Y0-BSN-508

CELL-MEDIATED IMMUNITY IN RUMINANTS

Edited by

Bruno M. L. Goddeeris
Laboratory of Physiology and Immunology of Domestic Animals
Catholic University of Leuven
Leuven, Belgium

W. Ivan Morrison
AFRC Institute for Animal Health
Compton Laboratory
Compton, Near Newbury
Berkshire, England

CRC Press
Boca Raton Ann Arbor London Tokyo

Library of Congress Cataloging-in-Publication Data

Cell-mediated immunity in ruminants / edited by Bruno M. L. Goddeeris,
William I. Morrison
p. cm.
Includes bibliographical references and index.
ISBN 0-8493-4952-4
1. Ruminants—Immunology. 2. Ruminants—Diseases. 3. Cellular immunity. I. Goddeeris, Bruno M. L. II. Morrison, W. Ivan.
SF757.2.C44 1994
636.2′0896079—dc20

93-42038
CIP

This book contains information obtained from authentic and highly regarded sources. Reprinted material is quoted with permission, and sources are indicated. A wide variety of references are listed. Reasonable efforts have been made to publish reliable data and information, but the author and the publisher cannot assume responsibility for the validity of all materials or for the consequences of their use.

Direct all inquiries to CRC Press, Inc., 2000 Corporate Blvd., N.W., Boca Raton, Florida 33431.

International Standard Book Number 0-8493-4952-4
Library of Congress Card Number 93-42038
Printed in the United States of America 2 3 4 5 6 7 8 9 0
Printed on acid-free paper

CONTENTS

Chapter 1
The Leukocytes: Markers, Tissue Distribution and Functional Characterization 1
C. J. Howard and W. I. Morrison

Chapter 2
Ontogeny of T Cells 19
W. R. Hein

Chapter 3
The Major Histocompatibility Complex 37
L. Andersson and C. J. Davies

Chapter 4
The T Cell Receptor 59
N. Ishiguro and W. R. Hein

Chapter 5
The Cytokines: Origin, Structure, and Function 75
D. M. Haig, C. J. McInnes, P. R. Wood, and H. F. Seow

Chapter 6
Induction of T Cell-Mediated Immune Responses in Ruminants 93
D. J. McKeever

Chapter 7
Lymphocyte Recirculation and Homing 109
W. G. Kimpton, E. A. Washington, and R. N. Cahill

Chapter 8
Neutrophils and Killer Cells 127
J. A. Roth

Chapter 9
East Coast Fever (*Theileria parva*): Cell-Mediated Immunity and Protection 143
B. M. Goddeeris, W. I. Morrison, E. L. Taracha, and D. J. McKeever

Chapter 10
Infectious Bovine Rhinotracheitis (Bovine Herpesvirus 1): Helper T Cells,
Cytotoxic T Cells and NK Cells 157
M. Denis, G. Splitter, E. Thiry, P.-P. Pastoret, and L. A. Babiuk

Chapter 11
Foot and Mouth Disease (Aphthovirus): Viral T Cell Epitopes 173
T. Collen

Chapter 12
Maedi-Visni (Lentivirus): Cell-Mediated Immune Responses in Pathogenesis 199
B. A. Blacklaws, P. Bird, and I. McConnell

Chapter 13
Cell-Mediated Responses against Gastrointestinal Nematode Parasites of Ruminants 213
S. J. McClure and D. L. Emery

Chapter 14
Cytokine Application in Infectious Diseases 229
M. Campos, D. L. Godson, H. P. A. Hughes, and L. A. Babiuk

Index 241

THE EDITORS

Bruno M. Goddeeris, DVM, Ph.D. is Professor at the Faculty of Agriculture and Applied Biological Sciences of the Catholic University of Leuven and Associate Professor at the Faculty of Veterinary Medicine of the University of Gent in Belgium. In 1973 he obtained his M.Sc. from the University of Leuven and in 1977 his DVM from the University of Gent. In 1978 he obtained the diploma of Tropical Veterinary Medicine and Animal Husbandry at the Prince Leopold Institute of Antwerp, Belgium. In 1987 he received his Ph.D. from the Free University of Brussels, Belgium. From 1978 to 1981 he was working as an Associate Expert of the FAO in Kenya and from 1982 to 1989 as a scientist of the Belgian Administration for Development Cooperation at the International Laboratory for Research on Animal Diseases in Kenya. Dr. Goddeeris was appointed Associate Professor in 1989 and Professor in 1993 at the Faculty of Agriculture and Applied Biological Sciences for the University of Leuven. There he is in charge of the Research Unit Immunology of Domestic Animals and is responsible for the following courses: Prophylaxis of Infectious Diseases of Domestic Animals, Animal Husbandry of the Monogastric Animals, Tropical Animal Husbandry, Biotechnology in Animal Production and Cell-Mediated Immunology. In 1992 he was appointed Associate Professor of Immunology at the Faculty of Veterinary Medicine of the University of Gent in charge of the Laboratory of Immunology and responsible for the Immunology and Immunopathology courses. His research interests cover cell-mediated immunity in cattle and lately also in swine and chicken.

Bruno Goddeeris is a member of the Scientific Board of the National Institute for Veterinary Research of Belgium and a member of the Veterinary Immunology Committee of the International Union of Immunological Societies. He is also a member of the Belgian Societies for Immunology, Protozoology and Parasitology. His publications include over 50 peer-reviewed scientific papers in international journals.

William Ivan Morrison, Ph.D., BVMS, MRCPath is Head of the Division of Immunology and Pathology, AFRC Institute for Animal Health, Compton, Berkshire, United Kingdom.

Dr. Morrison received his BVMS degree from the University of Glasgow in 1972, a Ph.D. from the same University in 1975, and was admitted to membership in the Royal College of Pathologists on the basis of his published work in 1986. He joined the International Laboratory for Research on Animal Diseases as a postdoctoral fellow in 1975, became a scientist in 1977 and eventually was appointed head of the Theileriosis research program in 1987. He took up his present position in 1990.

Dr. Morrison's research interests center around cellular immunology in ruminants with particular interests in function of the MHC and the role of cellular immune responses in immunity to parasitic infections. He has published over 120 research papers, has edited two books and has made numerous presentations at national and international scientific meetings. Dr. Morrison has served on a number of science review groups and is currently a member of several grants boards in the U.K. He received the Pfizer Award for Outstanding Contribution to the Advancement of Knowledge in the Origin and Treatment of Diseases of Cattle in 1990 and the Wellcome Trust Medal for Veterinary Research in 1991.

To Kenya and the African People

CONTRIBUTORS

Leif Andersson
Department of Animal Breeding and Genetics
Swedish University of Agricultural Sciences
Uppsala, Sweden

L. A. Babiuk
VIDO
University of Saskatchewan
Saskatoon, Saskatchewan, Canada

P. Bird
Department of Veterinary Pathology
University of Edinburgh
Summerhall, Edinburgh, Scotland

B. A. Blacklaws
Department of Veterinary Pathology
University of Edinburgh
Summerhall, Edinburgh, Scotland

R. N. P. Cahill
Laboratory of Fetal & Neonatal Immunology
University of Melbourne
Parkville, Victoria, Australia

M. Campos
SmithKline Beecham Animal Health
Lincoln, Nebraska

Trevor Collen
Department of Immunology and Pathology
AFRC Institute for Animal Health
Compton, Near Newbury
Berkshire, England

Christopher J. Davies
Department of Animal Breeding
Wageningen Agricultural University
Wageningen, The Netherlands
Department of Microbiology, Immunology, and Parasitology
College of Veterinary Medicine
Cornell University
Ithaca, New York

Martine Denis
Department of Virology & Immunology
Faculty of Veterinary Medicine
University of Liège
Liège, Belgium

D. L. Emery
CSIRO Division of Animal Health
McMaster Laboratory
Glebe, New South Wales, Australia

Bruno M. L. Goddeeris
Laboratory of Physiology and Immunology of Domestic Animals
Catholic University of Leuven
Leuven, Belgium

D. L. Godson
VIDO
University of Saskatchewan
Saskatoon, Saskatchewan, Canada

D. M. Haig
Department of Immunobiology
Moredun Research Institute
Edinburgh, Scotland

W. R. Hein
Basel Institute for Immunology
Basel, Switzerland

Chris J. Howard
Department of Immunology & Pathology
AFRC Institute for Animal Health
Compton, Near Newbury
Berkshire, England

Huw P. A. Hughes
VIDO
University of Saskatchewan
Saskatoon, Saskatchewan, Canada

N. Ishiguro
Department of Veterinary Public Health
Obihiro University of Agriculture and Veterinary Medicine
Obihiro, Hokkaido, Japan

W. G. Kimpton
Laboratory of Fetal & Neonatal Immunology
University of Melbourne
Parkville, Victoria, Australia

Susan J. McClure
CSIRO Division of Animal Health
McMaster Laboratory
Glebe, New South Wales, Australia

I. McConnell
Department of Veterinary Pathology
University of Edinburgh
Summerhall, Edinburgh, Scotland

C. J. McInnes
Department of Immunobiology
Moredun Research Institute
Edinburgh, Scotland

Declan J. McKeever
International Laboratory for Research on Animal Diseases
Nairobi, Kenya

W. Ivan Morrison
AFRC Institute for Animal Health
Compton Laboratory
Compton, Near Newbury
Berkshire, England

P.-P. Pastoret
Department of Virology & Immunology
Faculty of Veterinary Medicine
University of Liège
Liège, Belgium

James A. Roth
Department of Microbiology, Immunology, & Preventive Medicine
Iowa State University
Ames, Iowa

H. F. Seow
CSIRO Division of Animal Health
Parkville, Victoria, Australia

Gary Splitter
Department of Veterinary Science
University of Wisconsin
Madison, Wisconsin

Evans L. Taracha
International Laboratory for Research on Animal Diseases
Nairobi, Kenya

E. Thiry
Department of Virology & Immunology
Faculty of Veterinary Medicine
University of Liège
Liège, Belgium

E. A. Washington
Laboratory of Fetal & Neonatal Immunology
University of Melbourne
Parkville, Victoria, Australia

P. R. Wood
CSIRO Division of Animal Health
Parkville, Victoria, Australia

Chapter 1

The Leukocytes: Markers, Tissue Distribution and Functional Characterization

Chris J. Howard and W. Ivan Morrison

CONTENTS

I. Introduction 1
II. Nomenclature 2
III. Differentiation Antigens that Identify Subpopulations of Lymphocytes 2
A. Definitive Markers of T Cells 2
B. Molecules Defining the Major T Cell Subpopulations 2
C. Other Markers Useful in the Identification of T Cells 4
D. Adhesion Molecules and Homing Receptors 5
E. Lymphocyte Activation Antigens 6
F. The Leukocyte Common Antigen 7
G. Antigens Expressed on B Cells 8
IV. Tissue Distribution and Function of T Cell Subpopulations 8
A. Immature T Cells 8
B. Mature T Cells 9
1. CD4 T Cells 9
2. CD8 T Cells 9
3. WC1$^+$ $\gamma\delta$ T Cells 9
4. Subsets of CD4$^+$ and CD8$^+$ T Cells Defined by CD45 Isoforms 10
5. Subsets of CD4$^+$ and CD8$^+$ T Cells Defined by L-Selectin 11
V. Concluding Remarks 11
References 12

I. INTRODUCTION

The immune system exhibits a number of structural and functional features which distinguish it from other organ systems. The most unique feature is the capacity of both B and T lymphocytes to recognize and respond to a wide variety of foreign antigens. This is a consequence of the acquisition during differentiation of clonotypic receptors so that mature populations contain an extensive repertoire of antigen recognition specificities. The process by which lymphocytes become activated by specific antigen involves, in addition to antigen recognition, a number of other molecular events at the cell surface including interaction of adhesion molecules, binding of cytokines, and engagement of molecules that mediate signal transduction. Another distinctive feature of the immune system is its organization into separate primary and secondary compartments, the latter being populated by functionally competent cells that are constantly replenished from stem cells in the primary hemopoietic organs. The secondary lymphoid tissues are in a particularly dynamic structural state, in that lymphocytes are continuously recirculating through them via the lymphatic and blood vascular systems. The differential expression of cell surface homing and adhesion molecules enables subpopulations of cells to recirculate preferentially to particular lymphoid tissues and to particular sites within the tissues. Such a system of recirculation facilitates efficient recruitment of antigen-specific lymphocytes to sites of antigen incursion and also allows rapid dissemination of immune effector cells.

Monoclonal antibodies (MAb) have proved invaluable in characterizing leukocyte surface molecules involved in a variety of these cell functions and have allowed the identification of functionally distinctive subsets of lymphocytes and accessory cells. The detection of changes in expression of surface molecules during ontogeny and following activation or antigenic stimulation has also enabled a precise definition of the stages of differentiation within individual lineages. Herein, we will review the current state of

0-8493-4952-4/94/$0.00+$.50

knowledge on the phenotype and function of ruminant leukocytes. The major emphasis will be on the lymphocytic lineage since antigen presenting cells are discussed by D. McKeever in another contribution to this volume.

II. NOMENCLATURE

Investigations in man, rodents, ruminants, and other animal species have shown considerable homology between the molecules expressed by leukocytes and the function of the cells that they identify. Comparison of the reactivity of MAb to human leukocyte antigens on panels of different cells, together with biochemical analyses of the target antigens, has allowed MAb with the same specificity to be grouped in clusters of differentiation (CD). The allocation of CD numbers to these clusters by international workshops has resulted in the development of a common nomenclature for human antigens. A similar system has been adopted for the differentiation antigens in other species of animals, with the equivalent antigens in different species being allocated the same CD number.

Two international workshops have now been held to compare MAb to ruminant leukocyte differentiation antigens. Within these workshops, it was agreed that where possible the ruminant nomenclature should follow that adopted for human antigens. For an antigen to be given a ruminant CD number, there should be compelling evidence that the molecule identified is the homologue of a human CD antigen. This evidence might include similarities in the M_r and cellular and tissue distribution of the molecule, information on its function, and gene sequence data.[1]

Within the two ruminant workshops, several clusters of MAb were identified that were shown to define the same molecule, but either the human homologue was not evident or insufficient evidence was available to be certain of concluding a homology between the ruminant and human molecules. These MAb and the molecules that they identify have been given WC (workshop cluster) numbers prefixed with Bo, Ov, or Cp (bovine, ovine, caprine) as required. Other molecules have been identified by making use of MAb originally raised against human antigens that cross-react with ruminant leukocytes. In these instances, the identity of the antigen can be inferred from the specificity of the MAb in man. Examples are L-selectin, VLA-4, MECA-79, and VCAM-1.[2,3]

III. DIFFERENTIATION ANTIGENS THAT IDENTIFY SUBPOPULATIONS OF LYMPHOCYTES

A. DEFINITIVE MARKERS OF T CELLS

The CD3 molecule consists of several invariant polypeptides that are associated with the T cell receptor (TCR) and mediate signal transduction upon engagement of the TCR. Because of its association with both mature $\alpha\beta$ and $\gamma\delta$ TCR^+ cells and absence from other lineages of leukocytes, CD3 is the only true "pan T" marker. A murine MAb that recognizes the bovine CD3 antigen has recently been described.[4] This MAb recognizes an extracellular epitope on the antigen and immunoprecipitates polypeptides of 12, 16, 22, 32, 36, and 44 kDa under reducing conditions and an additional one of 96 kDa under nonreducing conditions. Of potential usefulness for the future is the observation that polyclonal rabbit antisera and murine MAb raised against a synthetic polypeptide corresponding to part of the cytoplasmic domain of human CD3 cross-react with ruminant CD3 and the CD3 molecule expressed by a number of other animal species.[5] MAb derived by this method may enable the identification of other ruminant homologues of human CD antigens for which it has been difficult to obtain specific MAb. Such reagents can then be used to purify antigens for production of further MAb to the extracellular domains or to clone the respective gene in an expression system.

A number of MAb have been produced that are specific for ruminant $\gamma\delta$ TCR and many cross-react with several ruminant species. However, most recognize only restricted, often overlapping subpopulations; e.g., the MAb 86D originally produced against ovine cells[6] and MAb CACTB6A and CACTB81A originally produced against bovine T cells, and called N6 and N7, respectively.[7–9] Others (e.g., MAb CACT61A, N12 of Davis),[7,10] appear to recognize most, if not all, $\gamma\delta$ TCR^+ cells. No MAb specific for $\alpha\beta$ TCR have been published.

B. MOLECULES DEFINING THE MAJOR T CELL SUBPOPULATIONS

Three major populations of mature T cells are evident in ruminants on the basis of expression of one of the three antigens CD4, CD8, or WC1. Expression of the CD4 or CD8 antigen identifies the MHC class

II- and class I- restricted subpopulations of αβ TCR+T cells, respectively. The WC1 molecule identifies a major population of T cells that is CD4− and CD8− and expresses the γδ TCR. Additional minor populations devoid of these markers exist in blood and, to a greater or lesser extent, in lymphoid tissues. In addition, a minor subpopulation of CD8+ cells express the γδ TCR.

The CD4 molecule consists of a single polypeptide belonging to the immunoglobulin gene superfamily. Although apparently restricted to T cells in ruminants and mice, human and rat CD4 is also expressed on monocytes and macrophages.[11] An M_r of about 55 kDa has been reported for the ovine molecule[12,13] and 50 kDa for the bovine molecule.[14,15] Bovine CD4 is polymorphic. Two alleles, codominantly expressed, were initially identified with MAb.[16] Subsequently, biochemical studies revealed that a proportion of one of the initially defined CD4 alleles coded for a larger molecule of 57 kDa which was shown to be due to additional N-linked glycosylation. This third allele was only found in *Bos indicus.*[17] The three alleles have been termed CD4.1, CD4.2, and CD4.3.

Homologues of the human CD8 molecule have been described in cattle,[18] sheep,[12] and goats.[1] CD8 is usually expressed as a noncovalently linked heterodimer consisting of α and β chains, although homodimers of the α chain can also be expressed on the cell surface. Under reducing conditions, two bands of 36 and 33 kDa were reported for the CD8 molecule in sheep[12,19] and two bands of 38 and 34 kDa for cattle[18,20], presumably representing the α and β chains. Recently, cloning of the α-chain of bovine CD8 has established a 75% homology at the nucleotide level and 62% at the protein level with the human CD8α chain sequence.[21] Three particularly conserved regions were identified in comparison with the human molecule, two of which are also highly conserved in the rat and mouse sequences. One of these sites, region 1, is involved in binding to MHC class I molecules. The cloned CD8α gene has been expressed in murine L cells and the resultant cells used to select a further transfectant expressing both the α and β chains.[22] This has allowed the precise specificity of MAb to BoCD8 to be examined and has identified two distinct specificities. One group of MAb reacts with the homodimeric and heterodimeric forms of the molecule and therefore must reconize the α chain. The second group of MAb reacts only with the heterodimeric form and therefore is specific either for the β chain or a determinant generated by the association of both chains. Thus, by using selected CD8-specific MAb, it is possible to distinguish cells expressing the heterodimeric or homodimeric forms of bovine CD8.

The majority of CD8+ cells express heterodimeric CD8. Cells expressing only the homodimeric form of the molecule are much less common but are present as a low percentage of the CD8+ cells in PBM and gut mucosa and as a higher percentage in spleens from young calves.[23] The homodimeric molecule has been reported on NK cells in man that are membrane CD3−.[24] This population of cells remains to be identified precisely in cattle; however, a recent report described a cloned bovine cell line that displayed non-specific cytolytic activity against lymphoblastoid targets and was CD2−,CD6−,CD4−,WC1−, expressed homodimeric CD8 but was negative for CD3 and TCR RNA transcripts in Northern blots.[25] Another situation in which homodimeric CD8 is expressed is on lineages of lymphoblasts infected with the protozoan parasite *Theileria parva* which, prior to infection, are CD8−. Thus, CD4+ T cells and WC1+,CD8−,γδ TCR+ T cells commonly express homodimeric CD8 following infection with the parasite.

The WC1 molecule, which is also called T19 in sheep, was originally identified as an antigen with a predominant M_r under reducing conditions of 215 kDa expressed on CD2−,CD4−,CD8−,CD5+ cells presumed to be T lymphocytes.[26,27] Following cloning of CD3 and TCR genes in cattle and sheep, analyses of RNA in Northern blots with specific probes clearly demonstrated that these cells express CD3 and TCR γδ. A number of MAb considered to be specific for subpopulations of TCR γδ also react with these cells. A large number of MAb (more than 20) specific for the WC1 antigen have been produced and these all cross-react between cattle, sheep, goats, and a number of wild ruminant species.[28,29] Although the MAb appear to immunoprecipitate similar molecules, the numbers of cells with which different MAb react in individual animals varies markedly.[8,30] Recent studies in which genes encoding the WC1 antigen have been cloned and characterized have helped to explain this anomaly. The nucleotide sequences of two cDNA clones predict a 1436 amino acid type I membrane protein with an extracellular part containing 11 tandem repeats of a cysteine-rich domain of 110 amino acids.[31] Although homologous genes have not been described in other species, similar cysteine-rich domains are found in a number of glycoproteins including CD5, CD6, epidermal growth factor, and the macrophage scavenger receptor. The two cDNA clones exhibited differences in their nucleotide sequences and, following transfection and expression in mouse L cells,[32] some WC1-specific MAb were shown to recognize epitopes common to both gene products whereas other MAb recognized epitopes that were restricted to one or the other gene product. Further studies have indicated that the two gene products (termed WC1.1 and WC1.2) have slightly

Table 1-1 **CD1 Expression in Cattle**

	M_r		PBM			Thymus	Skin		ALVC
	α	β	T	B	Mo		EpDC	DDC	
CD1w1	46	12	–	+	+	++	+	+	++/–
CD1w2	46	12	–	–	–	++	–	+	++/–
CD1w3	44	12	–	+	+	++	–	+	++/–

Note: T = T cells; B = B cells; Mo = monocytes; EpDC = epidermal dendritic cells; DDC = dermal dendritic cells; ALVC = afferent lymph veiled cells.

different molecular weights — 205 and 215 kDa, respectively.[30] These findings, together with Northern and Southern blot analyses, suggest the existence of a family of WC1 genes encoding at least three distinct molecules. The final picture is likely to be complex and definition of the specificities of MAb will depend on further studies of cloned genes.

The molecule is particularly important as it identifies a unique phenotype of γδ TCR⁺ T cells in ruminants that do not express the CD2, CD4, or CD8 molecules. The molecule was thought to be restricted to ruminants,[28] but recently an antigenically related molecule has been identified in pigs.[33]

C. OTHER MARKERS USEFUL IN THE IDENTIFICATION OF T CELLS

The CD1 antigens are a family of molecules each composed of a heavy chain (α) associated with a β2 microglobulin light chain. At least three distinct CD1 molecules are expressed on ruminant leukocytes (Table 1-1). In man, Southern blot analysis has revealed five CD1 genes; eight have been found in rabbits[34] and a number of genes are certain to be established in ruminants. However, the relation between ruminant and human gene products has yet to be precisely established. The second workshop proposed naming the three bovine CD molecules identified to date as ws (workshop) 1, 2, and 3; i.e., BoCD1ws1, CD1ws2, and CD1ws3.[35] A complication is that immunoprecipitation and flow cytometric data indicated different specificities for the BoCDws3 MAb in sheep from cattle.[36,37] Furthermore, it has been suggested that in sheep the MAb 20-27 (CD1ws1) identifies an epitope common to different CD1 antigens.[38]

In cattle, MAb 20-27 (CDws1), which was originally raised against ovine cells,[39] reconizes a CD1 molecule expressed on B cells and monocytes in PBM. Monoclonal antibody CC20 (CDws2), originally raised against bovine cells,[40] has been shown to reconize human CD1b cDNA expressed in Cos cells;[41] hence, the molecule detected by this MAb, and other MAb in this cluster, is likely to be the ruminant homologue of CD1b. A third bovine CD1 molecule (CD1ws3) has a heavy chain with M_r 44 kDa which distinguishes it from the other two CD1 molecules which had M_r 46 kDa.[36] None of the CD1 molecules are expressed on mature T cells. However, all three CD1 antigens are strongly expressed on immature cortical thymocytes and on dendritic cells in the thymic medulla.[13,39,40] These three CD1 antigens are also expressed on dendritic cell subsets in peripheral lymphoid tissues, but some variation is evident between expression of the different molecules.[36,38,41,42] In skin, CD1ws1 is expressed on Langerhans cells in the epidermis, but CD1ws2 is not. All three antigens are expressed on dermal dendritic cells and subsets of afferent lymph veiled cells. The CD1ws1 and CD1ws3 antigens are expressed by bovine B cells and monocytes in PBM. CD1ws2 is not expressed by cells in PBM. The CD1ws1 and CD1ws3 molecules both have features in common with CD1c in man. Establishment of the precise relationship of these antigens with human CD1 molecules will require definition at the molecular level.

Like MHC class I antigens, the α chain of CD1 is associated with β2 microglobulin but sequence data does not reveal a close homology of the heavy chains. Unlike classical MHC class I, the CD1 molecules are not polymorphic. A peptide presentation function has been proposed for CD1 based on homology with MHC class I, and γδ T cells have been shown to reconize CD1 although no evidence of presentation of peptides has been presented.[43]

The CD2 molecule is an intercellular adhesion molecule implicated in T cell activation and differentiation. The molecule has been identified in cattle, sheep, and goats. In cattle, immunoprecipitation has consistently revealed a molecule with M_r 50 to 60 kDa together with, in a number of experiments, a second entity of about 48 kDa.[44–46] The sheep CD2 molecule has M_r 50 to 55 kDa and is expressed on ovine T cells at a lower level than CD2 in cattle.[13,47,48] LFA-3 (CD58), which is broadly expressed on hemopoietic and nonhemopoietic cells, has been established as the ligand for CD2 although recently it has been reported that CD48 is also a counter receptor for murine CD2.[49] LFA-3 is expressed at a high level on sheep erythrocytes accounting for the binding to T cells of many species.[50] Inhibition of sheep erythrocyte binding to bovine T cells is a property of many MAb directed against the CD2 antigen.[51] In ruminants, the CD2 antigen is expressed by most thymocytes and most CD4+ and CD8+ T cells in PBM that are αβ TCR+. It is not expressed by WC1+,CD4−,CD8− γδ T cells, but is expressed by a small CD8+ γδ T cell subset and a CD4−,CD8−,CD6−subset of lymphocytes in peripheral blood.[52,53] Some of the CD2-specific MAb have been shown to stimulate bovine T cells to proliferate, thus indicating that, as in other species, CD2 can serve as a transducer of activation signals in T cells.[44,45] Expression of CD2 on some murine B cells has been described;[54] however, CD2 has not been detected on normal sheep or cattle B cells although expression on *Theileria annulata* transformed cells, presumed to be derived from B cells or monocytes, has been observed.[55] CD2 has also been observed on ovine macrophage cell lines.[56]

A homologue of the human CD5 antigen has been identified in sheep[39,57,58] and cattle.[59] Under reducing conditions, the dominant molecule in sheep and cattle has M_r 67 kDa,[39,59] but an additional 60 to 65-kDa entity of variable intensity has sometimes been observed in sheep.[57,58] It is not known whether the lower molecular weight molecule is the result of proteolysis or is due to differences in glycosylation. The human CD5 molecule has recently been shown to function as the receptor for CD72 which is expressed strongly on B cells and weakly on tissue macrophages.[60] The CD5 molecule is involved in T cell activation through the TCR, and MAb to the CD5 molecule have been found to activate protein kinase C and tyrosine kinase.[61,62]

The bovine CD5 molecule is expressed weakly on immature cortical thymocytes and at a higher level on mature medullary thymocytes. It is expressed on the majority of mature T cells in the peripheral circulation and secondary lymphoid tissues, but at a lower intensity on WC1+ γδ T cells compared to CD2+ αβ T cells. A subpopulation of B cells, 5 to 35%, also expresses the molecule,[1,63] albeit at a lower intensity than CD2+ T cells.

Two codominantly expressed alleles of the bovine CD5 antigen (designated CD5.1 and CD5.2) have been defined.[64] *B. taurus* cattle only express CD5.1, while the frequency of this allele in *B. indicus* was only 10%. Since the population of *B. indicus* studied was believed to contain some *B. taurus* genetic material, it was suggested that the two alleles are subspecies markers. Similar allelic differences have been described for the murine equivalent of CD5,Ly1.[65,66]

The bovine CD6 molecule was first described as a pan-T cell surface antigen[45] that stimulated *in vitro* proliferation of T cells and thus appeared to play a role in the transduction of signals involved in cell activation. Under reducing conditions, the molecule has M_r 110 kDa. The majority of mature CD4+ and CD8+ T cells express the antigen, while WC1+ γδ T cells do not. In the thymus, expression of CD6 is restricted to the more mature medullary thymocytes. The antigen is not expressed by monocytes, granulocytes, or B cells.[67]

D. ADHESION MOLECULES AND HOMING RECEPTORS

Members of the integrin gene superfamily are expressed on the surface of cells as heterodimeric molecules that are involved in a variety of cell adhesion phenomena. There are three major subgroups defined as having common β chains that can be associated with different α chains. The β2 group, known as the leukocyte integrins, are mostly involved in cell/cell interactions within the immune system,[68,69]

The ruminant leukocyte integrins have been identified on the basis of their similarity in cellular distribution and biochemical characteristics to the human molecules.[70–72] A common β chain of 95 kDa (CD18) is associated with α chains of 180 kDa (CD11a),160 kDa (CD11b), or 150 kDa (CD11c). The different antigens can be identified with MAb that are specific for the α chain, while MAb to the common β chain precipitate all three antigens.[73] The function of the ruminant molecules, where it has been investigated, has been found to be the same as the human homologues.

CD11a (LFA-1) is expressed on the majority of leukocytes. Its ligands are ICAM-1 (CD54) and ICAM-2. ICAM-1 is a 90-kDa molecule which is constitutively expressed on endothelial cells and weakly expressed or absent from resting lymphoid cells; expression on many cell types is upregulated on activation.[69,74] Inhibition of the interaction of these adhesion molecules can prevent the induction of proliferation and cytotoxicity by T cells. In sheep, it has been reported that CD11a is expressed on all leukocytes in PBM, but that memory CD4+ T cells expressed a higher level of the antigen than naive cells.[71] Memory in this case was defined in *in vitro* antigen-specific proliferation assays using T cells from sheep primed with rabbit Ig. Despite the perceived importance of LFA-1/ICAM-1 interaction in T cell stimulation, recent studies in cattle demonstrated that a subset of afferent lymph veiled cells (dendritic cells) that had little or no detectable surface CD11a were much more effective at presenting soluble antigen to antigen-specific T cell clones than the CD11a$^+$ subset.[75] CD11b (Mac-1) is expressed by monocytes and granulocytes, but not by afferent lymph veiled cells.[72,75] It is also present on a small subset of CD8$^+$ T cells[23] and a subset of B cells.[63] In addition to its function as an adhesion molecule, CD11b is a receptor for C3b. CD11c is expressed by monocytes, granulocytes, afferent lymph veiled cells,[75] and activated T cells.[1] Monoclonal antibodies to CD11c have been shown to inhibit primary proliferative responses of bovine T cells to alloantigen.[76]

Cattle with Leukocyte Adhesion Deficiency (BLAD) have a mutation in the β2 subunit (CD18) that prevents association with the CD11 subunits and leads to a failure of surface expression of β2 integrins.[77] The mutation is at a site similar to that in molecules from humans with LAD.[78] Calves with the defect suffer recurrent pneumonia, ulcerative and granulomatous stomatitis, enteritis with bacterial overgrowth, delayed wound healing, and early death.

Members of the β1 and β3 integrin subfamilies mostly mediate interactions of cells with extracellular matrix but one antigen, VLA-4 (α4,β1), is restricted to lymphoid and myeloid cells. This molecule binds to VCAM-1 which is present on vascular endothelium at sites of inflammation resulting in the recruitment of VLA-4$^+$ cells. Mackay et al.[3] identified VLA-4 in sheep and showed that expression was upregulated on memory cells. Immediately following antigen challenge of the peripheral node, there is a transient increase of VLA-4$^+$ T cells in efferent lymph as a result of recruitment from the blood of nonblasting lymphocytes. This appears to be a non-specific result of inflammation associated with upregulation of VCAM-1 on the endothelium in the node.[79]

L-selectin, also called LECAM-1 or LAM-1, is a member of the selectin family of molecules and functions as the peripheral lymph node homing receptor, binding to a ligand on high endothelial venules (HEV). In cattle, the molecule has an M_r of 90 kDa. Some epitopes are conserved on the L-selectin molecules expressed on human, bovine, caprine, and porcine cells.[2,3,80] The ligand on the lymph node HEV in man is identified with the MAb MECA-79 which also reacts with sheep.[3] The MECA-79 epitope is expressed to a lesser degree by HEV in Peyers patches, but not by vessels in the spleen.[81] Downregulation of L-selectin following activation of human cells has been reported. However, bovine lymphocytes appear not to lose expression of the molecule so readily upon activation.[80] L-selectin is expressed by bovine granulocytes, monocytes, WC1$^+$ γδ T cells and subpopulations of B cells and CD4$^+$,CD8$^+$ T cells in peripheral blood.

The CD44 antigen has been identified by specific MAb in sheep and cattle. The bovine gene has been cloned[82] and also transfected into Cos cells, allowing the recent identification of several MAb specific for bovine CD44.[22] The molecule functions as a homing receptor with distinct epitopes controlling binding to endothelium in different tissues. It is the principle cell surface receptor for hyaluronate and this interaction mediates binding to HEV.[83] Although expressed on the majority of leukocytes in ruminants, it has been reported to be expressed at a higher level on memory T cells and may therefore allow memory cells to bind to endothelia and extravasate more efficiently.[71]

E. LYMPHOCYTE ACTIVATION ANTIGENS

A number of antigens have been identified that are upregulated or expressed *de novo* on activated T cells in ruminants. For the most part, they are not well defined. One that has been partially characterized is the receptor for IL-2 (IL-2R). In mouse and man, the IL-2R comprises several polypeptides, two of which –gp55 and gp75– bind IL-2; gp55 and gp75 bind IL-2 with low and intermediate affinities, respectively, and can combine to form a high-affinity receptor. The gp55 component (CD25) of the bovine IL-2R has been identified with MAb and a cDNA clone has been isolated, sequenced, and expressed in mouse L cells. Bovine CD25 has been identified as a 55-kDa molecule expressed on Con-A-activated PBM.[84–86] The molecule is not expressed on resting T cells in man or mouse but in cattle a low level of expression is seen on resting WC1$^+$ γδ$^+$ T cells in PBM (Figure 1-1), which may explain in part the rapid response of these cells to IL-2 *in vitro*.[87] CD25 is highly expressed on T lymphocytes that have been activated in culture for 1 to 7 days with mitogen. The antigen appears within 10 h after lectin stimulation and reaches a maximum on days 3–4, after which expression slowly decreases.[84] Monocytes and B cells in stimulated bovine PBM cultures express the CD25 antigens at a lower intensity than T cells.[85]

A number of less well-defined activation antigens have been described. The WC5 molecule, which is detected as a 47-kDa polypeptide, is expressed on activated T cells and on some T cell lines. However, it is not restricted to T cells and is evident on resting B cells in PBM.[88] Another antigen that is expressed by T cells late after activation (6 days), but not by resting cells, is the BoWC8 molecule. Expression is induced by alloantigens and lectins; the M_r under reducing conditions is 150 kDa.[89]

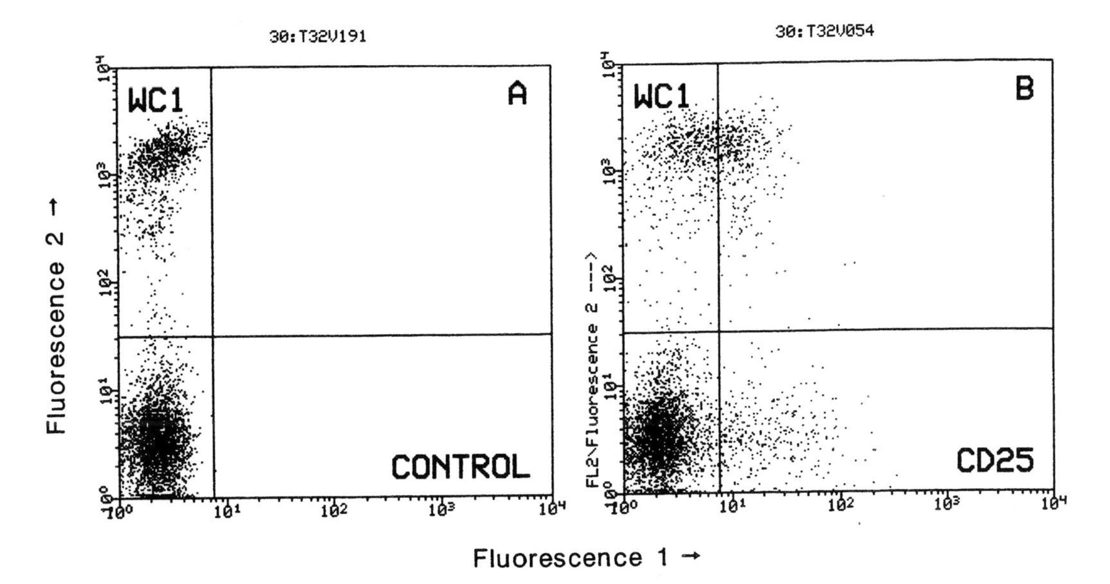

Figure 1-1 Expression of IL2-R (CD25) by WC1+ γδ T cells. Bovine PBM from an apparently normal healthy calf stained with MAb CC15 (WC1) and IL-A111 (CD25). A low level of expression of the CD25 antigen by a large proportion of the WC1+ T cells is evident. The animal is typical of >20 tested. Under the same conditions, only a minor subset of CD2+ cells were CD25+.

F. THE LEUKOCYTE COMMON ANTIGEN

The leukocyte common antigen, termed CD45, exists as a group of high molecular weight glycoproteins, the expression of which is restricted to leukocytes, and which make up some 10% of leukocyte surface proteins. Studies in rodents and man have established that the CD45 components represent different isoforms of the molecule that are generated as a result of differential splicing of RNA. Exons 4, 5, and 6 of the CD45 gene may be spliced out of mRNA individually or in combination, potentially producing eight isoforms of the molecule. Expression of some isoforms is restricted to subpopulations of leukocytes and hence can be used to identify cells with different functional properties and tissue distribution. The cytoplasmic domain of this molecule has protein tyrosine phosphatase activity and is believed to be involved in the modulation of signal transduction via the TCR.[90]

Monoclonal antibodies to epitopes on the CD45 molecule that are independent of exons 4, 5, and 6 identify all isoforms and react with all leukocytes. A number of these nonrestricted CD45 MAb identify polymorphic determinants within the genus *Bos.*[1] Glycoproteins with M_r 220, 205, and 180 kDa have been identified in ruminants with CD45-specific MAb.[71,91,92] However, isoforms with products of one or two of the variable exons included can have the same, or very similar, M_r. Moreover, there may also be variable glycosylation. Thus, M_r alone is only likely to identify isoforms with all three exon products present (termed ABC+) or absent (O isoform). A precise definition of the isoforms requires MAb of known isoform specificity. This in turn is dependent on characterization of the spliced RNA transcripts.

Mackay et al.[71] reported an MAb that reacted with restricted isoforms of the CD45 in sheep. This MAb immunoprecipitated molecules of 220 and, occasionally, 205 kDa. The antigen detected was present on B cells and subpopulations of CD4 and CD8 T cells, but not on WC1+ γδ T cells or monocytes. Based on the cellular distribution of the molecule and the functional properties of the positive T cells (see below), homology with human CD45RA was proposed.

In cattle, molecules of 220 and 205 kDa were precipitated from PBM by an MAb with specificity for restricted CD45 isoforms.[91] The determinant was expressed on B cells, subpopulations of CD4 and CD8 T cells, but not by monocytes or WC1+ γδ T cells. The tissue distribution, together with functional studies, indicate that this isoform is most similar to human CD45RB, but a final assignment of isoform specificity must await a molecular characterization.

Recent comparisons in sheep of the putative CD45RA MAb (73B) used in sheep by Mackay et al.[71] and the CD45RB MAb[91] used in cattle by Howard et al. (1991) showed that they reconize different isoforms;[93] the MAb 73B does not react with bovine cells.

The isoform resulting from RNA with all three exons spliced out is the smallest and in rodents and man has an M_r of 180 kDa.[90] A MAb reacting with the bovine CD45RO antigen has been identified recently.[92,94] This isoform is expressed on CD4+ and CD8+ subpopulations and on WC1+ γδ T cells and monocytes but not by B cells. The CD45R0+ positive CD4+ and CD8+ T cells include all of the CC76− (putative CD45RB) cells and a smaller component of the CC76+ cells.

G. ANTIGENS EXPRESSED ON B CELLS

Monoclonal antibodies to the bovine Ig isotypes — G1,G2,M, and A, but not IgD — have been described. The great majority of B cells in PBM are sIgM+ and a small percentage express IgG1, IgG2, or IgA.[95] Two allotypes of the μ heavy chain have been described.[96] A number of other antigens expressed predominantly on B cells have been identified, but most are not lineage specific.

The WC3 antigen was originally identified as a 145-kDa molecule expressed on B cells in PBM and in the follicles in lymph nodes, Peyer's patches, and the spleen.[88,97] An examination of stained frozen sections also indicated that the antigen was expressed on cells with a dendritic morphology in follicles. Several of the BoWC3-specific MAb were originally raised against alveolar macrophages which also express the WC3 antigen. The molecule has many features in common with CD21 in humans, a receptor for C3d, and it is likely that BoWC3 is the bovine homologue of this antigen.

Another molecule expressed on B cells in PBM and on cells within the organized follicles of lymph nodes is the BoWC4 antigen. Numerous cells in the domed area of the Peyer's patch also express this antigen. BoWC4 MAb precipitate a 90-kDa molecule from PBM, possibly homologous to human CD19. A further B cell antigen has been identified with MAb CC56 which precipitates a 180-kDa molecule from labeled PBM under reducing conditions.[88] This molecule appears restricted to B cells in PBM, but is expressed by other cells in tissues.

Two other molecules that are expressed on B cells but that are not lineage specific are the BoWC6 and BoWC10 molecules. The BoWC6 MAb immunoprecipitate an antigen of 210 to 220 kDa, which in preclearing experiments has been shown to be unrelated to CD45. It is expressed by B cells in PBM and at low levels by CD2+ T cells, but not by WC1+ γδ TCR+ T cells. The molecule is also expressed by immature cortical thymocytes and mature medullary thymocytes as well as being strongly expressed by veiled cells in afferent lymph.[75,98] The WC10 antigen is expressed at a low intensity by the majority of B cells and a subpopulation of CD2+ T cells in PBM. Monocytes and granulocytes and the majority of WC1+ T cells do not express the antigen. Cells other than leukocytes also express the antigen and staining of the brush border of the gut epithelium is particularly pronounced. Under reducing conditions, the BoWC10 MAb precipitate two molecules of 39 and 115 kDa.[99]

The CD5 and CD11b antigens are expressed on a largely overlapping B cell subset. In PBM from normal cattle, 5 to 35% of the B cells are CD5+, but expression is at a lower intensity than seen on CD2+ T cells. Most, but not all, of the CD11b+ B cells are CD5+.[1,63] In mice, CD5+ B cells are the source of most normal serum IgM and participate in T-independent antibody responses and production of autoantibodies. High numbers of CD5+,IgM+ cells are evident in cattle and sheep infected with bovine leukemia virus and in cattle infected with *Trypanosoma congolense*.[63,100]

Two B cell subsets are identified as being L-selectin positive or negative.[23] Within the CD5+ and CD5− B cell subsets the ratio of L-selectin+ to L-selectin− cells is 1:2 and 4:1, respectively. This implies that the B2 cells may have a greater propensity to recirculate through the lymph nodes than the more primitive B1 subset. The reported low level of B cells that express CD5 or CD11b in bovine lymph nodes and tonsil compared to blood[63] bear out this proposal.

IV. TISSUE DISTRIBUTION AND FUNCTION OF T CELL SUBPOPULATIONS

A. IMMATURE T CELLS

The thymus is the lymphopoietic tissue in which T cells are derived from stem cells, the TCR is rearranged, positive selection takes place, and self-reactive clones are removed or rendered anergic. Ruminants develop immunological competence about half way through gestation. All investigations to date indicate that, structurally and functionally, the ruminant thymus follows the general mammalian pattern. In calves, a layer of cells one or two cells thick is evident in the outer cortex that are CD1−, CD4−, CD8− but L-selectin+ and have high mitotic activity. These cells are the most immature thymocytes and some of them may represent recent thymic immigrants from the bone marrow.[23,101–103] The immature cells that comprise the bulk of the bovine thymus are sited in the cortex and are mostly CD1+, CD2+, CD4+, CD8+,

CD5$^+$, CD6$^-$, CD45R$^-$ (CC76$^-$).[23,40,52,91] Studies in mice have established this to be the area in which apoptosis predominates and the T cell repertoire is selected.[101] The medulla contains predominantly mature $\alpha\beta$ TCR$^+$ lymphocytes which are CD1$^-$, CD2$^+$, CD3$^+$, CD5$^+$, CD6$^+$, and either CD4$^+$ or CD8$^+$. WC1$^+$ $\gamma\delta$ T cells are evident in small numbers (<3% of total thymocytes) mostly in the medulla.[52,101–103] Studies of lambs thymectomized during fetal life have demonstrated that the peripheral WC1$^+$ $\gamma\delta$ T cell subpopulation is thymus-dependent (see Reference 102).

B. MATURE T CELLS

1. CD4 T Cells

The CD4 molecule functions by binding to an invariant region of the β chain of MHC class II. Its cytoplasmic domain is associated with a tyrosine kinase (p56^{lck} in man) which is believed to be involved in phosphorylation of components of CD3 during signal transduction. As in other species, CD4$^+$ T cells in ruminants have been shown to be MHC class II restricted.[14,52] *In vivo* depletion experiments with MAb specific for the CD4 antigen in both calves[104] and lambs[105] have demonstrated the role of these cells in providing help for antibody production to T-dependent antigens. CD4$^+$ T cells represent 25 to 35% of PBM. In the lymph nodes, a large percentage of cells in the paracortex are CD4$^+$ as are many cells in the interfollicular regions of the discrete Peyer's patches and tonsil and periarteriolar regions of the spleen. CD4$^+$ cells are also evident in secondary follicles in lymph nodes, tonsil, and discrete Peyer's patches where they form a distinct band on the luminal side of the follicle. In the nonorganized lymphoid tissue of the gut mucosa, CD4$^+$ T cells are evident in the lamina propria but not within the epithelium.[106]

2. CD8 T Cells

The second major T cell population expresses the CD8 antigen which functions in an analogous way to CD4 in class I-restricted T cells in that it binds to an invariant part of the α3 domain of the class I heavy chain and its cytoplasmic domain is associated with the same tyrosine kinase. Investigation of the function of these cells has shown that they mediate MHC class I-restricted cytolytic T cell responses against virus-infected cells and cells infected with protozoan parasites.[18,95,107–110] CD8 T cells make up 15 to 25% of PBM and are numerous within the paracortex of the lymph nodes, the interfollicular T-dependent areas of tonsils, and discrete Peyer's patches and periarteriolar regions in the spleen.[18,95] CD8$^+$ T cells are also seen in the red pulp of the spleen in young calves.[23] In the gut mucosa CD8$^+$ T cells are primarily within the epithelium.[106,111] In sheep, a subset of WC1$^-$,$\gamma\delta$ TCR$^+$ expressing cells is evident within the CD8$^+$ intraepithelial lymphocyte population.[112] The ratio of CD4:CD8 cells in mammary secretions was found to be 0.85:1, in contrast to the ratio of 1.53:1 in the blood, indicating a selective increase of the CD8 subset in secretions from this mucosal site.[113] Depletion studies *in vivo* have demonstrated a role for these cells in reducing the level of RSV infection in the bovine lung[114] and rotavirus infection in the bovine gut.[115] These studies indicate that, *in vivo,* CD8$^+$ T cells play a central role in the resolution of primary infection with these viruses at mucosal surfaces, probably as a result of their MHC class I-restricted cytolytic effector function, although this has yet to be confirmed *in vitro.*

In mice, a population of CD8$^+$ T cells in the gut epithelium has been identified that expresses only the homodimeric form of the molecule and that is thymus independent.[116] The great majority of CD8$^+$ cells in the gut epithelium of adult cattle express the heterodimeric molecule, cells expressing the homodimer alone being a minor population. However, a large percentage of splenic CD8$^+$ cells in neonatal calves express the homodimeric molecule.[23] Many of the CD2$^+$ cells in the spleens of calves, which may be the same cells, have also been shown to express the $\gamma\delta$ TCR.[117] At this site, they represent a major population of $\gamma\delta$ TCR$^+$ cells that is WC1$^-$ and are thus distinct from the WC1$^+$ $\gamma\delta$ TCR$^+$ T cells present in the circulation.

3. WC1$^+$ $\gamma\delta$ T Cells

Expression of the WC1 antigen identifies the third major T cell subpopulation in ruminants, the CD4$^-$,CD8$^-$,$\gamma\delta$ T cells.[27] The WC1$^+$ T cells in PBM appear relatively homogeneous with respect to expression of molecules that identify functionally distinct CD4 and CD8 subsets, the majority being L-selectin$^+$, CD45RO$^+$, and negative for the putative CD45RB(MAb CC76) and RA(MAb 73B) isoforms in cattle and sheep, respectively.[71,91,94] The WC1$^+$ T cells comprise a major proportion of the T cells in the circulation of young ruminants; a mean of 27% in PBM has been reported in calves aged 1 to 3 weeks, while in individual animals greater than 50% of PBM can be within the WC1$^+$ population.[27] In adults, the WC1$^+$ T cells comprise about 5% of PBM.

The percentage of WC1+ cells in suspensions of cells from lymph nodes and spleen is about 4%. The distribution of these cells in secondary lymphoid tissues is distinct from that of CD4+ and CD8+ T cells. In the lymph node, they are predominantly in the outer areas of the cortex adjacent to the subcapsular sinuses; a few cells are evident in the sinuses and paracortex. Positive cells are not usually seen in B cell follicles but have been noted in involuting secondary follicles of older animals. In the spleen, positive cells are concentrated mainly in the marginal zones.[26,27,104] In other tissues, WC1+ populations are prominent in the skin, where they are mostly found in the dermis with a few cells in the basal layers of the epidermis[27] and in the gut mucosa, where most are in the lamina propria near the base of the epithelium with a few cells evident in the epithelium.[27,104,106] In contrast to blood where the percentage of WC1+ cells decreases as animals age, an increase in the proportion of WC1+ T cells was noted in the gut mucosa of older calves.[106] A preliminary investigation of the tissue distribution of cells staining with MAb specific for the different WC1 gene products failed to show any differences in distribution.[30]

The presence of much larger numbers of WC1+ cells in blood than in lymph nodes would indicate that they do not readily recirculate directly from the blood. However, the WC1+ cells are uniformly L-selectin+, implying a capacity to enter the node from blood via the HEV.[80,118] A comparison of the distribution of WC1+ cells in sheep[71] showed 20% in afferent lymph, 6% in efferent lymph, and 13% in blood. Given that afferent lymph contributes 5 to 10% of the cell input into lymph nodes, the implication is that the majority of the WC1+ T cells in lymph nodes are extracted by endothelium in peripheral tissues rather than the HEV of lymph nodes, though there is some contribution from the blood directly. The location of the cells in sections of lymph nodes are consistent with them having entered the node from afferent lymph.[118] The increased number of WC1+ cells in afferent lymph and expression of the memory CD45 isoform (CD45RO) has been taken to imply that these cells are memory cells that selectively recirculate in a similar manner to CD4+ memory cells,[71] but the decrease in percentage of WC1+ cells with age is in marked contrast with CD4+ cells of memory phenotype which in several species have been shown to increase with age.[71,91,119]

The function of the WC1+ cells remains to be established, but a few relevant observations have been made. There is a disproportionate proliferation of the WC1+ cells in autologous MLR[27] and their depletion *in vivo* with MAb was noted to result in a higher antibody response to human type O erythrocytes injected intravenously.[104] The WC1+ T cells may therefore have a role in controlling the proliferation of other lymphocytes. A potential cytolytic activity for these cells has been proposed[13,120] but not clearly established.

4. Subsets of CD4+ and CD8+ T Cells Defined by CD45 Isoforms

Expression of different isoforms of CD45 has defined several subsets of both CD4 and CD8 T cells in ruminants. Mackay et al.[71] identified two subsets of CD4+ T cells in sheep on the basis of staining with the CD45R MAb 73B (a probable CD45RA homologue). Cells that proliferated in recall assays to soluble antigen, rabbit Ig, were within the CD45R− subset. The functional memory cells defined by this assay also expressed higher levels of CD2, CD58, CD44, and CD11a compared to the CD45R+ subset. The upregulation of these molecules involved in cellular interactions may explain in part the more rapid and extensive response of memory compared to naive T cells. In adult sheep, 30 to 40% of T cells in PBM were CD45R−(73B−), while the majority of T cells in efferent lymph were CD45R+(73B+) and those in afferent lymph were mostly CD45R−(73B−), indicating that the cells with the memory phenotype selectively recirculate through the tissues rather than via the HEV of peripheral nodes.

Studies in cattle[80,91] with a CD45R MAb CC76 (probably homologous to CD45RB) showed that the CD4+ T cells that responded in proliferation assays to soluble antigen, either VSG from *Trypanosoma brucei* or ovalbumin, were entirely within the CD45R− population. In contrast to these findings with the CD4+ cells, an examination of CTL precursor frequency in sorted CD8+CC76+ and CD8+CC76− PBM from cattle that were immune to *Theileria parva* failed to show an association between CD45R expression and CTL functional memory.[91] Subsequent studies with a MAb specific for bovine CD45RO[94] also allowed the identification of two CD4+ subsets and in experiments with sorted CD45RO+ and RO− cells proliferation to soluble antigen was limited to the CD45RO+ subset. However, three-color staining of CD4+ and CD8+ T cells revealed three subsets of each based on expression of the CD45R (MAb CC76) and CD45RO isoforms,[121] two subsets in which expression is mutually exclusive, and a third that expresses both isoforms.

It can be concluded from these observations that the CD45R+(CC76−),CD45RO−subset of bovine CD4+ T cells does not respond in proliferative assays to soluble antigen, whereas the CD45R−,CD45RO+ subset does. Whether the CD45R+(CC76+),CD45RO+ subset, which is a smaller subset of the CD45RO+

cells, responds in proliferation assays remains to be established. A potentially important possibility is that this double positive subset, which expresses two CD45 isoforms at a high level represents a functionally distinct CD4 memory cell population with the potential to express a different cytokine profile than that of the $CD45RO^+$,$CD45R^-$($CC76^-$) subset. Such a division into TH1 and TH2 subsets is suggested by studies of human T cells.[119] The results of the three-color staining with the CD8 cells also provides an alternative explanation for the results of CTLp analysis with *Theileria parva* noted above. The $CD45R^+$ ($CC76^+$) CD8 T cells that contained a high frequency of CTLp could have been within the RO^+ subset.

The tissue distribution of CD4 and CD8 T cells expressing different CD45 isoforms shows a marked preference of the different subsets for certain tissues (Table 1-2). In adult cattle, a higher proportion of $CD4^+$ and $CD8^+$ cells in lymph nodes were $CD45R^+$($CC76^+$) than in blood. In contrast, the great majority of $CD4^+$ cells in the gut mucosa were $CD45R^-$($CC76^-$), while a higher proportion of CD8 cells in the gut mucosa were also $CD45R^-$($CC76^-$) compared to blood.[91] Thus, for the CD4 cells, the cells in the gut were predominantly the subset that expressed the CD45 isoform associated with memory responses; if these are indeed memory cells, their presence at this site will ensure a rapid response to infection of the gut mucosa.

5. Subsets of $CD4^+$ and $CD8^+$ T Cells Defined by L-Selectin

Expression of L-selectin identifies two subsets of both the CD4 and CD8 populations. The L-selectin$^+$ subsets predominate in young calves, but the proportion of cells that are L-selectin$^-$ increases with age. Expression of L-selectin does not relate to memory status as determined by proliferation assays to soluble antigen; sorted L-selectin$^+$ and L-selectin$^-$ $CD4^+$ cells from cattle inoculated with ovalbumin both proliferated to a similar extent.[80] Three-color immunofluorescent staining of PBM divided the CD4 and CD8 cells each into four subsets, with L-selectin$^+$ and L-selectin$^-$ cells within the $CD45R^+$($CC76^+$) and $CD45R^-$ populations (Figures 1-2 and 1-3). The populations are not of equal size and a larger proportion of the L-selectin$^+$ cells are within the $CD45R^+$($CC76^+$) population. Expression of L-selectin identifies cells with a tissue distribution that is consistent with the function of the molecule as the peripheral node homing receptor. A large majority of the $CD4^+$ cells in prescapular lymph nodes and tonsil was L-selectin$^+$ — 56 and 66%, respectively — as was a majority of the $CD8^+$ cells — 79 and 80%, respectively. In contrast, only 4% of $CD4^+$ cells and 2% of $CD8^+$ cells in the gut mucosa were L-selectin$^+$. In the discrete Peyer's patches, 19% of CD4 cells and CD8 cells were L-selectin$^+$. In the mesenteric and bronchial nodes which drain mucosal tissues and in the spleen, more L-selectin$^-$ cells were evident than in the prescapular lymph nodes. The abundance of L-selectin$^+$ cells correlates with expression of the epitope identified by MAb MECA-79, a receptor for L-selectin, on HEV in the organized secondary lymphoid tissues.[79,81] The uneven expression of L-selectin by CD4 T cells with the naive or memory phenotype noted above is likely to affect the migration of these two populations through peripheral lymph nodes and result in more cells of the naive phenotype leaving the blood via the HEV.[3]

V. CONCLUDING REMARKS

The characterization of cell surface differentiation antigens in sheep and cattle has led to the identification of surface markers for the major leukocyte populations and has resulted in the identification of subpopulations of T lymphocytes. Evidence for functional heterogeneity within the CD4 and CD8 subsets

Table 1-2 **Expression of CD45R and L-Selectin by $CD4^+$ and $CD8^+$ T Cells**

	% of indicated populations that are CD45R$_+$		% of indicated populations that are L-selectin$^+$	
	CD4$_+$	$CD8^+$	$CD4^+$	$CD8^+$
PBM (calf)	85	87	90	83
PBM (adult)	41	68	39	68
Prescapular LN	51	75	56	78
Mesenteric LN	33	65	22	44
Peyer's patch	9	28	19	19
Gut mucosa	<5	35	<4	<2

Note: Mean % shown, taken from References 80 and 91. CD45R (MAb CC76); L-selectin (MAb CC32); LN, lymph node; Gut mucosa, from an area not containing any organized lymphoid tissue.

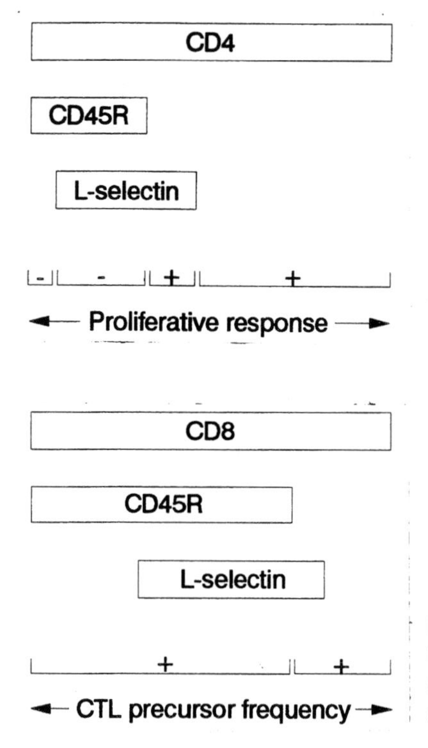

Figure 1-2 Expression of CD45R and L-selectin by bovine CD4+ T cells. Proliferative responses to soluble antigen and staining summarized from References 80 and 91. CD45R(MAb CC76). Antigen, ovalbumin injected subcutaneously with incomplete Freund's adjuvant or variable surface glycoprotein from *Trypanosoma brucei* injected intramuscularly in Freund's complete adjuvant.

Figure 1-3 Expression of CD45R and L-selectin by bovine CD8+ T cells. Staining and CTLp assays from References 80 and 91 as for Figure 1-2. Sorted PBM from calves that had recovered from infection with *Theileria parva.*

of T cells, as well as differences in their recirculation and tissue distribution, has been presented. A number of important questions remain to be answered. The full extent of the functional differences in subpopulations of CD4 and CD8 T cells, with respect to recirculation pathways, capacity to respond to different antigen presenting cells and cytokine production, has yet to be defined. The function of the WC1+ $\gamma\delta$ T cells also remains an important unanswered question of specific significance in ruminants and of general relevance to the immunobiology of $\gamma\delta$ T cells. Answers to these questions will create a level of knowledge of the immune system of ruminants that will allow in-depth investigations to characterize the immune response to pathogens and to understand the cellular events in the induction of immune responses with defined antigens.

REFERENCES

1. **Howard, C. J. and Morrison, W. I.,** Leukocyte antigens in cattle, sheep and goats, *Vet. Immunol. Immunopathol.*, 27, 1, 1991.
2. **Spertini, O., Kausas, G. S., Reimann, K. A., Mackay, C. R., and Tedder, T. F.,** Function and evolutionary conservation of distinct epitopes on the leukocyte adhesion molecule-1 (TQ-1, Leu 8) that regulate leukocyte migration, *J. Immunol.*, 147, 942, 1991.
3. **Mackay, C. R., Marston, W. L., Dudler, L., Spertini, O., Tedder, T. F., and Hein, W. R.,** Tissue-specific migration pathways by phenotypically distinct subpopulations of memory T cells, *Eur. J. Immunol.*, 22, 887, 1992.
4. **Davis, W. C., MacHugh, N. D., Park, Y. H., Hamilton, M. J., and Wyatt, C. R.,** Identification of a monoclonal antibody reactive with the bovine orthologue of CD3 (BoCD3), *Vet. Immunol. Immunopathol.*, 39, 85, 1993.
5. **Jones, M., Cordell, J. L., Beyers, A. D., Tse, A. G. D., and Mason, D. Y.,** The detection of T and B cells in many animal species using cross reactive antipeptide antibodies., *J. Immunol.*, 150, 5429, 1993.
6. **Mackay, C. R., Beya, M. F., and Matzinger, P.,** γ/δ T cells express a unique surface molecule appearing late during thymic development, *Eur. J. Immunol.*, 19, 1477, 1989.
7. **Davis, W. C., Hamilton, M. J., Park, Y.-H., Larsen, R. A., Wyatt, C. R., and Okada, K.,** Ruminant leukocyte differentiation molecules, in *MHC, Differentiation Antigens, and Cytokines in Animals and Birds. Monographs in Animal Immunology,* Barta, E. O., Ed., BAR-LAB Inc., Blacksburg, 1990, 47.

8. **Morrison, W. I. and Davis, W. C.,** 4.7 differentiation antigens expressed predominantly on CD4$^-$ CD8$^-$ T lymphocytes (WC1, WC2), *Vet. Immunol. Immunopathol.*, 27, 71, 1991.
9. **Sopp, P., Howard, C. J., and Parsons, K. R.,** Investigating monoclonal antibodies to bovine "null" cell antigens using two-colour immunofluorescence, *Vet. Immunol. Immunopathol.*, 27, 163, 1991.
10. **Parsons, K. R., Crocker, G., Sopp, P., Howard, C. J., and Davis, W. C.,** Identification of monoclonal antibodies specific for the γ/δ TCR, *Vet. Immunol. Immunopathol.*, 39, 161, 1993.
11. **Parnes, J. R.,** Molecular biology and function of CD4 and CD8, *Adv. Immunol.*, 44, 265, 1989.
12. **Maddox, J. F., Mackay, C. R., and Brandon, M. R.,** Surface antigens, SBU-14 and SBU-T8, of sheep T lymphocyte subsets defined by monoclonal antibodies, *Immunology*, 55, 739, 1985.
13. **Mackay, C.,** Sheep leukocyte molecules: a review of their distribution, structure and possible function, *Vet. Immunol. Immunopathol.*, 19, 1, 1988.
14. **Baldwin, C. L., Teale, A. J., Naessens, J. G., Goddeeris, B. M., MacHugh, N. D., and Morrison, W. I.,** Characterisation of a subset of bovine T lymphocytes that express BoT4 by monoclonal antibodies and function: similarity to lymphocytes defined by human T4 and murine L3T4[1], *J. Immunol.*, 136, 4385, 1986.
15. **Bensaid, A. and Hadam, M.,** Bovine CD4 (BoCD4), *Vet. Immunol. Immunopathol.*, 27, 51, 1991.
16. **Morrison, W. I., Howard, C. J., and Hinson, C. A.,** Polymorphism of the CD4 and CD5 differentiation antigens in cattle, *Vet. Immunol. Immunopathol.*, 27, 235, 1991.
17. **Morrison, W. I., Howard, C. J., Hinson, C. J., MacHugh, N. D., and Sopp, P.,** Identification of three distinct allelic forms of bovine CD4, submitted.
18. **Ellis, J. A., Baldwin, C. L., MacHugh, N. D., Bensaid, A., Teale, A. J., Goddeeris, B. M., and Morrison, W. I.,** Characterization by a monoclonal antibody and functional analysis of a subset of bovine T lymphocytes that express BoT8, a molecule analogous to human CD8, *Immunology*, 58, 351, 1986.
19. **Ezaki, T., Miyasaka, M., Beya, M.-F., and Trnka,** Z., A murine anti-sheep T8 monoclonal antibody, ST-8, that defines the cytotoxic T lymphocyte population, *Int. Arch. Allergy. Appl. Immunol.*, 82, 168, 1987.
20. **MacHugh, N. D. and Sopp, P.,** Bovine CD8 (BoCD8), *Vet. Immunol. Immunopathol.*, 27, 65, 1991.
22. **MacHugh, N. D., Taracha, E. L., and Toye, P. G.,** Reactivity of workshop antibodies on L cell and COS cell transfectants expressing bovine CD antigens, *Vet. Immunol. Immunopathol.*, 39, 61, 1993.
23. **Howard, C. J., Sopp, P., and Parsons, K. R.,** unpublished data, 1993.
24. **Lanier, L. L., Spits, H., and Phillips, J. H.,** The developmental relationship between NK cells and T cells, *Immunol. Today*, 13, 392, 1992.
25. **Goddeeris, B. M., Dunlap, S., Bensaid, A., MacHugh, N. D., and Morrison, W. I.,** Cell surface phenotype of two cloned populations of bovine lymphocytes displaying non-specific cytotoxic activity, *Vet. Immunol. Immunopathol.*, 27, 195, 1991.
26. **Mackay, C. R., Maddox, J. F., and Brandon, M. R.,** Three distinct subpopulations of sheep T lymphocytes, *Eur. J. Immunol.*, 16, 19, 1986.
27. **Clevers, H., MacHugh, N. D., Bensaid, A., Dunlap, S., Baldwin, C. L., Kaushal, A., Iams, K., Howard, C. J., and Morrison, W. I.,** Identification of a bovine surface antigen uniquely expressed on CD4$^-$ CD8$^-$ T cell receptor γ/δ^+ T lymphocytes, *Eur. J. Immunol.*, 20, 809, 1990.
28. **Hein, W. R., Dudler, L., Beya, M.-F., and Mackay, C. R.,** Epitopes of the T19 lymphocyte surface antigen are extensively conserved in ruminants, *Vet. Immunol. Immunopathol.*, 27, 173, 1991.
29. **Naessens, J., Olubayo, R. O., Davis, W. C., and Hopkins, J.,** Cross-reactivity of workshop antibodies with cells from domestic and wild ruminants, *Vet. Immunol. Immunopathol.*, 39, 283, 1993.
30. **Crocker, G., Sopp, P., Parsons, K., Davies, W. C., and Howard, C. J.,** Analysis of the γ/δ T cell restricted antigen WC1, *Vet. Immunol. Immunopathol.*, 39, 137, 1993.
31. **Wijngaard, P. L. J., Metzelaar, M. J., MacHugh, N. D., Morrison, W. I., and Clevers, H. C.,** Molecular characterization of the WC1 antigen expressed specifically on bovine CD4$^-$ CD8$^-$ T lymphocytes, *J. Immunol.*, 149, 3273, 1992.
32. **MacHugh, N. D., Wijngaard, P. L. J., Clevers, H. C., and Davis, W. C.,** Clustering of monoclonal antibodies recognizing different members of the WC1 gene family, *Vet. Immunol. Immunopathol.*, 39, 155, 1993.
33. **Carr, M. M., Howard, C. J., Sopp, P., Manser, J. M., and Parsons, K. R.,** Expression on porcine γ/δ lymphocytes of a phylogenetically conserved surface antigen previously restricted in expression to ruminant γ/δ T lymphocytes, *Immunology,* in press, 1993.

34. **Hughes, A. L.,** CD1 genes form a multigene family, *Mol. Biol. Evol.*, 8, 185, 1991.
35. **Howard, C. J. and Naessens, J.,** General summary of second bovine leukocyte workshop findings for cattle, *Vet. Immunol. Immunopathol.*, in press, 1993.
36. **Howard, C. J., Sopp, P., Parsons, K. R., Bembridge, G. P., and Hall, G.,** A new bovine leukocyte antigen cluster comprising two monoclonal antibodies, CC43 and CC118, possibly related to CD1, *Vet. Immunol. Immunopathol.*, 39, 69, 1993.
37. **Hopkins, J., Ross, A., and Dutia, B. M.,** Summary of second workshop findings for sheep leukocyte antigens, *Vet. Immunol. Immunopathol.*, in press, 1993.
38. **Hopkins, J. and Dutia, B. M.,** Workshop studies on the ovine CD1 homologue, *Vet. Immunol. Immunopathol.*, 27, 97, 1991.
39. **Mackay, C. R., Maddox, J. F., Gogolin-Ewens, K. J., and Brandon, M. R.,** Characterization of two sheep lymphocyte differentiation antigens, SBU-T1 and SBU-T6, *Immunology*, 55, 729, 1985.
40. **MacHugh, N. D., Bensaid, A., Davis, W. C., Howard, C. J., Parsons, K. R., Jones, B., and Kaushal, A.,** Characterization of a bovine thymic differentiation antigen analogous to CD1 in the human, *Scand. J. Immunol.*, 27, 541, 1988.
41. **Howard, C. J., Sopp, P., Bembridge, G., Young, J., and Parsons, K. R.,** Comparison of CD1 monoclonal antibodies on bovine cells and tissues, *Vet. Immunol. Immunopathol.*, 39, 77, 1993.
42. **Parsons, K. R., Howard, C. J., and Sopp, P.,** Immunohistology of workshop monoclonal antibodies to the bovine homologue of CD1, *Vet. Immunol. Immunopathol.*, 27, 201, 1991.
43. **Porcelli, S., Brenner, M. B., Greenstein, J. L., Tierhorst, C., and Bleicher, P. A.,** Recognition of cluster of differentiation 1 antigens by human $CD4^-$ $CD8^-$ cytolytic T lymphocytes, *Nature*, 314, 447, 1989.
44. **Davis, W. C., Ellis, J. A., MacHugh, N. D., and Baldwin, C. L.,** Bovine pan T-cell monoclonal antibodies reactive with a molecule similar to CD2, *Immunology*, 63, 165, 1988.
45. **Baldwin, C. L., MacHugh, N. D., Ellis, J. A., Naessens, J., Newson, J., and Morrison, W. I.,** Monoclonal antibodies which react with bovine T-lymphocyte antigens and induce blastogenesis: tissue distribution and functional characteristics of the target antigens, *Immunology*, 63, 439, 1988.
46. **Davis, W. C. and Splitter, G. S.,** Bovine CD2 (BoCD2), *Vet. Immunol. Immunopathol.*, 27, 43, 1991.
47. **Mackay, C. R., Hein, W. R., Brown, M. H., and Matzinger, P.,** Unusual expression of CD2 in sheep: implications for T cell interactions, *Eur. J. Immunol.*, 18, 1681, 1988.
48. **Giergerich, G. W., Hein, W. R., Miyasaka, M., Tiefenthaler, G., and Hönig, T.,** Restricted expression of CD2 among subsets of sheep thymocytes and T lymphocytes, *Immunology*, 66, 354, 1989.
49. **Kato, K., Koyanagi, M., Okada, H., Takanashi, T., Wong, Y. W., Williams, A. F., Okumura, K., and Yagita, H.,** CD48 is a counter-receptor for mouse CD2 and is involved in T cell activation, *J. Exp. Med.*, 176, 1241, 1992.
50. **Selvaraj, P., Plunkett, M. L., Dustin, M., Sanders, M. E., Shaw, S., and Springer, T. A.,** The T lymphocyte glycoprotein CD2 binds the cell surface ligand LFA-3, *Nature*, 326, 400, 1987.
51. **Davis, W. C. and Splitter, G. S.,** Bovine CD2 (BoCD2), *Vet. Immunol. Immunopathol.*, 27, 43, 1991.
52. **Baldwin, C. L., Morrison, W. I., and Naessens, J.,** Differentiation antigens and functional characteristics of bovine leukocytes, in *Differentiation Antigens in Lymphohemopoietic Tissues,* Miyasaka, M. and Trnka, Z., Eds., Marcel Dekker, New York, 1988, 455.
53. **Sopp, P., Howard, C. J., and Parsons, K. R.,** Investigating monoclonal antibodies to bovine "null" cell antigens using two-colour immunofluorescence, *Vet. Immunol. Immunopathol.*, 27, 163, 1991.
54. **Driscoll, P. C., Cyster, J. G., Campbell, I. D., and Williams, A. F.,** Structure of domain 1 of rat T lymphocyte CD2 antigen, *Nature*, 353, 762, 1991.
55. **Howard, C. J., Sopp, P., Preston, P. M., Jackson, L. A., and Brown, C. G. D.,** Phenotypic analysis of bovine leukocyte cell lines infected with *Theileria annulata*, *Vet. Immunol. Immunopathol.*, 39, 275, 1993.
56. **Haig, D. M., Thomson, J., and Dawson, A.,** Reactivity of the workshop monoclonal antibodies with ovine bone marrow cells and bone marrow-derived monocyte/macrophage and mast cell lines, *Vet. Immunol. Immunopathol.*, 27, 135, 1991.
57. **Beya, M.-F. and Miyasaka, M.,** Studies on the differentiation of T lymphocytes in sheep. III. Preliminary characterization of an antigen recognized by two anti-pan T-cell monoclonal antibodies, *Immunology*, 58, 71, 1986.

58. **Beya, M.-F., Miyasaka, M., Dudler, L., Ezaki, T., and Trnka, Z.,** Studies on the differentiation of T lymphocytes in sheep. II. Two monoclonal antibodies that recognize all ovine T lymphocytes, *Immunology*, 57, 115, 1986.
59. **Howard, C. J., Parsons, K. R., Jones, B. V., Sopp, P., and Pocock, D. H.,** Two monoclonal antibodies (CC17, CC29) recognizing an antigen (Bo5) on bovine T lymphocytes, analogous to human CD5, *Vet. Immunol. Immunopathol.*, 19, 127, 1988.
60. **van deVelde, H., von Hoegen, I., Luo, W., Parnes, J. R., and Thielemans, K.,** The B-cell surface protein CD72/Lyb-2 is the ligand for CD5, *Nature*, 351, 662, 1991.
61. **McAteer, M. J., Lagarde, A.-C., Georgiou, H. M., and Bellgrau, D.,** A requirement for the CD5 antigen in T cell activation, *Eur. J. Immunol.*, 18, 1111, 1988.
62. **Alberola-Ila, J., Places, L., Cantrell, D. A., Vives, J., and Lozano, F.,** Intracellular events involved in CD5-induced human T cell activation and proliferation, *J. Immunol.*, 148, 1287, 1992.
63. **Naessens, J. and Williams, D. J. L.,** Characterization and measurement of $CD5^+$ B cells in normal and *Trypanosoma congolense*-infected cattle, *Eur. J. Immunol.*, 22, 1713, 1992.
64. **Howard, C. J., Morrison, W. I., Brown, W. C., Naessens, J., and Sopp, P.,** Demonstration of two allelic forms of the bovine T cell antigen Bo5 (CD5) and studies of their inheritance, *Anim. Genet.*, 20, 351, 1989.
65. **Boyse, E. A., Miyazawa, M., Aoki, T., and Old, L. J.,** Ly-A and Ly-B: two systems of lymphocyte isoantigens in the mouse, *Proc. Roy. Soc. B*, 170, 175, 1970.
66. **Mathieson, B. J. and Sharrow, S. D.,** Cell surface differentiation antigens expressed on thymocytes and T cells in the mouse, in *Differentiation Antigens in Lymphohemopoietic Tissues*, Miyasaka, M. and Trnka, Z., Eds., Marcel Dekker, New York, 1988, 65.
67. **Letesson, J. J. and Bensaid, A.,** Bovine CD6 (BoCD6), *Vet. Immunol. Immunopathol.*, 27, 61, 1991.
68. **Lobb, R., Hession, C., and Osborn, L.,** Vascular cell adhesion molecule-1, *Cell Mol. Mech. Inflamm.*, 2, 151, 1991.
69. **Uciechowski, P. and Schmidt, R. E.,** N1 cluster report: CD11, in *Leukocyte Typing IV,* Knapp, W. et al.. Eds., Oxford University Press, New York, 1989, 543.
70. **Hein, W. R. and Mackay, C. R.,** Other surface antigens identified on sheep leukocytes, *Vet. Immunol. Immunopathol.*, 27, 115, 1991.
71. **Mackay, C. R., Marston, W. L., and Dudler, L.,** Naive and memory T cells show distinct pathways of lymphocyte recirculation, *J. Exp. Med.*, 171, 801, 1990.
72. **Splitter, G. and Morrison, W. I.,** Antigens expressed predominantly on monocytes and granulocytes: identification of bovine CD11b and CD11c, *Vet. Immunol. Immunopathol.*, 27, 87, 1991.
73. **Letesson, J. J. and Delcommenne, M.,** Production of a monoclonal antibody to the light chain of the bovine β_2-integrin family (BoCD18), *Vet. Immunol. Immunopathol.*, 39, 103, 1993.
74. **Boyd, A. W., Wicks, I. P., Wilkinson, D., Novotny, J. R., Campbell, I., Wawryk, S. O., Harrison, L. C., and Burns, G. F.,** N14.1 Intercellular adhesion molecule 1 (ICAM-1): regulation and role in cell contact-mediated lymphocyte function, in *Leukocyte Typing IV*, Knapp, W. et al., Eds., Oxford University Press, New York, 1989, 684.
75. **McKeever, D. J., MacHugh, N. D., Goddeeris, B. M., Awino, E., and Morrison, W. I.,** Bovine afferent lymph veiled cells differ from blood monocytes in phenotype and accessory function, *J. Immunol.*, 147, 3703, 1991.
76. **Eskra, L., O'Reilly, K. L., and Splitter, G. A.,** The bovine p150/95 molecule (CD11c/CD18) functions in primary cell-cell interaction, *Vet. Immunol. Immunopathol.*, 29, 213, 1991.
77. **Kehrli, M. E. K., Jr., Ackermann, M. R., Shuster, D. E., van der Maaten, M. J., Schmalstieg, F. C., Anderson, D. C., and Hughes, B. J.,** Animal model of human disease: bovine leukocyte adhesion deficiency: β_2 integrin deficiency in young holstein cattle, *Am. J. Pathol.*, 140, 1489, 1992.
78. **Hogg, N.,** Roll, roll, roll your leucocyte gently down the vein, *Immunol. Today*, 13, 113, 1992.
79. **Mackay, C. R., Marston, W., and Dudler, L.,** Altered patterns of T cell migration through lymph nodes and skin following antigen challenge, *Eur. J. Immunol.*, 22, 2205, 1992.
80. **Howard, C. J., Sopp, P., and Parsons, K. R.,** L-selectin expression differentiates T cells isolated from different lymphoid tissues in cattle but does not correlate with memory, *Immunology*, 77, 228, 1992.
81. **Berg, E. L., Picker, L. J., Robinson, M. K., Streeter, P. R., and Butcher, E. C.,** Vascular addressins: tissue selective endothelial cell adhesion molecules for lymphocyte homing, *Cell. Mol. Mech. Inflamm.*, 2, 111, 1991.

82. **Bosworth, B. T., St. John, T., Gallatin, W. M., and Harp, J. A.,** Sequence of the bovine CD44 cDNA: comparison with human and mouse sequences, *Mol. Immunol.*, 28, 1131, 1991.
83. **Gallatin, W. M., Rosenman, S. J., Ganji, A., and St. John, T. P.,** Structure-function relationships of the CD44 class of glycoproteins, *Cell. Mol. Mech. Inflamm.*, 2, 131, 1991.
84. **Naessens, J., Sileghem, M., MacHugh, N., Park, Y. H., Davis, W. C., and Toye, P.,** Selection of BoCD25 monoclonal antibodies by screening mouse L cells transfected with the bovine p55-interleukin-2 (IL-2) receptor gene, *Immunology*, 76, 305, 1992.
85. **Taylor, B. C., Stott, J. L., Scibienski, R. J., and Redelman, D.,** Development and characterization of a monoclonal antibody specific for the bovine low-affinity interleukin-2 receptor, BoCD25, *Immunology*, 77, 150, 1992.
86. **Weinberg, A. D., Shaw, J., Paetkau, V., Bleackley, R. C., Magnuson, N. S., Reeves, R., and Magnuson, J. A.,** Cloning of cDNA for the bovine IL-2 receptor (bovine Tac antigen), *Immunology*, 63, 603, 1988.
87. **Bujdoso, R., Lund, B. T., Evans, C. W., and McConnell, I.,** Different rates of IL-2 receptor expression by ovine γ/δ and α/β T cells, *Vet. Immunol. Immunopathol.*, 39, 109, 1993.
88. **Naessens, J. and Howard, C. J.,** Monoclonal antibodies reacting with bovine B cells (BoWC3, BoWC4 and BoWC5), *Vet. Immunol. Immunopathol.*, 27, 77, 1991.
89. **Nthale, J. M. and Naessens, J.,** Characterization of a late activation antigen defined by monoclonal antibodies of cluster BoWC8 (TC23), *Vet. Immunol. Immunopathol.*, 39, 201, 1993.
90. **Thomas, M. L.,** The leukocyte common antigen family, *Annu. Rev. Immunol.*, 7, 339, 1989.
91. **Howard, C. J., Sopp, P., Parsons, K. R., McKeever, D. J., Taracha, E. L. N., Jones, B. V., MacHugh, N. D., and Morrison, W. I.,** Distinction of naive and memory BoCD4 lymphocytes in calves with a monoclonal antibody, CC76, to a restricted determinant of the bovine leukocyte-common antigen, CD45, *Eur. J. Immunol.*, 21, 2219, 1991.
92. **Bembridge, G. P., Parsons, K. R., Sopp, P., MacHugh, N. D., and Howard, C. J.,** Comparison of mAb with potential specificity for restricted isoforms of the leukocyte common antigen (CD45R), *Vet. Immunol. Immunopathol.*, 39, 129, 1993.
93. **Dutia, B. M., Ross, A. J., and Hopkins, J.,** Comparison of workshop CD45R monoclonal antibodies with OvCD45R monoclonal antibodies in sheep, *Vet. Immunol. Immunopathol.*, 39, 121, 1993.
94. **MacHugh, N. D., Awino, E. O., and McKeever, D. J.,** A monoclonal antibody specific for bovine CD45RO differentiate memory from naive $CD4^+$ bovine T cells, *Proc. 3rd IVIS, Budapest*, 548, 1992.
95. **Morrison, W. I., Baldwin, C. L., MacHugh, N. D., Teale, A. J., Goddeeris, B. M., and Ellis, J.,** Phenotypic and functional characterisation of bovine lymphocytes, *Prog. Vet. Microbiol. Immunol.*, 4, 134, 1988.
96. **Williams, D. J. L., Newson, J., and Naessens, J.,** Quantitation of bovine immunoglobulin isotypes and allotypes using monoclonal antibodies, *Vet. Immunol. Immunopathol.*, 24, 267, 1990.
97. **Naessens, J., Newson, J., MacHugh, N., Howard, C. J., Parsons, K. R., and Jones, B. V.,** Characterization of a bovine leucocyte differentiation antigen of 145,000 MW restricted to B lymphocytes, *Immunology*, 69, 525, 1990.
98. **Parsons, K. R., Bembridge, G., Sopp, P., and Howard, C. J.,** Studies of monoclonal antibodies identifying two novel bovine lymphocyte antigen differentiation clusters: workshop clusters (WC) 6 and 7, *Vet. Immunol. Immunopathol.*, 39, 187, 1993.
99. **Sopp, P., Howard, C. J., and Parsons, K. R.,** A new non-lineage specific antigen with an M_r of 115 kDa and 39 kDa present on bovine leukocytes identified by monoclonal antibodies within BoWC10, *Vet. Immunol. Immunopathol.*, 39, 209, 1993.
100. **Letesson, J. J., Mager, A., Mammerickx, A., Burny, A., and Depelchin, A.,** B cells from bovine leukemia virus (BLV) infected sheep with hematologic disorders express the CD5 B cell marker, *Leukaemia*, 4, 377, 1990.
101. **Ritter, M. A., Crispe, I. N., and Male, D.,** *The Thymus,* Oxford University Press, New York, 1992, 85.
102. **Hein, W. R.,** Ontogeny of T cells, this publication, Chapter 2, 1993.
103. **Kansas, G. S., Spertini, O., and Tedder, T. F.,** Leukocyte adhesion molecule-1 (LAM-1): structure, function, genetics and evolution, *Cell. Mol. Mech. Inflamm.*, 2, 31, 1991.
104. **Howard, C. J., Sopp, P., Parsons, K. R., and Finch J.,** *In vivo* depletion of BoT4 (CD4) and of non-T4/T8 lymphocyte subsets in cattle with monoclonal antibodies, *Eur. J. Immunol.*, 19, 757, 1989.

105. **Gill, H. S., Watson, D. L., and Brandon, M. R.,** *In vivo* inhibition by a monoclonal antibody to CD4+ T cells of humoral and cellular immunity in sheep, *Immunology*, 77, 38, 1992.
106. **Parsons, K. R., Howard, C. J., Jones, B. V., and Sopp, P.,** Investigation of bovine gut associated lymphoid tissue (GALT) using monoclonal antibodies against bovine lymphocytes, *Vet. Pathol.*, 26, 396, 1989.
107. **Goddeeris, B. M., Morrison, W. I., and Teale, A. J.,** Generation of bovine cytotoxic cell lines, specific for cells infected with the protozoan parasite *Theileria parva* and restricted by products of the major histocompatibility complex, *Eur. J. Immunol.*, 16, 1243, 1986.
108. **Preston, P. M., Brown, C. G. D., and Spooner, R. L.,** Cell-mediated cytotoxicity in *Theileria annulata* infection of cattle with evidence for BoLA restriction, *Clin. Exp. Immunol.*, 53, 88, 1983.
109. **Cook, C. G. and Splitter, G. A.,** Comparison of bovine mononuclear cells with other species for cytolytic activity against virally-infected cells, *Vet. Immunol. Immunopathol.*, 20, 239, 1989.
110. **Denis, M., Slaoui, M., Keil, G., Babiuk, L. A., Ernst, E., Pastoret, P.-P., and Thiry, E.,** Identification of different target glycoproteins for bovine herpes virus type 1-specific cytotoxic T lymphocytes depending on the method of *in vitro* stimulation, *Immunology*, 78, 7, 1993.
111. **Nagi, A. M. and Babiuk, L. A.,** Characterization of surface markers of bovine gut mucosal leukocytes using monoclonal antibodies, *Vet. Immunol. Immunopathol.*, 22, 1, 1989.
112. **Gyorffy, E. J., Glogauer, M., Kennedy, L., and Reynolds, J. D.,** T-cell receptor association with lymphocyte populations in sheep intestinal mucosa, *Immunology*, 77, 25, 1992.
113. **Park, Y. H., Fox, L. K., Hamilton, M. J., and Davis, W. C.,** Bovine mononuclear leukocyte subpopulations in peripheral blood and mammary gland secretions during lactation, *J. Dairy. Sci.*, 75, 998, 1992.
114. **Taylor, G., Howard, C. J., Thomas, L. H., Wyld, S. G., Furze, J., Sopp, P., and Gaddum, R.,** Effect of T cell depletion on RSV infection in calves, *Proc. 3rd IVIS, Budapest*, 1992, 8.
115. **Oldham, G., Bridger, J. C., Howard, C. J., and Parsons, K. R.,** Effect of *in vivo* depletion with monoclonal antibody on infection and immune response of calves to rotavirus, *J. Virol.*, 67, 5012, 1993.
116. **Rocha, B., Vassalli, P., and Guy-Grand, D.,** The extrathymic T-cell development pathway, *Immunol. Today*, 13, 44, 1992.
117. **Wyatt, C. R., Madruga, C., Cluff, C., Parish, S., Hamilton, M. J., Goff, W., and Davis, W. C.,** Differential distribution of γ/δ T cell receptor lymphocyte subpopulations in blood and spleen of young and adult cattle, *Vet. Immunol. Immunopathol.*, in press, 1993.
118. **Mackay, C. R., Kimpton, W. G., Brandon, M. R., and Cahill, R. N. P.,** Lymphocyte subsets show marked differences in their distribution between blood and the afferent and efferent lymph of peripheral lymph nodes, *J. Exp. Med.*, 167, 1755, 1988.
119. **Mason, D. and Powrie, F.,** Memory CD4+ T cells in man form two distinct subpopulations, defined by their expression of isoforms of the leucocyte common antigen, CD45, *Immunology*, 70, 427, 1990.
120. **O'Reilly, K. L. and Splitter, G. A.,** $CD5^{dim}$ peripheral blood lymphocytes proliferate to exogenous IL-2 in the absence of antigen and kill in an NK-like manner, *Cell. Immunol.*, 130, 389, 1990.
121. **Bembridge, G., and Howard, C. J.,** unpublished 1993.

Chapter 2

Ontogeny of T Cells

W. R. Hein

CONTENTS

I. Introduction 19
II. Development of the Thymus 19
  A. Morphogenesis and Colonization of the Fetal Thymus 19
  B. Growth and Regression of the Thymus 21
III. Intrathymic Maturation of αβ T Cells 22
  A. Pathways of Thymocyte Development 23
  B. T Cell Receptor Gene Rearrangement and Expression 24
  C. Intrathymic Selection 25
IV. Intrathymic Maturation of γδ T Cells 26
V. Development of Peripheral T Cells 27
  A. Emigration from Thymus 27
  B. Expansion of the Peripheral T Cell Pool 28
  C. Ontogeny of the Peripheral T Cell Repertoire 30
  D. Extrathymic T Cell Development 32
Acknowledgments 33
References 33

I. INTRODUCTION

In the 30 years since the landmark experiments of Miller,[1] immunologists have uncovered a wealth of information about the complex series of events that occur in the thymus and confer to this organ such a central role in the immune system — the production of T lymphocytes. In terms of the actual movement of cells, the process might seem straightforward enough: immature precursor cells migrate into the thymus; once in the thymus, they proliferate, progress through a series of maturational stages, and, after appropriate selection, the surviving mature thymocytes then emigrate to the periphery to join other T cells in the secondary lymphoid organs. However, each stage in this process, namely the entry, maturation, selection, and exit of cells, is regulated by complex mechanisms that are not yet completely understood. Furthermore, the thymic development of T cells may be modulated differently in animals of different ages and some T cells also develop by extrathymic pathways. Keeping these points in mind, the current level of knowledge about T cell ontogeny in ruminant animals will be reviewed in this chapter. Of course, the vast majority of our present understanding of T cell ontogeny is derived from studies done in other species, chiefly laboratory rodents. While it is not the purpose of the present chapter to review work in other species in great detail, a coherent account of T cell ontogeny cannot be achieved without considering some of those studies.

II. DEVELOPMENT OF THE THYMUS

A. MORPHOGENESIS AND COLONIZATION OF THE FETAL THYMUS

The thymic anlage is first visible in 27-to 30-day-old sheep embryos (fetal age is described throughout as days post-coitus) as an epithelial chord growing from the ventral diverticulum of the third pharyngeal pouch.[2] Dorsally it is related closely to the third aortic arch and ventrally it is associated with the pericardium. As the fetal neck grows and extends, the epithelial components of the primordium proliferate as a chord of tissue which by day 33 has developed a characteristic lobular structure (Figure 2-1B). By day 40, and probably earlier, the thymus is divided into distinct cervical and mediastinal regions that are connected at the thoracic inlet by a thin isthmus of tissue (Figure 2-1A). The cervical thymus occurs as two elongated chords, fused at their caudal extremities, lying close to the jugular veins beneath the ventral

0-8493-4952-4/94/$0.00+$.50

cervical muscles (Figure 2-1C). At day 41, the mediastinal component also has a dual nature (Figure 2-1C) but the two parts fuse shortly afterwards.[2] This basic anatomical arrangement persists from this time onward and further thymic development is associated mainly with an increase in size and cellularity.

The thymic rudiment of sheep embryos consists mainly of epithelial cells, derived from both ectodermal and endodermal components, as shown by the expression of characteristic markers such as keratins.[3] Colonization by hemopoietic precursor cells probably begins within 2 to 3 days of the appearance of the thymic anlage since large, basophilic cells can already be seen within the epithelial framework in sections derived from a 30-day-old fetus.[2] Examination of fresh explants of 33-day-old fetal thymus by transmission light microscopy reveals a few large, round, lymphoid-like cells scattered within and sometimes adherent to a translucent network of epithelial and mesenchymal cells (Figure 2-1B).

The origin of the precursor cells that migrate into the fetal lamb thymus is not known with certainty.

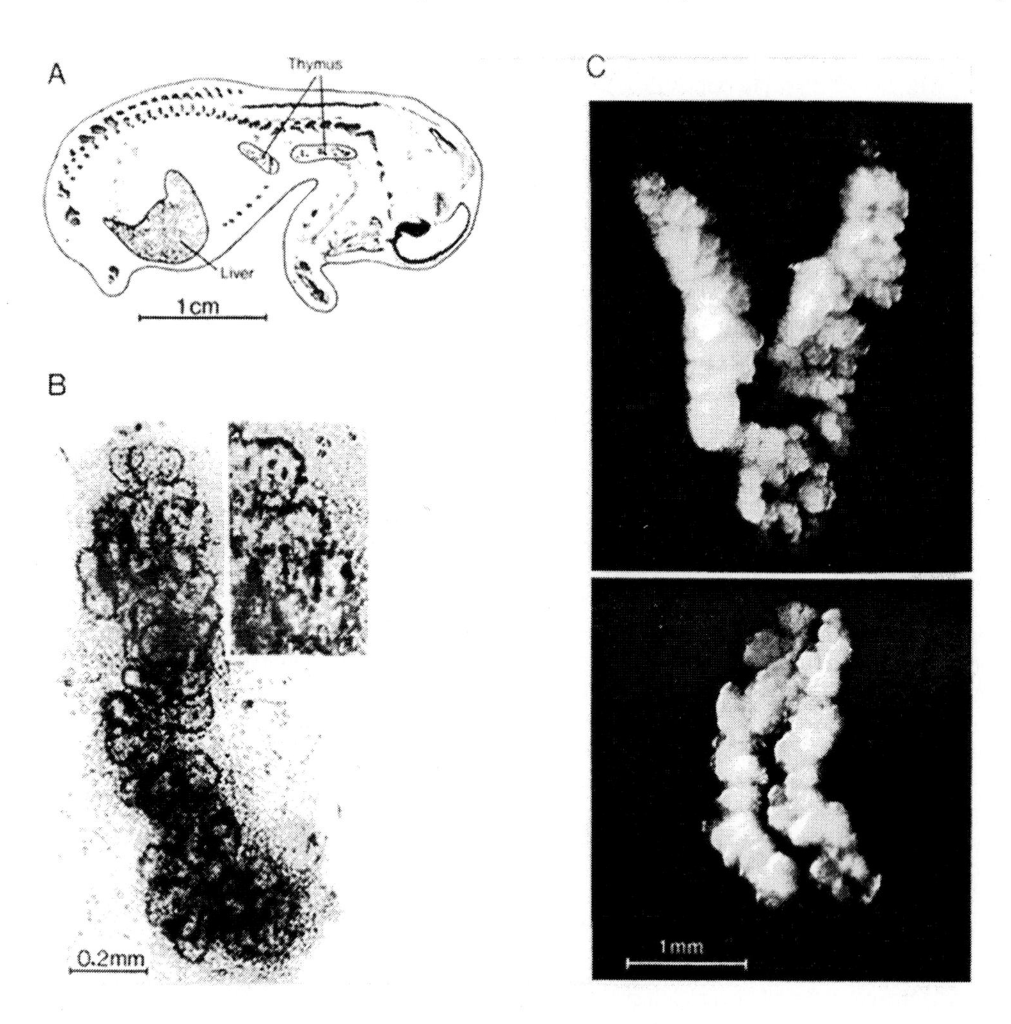

Figure 2-1 Gross morphology of the thymus in sheep embryos and early fetuses. (A) Outline of a whole-body section from a 41-day-old fetus showing the location of the mediastinal and cervical thymus and the fetal liver. (B) Thymus from a 33-day-old embryo after overnight *in vitro* culture, photographed by transmitted light. Note the lobulated appearance of the epithelial and mesenchymal framework. The inset shows the top right-hand edge of the thymus at higher magnification. Dense, round lymphoid-like cells are indicated within the epithelial stroma (arrows). (C) Cervical (top) and mediastinal (bottom) thymus from a 40-day-old fetus photographed by reflected light. At this stage of development, both parts of the thymus have a dual nature.

Surface markers associated with the lymphoid lineage, such as the leucocyte common antigen (CD45) and MHC class I antigens, are expressed first by a few cells in the yolk sac and at other sites scattered throughout day 19 embryos.[4,5] This is just after the time that the vitelline and embryonic blood vessels fuse, so that these cells may have traveled from the yolk sac to other sites via the circulation. Hemopoiesis continues in the yolk sac until around 27 days of gestation, when the organ disappears,[6] and the yolk sac therefore is unlikely to be the direct source of the first immigrants entering the thymus 2 to 3 days later. From the earliest stages examined (20 to 25 days), the fetal liver contains cells that express CD45, MHC class I and CD1 molecules, as well as myeloid markers,[5,7] and it constitutes the major hemopoietic organ from this time at least until when the bone marrow starts to develop at around 70 to 75 days of gestation.[7] Therefore, the fetal liver is the most likely immediate source of thymic immigrants during this period of gestation, but it cannot be excluded that other populations of precursor cells dispersed as foci in other embryonic sites might have a specialized role in the initial colonizing events.

Studies in chicken and rat embryos have given the clearest insights obtained so far as to the actual process of initial thymic colonization. Hemopoietic precursors extravasate from adjacent blood vessels, migrate in the mesenchyme, traverse the perithymic basement membrane, and migrate into the thymic epithelial primordium.[8] The invasive process is regulated by at least two mechanisms. First, the invading cells express receptors for structural components of basement membranes and the extracellular matrix, such as fibronectin and laminin, and colonization is inhibited by antibodies reacting with the receptors.[8] Second, the migration of the invading cells is a chemotactic process induced by specific peptides secreted by the thymic epithelium, originally termed "thymotaxin", but later shown to be soluble β2-microglobulin.[9-13] Immigration of precursors into the thymus of chick embryos occurs as three distinct waves at different times in development and available evidence suggests that a similar phenomenon involving two waves of migration may also occur in rodents.[14,15] It is likely that some aspects of the invasion mechanism will vary during the later colonization waves since the architecture, blood supply, and overall physiology of the thymus probably changes rapidly after the first wave.

Some preliminary studies have been done to determine whether a similar overall mechanism might operate in the sheep fetus. Using an in vitro Boyden chamber assay, thymocytes and mononuclear cells isolated from the liver of 39- to 49-day-old sheep fetuses were tested for their ability to show directional migration in response to chemotactic peptides produced by a rat thymic epithelial cell line. None of the fetal liver cell preparations showed a detectable migratory response, perhaps because the frequency of precursor cells was too low. However, during a short window of development (days 40 to 41), fetal thymocytes showed significant migration, while at other times they did not.[16] Although more studies need to be done, these results support the idea that the mechanism of chemotactic migration of thymic precursors is well conserved between different mammals and suggest that the colonization of the fetal sheep thymus may also occur in distinct waves.

Surprisingly perhaps, considering the difference in the length of gestation between sheep and cattle, the time course of thymic morphogenesis in the fetal calf seems to be quite similar to that of the fetal lamb. Thus, a primordial thymus can be detected in the region of the third pharyngeal pouch at approximately 30 days of gestation,[17] and lymphoid development occurs within the fetal bovine thymus by day 42.[18] By comparison, in humans, where the length of gestation is similar to that in cattle, the fetal thymus is not colonized with hemopoietic cells until around day 55 to 60.[19]

B. GROWTH AND REGRESSION OF THE THYMUS

In the first few days after colonization, there is an exponential increase in the total number of thymocytes that can be recovered from the fetal lamb thymus; but by 49 days of gestation, the rate of increase becomes slower (Figure 2-2A). Nonetheless, the thymus continues to increase in size throughout fetal development and post-natally, reaching its maximum absolute weight in the first 2 months after birth. From this time onward, there is a gradual decrease in the mass of the thymus (Figure 2-2B). When expressed as a percentage of body weight, the thymic index reaches its maximum value in the fetal lamb at around 120 days of gestation and decreases rapidly after birth (Figure 2-2B). Essentially similar relationships exist between the size of the thymus and body weight of the bovine fetus.[18]

Most of the increase in the mass of the thymus is almost certainly due to the intrathymic proliferation of thymocytes rather than the continuous ingress of precursor cells. In a study of cell division in sheep thymocytes, the highest rate of division was measured in 40-day-old fetuses, when 30% of thymocytes

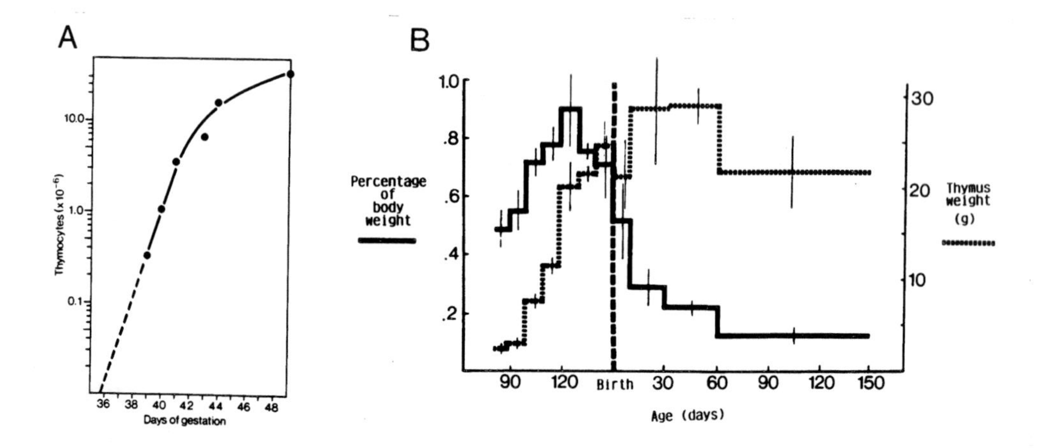

Figure 2-2 Growth of the sheep thymus. (A) Number of thymocytes isolated from fetal thymuses at different stages of gestation. In early fetuses, there is an exponential increase in thymocyte numbers. (B) Changes in the absolute weight of thymus and in the thymic index (percent of body weight) during fetal and early post-natal development. The graph is derived from observations made on Australian[20] and Swiss[22] sheep.

incorporated BrdU over a 1-h labeling period.[21] The overall rates of division at later stages of development were lower, with incorporation levels of 13 and 9% being measured in 125-day-old fetuses and 8-week-old lambs, respectively.[21]

Although the thymus begins to regress from about the time of puberty, it does not necessarily involute completely and older animals may contain a substantial thymus. For example, Swiss sheep aged 3 to 6 years invariably contain a 10- to 20-g cervical thymus which is essentially normal on histological examination, although there can be variable degrees of fatty infiltration in the outer cortex of some lobules in the older animals. The mediastinal thymus is not always present and appears to regress earlier and more completely. The oldest animal of confirmed age that we have examined was a 13-year-old ewe. At post-mortem examination, the cervical thymus appeared to have been replaced completely by serous fatty tissue. However, on histological examination, small islands of lymphoid tissue that were positive for a number of lymphocyte differentiation markers could still be detected, although the thymic architecture was abnormal.[22] The rate of thymic regression is influenced by a number of factors and is usually accelerated by stress. The extent of thymic regression at a given age is therefore likely to vary widely between animals exposed to different husbandry and environmental conditions.

III. INTRATHYMIC MATURATION OF $\alpha\beta$ T CELLS

The maturational changes that occur in thymocytes from the time they enter the thymus as a prothymocyte stem cell until their exit as mature T cells are complex and have been studied in a number of different ways using various phenotypic and genotypic markers to identify intermediate stages in this process. In the invading stem cells, the T cell receptor (TCR) loci are in germline configuration, but these genes begin to rearrange in a defined sequence shortly after entry to the thymus. In addition, the nature and density of molecules expressed at the cell surface changes in defined ways as thymocytes become progressively more mature and their functional properties also change. All of these processes may partly reflect a developmental potential that is intrinsic to each stem cell, but they are also controlled and regulated at many stages by interactions between the thymocytes and other cells in the thymus, at the level of both cell-cell contact and through soluble mediators such as cytokines. Two distinct lineages of T cells that are distinguished by the type of TCR expressed on the cell surface develop within the thymus. The pathways involved in the maturation of $\alpha\beta$ T cells are generally better understood, and a number of detailed reviews have been published recently.[23-26] The main cellular and molecular events involved in the development of this lineage will be discussed briefly here, following the schematic outline shown in Figure 2-3. Some reference is also made to the development of $\gamma\delta$ T cells, although this lineage is considered in more detail in a later section.

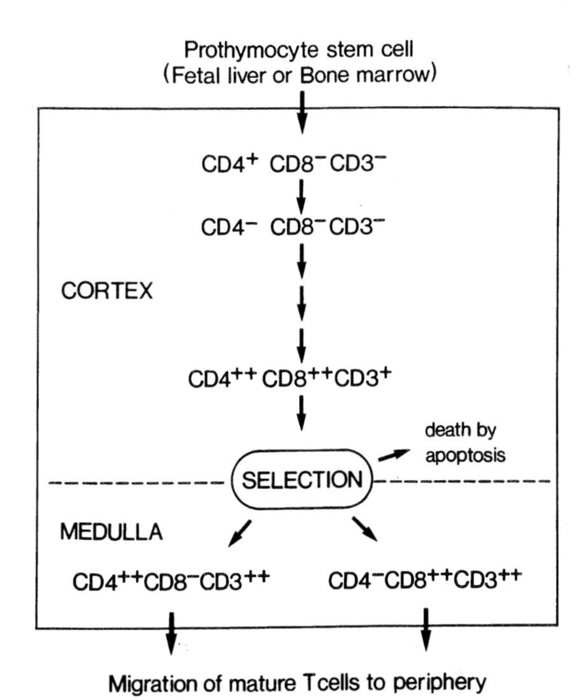

Figure 2-3 Schematic outline of the intrathymic development of TCR αβ T cells. The major stages of thymocyte development are shown in terms of the expression of the CD4, CD8, and CD3 molecules. For a detailed discussion, see text.

A. PATHWAYS OF THYMOCYTE DEVELOPMENT

There is no general agreement as to the nature of the stem cells that migrate into the thymus and no single surface phenotype by which these cells can be reliably identified.[27] During early fetal development, these cells arise in the liver and during later fetal and adult stages in the bone marrow, but it remains unclear whether they remain multipotent or are restricted in their developmental potential by the time they enter the thymus. It is possible that these properties differ between fetal liver- and adult bone marrow-derived stem cells. In later fetal and post-natal stages, the thymus is extensively vascularized and the mechanism of invasion of stem cells will be quite different from the initial colonizing events. The migration of pro-T cells through the vascular endothelium is clearly a necessary event and several members of distinct families of adhesion receptor molecules are thought to be involved in this process; the overall mechanism of stem cell homing to the thymus has been reviewed recently.[28] Reconstitution studies in adult mice show that the thymus can accept only a small number of stem cells and that new immigrants undergo about a week of development before recognizable subpopulations of $CD4^-CD8^-$ progeny can be detected.[29] The earliest recognizable intrathymic precursor cells have recently been shown to express moderate levels of the CD4 molecule ($CD4^+CD8^-CD3^-$).[30] These cells have their TCR genes in germline configuration and can give rise to both αβ and γδ T cells,[31] so presumably the divergence between these two lineages could occur quite soon after stem cell entry. At this time, the thymocytes downregulate CD4 expression and become so-called *double-negative* thymocytes ($CD4^-CD8^-CD3^-$). These immature cells localize as a narrow margin in the outer cortex (see Figure 2-4 d-f).

Progression from the double-negative state to the *double-positive* phenotype ($CD4^+CD8^+CD3^+$) probably occurs by one of several pathways since an increasing number of distinct intermediate stages are now being identified.[23,24] During this interval, there are many changes in thymocyte metabolism. TCR gene rearrangement commences in a defined temporal sequence (see below) and culminates in the eventual expression of low levels of surface CD3/TCR complexes. The expression of CD4 and CD8 molecules is also regulated, and intermediate *single-positive* stages have been described. There are also marked changes in the secretion of different cytokines and in the expression of cytokine receptors.[25] Once the developing thymocytes express both CD4 and CD8 and have a functional surface TCR, they are exposed to two types of selection, positive and negative, which have a critical role in first determining whether they survive or are destined for cell death, and second, whether they will become either CD4 or CD8 mature *single-positive* T cells (see below). The selected thymocytes upregulate the expression of the CD3/TCR complex and localize predominantly in the thymic medulla, from where they are considered to emigrate to the periphery.

There have been no detailed studies in ruminants of precursor-product relationships between thymocyte subpopulations, and the scheme outlined above is based primarily on rodent studies. Nonetheless, the patterns of expression of differentiation antigens on fetal and adult sheep thymocytes indicate that the overall pathways of thymocyte development are likely to be essentially similar to those in the mouse, although the composition of ovine thymocyte subsets changes to the adult pattern quite early in ontogeny.[5,32] Lymphoid cells found in the 35-day-old fetal lamb thymus are negative for all T cell markers except CD5. From day 40 to 80 of fetal development, the CD8 and CD4 antigens are expressed on an increasing number of thymocytes although the number of $CD8^+$ cells always exceeds the number of $CD4^+$ thymocytes. From day 80 of fetal development, the thymus resembles that of an adult animal in terms of the proportion of different subsets defined by the CD4 and CD8 molecules (Figure 2-4 d,e). In the postnatal thymus, the proportions of the different subsets are as follows: $CD4^-CD8^-$ (12%), $CD4^+CD8^+$ (74%), $CD4^+CD8^-$ (10%), and $CD4^-CD8^+$ (4%).[32] The double-negative subset is therefore more numerous in sheep than in rodents and humans, where it comprises 3 to 6% of thymocytes. The localization of cells expressing the CD4 and CD8 molecules in the bovine thymus is similar to that in sheep.[33,34] Some other differentiation molecules show variable expression in different parts of the thymus and again have a similar distribution in the sheep and bovine thymus. CD1 is expressed predominantly on cortical cells with only scattered positive thymocytes in the medulla,[5,35] whereas MHC class I antigens are expressed mainly on medullary thymocytes.[5] In the adult thymus, the "homing" molecule CD44 is expressed on medullary cells and a limited number of cortical cells. A similar pattern is evident in the fetal thymus although in early stages there is a relatively higher level of expression on outer cortical thymocytes.[36] All thymocytes express the CD45 leucocyte common antigen but some restricted determinants of this molecular complex that may be associated with a particular maturation stage are expressed by a minority of thymocytes. For example, only 2% of cells in the bovine thymus express the CD45 determinant detected by MAb CC76. The positive cells occur mainly in the medulla and nearly all of them co-express other markers typical of mature T cells, suggesting that the expression of this restricted CD45 epitope may be aquired just prior to emigration to the periphery.[37]

B. T CELL RECEPTOR GENE REARRANGEMENT AND EXPRESSSION

During intrathymic development, the genes encoding the TCR V regions are assembled in a series of ordered gene rearrangements. The analysis of both fetal thymus and immature adult thymocytes indicates that gene rearrangement at the four TCR loci occurs in the order γ, δ, β, α. Moreover, the earliest rearrangement events are targeted to the most proximal V and J gene segments within each locus so that the early repertoire produced by rearrangement seems to reflect the physical location of the genes. The overall process of TCR gene rearrangement is described in Chapter 4. In this section, we will briefly consider other issues, including the divergence of the $\alpha\beta$ and $\gamma\delta$ lineages and the regulation of allelic exclusion and TCR gene transcription during T cell development.

When the immature $CD4^-CD8^-$ thymocytes begin to rearrange their TCR genes, two critical levels of control must be imposed to ensure that each mature T cell will eventually express only one TCR at the cell surface. First, there has to be a mechanism that will direct the maturing thymocyte to become either a $\gamma\delta$ or an $\alpha\beta$ T cell; i.e., there has to be a point of divergence between the two T cell lineages. Because the γ and δ loci undergo rearrangement first, a progressive model of T cell development was initially proposed whereby all immature thymocytes attempt rearrangement of these two loci. According to this scheme, productive γ and δ gene rearrangement would result in further differentiation into $\gamma\delta$ T cells, whereas nonproductive rearrangements at either γ or δ loci would signal the thymocytes to rearrange TCR α and β genes instead.[38] However, more recent observations are not compatible with a simple stochastic model of this type (reviewed in Reference 39). For example, transgenic mice containing functionally rearranged γ and δ genes still produce normal numbers of $\alpha\beta$ T cells which contain the rearranged transgenes but these are not transcribed. Also, mice that have been depleted of the TCR α locus and cannot produce $\alpha\beta$ T cells do not have a compensatory increase in the number of $\gamma\delta$ T cells; only a low number of these cells are produced, comparable to the level in normal mice.[40] Therefore, other mechanisms that are not dependent on functional TCR rearrangement seem to regulate both the divergence between the two lineages and control the numbers of each type of T cell produced. Exactly how this is achieved remains unclear, although a good deal of evidence indicates a critical role for specific silencing elements that selectively block gene expression at the transcriptional level.[38,39,41]

T cells that become committed to the $\alpha\beta$ lineage first rearrange the TCR β locus, followed by TCR α. It is becoming increasingly clear that at least the earliest rearrangement events in immature thymocytes are not strictly random since the joining events seem to be targeted. The Vα gene segments located nearest the 3' end of the gene cluster are frequently associated with the Jα segments nearest the 5' end; there is also evidence, albeit less striking, for targeted rearrangement of TCR β segments (reviewed in Reference 42). Following productive rearrangement at both loci, a mature TCR transcript can be produced, translated into protein, assembled into the CD3/TCR complex, and expressed at the cell surface. Since thymocytes are diploid cells and contain two alleles of each locus, the rearrangement process could, at least in principle, lead to two functional rearrangements at the TCR α and/or β chain loci, giving the possibility of more than one TCR $\alpha\beta$ heterodimer combination. To prevent this, a second level of control must exist during the rearrangement process to achieve allelic exclusion; several mechanisms which could account for this have been reviewed recently.[42,43] About half of all developing T cells undergo VDJ rearrangements at both TCR β alleles, although usually only one of these is productive. However, a surprisingly high frequency of T cell clones have productive rearrangements at both α alleles and this may reflect an elevated level of secondary rearrangements at this locus. Most models of allelic exclusion predict that the surface expression of a functional TCR is a critical step and leads to signals which prevent subsequent rearrangements, although the exact way this is mediated remains controversial.[43] TCR β chain homodimers have recently been detected on immature T cells and this receptor may play a role in mediating β chain allelic exclusion and initiating TCR α rearrangement.[44]

To date, only a single study has been done on TCR gene expression in the sheep thymus.[45] Thymocytes were separated into subsets defined by the CD4 and CD8 molecules and analyzed for TCR transcripts by Northern hybridization. The double-negative thymocytes, which localize mainly in the outer cortex, expressed mature message for TCR γ and δ chains and strong but predominantly truncated β chain transcripts, whereas α chain transcripts were barely detectable. Inner cortical and medullary thymocytes expressed high levels of mature $\alpha\beta$ transcripts, low levels of mature δ chain transcripts, but no detectable γ chain message. These patterns of TCR expression are essentially similar to those detected in other animals and suggest that (1) TCR gene rearrangement commences in sheep outer cortical thymocytes and that this population contains intrathymic precursors for both the $\alpha\beta$ and $\gamma\delta$ lineages; and (2) there is a gradient of increasing maturity of $\alpha\beta$ T cells from the cortex to the medulla leading to the expression of a functional TCR in the $CD4^+CD8^+$ subset. As described in the next section, the specificity of the TCR expressed at the surface of each medullary thymocyte plays a crucial role in determining the ultimate fate of the developing T cells.

C. INTRATHYMIC SELECTION

The process of TCR gene rearrangement results in an enormous number of different receptor specificities, including by chance some that are directed against self-antigens. The need to remove these potentially self-reactive T cells has long been recognized, but it is only recently that significant insights have been gained into the nature of the selection processes.[46-48] Two distinct selection events are now recognized, termed *negative* and *positive selection,* respectively. In both selection events, the TCR expressed on the surface of developing thymocytes must engage MHC molecules expressed on thymic stromal cells. The outcome of this recognition event depends on the precise specificity of the TCR. Thymocytes expressing a receptor that is able to recognize a self-peptide bound within the presenting groove of self-MHC molecules are deleted by negative selection and diverted to a pathway of programmed cell death.[47] Those thymocytes which either recognize *empty* self-MHC molecules in the absence of self-peptide or bind only weakly to an MHC-peptide complex are positively selected for further differentiation. Subsequently, these T cells become restricted to the recognition of foreign antigens in the context of self-MHC.

Exactly how and when these two processes are integrated into the overall pathway of $\alpha\beta$ T cell development remains an unsolved puzzle. The consensus view is that selection operates at the $CD4^+CD8^+$ stage of thymocyte development, although a recent report suggests that negative selection can occur before this stage.[49] Two competing theories have been proposed to explain how thymocytes distinguish between these two processes. One theory holds that negative and positive selection is mediated by different subsets of thymic stromal cells and this results in qualitatively different signals, i.e., negative selection is mediated by bone marrow-derived stromal cells, whereas positive selection results from engagement of self-MHC molecules expressed on thymic epithelial cells. Another idea is that differences in the affinity of TCR binding affects the signals transduced into thymocytes and directs their subsequent development; a strong interaction, due to binding of self-peptide plus self-MHC, leads to deletion (negative selection), whereas a weak interaction due to binding of self-MHC alone results in positive selection.[47] The selection process also influences the subsequent phenotype of thymocytes since binding of the TCR to either class I or class II thymic MHC molecules determines whether the cell will become a $CD4^-CD8^+$ killer, or a $CD4^+CD8^-$ helper T cell, respectively.[46]

IV. INTRATHYMIC MATURATION OF γδ T CELLS

Although some mention has been made of γδ T cell development in the preceding sections, the ontogeny of this lineage of T cells warrants more detailed consideration because these T cells are unusually prominent in the immune system of ruminants. In mice and chickens, TCR $\gamma\delta^+$ thymocytes are the first lineage to arise in the fetal thymus and their numbers far exceed those of TCR $\alpha\beta^+$ thymocytes during the early stages of thymic development. To date, similar comparisons cannot be made with any certainty in ruminants due to the continuing lack of monoclonal antibodies that identify the αβ TCR. However, two sets of monoclonal antibodies have been used to identify γδ T cells; some of these are directed against the receptor itself, while others identify the WC1 (T19) differentiation antigen specifically expressed by ruminant γδ T cells. Staining of sheep thymuses with these reagents suggests a different relationship between these two lineages in that γδ T cells appear to constitute only a minor population of thymocytes, from the earliest times of fetal thymic development until adult stages.[50,51]

In 30- to 35-day-old fetal lamb thymus, all thymocytes are negative for CD4, CD8, γδ TCR, and T19.[5,31,50] From about day 40 onward, a few TCR $\gamma\delta^+$ cells are evident and these are scattered throughout the thymus; at this time, they are $T19^-$ and the T19 antigen is first detected on fetal thymocytes at around 55 days of gestation.[50] A characteristic localization and staining pattern of TCR $\gamma\delta^+$ thymocytes develops from about 70 to 80 days of gestation onward and persists into post-natal life. Throughout this period, $\gamma\delta^+$ thymocytes constitute a minor thymus population (1 to 4% of thymocytes). Most of them occur in the thymic medulla where they have a pronounced tendency to localize in the vicinity of Hassal's corpuscles (Figure 2-4a,c). A similar pattern of localization is seen in the bovine thymus (Figure 2-4b).[52] Hassal's corpuscles are formed by the differentiation of epithelial elements within the thymus and are unusually prominent in the ruminant thymus, especially in young animals. Medullary $\gamma\delta^+$ cells also express the T19 and MHC class I antigens.[50] A few $\gamma\delta^+$ thymocytes are scattered throughout the thymic cortex (Figure 2-4f) and some of these, particularly in the outer cortex, are negative for T19 and MHC class I antigens. These staining patterns have led to the proposal that the T19 antigen is a maturation marker that is only expressed on mature γδ T cells.[50]

However, in the light of more recent developments, some important caveats should be added to the proposal mentioned above. Studies in mice clearly show a sequential pattern to the earliest gene rearrangement events at TCR γ and δ loci so that the repertoire of fetal thymocytes changes during ontogeny and is distinct to the adult repertoire.[53] Although the thymic expression of Vγ and Vδ genes has not yet been examined in great detail in ruminants, a recent study clearly shows that the peripheral repertoire in sheep is highly diverse and changes during ontogeny; this may be due, at least in part, to regulated rearrangement in thymus.[54] In view of such diversity, and since all available monoclonals directed against the sheep γδ TCR were made by immunizing mice with peripheral γδ T cells from adult animals, there is a distinct possibility that they may not recognize all forms of the receptor, especially in fetuses. Also, emerging evidence shows that the WC1 (T19) antigen is polymorphic and is encoded by a multi-gene family.[55] The large number of monoclonals raised against this antigen exhibit different and complex patterns of cross-reactivity and it remains to be definitively established whether the antibodies used in the studies mentioned above identify all possible forms of this marker. Therefore, rather than indicating a maturation-related expression of T19, the patterns of TCR γδ and T19 staining detected in sheep thymus might instead reflect the ontogeny of different lineages of γδ T cells.[45]

At present, there is no general agreement as to whether γδ T cells are exposed to selective pressures during development in the thymus in a way that is analogous to positive and negative selection of αβ T cells (reviewed in Reference 39). Two recent studies which examined the role of MHC class I in γδ T cell selection in mice using gene deletion techniques resulted in directly conflicting conclusions.[56,57] The development and maturation of γδ T cells has also been difficult to study in *in vitro* systems due to the scarcity of these cells in thymus, their poorly characterized growth requirements, and continuing uncertainty about the nature of their physiological ligands. No attempt has yet been made to examine these features in either sheep or cattle. The close association seen between TCR $\gamma\delta^+$ thymocytes and Hassal's corpuscles in the thymus of ruminants suggests that these epithelial elements may play some role in the development process, perhaps by providing an important signal needed for continued growth or differentiation or by mediating some form of selection.

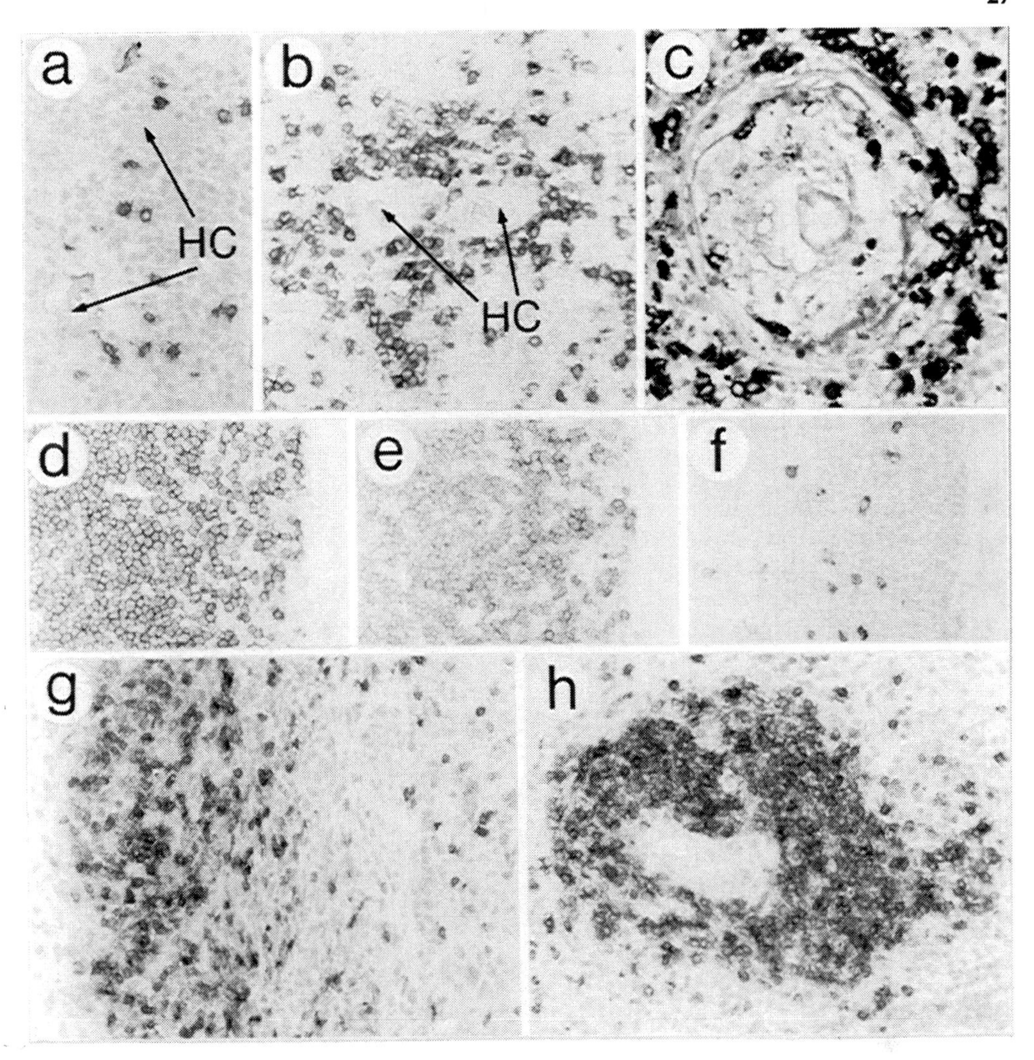

Figure 2-4 Development and tissue localization of different T cell subsets. Panels (a) through (c) show the localization of γδ T cells in the thymic medulla. (a) In a 68-day-old fetal lamb, γδ thymocytes are relatively scarce and are just beginning to associate with the Hassal's corpuscles (HC) that differentiate from medullary epithelial cells. (b) In a post-natal calf thymus, γδ thymocytes are more numerous and are tightly clustered around the Hassal's corpuscles. (c) An enlargement of a Hassal's corpuscle from a 1-day-old lamb thymus. γδ thymocytes are closely adherent to the outer epithelial whorls and a few positive cells can be seen within the Hassal's corpuscle. Panels (d) through (f) show the localization of T cell specific molecules in serial sections of the outer cortex of a 4-month-old lamb thymus; (d), CD8; (e) CD4; and (f), γδ TCR. The region immediately beneath the thymic capsule, at the right of each panel, contains some thymocytes that are negative for all of these markers. Note the relatively low frequency of $\gamma\delta^+$ thymocytes. The two bottom panels show the localization of T lymphocytes, as detected with a CD5 monoclonal antibody, in (g) prescapular lymph node and (h) spleen of a 76-day-old fetal lamb. In the lymph node, T cells occur mainly as a narrow band in the outer cortex and only a few cells are scattered in other regions. In the 76-day-old fetal spleen, T cells already show a characteristic localization to the periarteriolar lymphoid sheath region. (Panel (c) ×600. All other panels ×200.)

V. DEVELOPMENT OF PERIPHERAL T CELLS

A. EMIGRATION FROM THYMUS

During their development in the thymus, T cells show a largely sessile behavior although there must clearly be some limited movement within the confines of the thymic architecture. Throughout their subsequent life history, one of the most important properties of T cells is their ability to circulate freely between various compartments of the secondary lymphoid organs. Of necessity, the first step toward this metastatic lifestyle involves the emigration of mature T cells from the thymus. In this section, the route of migration as well as quantitative and qualitative aspects of this process are considered.

Numerous studies of thymic emigration in rodents, which have been reviewed recently,[58] indicate that only a small fraction of thymocytes are destined to emigrate. In quantitative terms, it is estimated that in mature post-natal animals about 1% of total thymocytes emigrate each day and that this represents 3 to 4% of the daily production of thymocytes. Therefore, only a very small fraction of the cells produced each day in the thymus are selected to enter the secondary lymphoid compartments and the remainder evidently die *in situ*. There are indications that the relative rate of migration may be higher in fetal and newborn animals.[58] Similar overall results have been recorded in studies of thymic emigration in post-natal and fetal lambs using different techniques. The emigration of T cells from the post-natal thymus was studied directly by fluorescent labeling of thymocytes by *in situ* perfusion and then monitoring the appearance of emigrants in the thymic vein. Around 0.5% of all thymocytes were estimated to exit from the thymus each day and this corresponded to 4 to 5% of newly formed cells.[59] An earlier study compared the expansion of T cells in normal and thymectomized fetuses and inferred that the emigration rate in 120-day-old fetal lambs is probably higher in that the thymus contributed about 4×10^7 cells/hour to the recirculating pool;[60] this would constitute about 5% of total thymocytes. More recently, elevated rates of thymic emigration have also been measured in fetal lambs using direct labeling techniques.[61]

T cells may emigrate from the thymus by one of two pathways: via the thymic vein or via the lymphatics. Cells emigrating through the lymphatic route have been collected by cannulating the thymic lymphatics in sheep, and it has been suggested that around 8% of newly formed cells might leave by this pathway.[59,62] These cells would be delivered first to the regional parathymic lymph node, where it is proposed they may undergo further post-thymic maturation.[62] Interestingly, the parathymic lymph node seems to be the first to develop during fetal ontogeny, probably because it is directly seeded by thymic emigrants; but continued maturation at this site has not yet been established by functional or phenotypic criteria.[63]

Thymic emigrants in sheep differed from peripheral T cells by some morphological and phenotypic criteria.[59] Lymphatic emigrants had an irregular profile with multiple membrane protrusions, suggesting that they were actively motile cells. The emigrants expressed variable amounts of MHC class I molecules, whereas peripheral T cells were more homogeneous. Also, an increased proportion of thymic emigrants expressed MHC class II molecules. Curiously, the lymphatic emigrants contained a notably elevated proportion of γδ T cells and a high number of mitotic bodies, although the exact phenotype of the cells in cycle was not determined.[59,62] Collectively, these results suggest that some thymic emigrants have been recently activated and that their phenotype is not typical of mature resting T cells. More recent studies of thymic emigration in sheep of different fetal and post-natal ages have shown distinct changes in the proportion of different subsets that migrate to the periphery, suggesting that there is some sort of developmental regulation of the rate at which different T cell types are added to the peripheral pool.[61]

B. EXPANSION OF THE PERIPHERAL T CELL POOL

Throughout fetal ontogeny, there is a steady increase in the number of lymphocytes present in the periphery, as shown by the growth of the solid lymphoid organs and the expansion of the recirculating lymphocyte pool. The vast majority of peripheral lymphocytes in the fetal lamb and calf are T cells, and earlier experiments which examined their general development in the secondary lymphoid organs have been described in a number of reviews (References 7, 64, and 65). More recently, the colonization of lymph nodes and spleen by different T cell subsets has been followed at different intervals of gestation in fetal lambs.[66,67] Isolated and scattered lymphocytes expressing T cell markers were detected in 43- to 44-day-old fetal spleens, but there was no clear organization of T cell areas in the periarteriolar region until about day 55. From this time onward, there was a rapid expansion of the T cell components in the spleen (Figure 2-4h). $CD8^+$ cells appeared before the $CD4^+$ subset while $T19^+$ γδ T cells were not found in fetal spleen until later in gestation, around day 57. During mid-gestation, there was a greater number of $CD8^+$ than $CD4^+$ T cells in the spleen, whereas in older fetuses and after birth, this ratio was reversed.[66] In the earliest lymph nodes examined, lymphocytes localized first in the outer cortex, while the medullary region contained extremely few lymphocytes and consisted of a loose reticulum framework (Figure 2-4g). With advancing fetal age, the size and cellularity of the cortex increased, but did not reach adult-like proportions until after birth; there was also an increase in the cellularity of the medulla. As in the fetal spleen, $CD8^+$ T cells were the first subset detected in lymph nodes, followed soon by $CD4^+$ cells. The $T19^+$

γδ T lymphocytes again appeared later in gestation and were not seen in fetal lymph nodes until day 69 of gestation.[67]

In addition to localizing in solid lymphoid organs such as lymph nodes and spleen, T lymphocytes in the fetal lamb and calf acquire from an early time in ontogeny the capacity to recirculate through the lymphatic apparatus, and the size of the recirculating pool increases throughout ontogeny.[60,68] General aspects of lymphocyte recirculation are described in greater detail elsewhere in this volume (see Chapter 7). In the context of the ontogeny of various T cell compartments in the body, the movement of lymphocytes may clearly play an important role, one which appears to be modulated differently in different organs and at different times of development. For example, there would seem to be two quite different routes whereby colonizing T cells enter the fetal spleen and lymph nodes. In the spleen, T cells must enter from blood vessels, via a route that does not involve high endothelial venules (HEV) since splenic blood vessels do not have these structures, and this pattern persists into adult life. The predominantly subcapsular localization of lymphocytes in early fetal lymph nodes, however, is consistent with their entry from afferent lymphatic vessels and suggests that the peripheral tissue pathway of migration operates first in ontogeny and plays an important role in early lymph node development.[69] Subsequently, fetal T cells begin to recirculate via HEV or analogous structures in lymph nodes and this pathway predominates after birth. It is likely that the orderly ontogeny of these different migratory pathways requires the regulated expression of specific adhesion-related molecules and their ligands on fetal T cells and various types of blood vessel endothelium, respectively.

A number of experimental studies involving thymectomy of the fetal lamb indicate that continued migration from the thymus and cell division in the periphery are both important mechanisms that account for the expansion of the peripheral T cell pool. Earlier studies established the critical importance of the thymus by showing that its surgical removal from fetuses from around day 70 onward resulted in significant lymphopoenia at different times after birth due to a deficit of T cells.[70,71] Nonetheless, some of the T cells that had migrated from the fetal thymus before its removal were able to persist for long periods of time and the peripheral pool expanded even in the absence of a thymus. After birth, such expansion might be due to antigen-stimulated cell division. However, a significant number of lymphocytes are also produced in extrathymic tissues in the fetus in the absence of any obvious antigenic stimulation. For example, by comparing the size of peripheral lymphocyte pools in normal and thymectomized 120-day-old fetal lambs, it has been estimated that extrathymic production amounts to as much as 10% of the thymic contribution.[60] However, the phenotype of lymphocytes was not determined in those experiments and it is likely that the extrathymic contribution included non-T cell lineages.

More recently, we have used monoclonal antibodies to monitor the post-natal expansion of different lymphocyte subsets in the blood of both normal sheep (Ti) and in animals which were thymectomized in the first half of gestation (Tx). At birth, the Tx animals were severely depleted of T cells (around 10% of Ti values) and our earlier observations on the relative numbers of different lymphocytes over the first year of life have been published.[72,73] We have continued to monitor these animals, some of which have now reached 4 years of age, and the changes in lymphocyte subsets over this time is shown (in Figure 2-5). In the normal animals, there was a rapid increase in the numbers of all circulating T cells ($CD4^+$, $CD8^+$, TCR $\gamma\delta^+$), after birth reaching peak levels at around 8 months of age. After this time, the $CD4^+$ and $CD8^+$ subsets declined quite quickly to a relatively stable plateau level, while the number of γδ T cells in blood continued to fall. In quantitative terms, γδ T cells were a major subset in normal neonatal animals and continued to be prominent until around 2 years of age. In the Tx animals, the $CD4^+$ and $CD8^+$ T cells were able to expand, albeit much less rapidly, and also reached their maximum concentration within 8 to 12 months, at levels of 50 to 70% of those measured in Ti animals; subsequently, their numbers fell as in the normal animals, although there was an indication of another phase of increasing numbers of $CD4^+$ and $CD8^+$ T cells in the older Tx animals. In contrast to the behavior of αβ T cells, the Tx animals remained severely and consistently depleted of γδ T cells throughout their life. However, these cells were present in blood as a scarce population at all times and a small increase in their numbers was detected in the older animals (Figure 2-5).

These results have a bearing on two main aspects of T cell ontogeny. First, they suggest that, in quantitative terms, the peripheral pools of the two T cell lineages are dependent on continued thymic production to very different degrees. Thus, αβ T cells emigrating from the thymus, even from early stages in fetal development, are capable of very considerable extrathymic expansion while the early γδ T cells have no such capacity. While it remains possible that γδ T cells emigrating later in development could have a greater potential for peripheral expansion, experiments in swine suggest this is not the case since

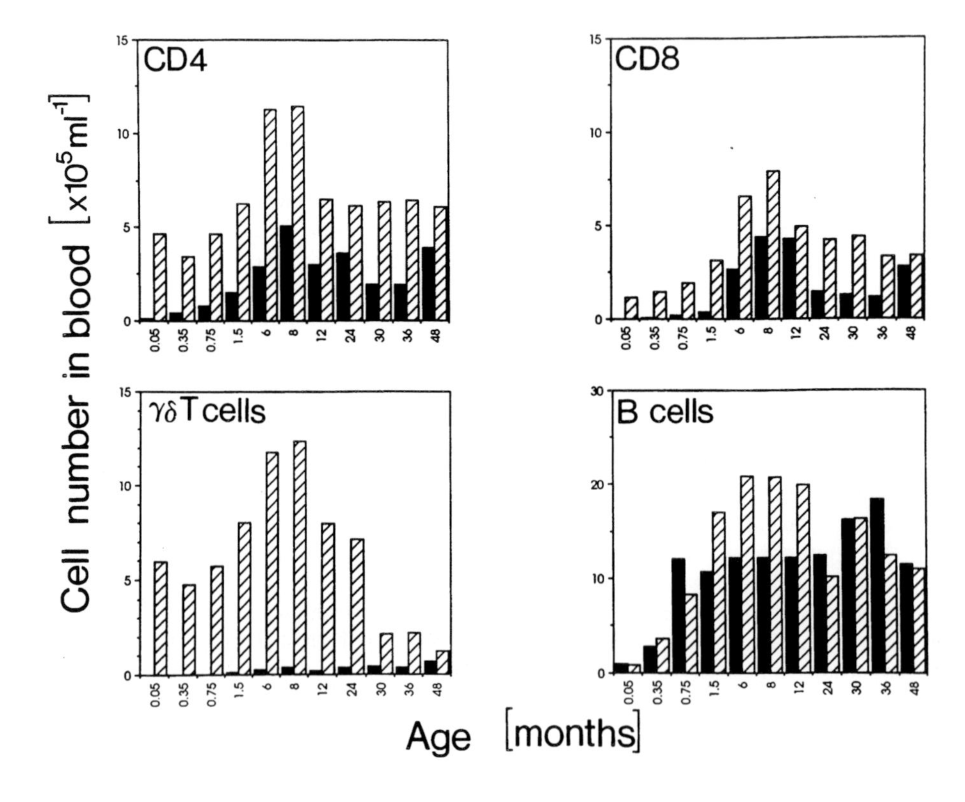

Figure 2-5 Concentration of lymphocyte subsets in the blood of normal sheep (hatched columns) and animals which had been thymectomized in the first half of gestation (solid columns). The graphs show the absolute numbers of CD4+ and CD8+ T cells, γδ T cells, and B cells in blood at different times after birth. The number of animals examined at each time point ranged from 2 to 8.

neonatal thymectomy in this species also results in a persistent depletion of γδ T cells.[74] Therefore, although all recirculating T cells are probably dependent on the thymus for their initial ontogeny, the γδ lineage appears to have a special requirement for the presence of this organ throughout development and suffers a more severe distortion if it is removed. Secondly, the similarity between the overall pattern of αβ T cell numbers in Ti and Tx animals indicates that not only does some sort of homeostatic mechanism control the numbers of T cells in the blood at different times of development, but that this mechanism operates even in the absence of a continued output of T cells from the thymus.

The information discussed so far has outlined the major changes that occur in the composition and behavior of T cells in the thymus and some peripheral compartments during ontogeny. Figure 2-6 shows an overall summary of the relative timing of these events in sheep.

C. ONTOGENY OF THE PERIPHERAL T CELL REPERTOIRE

As indicated in earlier sections, the rearrangement of TCR V genes occurs in a temporal order, especially during early fetal development, and the TCR repertoire expressed in the fetal thymus is distinct to that at an adult stage. This suggests that the peripheral repertoire which becomes established by thymic emigrants should also vary at different times in development and there is increasing evidence that this is the case. Studies in mice show that the usage of Vα gene segments and the level of N nucleotide diversification at junctional regions differs significantly between neonatal and adult αβ T cells.[42,75] In the case of the γδ lineage, there are numerous reports of differences between fetal and adult thymic repertoires. These features are again taken to indicate that the peripheral repertoire changes during development, although there is remarkably little direct evidence since these cells are scarce and difficult to sample from the peripheral immune system of fetal mice.

Recent experiments have examined the ontogeny of the TCR γδ repertoire in sheep by directly cloning expressed V gene segments from peripheral lymphocytes at different stages of development.[54] The

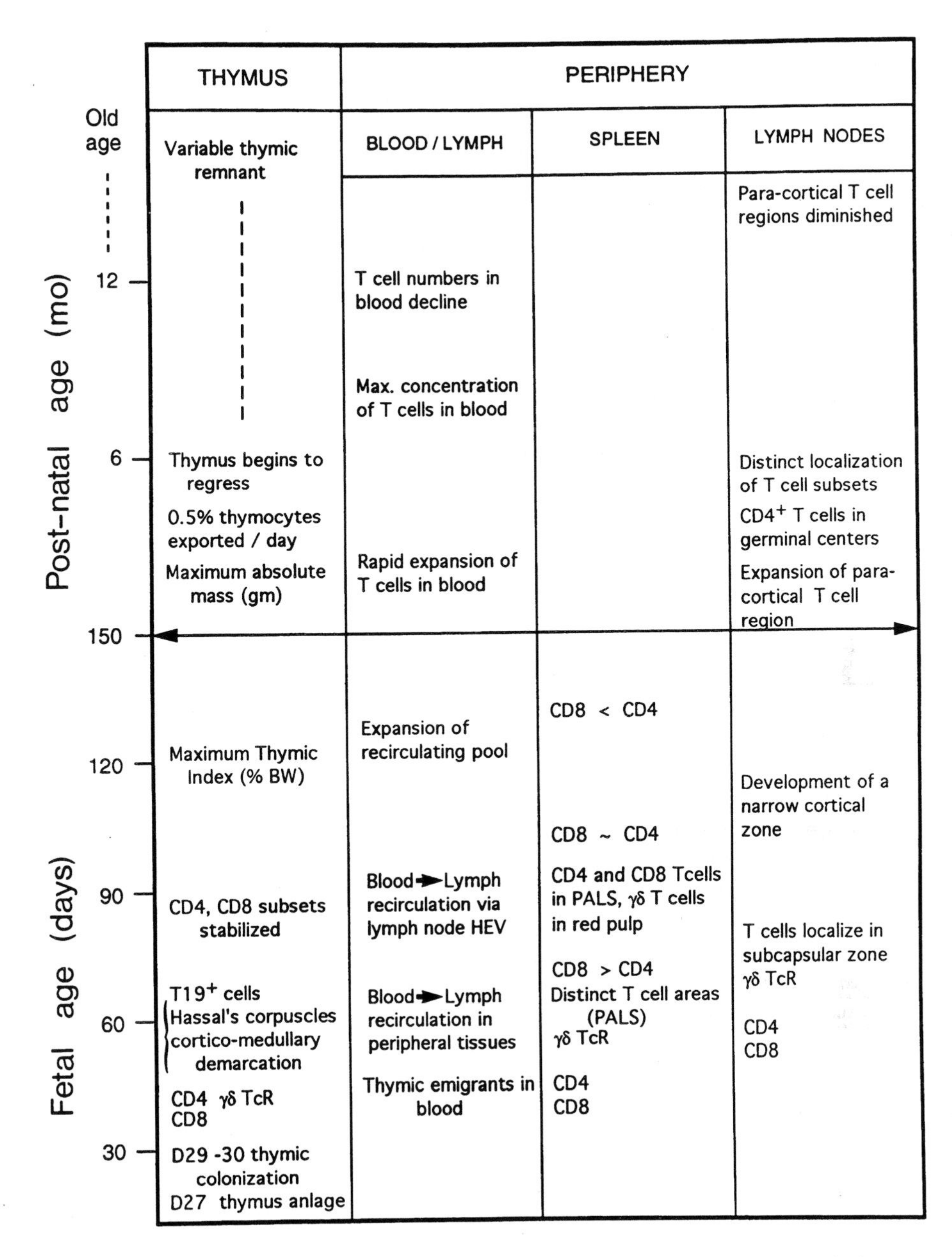

Figure 2-6 A summary of the main cellular and physiological events associated with the ontogeny of T cells in sheep.

frequency of usage of both Vγ and Vδ genes changed during fetal ontogeny and differed between fetuses and adults (Figure 2-7). To a large extent, the fetal and adult peripheral repertoires were nonoverlapping, and some V gene segments were used exclusively at a particular stage in development. The frequency of usage of other gene segments also differed since there was a nearly invariant pattern of rearrangement between particular Vγ-Jγ segments followed by splicing to distinct Cγ regions. Since the different sheep Cγ segments vary quite markedly in the structure of their hinge regions, this means that distinct isotypic forms of the γδ TCR are expressed at different times in ontogeny. The peripheral repertoire in two animals that had been thymectomized *in utero* was markedly different from the normal pattern and in many respects had retained a fetal character, emphasizing once more the critical role of the thymus in the development of γδ T cells.[54] To date, neither the development of the recirculating TCR αβ repertoire nor the ontogeny of tissue-specific repertoires have been studied in ruminants.

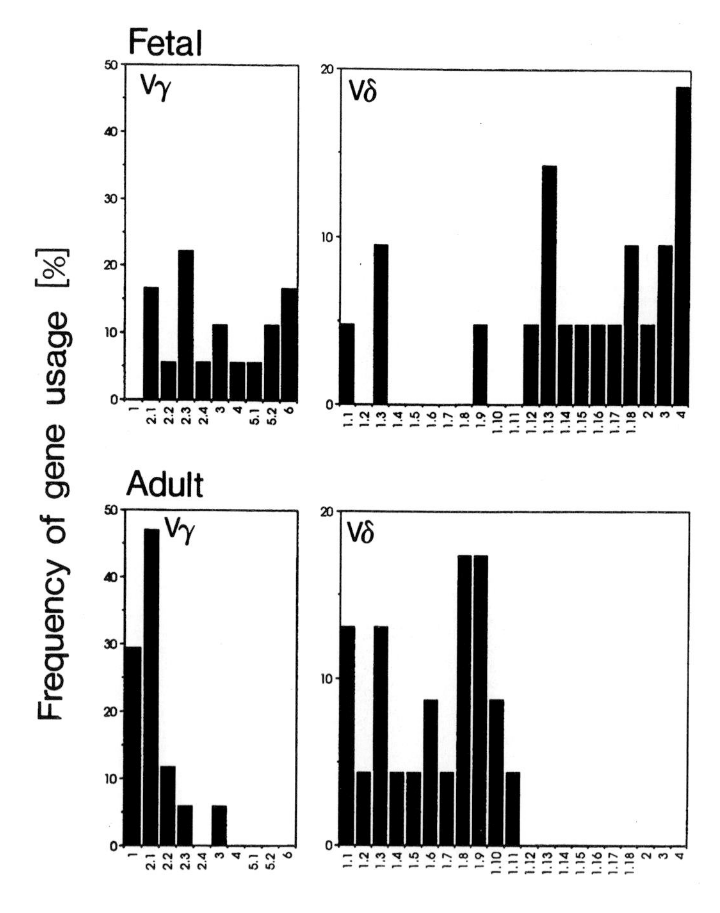

Figure 2-7 Development of the TCR γδ repertoire in sheep. The graphs show the relative frequency of expression of different V gene segments by peripheral γδ T cells in normal fetal (61 to 146 days of gestation) and adult animals (1 year old) and are derived from the sequence analysis of cDNA clones. For more detailed discussion of changes in V, J, and C-region gene usage during fetal ontogeny, see Reference 54.

D. EXTRATHYMIC T CELL DEVELOPMENT

The possibility that T cells might develop *de novo* from immature precursors using extrathymic maturation pathways has been a subject of continuing debate almost since the time of Miller's experiments which demonstrated the critical role of the thymus. However, there is now a large body of convincing evidence to show, at least in the mouse, that extrathymic pathways do exist and that these make an important contribution to the total T cell pool (reviewed in Reference 76). Murine T cells that developed inside or outside the thymus can be differentiated by a number of criteria, including TCR repertoires, the expression of either the CD8αβ or CD8αα dimeric chains and their functional properties. The two pathways of development appear to contribute differentially to different compartments of the peripheral T cell pool. Thus, the lymphoid compartment seems to be predominantly thymus derived, whereas many — even the major proportion — of lymphocytes in some peripheral tissues such as the gut, have features suggesting they developed extrathymically. In overall quantitative terms, extrathymically derived T cells could comprise as much as 25 to 30% of all T cells in the body.[76] Since these cells have escaped the selection procedures operating in the thymus, there must be other mechanisms that maintain peripheral tolerance; however, the nature of these events and the ligand specificity of extrathymically derived T cells remain unclear.[76,77]

The above findings should provoke a careful search for corresponding pathways of T cell development in other species. As already indicated, similar to the situation in the mouse, most evidence argues that T cells in the recirculating lymphoid compartment of ruminants are mainly thymus derived since thymectomized animals maintain a permanent deficit of T cells in blood and lymph. The outgrowth of αβ T cells in sheep that have been thymectomized to date can be explained by the expansion of cells which emigrated from the thymus before its removal; it is not yet feasible to thymectomize fetal lambs before thymic emigration begins. Also, the repertoire of γδ T cells in thymectomized adult animals had retained fetal characteristics, supporting the idea that they were either survivors or progeny of fetal emigrants.[54] At the same time, however, there are indications that some γδ T cells in blood might develop extrathymically since there was a higher frequency of unusual Vγ-Jγ-Cγ combinations in Tx animals and this might reflect different levels of control over gene rearrangement and RNA splicing in T cells maturing outside the thymus.[54] Unfortunately, there is no phenotypic marker which distinguishes between T cells which developed either within or outside the thymus in sheep; thus, the relative quantitative importance of these two pathways cannot be definitively resolved, although for those T cells destined to become part of the circulating pool in blood, extrathymic pathways would seem to play a minor role.

However, in view of the evidence from mice, this may not apply at other body sites, and the development of defined populations of T cells resident in specific tissues such as the gut, lung, or skin of ruminants should be re-examined. Although these sites remained severely depleted of all T cells in 4 to 5-month-old Tx lambs,[72] it is not yet known whether there was a subsequent increase in older animals. In normal adult cattle and sheep, T cells are prominent in a number of different tissues that have either an endodermal or ectodermal component. The primordial thymus is derived from the same embryonic layers, so it is not too implausible that these other tissues could have a residual capacity to promote T cell differentiation. Many of these tissue sites form a barrier between the internal and external environment; the phenotype, repertoire, and functional properties of T cells populating them deserve close study.

ACKNOWLEDGMENTS

Some of the previously unpublished data shown in Figures 2-1 and 2-2 resulted from work done in collaboration with Dr. Beat Imhof. I thank him both for allowing the results to be presented here and for critically reading the manuscript. I also thank Lisbeth Dudler for excellent technical assistance. The Basel Institute for Immunology was founded and is supported by F. Hoffmann-La Roche and Co., Basel, Switzerland.

REFERENCES

1. **Miller, J. F. A. P.,** Immunological function of the thymus, *Lancet,* i, 748, 1961.
2. **Jordan, R. K.,** Development of sheep thymus in relation to *in utero* thymectomy experiments, *Eur. J. Immunol.,* 6, 693, 1976.
3. **White, T. R., Lee, C. S., French, P., Hewish, D., and Brandon, M. R.,** Characterization of a monoclonal antibody (SBU-1) made to the thymic rudiment of sheep, *J. Histochem. Cytochem.,* 8, 785, 1985.
4. **Landsverk, T.,** personal communication, 1992.
5. **Maddox, J. F., Mackay, C. R., and Brandon, M. R.,** Ontogeny of ovine lymphocytes. I. An immunohistological study on the development of T lymphocytes in the sheep embryo and fetal thymus, *Immunology,* 62, 97, 1987.
6. **Al Salami, M., Simpson-Morgan, M. W., and Morris, B.,** Haemopoiesis and the development of immunological reactivity in the sheep foetus, in *Immunology of the Sheep,* Morris, B. and Miyasaka, M., Eds., Editiones Roche, Basel, 1985, 19.
7. **Miyasaka, M. and Morris, B.,** The ontogeny of the lymphoid system and immune responsiveness in sheep, *Prog. Vet. Microbiol. Immunol.,* 4, 21, 1988.
8. **Savagner, P., Imhof, B. A., Yamada, K. M., and Thiery, J.-P.,** Homing of hemopoietic precursor cells to the embryonic thymus: characterization of an invasive mechanism induced by chemotactic peptides, *J. Cell Biol.,* 103, 2715, 1986.
9. **Champion, S., Imhof, B. A., Savagner, P., and Thiery, J.-P.,** The embryonic thymus produces chemotactic peptides involved in the homing of hemopoietic precursors, *Cell,* 44, 781, 1986.
10. **Imhof, B. A., Deugnier, M.-A., Girault, J.-M., Champion, S., Damais, C., Itoh, T., and Thiery, J.-P.,** Thymotaxin: a thymic epithelial peptide chemotactic for T-cell precursors, *Proc. Natl. Acad. Sci. U.S.A.,* 85, 7699, 1988.
11. **Deugnier, M.-A., Imhof, B. A., Bauvois, B., Dunon, D., Denoyelle, M., and Thiery, J.-P.,** Characterization of rat T cell precursors sorted by chemotactic migration toward thymotaxin, *Cell,* 56, 1073, 1989.
12. **Dargemont, C., Dunon, D., Deugnier, M.-A., Denoyelle, M., Girault, J.-M., Lederer, F., Kim, H., Godeau, F, Thiery, J.-P., and Imhof, B. A.,** Thymotaxin, a chemotactic protein, is identical to β2-microglobulin, *Science,* 246, 803, 1989.

13. **Dunon, D., Kaufman, J., Salomonsen, J., Skjoedt, K., Vainio, O., Thiery, J.-P., and Imhof, B. A.,** T cell precursor migration towards β2-microglobulin is involved in thymus colonization of chicken embryos, *EMBO J.,* 9, 3315, 1990.
14. **Le Douarin, N. M., Dieterlen-Lievre, F., and Oliver, P. D.,** Ontogeny of primary lymphoid organs and lymphoid stem cells, *Am. J. Anat.,* 170, 261, 1984.
15. **Jotereau, F., Heuze, F., Salomon-Vie, V., and Gascan, H.,** Cell kinetics in the fetal mouse thymus: precursor cell input, proliferation, and emigration, *J. Immunol.,* 138, 1026, 1987.
16. **Imhof, B. A. and Hein, W. R.,** unpublished data.
17. **Anderson, E.L.,** Pharyngeal derivatives in the calf, *Anat. Rec.,* 24, 25, 1922.
18. **Schultz, R. D., Dunne, H. W., and Heist, C. E.,** Ontogeny of the bovine immune response, *Infect. Immun.,* 7, 981, 1973.
19. **Lobach, D. F. and Haynes, B. F.,** Ontogeny of the human thymus during fetal development, *J. Clin. Immunol.,* 7, 81, 1987.
20. **Reynolds, J. D.,** The Development and Physiology of the Gut-Associated Lymphoid Tissue in Lambs, Ph.D. thesis, Australian National University, Canberra, 1976.
21. **McClure, S., Dudler, L., Thorpe, D., and Hein, W. R.,** Analysis of cell division among subpopulations of lymphoid cells in sheep. I. Thymocytes, *Immunology,* 65, 393, 1988.
22. **Hein, W. R. and Dudler, L.,** unpublished data.
23. **Shortman, K.,** Cellular aspects of early T-cell development, *Curr. Opinion Immunol.,* 4, 140, 1992.
24. **Scollay, R.,** T-cell subset relationships in thymocyte development, *Curr. Opinion Immunol.,* 3, 204, 1991.
25. **Carding, S. R., Hayday, A. C., and Bottomly, K.,** Cytokines in T-cell development, *Immunol. Today,* 12, 239, 1991.
26. **Nikolic-Zugic, J.,** Phenotypic and functional stages in the intrathymic development of αβ T cells, *Immunol. Today,* 12, 65, 1991.
27. **O'Neill, H. C.,** Prothymocyte seeding in the thymus, *Immunol. Lett.,* 27, 1, 1991.
28. **Dunon, D. and Imhof, B. A.,** Mechanisms of thymus homing, *Blood,* 81, 1, 1993.
29. **Spangrude, G. J. and Scollay, R.,** Differentiation of hematopoietic stem cells in irradiated mouse thymic lobes: kinetics and phenotype of progeny, *J. Immunol.,* 145, 3661, 1990.
30. **Wu, L., Scollay, R., Egerton, M., Pearse, M., Spangrude, G. J., and Shortman, K.,** CD4 expressed on earliest T-lineage precursor cells in the adult murine thymus, *Nature,* 349, 71, 1991.
31. **Shortman, K., Wu, L., Kelly, K. A., and Scollay, R.,** The beginning and the end of the development of TCR γδ cells in the thymus, *Curr. Topics Microbiol. Immunol.,* 173, 71, 1991.
32. **Mackay, C. R., Maddox, J. F., and Brandon, M. R.,** Thymocyte subpopulations during early fetal development in sheep, *J. Immunol.,* 136, 1592, 1986.
33. **Baldwin, C. L., Teale, A. J., Naessens, J. G., Goddeeris, B., MacHugh, N. D., and Morrison, W. I.,** Characterization of a subset of bovine T lymphocytes that express BoT4 by monoclonal antibodies and function: similarity to lymphocytes defined by human T4 and murine L3T4, *J. Immunol.,* 136, 4385, 1986.
34. **Ellis, J. A., Baldwin, C. L., MacHugh, N. D., Bensaid, A., Teale, A. J., Goddeeris, B. M., and Morrison, W. I.,** Characterization by a monoclonal antibody and functional analysis of a subset of bovine T lymphocytes that express BoT8, a molecule analogous to human CD8, *Immunology,* 58, 351, 1986.
35. **MacHugh, N. D., Bensaid, A., Davis, W. C., Howard, C. J., Parsons, K. R., Jones, B., and Kaushal, A.,** Characterization of a bovine thymic differentiation antigen analogous to CD1 in the human, *Scand. J. Immunol.,* 27, 541, 1988.
36. **Mackay, C. R., Maddox, J. F., Wijffels, G. L., Mackay, I. R., and Walker, I. D.,** Characterization of a 95,000 molecule on sheep leucocytes homologous to murine Pgp-1 and human CD44, *Immunology,* 65, 93, 1988.
37. **Howard, C. J., Sopp, P., Parsons, K. R., McKeever, D. J., Taracha, E. L. N., Jones, B. V., MacHugh, N. D., and Morrison, W. I.,** Distinction of naive and memory BoCD4 lymphocytes in calves with a monoclonal antibody, CC76, to a restricted determinant of the bovine leukocyte-common antigen, CD45, *Eur. J. Immunol.,* 21, 2219, 1991.
38. **Winoto, A.,** Regulation of the early stages of T-cell development, *Curr. Opinion Immunol.,* 3, 199, 1991.
39. **Haas, W. and Tonegawa, S.,** Development and selection of γδ T cells, *Curr. Opinion Immunol.,* 4, 147, 1992.

40. **Philpott, K. L., Viney, J. L., Kay, G., Rastan, S., Gardiner, E. M., Chae, S., Hayday, A. C., and Owen, M. J.,** Lymphoid development in mice congenitally lacking T cell receptor αβ-expressing cells, *Science,* 256, 1448, 1992.
41. **Leiden, J. M.,** Transcriptional regulation during T-cell development: the α TCR gene as a molecular model, *Immunol. Today,* 13, 22, 1992.
42. **Benoist, C. and Mathis, M.,** Generation of the αβ T-cell repertoire, *Curr. Opinion Immunol.,* 4, 156, 1992.
43. **Malissen, M., Trucy, J., Jouvine-Marche, E., Cazenave, P.-A., Scollay, R., and Malissen, B.,** Regulation of TCR α and β gene allelic exclusion during T-cell development, *Immunol. Today,* 13, 315, 1992.
44. **Groettrup, M., Baron, A., Griffiths, G., Palacios, R., and von Boehmer, H.,** T cell receptor (TCR) β chain homodimers on the surface of immature but not mature α, γ, δ chain deficient T cell lines, *EMBO J.,* 11, 2735, 1992.
45. **Hein, W. R., Dudler, L., Beya, M.-F., Marcuz, A., and Grossberger, D.,** T cell receptor gene expression in sheep: differential usage of TcR1 in the periphery and thymus, *Eur. J. Immunol.,* 19, 2297, 1989.
46. **von Boehmer, H.,** Positive and negative selection of the αβ T-cell repertoire *in vivo, Curr. Opinion Immunol.,* 3, 210, 1991.
47. **Pardoll, D. and Carrera, A.,** Thymic selection, *Curr. Opinion Immunol.,* 4, 162, 1992.
48. **von Boehmer, H.,** Thymic selection: a matter of life and death, *Immunol. Today,* 13, 454, 1992.
49. **Takahama, Y., Shores, E. W., and Singer, A.,** Negative selection of precursor thymocytes before their differentiation into $CD4^+CD8^+$ cells, *Science,* 258, 653, 1992.
50. **Mackay, C. R., Beya, M.-F., and Matzinger, P.,** γ/δ T cells express a unique surface molecule appearing late during thymic development, *Eur. J. Immunol.,* 19, 1477, 1989.
51. **McClure, S. J., Hein, W. R., Yamaguchi, K., Dudler, L., Beya, M.-F., and Miyasaka, M.,** Ontogeny, morphology and tissue distribution of a unique subset of $CD4^-CD8^-$ sheep T lymphocytes, *Immunol. Cell Biol.,* 67, 215, 1989.
52. **Mackay, C. R. and Hein, W. R.,** A large proportion of bovine T cells express the γδ T cell receptor and show a distinct tissue distribution and surface phenotype, *Int. Immunol.,* 5, 540, 1989.
53. **Havran, W. L. and Allison, J. P.,** Developmentally ordered appearance of thymocytes expressing different T-cell antigen receptors, *Nature,* 335, 443, 1988.
54. **Hein, W. R. and Dudler, L.,** Divergent evolution of T cell repertoires: extensive diversity and developmentally regulated expression of the sheep γδ T cell receptor, *EMBO J.,* 12, 715, 1993.
55. **Wijngaard, P. L. J., Metzelaar, M. J., MacHugh, N. D., Morrison, W. I., and Clevers, H. C.,** Molecular characterization of the WC1 antigen expressed specifically on bovine $CD4^-CD8^-$ γδ T lymphocytes, *J. Immunol.,* 149, 3273, 1992.
56. **Periera, P., Zijlstra, M., McMaster, J., Loring, J. M., Jaenisch, R., and Tonegawa, S.,** Blockade of transgenic γδ T cell development in β2-microglobulin deficient mice, *EMBO J.,* 11, 25, 1992.
57. **Correa, I., Bix, M., Liao, N.-S., Zilstra, M., Jaenisch, R., and Raulet, D.,** Most γδ T cells develop normally in β2-microglobulin-deficient mice, *Proc. Natl. Acad. Sci. U.S.A.,* 89, 653, 1992.
58. **Scollay, R.,** Migration of cells from the thymus to the secondary lymphoid organs, in *Migration and Homing of Lymphoid Cells,* Vol. I., Husband, A. J., Ed., CRC Press, Boca Raton, FL, 1988, 51.
59. **Miyasaka, M., Pabst, R., Dudler, L., Cooper, M., and Yamaguchi, K.,** Characterization of lymphatic and venous emigrants from the thymus, *Thymus,* 16, 29, 1990.
60. **Pearson, L. D., Simpson-Morgan, M. W., and Morris, B.,** Lymphopoiesis and lymphocyte recirculation in the sheep fetus, *J. Exp. Med.,* 143, 167, 1976.
61. **Kimpton, W. G.,** personal communication, 1992.
62. **Yamashita, A., Miyasaka, M., and Trnka, Z.,** Early post-thymic T cells: studies on lymphocytes in the lymph coming from the thymus of sheep, in *Immunology of the Sheep,* Morris, B. and Miyasaka, M., Eds., Roche, Basel, 1985, 162.
63. **Hein, W. R. and Dudler, L.,** unpublished data.
64. **Hein, W. R., Simpson-Morgan, M. W., and Morris, B.,** The traffic of lymphocytes in fetal and adult ruminants, *Expl. Biol. Med.,* 10, 231, 1985.
65. **Morris, B.,** The ontogeny and comportment of lymphoid cells in fetal and neonatal sheep, *Immunol. Rev.,* 91, 219, 1986.

66. **Maddox, J. F., Mackay, C. R., and Brandon, M. R.,** Ontogeny of ovine lymphocytes. II. An immunohistological study on the development of T lymphocytes in the sheep fetal spleen, *Immunology*, 62, 107, 1987.
67. **Maddox, J. F., Mackay, C. R., and Brandon, M. R.,** Ontogeny of ovine lymphocytes. III. An immunohistological study on the development of T lymphocytes in sheep fetal lymph nodes, *Immunology,* 62, 113, 1987.
68. **Hein, W. R., Shelton, J. N., Simpson-Morgan, M. W., and Morris, B.,** Traffic and proliferative responses of recirculating lymphocytes in fetal calves, *Immunology,* 64, 621, 1988.
69. **Morris, B. and Al Salami, M.,** The blood and lymphatic capillaries of lymph nodes in the sheep foetus and their involvement in cell traffic, *Lymphology,* 20, 244, 1987.
70. **Cole, G. J. and Morris, B.,** The growth and development of lambs thymectomized *in utero, Aust. J. Exp. Biol. Med. Sci.,* 49, 33, 1971.
71. **Fahey, K. J., Outteridge, P. M., and Burrells, C.,** The effect of pre-natal thymectomy on lymphocyte sub-populations in the sheep, *Aust. J. Exp. Biol. Med. Sci.,* 58, 571, 1980.
72. **Hein, W. R., Dudler, L., and Morris, B.,** Differential peripheral expansion and *in vivo* antigen reactivity of α/β and γ/δ T cells emigrating from the early fetal lamb thymus, *Eur. J. Immunol.,* 20, 1805, 1990.
73. **Hein, W. R. and Mackay, C. R.,** Prominence of $\gamma\delta$ T cells in the ruminant immune system, *Immunol. Today,* 12, 30, 1991.
74. **Binns, R. M., Pallares, V., Symons, D. B. A., and Sibbons, P.,** Effect of thymectomy on lymphocyte subpopulations in the pig. Demonstration of a thymus-dependent 'null' cell, *Int. Archs. Allergy Appl. Immunol.*, 55, 96, 1977.
75. **Bogue, M., Candeias, S., Benoist, C., and Mathis, D.,** A special repertoire of α:β T cells in neonatal mice, *EMBO J.*, 10, 3647, 1991.
76. **Rocha, B., Vassalli, P., and Guy-Grand, D.,** The extrathymic T-cell development pathway, *Immunol. Today,* 13, 449, 1992.
77. **Arnold, B., Schönrich, G., and Hämmerling, G. J.,** Extrathymic T-cell selection, *Curr. Opinion Immunol.,* 4, 166, 1992.

Chapter 3

The Major Histocompatibility Complex

Leif Andersson and Christopher J. Davies

CONTENTS

I. Introduction 37
II. Structure of MHC Molecules 38
III. The Bovine Major Histocompatibility Complex 41
A. General Organization of the Bovine MHC 41
1. The Class I Region 41
2. The Class II Regions 42
a. The Class IIa Region 42
b. The Class IIb Region 42
3. The Class III Region 43
B. Polymorphism of the Class I and Class II Genes 43
1. Class I Genes 43
a. Polymorphism Revealed by Serological Methods 43
b. Polymorphism Revealed by Cellular Methods 43
c. Polymorphism Revealed by Isoelectric Focusing 44
d. Restriction Fragment Length Polymorphism (RFLP) 44
2. Class II Genes 44
a. Polymorphism Revealed by Cellular Methods 44
b. Polymorphism Revealed by Serological Methods 45
c. Polymorphism Revealed by Isoelectric Focusing 45
d. Restriction Fragment Length Polymorphism (RFLP) 45
e. Polymorphism Revealed by DNA Sequencing 46
f. Polymorphism Revealed by PCR-Based Methods 46
g. Correlation Between Class II Typing Methods 47
IV. The Ovine and Caprine Major Histocompatibility Complexes 47
A. General Organization of the Ovine and Caprine MHC 47
B. Polymorphism of the Class I and Class II Genes 47
V. Significance of MHC Polymorphism 48
A. MHC Restriction 48
B. Immune Response Associations 48
C. Disease Associations 48
1. Bovine Disease Association Studies 49
2. Ovine and Caprine Disease Association Studies 49
D. Associations with Physiological Traits 49
E. Reproduction and the Fetal Allograft 50
References 50

I. INTRODUCTION

The major histocompatibility complex (MHC) was first discovered by Gorer more than 50 years ago as a polymorphic genetic system controlling the rejection of foreign tissue grafts in the mouse (see Reference 1 for review). The critical role of MHC molecules in the immune system was revealed much later. During the 1970s, it was clearly established that genes in the MHC control antibody responses to well-defined antigens such as short peptides.[2,3] A major breakthrough in the understanding of MHC function was the finding of Zinkernagel and Doherty[4] that T lymphocytes are MHC restricted; i.e., they recognize antigens only in association with self-histocompatibility molecules. It has now been clearly established that a major function of the molecules encoded in the MHC is to present self and foreign antigens to T cells. Antigens are degraded intracellularly and presented by MHC molecules in the form

0-8493-4952-4/94/$0.00+$.50

of short peptides.[5] The presentation of antigen in this way is a prerequisite for an efficient immune response and hence the MHC molecules involved are essential in the defense against pathogens.

There are two classes of MHC molecules involved in antigen presentation, namely class I and class II. They differ in function in that class I molecules present endogenous peptides, (i.e., peptides synthesized within the cell), whereas class II molecules present exogenous peptides, (i.e., derived from antigens synthesized outside the cell) (Figure 3-1). This means that presentation of foreign peptides by class I and class II molecules gives strikingly different signals to the immune system. Class I presentation implies that something is wrong with the cell; e.g., a pathogen is replicating within the cell. Consequently, class I molecules present antigens to cytotoxic T cells, which in turn eliminate the infected cell. In contrast, class II presentation of exogenous peptides signals that pathogens are replicating in other cells or in extracellular spaces. In this case, the immune response should not be directed against the antigen presenting cell. Thus, class II molecules present foreign peptides to T helper cells which induce the activity of other cells like the production of antibodies by B lymphocytes. The MHC specificity of T cells is determined not only by the T cell receptor, but also by the expression of the CD4 and CD8 molecules on cytotoxic T cells and T helper cells, respectively (Figure 3-1). The function of class I and class II molecules is also reflected in their tissue distribution. While class I molecules are expressed on a wide variety of cell types, the expression of class II molecules is restricted primarily to cells of the immune system such as macrophages, B cells, and dendritic cells that are involved in presentation of exogenous antigens.

The reason why the class I and class II MHC genes and molecules have been so extensively studied, compared with other important molecules of the immune system, is their extreme genetic polymorphism and the functional significance of this polymorphism with regard to immune responses and disease susceptibility. Except for the immunoglobulin and T cell receptor molecules, whose diversity is based on somatic recombination rather than germline polymorphism, the MHC molecules are the most polymorphic proteins which have been found in vertebrates. The MHC polymorphism deviates from most other known protein polymorphisms in several respects. First, the number of alleles at a single MHC locus is often very large and may exceed 50. Second, the number of amino acid substitutions between alleles is very high as two alleles may differ by more than 10% in their amino acid sequence. Third, the allele frequency distribution at MHC loci is often more even than expected for selectively neutral alleles.[6] Fourth, the frequency of replacement substitutions is significantly higher than the frequency of silent substitutions at codons encoding the peptide binding site.[7] In conclusion, there is now compelling evidence that MHC polymorphism is maintained by natural selection. The common view is that the selection pressure favoring MHC polymorphism is related to defense against foreign pathogens.

II. STRUCTURE OF MHC MOLECULES

Class I molecules are composed of a light chain, denoted β_2 microglobulin (β_2m), and a heavy chain with molecular weights of about 12,000 and 45,000, respectively; β_2m is encoded by a single gene not located in the MHC. The class I heavy chain is composed of three extracellular domains (α1-α3), each of about 90 amino acids, a transmembrane region, and a cytoplasmic domain (Figure 3-2). The polymorphic amino acids of class I molecules are primarily located in the α1 and α2 domains which form the peptide binding site. The class II molecules are composed of an α and a β chain with molecular weights of approximately 35,000 and 28,000, respectively. The two chains are encoded by very closely linked MHC genes. The structure of class II molecules is very similar to that of class I molecules (Figure 3-2) but with the difference that both the α and β chains have two extracelluar domains. In a similar fashion to the class I molecules, the genetic polymorphism of class II molecules occurs predominantly in the α1 and β1 domains which form the peptide binding site. The striking structural similarities between class I and class II molecules strongly suggest that they have evolved from a common ancestral gene by gene duplication.

The three-dimensional structure of the extracellular part of class I molecules has been determined by high-resolution X-ray analysis of a crystallized human class I molecule.[8] The carboxy-terminal α3 domain and β_2 microglobulin show sequence and structural homology to the immunoglobulin constant region domains. The peptide binding site, formed by the α1 and α2 domains, consists of a pocket with two α-helices forming the sides on a floor of eight strands of antiparallel β sheets (Figure 3-3). Short peptides are bound in the pocket, and amino acid residues in the β-strands and α-helical regions facing the pocket are involved in peptide binding. The fact that many of these residues are highly polymorphic

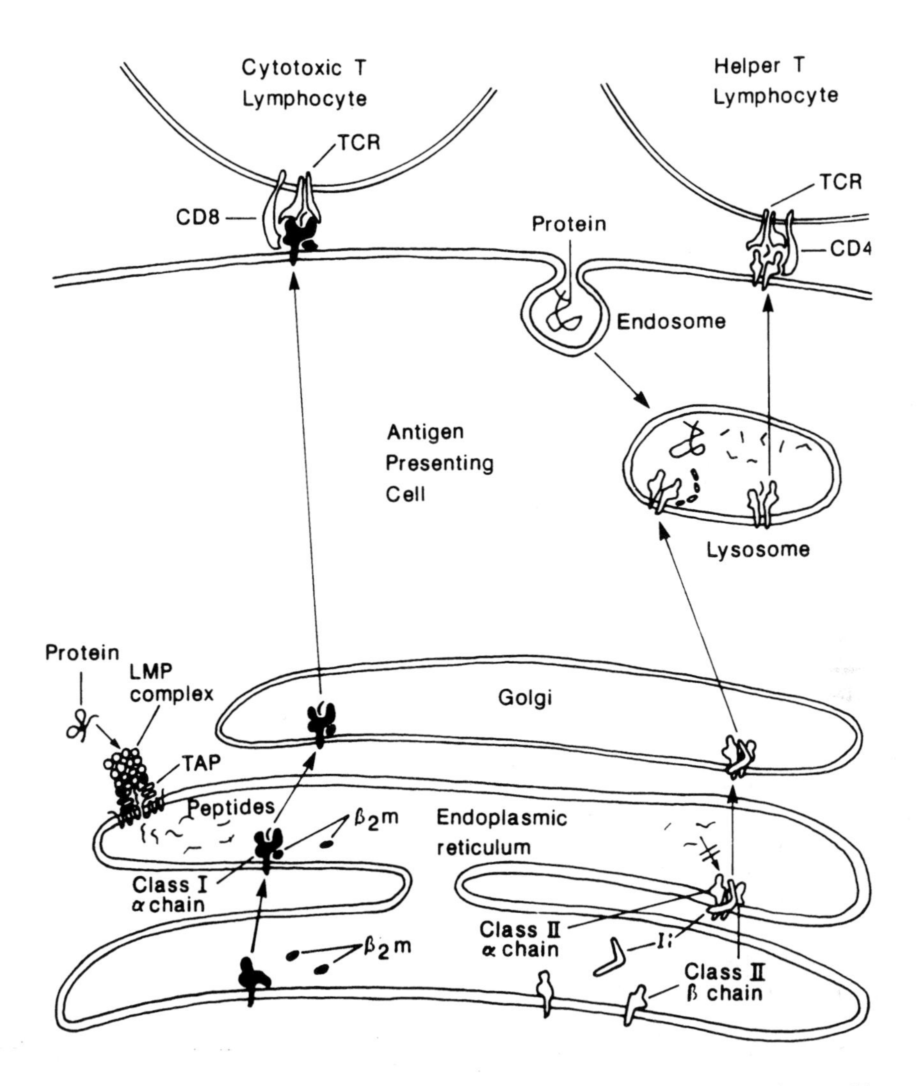

Figure 3-1 Endogenous class I and exogenous class II antigen presentation pathways (drawn after DeMars and Spies[146]). Class I molecules are assembled in the endoplasmic reticulum from an MHC-encoded α chain and a non-MHC encoded chain known as β_2m. Binding of peptide at the time of assembly is essential for the stability of the molecule and subsequent expression at the cell surface. Peptides are transported into the endoplasmic reticulum by the MHC-encoded Transporter associated with Antigen Processing (TAP).[147,148] The peptides presented by class I molecules are generated from cytoplasmic proteins by proteasome complexes. Two genes encoding proteasome or Large Multifunctional Protease (LMP) components are located in the MHC adjacent to the TAP genes. The MHC-encoded LMP components are only present in a small portion of cytoplasmic proteasomes[149] and these genes are apparently not essential for normal class I-mediated antigen presentation.[150,151] One hypothesis, shown here, is that the MHC-encoded LMP components (represented by filled circles) link certain proteasomes directly to the TAP transporters.[146] Following transport to the cell surface, class I molecules present peptides to T cell receptors (TCR) on CD8+ cytotoxic T lymphocytes. Class II molecules are assembled in the endoplasmic reticulum from the MHC-encoded α and β chains and the non-MHC-encoded invariant chain (Ii). The invariant chain stabilizes the class II molecule, blocks the binding of peptides, and directs transport of the molecule to the endocytic pathway.[151] In phagolysosomes, the invariant chain is degraded and peptides generated from proteins brought into the cell by phagocytosis are bound to the class II molecule. The class II molecule peptide complex is then transported to the cell surface where the peptides are presented to TCR on CD4+ helper T lymphocytes.

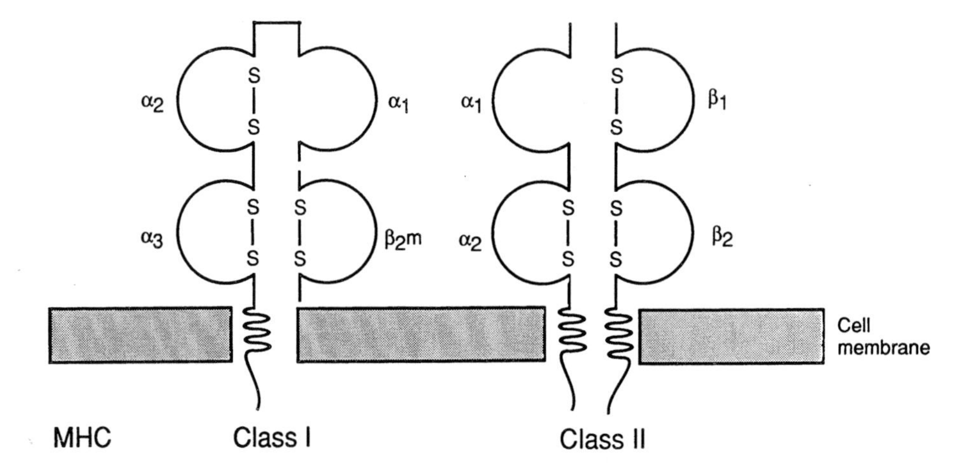

Figure 3-2 Schematic representation of the MHC class I and class II molecules.

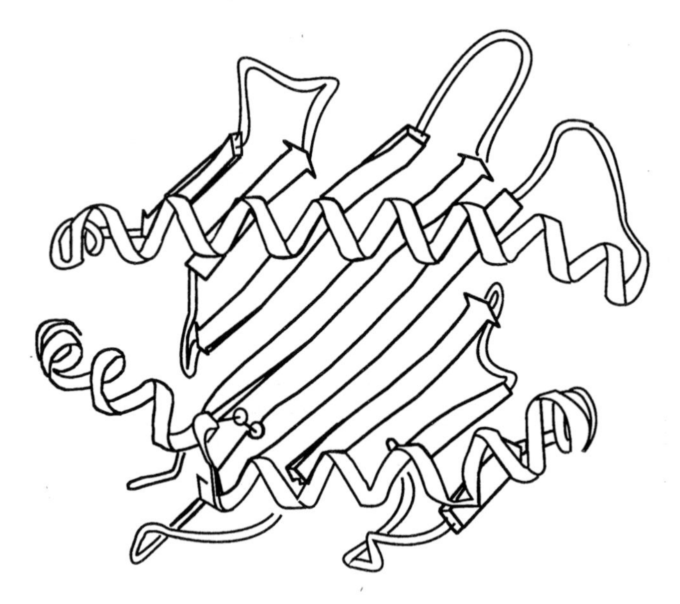

Figure 3-3 Schematic representation of the peptide binding site of a MHC class I molecule (drawn after Bjorkman et al.[8]).

strongly suggests that the polymorphism has a significant effect on antigen presentation. Residues in the α-helical region that point upward presumably interact with the T cell receptor.

The structure of the class II molecule has not yet been determined at the same high resolution as that of the class I molecule; however, a hypothetical model has been constructed by comparing available X-ray data for class II with the class I data and analyzing the pattern of polymorphic and conserved residues in class I and class II sequences.[9] The data indicate that the peptide binding site of the class II molecule has the same basic structure as the class I binding site. The class II α and β chains contribute equally to the peptide binding site.

The description of the three-dimensional structure of the peptide binding site has given new insight into the interaction between the MHC, peptide, and T cell receptor. The immune system is able to evoke an immune response against an enormous number of different peptides and, consequently, it has been a puzzle how MHC molecules are able to bind such a wide variety of peptides with high affinity. This question appears to have been solved by the elucidation of the X-ray structure of a murine class I molecule in complex with two different viral peptides.[10] The results show that peptide binding by a particular MHC molecule focuses on features that are common among all peptides, namely the main-chain atoms and the amino- and carboxy-terminal ends.

III. THE BOVINE MAJOR HISTOCOMPATIBILITY COMPLEX

A. GENERAL ORGANIZATION OF THE BOVINE MHC

The bovine MHC, denoted the BoLA (for Bovine Lymphocyte Antigen) complex, is located on chromosome 23[11] and is divided into four regions: class I, class IIa, class IIb, and class III. The class I MHC molecules are encoded in the class I region, the class IIa region encodes the functional class II MHC molecules, the class IIb region harbors a number of class II genes whose function is yet unknown, and the class III region contains a large number of non-antigen-presenting molecule genes several of which encode proteins that have immunological functions. A linkage map of the BoLA region is shown in Figure 3-4. It should be noted that the exact order of all genes has not yet been determined, so the map has been drawn partly on the basis of the order established in man and mouse. The recombination distance in the interval including the subregions I, IIa, and III is low (less than 3% recombination), whereas the IIb region is separated from the other regions by a recombination distance of about 17 cM.[12,13] The orientation of the IIb region in relation to the I and IIa regions has not yet been established and it may be located to the "right" of the class I genes. Such a high recombination frequency between class II genes has not been observed in any other mammalian species and it is still an open question whether the result reflects a chromosomal rearrangement or a recombination hotspot.

1. The Class I Region

Southern blot analysis using a human class I cDNA probe indicated the presence of a large number (at least ten) of class I genes in the bovine genome.[14] The number of expressed class I genes in cattle is not yet known. However, biochemical studies indicate that there are at least two expressed class I genes.[15] In addition, Kemp and co-workers[16] have produced monoclonal antibodies specific for polymorphic determinants on two class I molecules encoded by a single haplotype. The class I molecules recognized by these monoclonal antibodies are encoded at the BoLA-A and BoLA-B loci. The independent nature of the two loci was definitively demonstrated by the differential reactivity of the monoclonal antibodies with BoLA-A and BoLA-B locus L cell transfectants produced using DNA from a MHC homozygous cow.[17]

The presence of two or three expressed class I genes is also supported by cDNA cloning experiments.[18,19,20a] The 3′ untranslated regions of human and murine class I alleles encoded at a single locus generally have homology of 93 to 99%,[18] whereas three bovine cDNA clones with homology of only 80 to 87% have been isolated.[19] Furthermore, evidence for at least two expressed class I loci in cattle has been obtained, as up to four expressed class I alleles have been detected in a single individual by cDNA sequencing.[20a] The cDNA clones described by Bensaid and co-workers[19] were demonstrated to be BoLA-A and BoLA-B locus products by differential hybridization of their 3′ untranslated regions to the L cell transfectants mentioned above. The 3′ untranslated regions from these cDNA clones were also used as locus-specific probes to physically map the two loci by field inversion gel electrophoresis. The two genes are less than 210 kb apart.

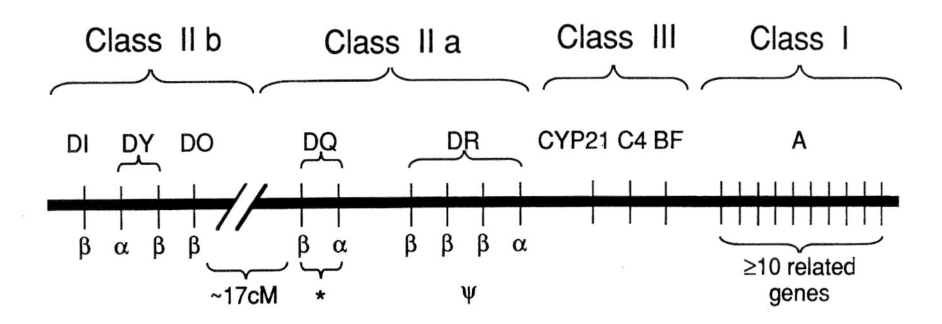

Figure 3-4 A genetic map of the BoLA region. * indicates the presence of a polymorphism in the number of DQ genes between different BoLA haplotypes. Ψ indicates that at least one of the DRB genes is a pseudogene.

2. The Class II Regions

It has been more straightforward to unravel the organization of the bovine class II regions than the class I region. The explanation for the difference is the contrasting evolutionary histories of the two classes of MHC genes in mammals. Both the class I and class II regions include multiple genes in all mammalian species studied so far. The difference is that class I genes are most closely related to other class I genes in the same species, whereas class II genes can be assigned to a set of distinct class II loci which evolved prior to the divergence of mammalian families. This difference implies a more dynamic evolution of the class I region involving expansions and contractions, as well as some form of concerted evolution, i.e., exchange of sequence motifs between nonallelic genes.

In man, there are three major subtypes of class II molecules denoted DQ, DR, and DP, each composed of an α and a β chain encoded by separate A and B genes, respectively. Two additional human class II genes, DOB and DNA, have been described but their function is obscure. Bovine class II genes have been investigated using human probes in Southern blot analyses.[12,20b-23] Human probes have also been used to clone bovine class II genes.[24-29] These studies have given us the knowledge indicated below and summarized in Figure 3-4.

a. Class IIa Region

Evidence suggests that there is only a single bovine DRA gene. However, Southern blot experiments have indicated the presence of at least three bovine DRB genes and genomic clones for three DRB genes (DRB1,[24,25] DRB2,[24] and DRB3[27]) have been isolated. DRB1 is evidently a pseudogene and only the DRB3 gene is expressed at a high level by peripheral lymphocytes.[30] The DRB3 gene is also well conserved compared with human DRB genes, whereas the second exons encoding the β1 domains of the DRB1 and DRB2 genes are quite divergent.[31] Thus, it appears that the DRB3 gene is the most functionally important bovine DRB gene characterized so far. This conclusion is supported by the finding of an extreme genetic polymorphism at this locus (see below).

Southern blot analyses have clearly shown that the number of DQ genes varies between MHC haplotypes.[22] Three types of haplotypes have been found: those carrying single DQA and DQB genes, those carrying two DQA genes and a single DQB gene, and finally, those carrying two genes of each type. The variation in the number of DQB genes has recently been confirmed by PCR amplification and cloning experiments.[32] The expression of bovine DQ genes is strongly suggested by the isolation of cDNA clones[33] and by their extensive genetic polymorphism revealed by RFLP and DNA sequence analyses (see below).

b. Class IIb Region

DOB and DNA are two class II genes which in humans show no or limited polymorphism and which are poorly expressed in those cells where the polymorphic DR, DQ, and DP genes are expressed. Southern blot analyses using human probes have revealed the presence of bovine DOB and DNA genes in the class IIb region.[12,22] There is no evidence for their expression and they are not highly polymorphic.

DYA, DYB, and DIB are three novel bovine class II genes. DYA and DYB were first defined by Southern blot analyses as restriction fragments cross-hybridizing to human DQA and DRB probes, respectively.[12] They were identified as distinct from the bovine DQA and DRB genes by the observation of a high recombination frequency between the DQ-DR region and the DY genes. A genomic clone for DYA was subsequently isolated.[28] Similarly, the DIB gene was first isolated as a slightly divergent bovine DQB gene.[29] Subsequent linkage analysis showed that DIB is located in the class IIb region.[34,35] The DY and DIB genes do not appear to be highly polymorphic and there is no evidence for their expression.

No evidence for the presence of bovine DP genes has been found.[22] As the same strategy used to search for the DQ and DR genes was used for the DP genes, it seems that if there are bovine DP genes they must have diverged considerably from their human counterparts.

A number of new genes have recently been found in the class II region of man, mouse, and rat.[36] These new genes include several functionally interesting genes such as two divergent class II-related genes (DMA and DMB), two genes (TAP1 and TAP2) encoding a peptide transporter that transports peptides generated from degraded cytoplasmic proteins into the endoplasmic reticulum for class I antigen presentation, and two genes that encode proteasome components (LMP2 and LMP7). Recently, bovine TAP1 and LMP7 transcripts have been detected, bovine LMP7 clones have been isolated,[37] and the bovine LMP7 gene has been mapped to the class IIb region by RFLP analysis (C. J. Davies, unpublished results). Since the TAP and LMP genes are located adjacent to each other in the HLA complex, it is exceedingly likely that the bovine homologues of these genes are all located in the BoLA class IIb region.

Recently, it was reported that genetic polymorphism in one of the MHC-linked peptide transporter genes in the rat (TAP-2) has a dramatic influence on the repertoire of peptides presented by class I molecules.[38] This raises the question of whether certain haplotypic combinations of the proteasome, peptide transporter, and class I genes function better together than other combinations. Favorable haplotypic combinations might be kept at a high frequency in the population due to selection. The significant recombination interval between the bovine class IIb region, which most likely contains the TAP and LMP genes, and the rest of the BoLA complex should make it possible to address the question of how TAP and LMP polymorphism affects class I antigen presentation.

3. Class III Region

This region is located between the class I and class II regions in all mammalian species studied so far. MHC class I and class II genes have evolved from the same progenitor and both encode cell surface molecules that present peptides to T cells. In contrast, the numerous genes mapped to the class III region have no relationship to the class I and class II genes except that several of them are involved in the immune system. The class III genes identified in humans include genes for the complement components C2, C4, and Bf, tumor necrosis factors A and B (TNFA and TNFB), 21 hydroxylase (CYP21), heat shock protein 70 (HSP70), and a number of genes with unknown functions.[36]

So far, only limited efforts have been made to characterize the bovine class III region but the C4, Bf, and CYP21 genes have been mapped to the MHC region by linkage analysis[12,39] or analysis of somatic cell hybrids.[40]

B. POLYMORPHISM OF THE CLASS I AND CLASS II GENES

1. Class I Genes

a. Polymorphism Revealed by Serological Methods

BoLA class I antigens were first identified using alloantisera collected from naturally immunized cows, cows immunized against fetal histocompatibility antigens during pregnancy, or cows specifically immunized by skin grafting or lymphocyte inoculation.[41-43] When used in the complement-mediated lymphocyte microcytotoxicity dye exclusion test, these antisera appeared to define alleles at a single highly polymorphic genetic locus denoted BoLA-A.[41,44] Amorena and Stone[45] established that BoLA-A was part of the MHC of cattle by showing that matching of skin graft donors and recipients for BoLA-A alleles resulted in prolonged graft survival time.

The complexity of many BoLA class I reagents has stimulated considerable debate about whether BoLA class I typing sera detect alleles at a single locus or are haplotype specific and react with the products of more than one polymorphic class I gene.[46] The tight genetic linkage between the BoLA-A and -B loci (discussed above) explains why it has been exceedingly difficult to identify alloantisera that define a second, independent segregant series of BoLA class I antigens, that is, the B locus products. To date, there is no evidence for a third "classical", polymorphic class I locus in cattle.

Five international BoLA workshops have been organized under the auspices of the International Society for Animal Genetics to standardize BoLA typing reagents and nomenclature.[44,47-51] In total, 53 class I workshop specificities have been defined. Twenty-seven specificities have been assigned to the A locus, while 26 are still classified as workshop (denoted by "w") specificities. One "w" specificity is believed to recognize a B locus epitope. There are seven "full house" public specificities, i.e., specificities that have been split and for which all animals tested thus far carry one of the subtypic specificities. Consequently, the workshop specificities presumably define 45 A locus alleles and 1 B locus allele. Many additional putative A locus and a few putative B locus alleles are recognized in individual laboratories. Both at the workshop and local level, a large number of broad public specificities have also been identified.

Despite behaving in a locus-specific manner, many of the alloantisera used to define BoLA-A locus alleles may be haplotype specific and may contain antibodies against both A and B locus products. However, because the linkage between the A and B loci is exceedingly tight and the reagents define a single segregant series of alleles, the reagents are functionally A locus specific.

In most cattle populations, there are from 10 to 20 BoLA-A alleles segregating. The gene frequencies for these alleles generally vary from less than 1% to as high as 15 to 50% (usually less than 30%).[52-55]

b. Polymorphism Revealed by Cellular Methods

Receptors on cytotoxic T lymphocytes (CTLs) react with peptides that are bound in the antigen presentation groove of class I molecules. Because of clonal deletion and anergy, animals normally do not have CTLs that react with self peptides (minor histocompatibility antigens) in combination with self class I antigens (major histocompatibility antigens). However, if either the peptides or the class I MHC molecules are "foreign" to the responder lymphocyte population, then clones of CTLs that recognize the foreign class I – peptide combinations will respond, i.e., proliferate and lyse the target cells. Bovine alloreactive CTLs are generated in one-way mixed lymphocyte culture (MLC) reactions and these CTLs have specificity for particular class I alleles, and presumably particular endogenous cattle peptides, when used in chromium release assays.[56,57] It is possible to generate CTLs lines that are highly specific. This was demonstrated by Spooner et al.[57] who generated lines that distinguished four different BoLA-A6 subgroups (serological reagents that distinguish these subgroups have also been developed). Despite the good specificity of CTL lines, the use of CTLs for routine MHC typing is not practical.

c. Polymorphism Revealed by Isoelectric Focusing

One-dimensional isoelectric focusing (1D-IEF) is a powerful technique for studying expressed polymorphisms. Several investigators have utilized either immunoprecipitation and 1D-IEF[58-60] or 1D-IEF and immunoblotting[61] to characterize serologically defined BoLA class I haplotypes. Using 1D-IEF, these investigators have identified serologically undetected (blank) haplotypes, BoLA-A locus polymorphisms (splits) that were not resolved using the available serological reagents, and products of more than one expressed class I gene. Unfortunately, because there are no commercially available standards for 1D-IEF gels, because there can be considerable gel to gel variation in the pH gradient, because the available monoclonal antibodies may not react with all alleles at all polymorphic class I loci, because the intensity of bands can vary from sample to sample, and because the banding patterns are complex, it is exceedingly difficult to use 1D-IEF as a primary class I typing technique. Nevertheless, class I 1D-IEF is a valuable adjunct to serological class I typing. In the Fifth International BoLA Workshop, a number of discrepancies were resolved and several class I haplotypes were split on the basis of 1D-IEF data contributed by four laboratories.[50]

d. Restriction Fragment Length Polymorphism (RFLP)

When Southern blots of restriction enzyme-digested bovine genomic DNA are probed with human or bovine class I cDNA probes that include some or all of the extracellular domains of class I molecules, a large number of monomorphic and polymorphic fragments are detected.[14,60] These fragments carry portions of at least ten bovine class I genes. Despite the fact that the patterns are complex, it is possible to follow the segregation of class I RFLP patterns in family studies and it has been found that RFLP patterns and class I serotypes are highly correlated.[14] As a class I typing technique, RFLP typing has many of the same strengths and weaknesses as 1D-IEF typing: both techniques are expensive and laborious; the banding patterns detected with both techniques are complex and, therefore, difficult to interpret; neither technique is very good as a primary class I typing techniques; and both techniques provide a wealth of information when used in conjunction with class I serotyping. As more bovine class I sequence information becomes available, it may be possible to identify locus-specific probes, corresponding to untranslated portions of class I genes. With locus specific probes it would undoubtedly be possible to do reliable class I RFLP typing in outbred populations. The availability of sequence information may also make it possible to develop other less laborious DNA typing techniques; for example, PCR-based techniques similar to those discussed below for class II typing.

2. Class II Genes

a. Polymorphism Revealed by Cellular Methods

At approximately the same time that the BoLA class I antigens were identified, Usinger et al.[62] detected "lymphocyte-defined" or class II loci using the mixed lymphocyte culture (MLC) test. The MLC assay is a lymphocyte proliferation assay and the majority of the proliferation is due to $CD4^+$ helper T lymphocytes responding to class II differences.[63] Receptors on helper T lymphocytes recognize peptides that are bound in the antigen presentation groove of particular MHC class II molecules. Because of clonal deletion and anergy, animals normally do not have helper T lymphocytes that react with self-peptides (minor histocompatibility antigens) in combination with self class II antigens (major histocompatibility antigens). However, if the class II molecules are "foreign" to the responder lymphocyte population, then clones of helper T lymphocytes that recognize particular endogenous cattle peptides bound in the antigen presentation grooves of the foreign class II molecules will respond, i.e., proliferate. The MLC test detects foreign class II molecules on stimulator cells. Consequently, a negative response indicates that the stimulator cells do not express any class II molecules that are foreign to the responder cells.

Several studies have demonstrated that the lymphocyte-defined loci of cattle, the class II loci, and the serologically defined class I loci segregate together in full sibling embryo transfer families.[55,64,65] Davies and Antczak[55] also identified six BoLA complex homozygous, homozygous typing cell (HTC) donors, and using serology for class I typing and the MLC assay for class II typing demonstrated the presence of class I identical – class II disparate and class I disparate – class II identical BoLA haplotypes in a herd of Holstein Friesian cattle. The identification of cattle carrying BoLA haplotypes with different BoLA-A alleles but identical HTC-defined class II haplotypes implies very strong linkage between all of the expressed, polymorphic class II loci of cattle.

Since the biological function of class II molecules is to present peptides to T cell receptors on $CD4^+$ lymphocytes, alloreactive $CD4^+$ T lymphocyte clones are the ultimate tool for identifying functionally relevant class II polymorphism.[66,67] Teale and Kemp[66] isolated both cytolytic and noncytolytic $CD4^+$ alloreactive T lymphocyte clones from an MLC culture. Unfortunately, T lymphocyte clones have a limited life span. Teale and Baldwin[68] tried to immortalize alloreactive T lymphocyte clones with *Theileria parva*; however, they were not successful.

b. Polymorphism Revealed by Serological Methods

Generation of alloantisera specific for bovine class II molecules is quite difficult. Cows immunized with leucocytes or skin carrying a combination of foreign class I and class II molecules tend to make antibody responses directed primarily against the class I molecules. One explanation for this is that the dose of class II molecules is much lower than the dose of class I molecules. While class I molecules are expressed on all cells, under normal circumstances class II molecules are only expressed on antigen presenting cells, e.g., macrophages and B lymphocytes. The fact that class II molecules are not expressed on resting T lymphocytes also makes it more difficult to use the lymphocyte cytotoxicity assay for serological class II typing. When the standard lymphocyte microcytotoxicity assay is performed with unfractionated peripheral blood lymphocytes, class II-specific antisera give inconsistent results because they only kill the B lymphocytes. Consequently, to do reliable class II typing it is necessary to use purified B lymphocytes or a two color immunofluorescence test[69] in which the B and T lymphocytes can be distinguished.

Two approaches have been used to produce alloantisera for bovine class II typing: (1) antisera with activity against both class I and class II antigens have been absorbed with platelets to remove class I-specific antibodies,[70-72] and (2) specific immunizations have been done using class I compatible – class II disparate cattle selected by doing class I serology in conjunction with MLC testing or class II 1D-IEF.[73-75] Both the two-color immunofluorescence test[70,71] and enrichment for B lymphocytes by panning, followed by testing in a standard lymphocyte microcytotoxicity assay,[72-74] have been used for bovine class II typing. The results obtained with the two methods are comparable.

c. Polymorphism Revealed by Isoelectric Focusing

Studies involving the immunoprecipitation of BoLA class II molecules with polyclonal rabbit anti HLA-DR antisera or monoclonal antibodies specific for bovine class II molecules, followed by the separation of the molecules by 1D-IEF, have been very informative. In the initial studies, in which rabbit anti HLA-DR antisera were used, a single monomorphic class II α chain band and pairs of bands representing a polymorphic β chain were identified.[76,77] Thus, a heterozygous individual shows two pairs of β chain bands. All indications are that the two β chain bands represent post-translational modifications of the product of a single B gene.[76] It has now been established that these are bovine DR genes products.[78] Because of the relative simplicity of the DRB focusing patterns, DRB 1D-IEF typing does not suffer from the interpretation problems that limit the usefulness of 1D-IEF class I typing.

BoLA-DQ products can also be separated using 1D-IEF.[78] DQ molecules immunoprecipitated with "DQ-specific" monoclonal antibodies and separated by 1D-IEF have polymorphic α and β chains. Bissumbhar et al.[78] also found that the "DQ-specific" monoclonal antibodies used in their study reacted with polymorphic DQ epitopes and also with some DR alleles. Moreover, some haplotypes appeared to express more than one DQ product. Because DQ α chains are polymorphic and some BoLA haplotypes express two different DQ products, DQ 1D-IEF patterns are considerably more complex than DR 1D-IEF patterns.

d. Restriction Fragment Length Polymorphism (RFLP)

This was the first typing method that became available for the routine definition of the allelic polymorphism of bovine class II genes.[20b,21] RFLPs have now been reported for each of the nine classes of class II genes found in cattle.[12,20b-23,34] RFLP analysis revealed that there are two groups of bovine class II genes: (1) genes with limited polymorphism, including DRA, DOB, DNA, DYA, DYB, and DIB, and (2) extremely polymorphic genes, including DQA, DQB, and DRB. Two to five different RFLP types have been identified for each locus in the first group, whereas more than 20 alleles have been identified for each locus in the latter group. Interestingly, a similar dichotomy of monomorphic DRA, DOB, and DNA genes, and highly polymorphic DQA, DQB, and DRB genes is found in man and the mouse.

The RFLP studies have revealed an extremely strong linkage disequilibrium in the bovine DQ-DR region which proved very beneficial for the development of RFLP typing for DQ-DR polymorphism. Very few exceptions were found to the rule that a given DQA allele is always associated with the same DQB allele and many of the DQ haplotypes first found in Swedish Red and White cattle appear to be conserved in most European (*Bos taurus*) cattle breeds. Moreover, the strong linkage disequilibrium between DQ and DRB alleles is exemplified by the fact that Sigurdardóttir et al.[23] found 20 DQ types and 28 DRB types, but only 29 DQ-DR haplotypes in one population of Swedish cattle.

e. Polymorphism Revealed by DNA Sequencing

Polymorphism of bovine class II genes has been studied by sequencing exon 2, which encodes the highly polymorphic residues of the peptide binding site, following polymerase chain reaction (PCR) amplification. Thus far, allelic series for the DRB3 and DQB genes have been reported.[31,32,79]

DRB3 exon 2 sequences were obtained for 14 major DQ-DR haplotypes defined by RFLP analysis and it was found that each haplotype carries a unique allele at the DRB3 locus.[31] The alleles differ by multiple amino acid substitutions, from 5 to 22 substitutions for the 88 positions compared (Figure 3-5). It is clear from this analysis that the bovine DRB3 locus is highly polymorphic both with regard to the number of alleles and the degree of divergence between alleles. The degree of polymorphism is comparable to that of the human DRB1 locus, which is the most polymorphic class II locus in man.

PCR-based sequence analysis of the bovine DQB genes has given a more complex picture than that obtained for the DRB genes.[32] The polymorphism in the copy number of DQB genes, first revealed by RFLP analysis,[22] was confirmed since two DQB sequences were obtained for certain haplotypes. Analysis of the DQB sequences indicated that they could be divided into four subtypes (DQB1, DQB2, DQB3, and DQB4). Extensive genetic polymorphism was revealed as ten DQB1 alleles and four DQB2 alleles were identified by analyzing DNA samples representing 14 major DQ-DR haplotypes defined by RFLP analysis.

f. Polymorphism Revealed by PCR-Based Methods

It is very likely that routine class II typing will in the future be carried out using DNA typing methods based on the sequence polymorphism of the expressed class II genes. Sequence-based typing will probably involve PCR amplification of polymorphic parts of the genes. Thus far, two PCR-based methods for class II typing have been reported. One method involves PCR amplification and RFLP analysis, using three restriction enzymes, of the second exon of the DRB3 gene;[80] about 30 alleles were defined. Another PCR-based method was reported by Ellegren et al.[81] This method involves PCR

```
        10         20         30         40         50         60         70         80          90
BoDRB3  +  + +               + +       ++                    +    +     +  + +  +    +   +   +
4A      EY STSECHFFNG TERVRFLDRY FYNGEEYVRF DSDWGEFQAV TELGRPDAKY WNSQKDILER ERAAVDTYCR HNYGGVESFTV
1A      -- -K-------- ---------- YT----T--- -------R-- -----Q--E- ------F--E K--E--RV-- -----M-----
1B      Q- HKG------- -----L---H ------F--- ----D--R-- ------A-EQ ------F--Q K--E--RV-- -----------
2A      -- ---------- ---------- -H----F--- ------YR-- --------E- ---- E---- A--------- -----------
3       -- CK-------- -------E-S ------F--- ------YR-- ---------- ------L--Q K--N------ ----VG-----
5       -- HK-------- ---L-Y---- ---------- ------YR-- ---------- -----E---- K--N------ ----V------
6       -- CKR------- -----L---C -H----F--- -------R-- -----RV-EH L----E---- K--E---V-- ----VG-----
7A      -- CKR------- ---------C -H----F--- -------R-- -----RV-EQ ------F--- R--E--RV-- ----V------
8A      -- A--------- ---------- -H----L--- -------R-- ------S-VH L-----F--D ---S--T--- ----V------
9A      -- -K-------- -------E-S ------N--- ------YR-- --------E- -----E---- K--N--RV-- ----VG-----
10      -- -K-------- ---------- -H-------- ------YR-- ----QRV-E- C-----F--- A--------- ----VG-----
11      Q- HKG------- -----L---H ---------- ----D--R-- ------S-E- ------F--- R--E---V-- ----V------
12      -- TKK------- -------N-- -H----F--- ------YR-- --------E- -----E---- A--------- ----VG-----
13A     -- LK-------- -------E-- ---------- ------YR-- ---------- ------L--- K--N------ ----V------
```

Figure 3-5 The amino acid sequence of the first domains of 14 bovine DRB3 alleles (based on data presented by Sigurdardóttir et al.[31]). + indicates positions in the peptide binding site. A dash indicates identity to the DRB3*4A allele. A blank in position 65 of DRB3*2A indicates the presence of a deleted codon in this allele.

amplification and size analysis of a microsatellite located in intron 2 of the DRB3 gene very close to the polymorphic exon 2. The microsatellite locus is highly polymorphic, comprising at least 14 alleles. It is a valuable marker because there is a very strong correlation between the microsatellite alleles and the expressed polymorphism in exon 2.

g. *Correlation Between Class II Typing Methods*

Very good correlations between serological, cellular, IEF, RFLP, and PCR-based typing have been reported in several studies.[49,51,82,83] The serological, cellular, and IEF typings, which detect polymorphism at the product level, all correlate closely with DQ-DR RFLP types. This shows that the most important class II polymorphism is encoded in the IIa and not the IIb region (Figure 3-4). The four standard typing methods all have their advantages and limitations. The serological method is excellent for large population screenings, but so far a rather limited panel of serological reagents has been developed. The application of cellular typing is limited by the paucity of a good panel of homozygous typing cells. The IEF and RFLP typing methods both give very good resolution of class II polymorphism but are rather laborious and, therefore, are difficult to apply to large population screenings. As a consequence of these limitations, locus-specific PCR-based typing will probably be the method of choice in the future.

IV. THE OVINE AND CAPRINE MAJOR HISTOCOMPATIBILITY COMPLEXES

A. GENERAL ORGANIZATION OF THE OVINE AND CAPRINE MHC

The ovine MHC, known as the ovine lymphocyte antigen (OLA) complex, and the caprine MHC, known as the caprine lymphocyte antigen (CLA) complex, have not been as thoroughly characterized as the BoLA complex. However, the data that are available suggest that the OLA and CLA complexes are similar in structure to the BoLA complex.

RFLP analysis using a human cDNA probe suggested that the caprine class I region contains 10 to 13 class I genes.[84] A cDNA cloning study[85] and experiments involving the sequential immunoprecipitation of ovine class I molecules with monoclonal antibodies[86] indicated that the OLA class I region contains at least three expressed class I genes. However, it is not known if all three genes are highly polymorphic. A 1D-IEF study has provided evidence for two expressed, polymorphic class I loci in the CLA complex.[87]

Hybridization and cloning experiments indicate that the OLA complex contains single copy DRA, DNA, and DOB genes and duplicated DRB, DQA, and DQB genes.[88-90] There is no evidence for ovine homologues of the human DPA and DPB genes.[88,91] However, the CLA complex apparently contains a homologue of the bovine DYA gene.[92] It has been established that some OLA haplotypes carry duplicated DQ genes and that in haplotypes with duplicated genes both sets can be transcribed.[89] In addition, transfection experiments have demonstrated that the OLA complex contains expressible DRA and DRB genes.[91,93] Immunoprecipitation studies suggest that sheep can express four or more distinct class II molecules.[94]

Because of a pulse field gel electrophoresis study conducted by Cameron and co-workers,[84] more is known about the class III region of goats than of other ruminants. The CLA class III region contains duplicated CYP21 and C4 genes as well as C2, HSP70, and TNF genes. All of these genes are located within 800 kb of the closest class I gene and the gene order is the same as that established for the HLA complex except that the C4 and CYP21 genes may be arranged with the two CYP21 genes between the C4 genes. If the C4 and CYP21 genes are arranged in this fashion, it would imply unique duplication events for the CLA, HLA, and H-2 complexes. A moderate degree of RFLP was associated with the C2 and C4 genes.

B. POLYMORPHISM OF THE CLASS I AND CLASS II GENES

All of the techniques used to study BoLA polymorphism can also be used for sheep and goats. To date, the following methods have been used to characterize OLA and CLA polymorphism: OLA class I-serology,[95-98] -RFLP;[99] OLA class II-RFLP;[88,99] CLA class I-serology,[100-102] -IEF,[87] -RFLP;[84] and CLA class II-MLC,[103] -RFLP.[84]

There is general agreement that serologically defined OLA and CLA class I antigens are encoded at two closely linked loci.[96-98,101,104] This conclusion is also supported by CLA lysostrip experiments[105] and CLA class I IEF.[87] In the latter study, class I IEF typing proved to be a valuable adjunct to serology as it helped to define splits in serotypes. Because of the limited number of laboratories doing OLA and CLA typing, only small comparison tests have been organized.[102] Consequently, there is no internationally accepted MHC nomenclature for sheep and goats.

It is not clear why it has been relatively easy to define two independent segregant series of OLA and CLA class I antigens, while it has been very difficult to define a second segregant series of BoLA class I antigens. One plausible explanation is that there is a higher recombination frequency between the loci encoding polymorphic class I molecules in sheep and goats than in cattle.

Class II typing in sheep and goats is still at a very rudimentary stage. Although RFLP bands have been shown to segregate in families,[99] alleles at different loci have not been defined and class II haplotypes have not been established.

V. SIGNIFICANCE OF MHC POLYMORPHISM

A. MHC RESTRICTION

Both class I- and class II-mediated MHC restriction of bovine T lymphocyte responses have been reported. Bovine class I molecules have been shown to act as MHC restriction elements for $CD8^+$ cytotoxic T cells specific for target cells infected with: *Theileria parva*[106-108] (see Chapter 9), *Theileria annulata,*[109] and bovine herpesvirus-1[110] (see Chapter 10). Bovine class II molecules have been shown to act as MHC restriction elements for $CD4^+$ helper T cells specific for ovalbumin,[111,112] foot-and-mouth disease virus (FMDV) peptides,[113] and *Oesophagostomum radiatum* antigens.[114] In the FMDV study, one immunodominant peptide was identified and nonresponder animals, animals carrying BoLA haplotypes incapable of mounting *in vitro* T cell proliferative responses to this peptide, were identified.[113]

B. IMMUNE RESPONSE ASSOCIATIONS

The MHC genes of ruminants can function as immune response (Ir) genes. In one study, Lie et al.[115] showed that BoLA type influenced the magnitude of antibody responses to human serum albumin and a synthetic polypeptide. In another study, Glass et al.[116] found that following immunization with ovalbumin, the magnitude of *in vitro* ovalbumin-specific, class II-dependant T lymphocyte proliferation, segregated with paternal BoLA haplotypes in two half-sibling families. In this study, BoLA type did not appear to influence antibody responses to ovalbumin and an association between BoLA class I alleles and T lymphocyte proliferation was not detected in outbred cattle. In outbred populations, there are usually haplotypes with different combinations of class I and class II alleles. Consequently, it is possible that there was a class II effect that was not detected.

C. DISEASE ASSOCIATIONS

It is a fundamental tenet of immunology that the immune system must be able to distinguish self from non-self. A number of MHC genes, including the “classical” class I and class II genes, are involved in immunoregulation; i.e., these genes are involved in self, non-self discrimination. Animals with different sets of MHC genes, and different endogenous peptides or “background genes”, become tolerant to different self-antigens, antigen presentation molecule-peptide combinations, and develop different T cell receptor repertoires. Consequently, animals respond differently to “foreign” organisms or particulate antigens. The MHC alleles carried by an animal influence whether the animal mounts an appropriate or inappropriate immune response to a pathogen or particulate antigen. A particular MHC haplotype is likely to confer resistance to most diseases and susceptibility to a few diseases.

An association between a polymorphic MHC gene and a disease may either be due to a direct effect of the gene or to the effect of a linked gene. Consequently, many MHC disease associations are ultimately mapped to genes other than the gene with which they were initially associated. For example, many diseases that were originally associated with HLA class I alleles have now been found to be more strongly associated with polymorphisms in the class II region. Nevertheless, the class I and class II genes are often viewed as candidate genes which can have a direct effect on disease susceptibility.

It only makes sense to conduct a disease association study when there is evidence for genetic control of the disease process. Furthermore, except in the case of candidate gene studies, it is important to use multiple markers to define MHC haplotypes or to work with genetically defined populations. In addition, disease association studies should be regarded as just the first step in the process of unraveling the disease process or mechanism of immunity.

Many infectious and autoimmune diseases have been shown to be associated with the presence of particular HLA alleles or haplotypes (reviewed by Tiwari and Terasaki[117]). Information on MHC disease associations in domestic animals is considerably more limited. Nevertheless, it has been established that MHC polymorphism can have a profound effect on disease susceptibility in domestic animals. One of the strongest MHC disease associations found in any species is the association between a particular equine MHC class II allele and the development of bovine papilloma virus induced sarcoid tumors.[118]

1. Bovine Disease Association Studies

MHC disease association studies in cattle include studies on bovine leukemia virus (BLV) infection,[119,120] bovine virus diarrhea,[121] mastitis,[122-124] endo and ectoparasites,[125,126] and ocular squamous cell carcinoma.[127] Several of these studies involved the assessment of MHC effects on quantitative disease traits. Quantitative trait studies are considerably more complicated than case control studies and this may explain why the studies have indicated weak associations that have been difficult to confirm. Different models that can be used to assess MHC effects on disease resistance have been reviewed by Østergård et al.[128] Lundén et al.[123] have used a novel approach, the correlation of bull breeding values for disease traits with bulls class II types, to assess the relationship between BoLA genes and disease resistance. Because the studies involving other diseases have been inconclusive or unsubstantiated only the work on BLV will be discussed further.

Lewin and Bernoco[119] were the first to provide evidence that the subclinical progression of BLV infection is influenced by genes in the BoLA complex. At the population (herd) level, they found class I alleles that were associated with low and high frequencies of polyclonal B cell expansion in BLV-infected seropositive cows. In addition, in a half-sibling family, resistance to lymphocytosis and normal numbers of B lymphocytes were found to segregate with one paternal BoLA haplotype. In a subsequent study, Lewin et al.[120] found that one class I allele was associated with delayed seroconversion. The authors hypothesized that delayed seroconversion might be due to a BoLA complex gene that influences the production of the plasma BLV blocking factor previously described.[129] Because different class I alleles have been associated with increased or decreased polyclonal B cell expansion in different breeds, it is likely that the class I genes are marker genes and do not have a direct effect. Since there is downregulation of BoLA-DR expression on B lymphocytes during polyclonal B cell expansion,[130] it has been suggested that class II polymorphism might have a direct influence.[131]

2. Ovine and Caprine Disease Association Studies

Very few ovine and caprine MHC disease association studies have been conducted. Nevertheless, the existence of an OLA complex-linked scrapie susceptibility gene[132,133] and a CLA-linked gene that influences the development of caprine arthritis-encephalitis (CAE) virus-induced arthritis[134] have been suggested.

Millot and colleagues[132,133] have postulated the existence of an autosomal recessive scrapie susceptibility gene. Their first study was a case control study that involved two breeds of sheep.[132] In this study, they found that certain OLA class I antigens were associated with an increased or decreased relative risk for the development of scrapie. As different alleles were associated with resistance/susceptibility in the two breeds, they concluded that susceptibility was controlled by a class I-linked gene and not by the class I genes themselves. In their second study, they looked at segregation of OLA class I alleles and the scrapie susceptibility locus in a single breed of sheep.[133] Based on the family studies, they estimated that the scrapie susceptibility locus was 11 to 16 centiMorgans from the OLA class I genes.

While a case control disease association study involving 546 diseased and 402 healthy controls did not provide convincing evidence for an association between the CLA class I genes and CAE virus-induced arthritis, segregation analysis in multiple case paternal half-sibling families revealed that there is a MHC-linked gene in goats that influences the development of arthritis.[134] The recombination rate between the arthritis susceptibility locus and the class I loci was estimated at 15%.

D. ASSOCIATIONS WITH PHYSIOLOGICAL TRAITS

In contrast to disease association studies where it is quite reasonable to view the class I and class II genes as candidate genes, in studies involving associations between the MHC and physiological traits, the situation is quite different. In these studies, it is more appropriate to view the class I and class II genes as marker genes for a particular chromosome or linkage group. If there are cases where class I or class II polymorphism is responsible for variation in physiological traits, it is undoubtedly an indirect effect.

There are several reports of significant associations between serologically defined class I polymorphism and growth,[135,136] milk production,[124,137,138] and fertility. However, there is no consistent pattern regarding the effect of specific alleles. Furthermore, Lundén et al.[139] revealed no or only weak associations between class II polymorphism and milk production or fertility traits. The results strongly suggest that MHC polymorphism does not have a direct effect on these traits and that MHC linked genes are responsible for the observed associations.

E. REPRODUCTION AND THE FETAL ALLOGRAFT

One of the great paradoxes of immunology is the survival of the "fetal allograft" in an immunologically hostile maternal environment. Since class I and class II antigens act as major transplantation barriers, their expression on placental tissues that come into contact with the maternal immune system could result in rejection of the fetus, i.e., abortion. One adaptation that probably helps protect the ruminant fetus from rejection is that the cotyledonary villi that are intimately associated with the maternal caruncular tissue do not express "classical" class I or class II molecules.[140,141] However, although a study with sheep did not detect class I expression on interplacentomal trophoblast,[140] it has been reported that class I antigens are expressed in the interplacentomal areas in cattle.[141] Furthermore, although high titered responses usually do not occur until around the time of parturition, pregnant ruminants frequently produce antibodies against their fetuses' paternally inherited class I antigens.[142,143]

There is now mounting evidence that during pregnancy, MHC incompatibility may actually be beneficial. A recent study by Joosten et al.[144] has revealed that class I compatibility between a cow and her calf can result in retention of the placenta. Evidence was also presented that in cases of retained placenta in which the cow and calf were incompatible, tolerance of the cow to noninherited maternal antigens, antigens carried by both the cow's dam and calf, could be involved. These authors hypothesize that during a normal, MHC incompatible, pregnancy expression of fetal class I antigens, encoded either at classical or nonclassical class I loci, induces an allogeneic immune response by the mother which results in the production of specific lymphokines that are required for normal placental maturation.[145] Consequently, lack of allorecognition should result in abnormal placental development and retention of the placenta.

REFERENCES

1. **Klein, J.,** *Natural History of the Major Histocompatibility Complex,* John Wiley & Sons, New York, 1986.
2. **Benacerraf, B. and McDevitt, H. O.,** Histocompatibility- linked immune response genes. A new class of genes that controls the formation of specific immune responses has been identified, *Science,* 175, 273, 1972.
3. **Benacerraf, B.,** Role of MHC gene products in immune regulation, *Science,* 212, 1229, 1981.
4. **Zinkernagel, R. M. and Doherty, P. C.,** Restriction of in vitro T cell-mediated cytotoxicity in lymphocytic choriomeningitis within a syngeneic or semiallogeneic system, *Nature,* 248, 701, 1974.
5. **Lanzavecchia, A.,** Antigen-specific interaction between T and B cells, *Nature,* 314, 537, 1985.
6. **Hedrick, P. W. and Thomson, G.,** Evidence for balancing selection at HLA, *Genetics,* 104, 449, 1983.
7. **Hughes, A. L. and Nei, M.,** Pattern of nucleotide substitution at major histocompatibility complex class I loci reveals overdominant selection, *Nature,* 335, 167, 1988.
8. **Bjorkman, P. J., Saper, M. A., Samraoui, B., Bennett, W. S., Strominger, J. L., and Wiley, D. C.,** Structure of the human class I histocompatibility antigen, HLA-A2, *Nature,* 329, 506, 1987.
9. **Brown, J. H., Jardetzky, T., Saper, M. A., Samraoui, B., Bjorkman, P. J., and Wiley, D. C.,** A hypothetical model of the foreign antigen binding site of class II histocompatibility molecules, *Science,* 332, 845, 1988.
10. **Fremont, D. H., Matsumara, M., Stura, E. A., Peterson, P. A., and Wilson, I. A.,** Crystal structures of two viral peptides in complex with murine MHC class I H-2K^b, *Science,* 257, 919, 1992.
11. **Fries, R., Hediger, R., and Stranzinger, G.,** Tentative chromosomal localization of the bovine major histocompatibility complex by in situ hybridization, *Anim. Genet.,* 17, 287, 1986.
12. **Andersson, L., Lundén, A., Sigurdardóttir, S., Davies, C. J., and Rask, L.,** Linkage relationships in the bovine MHC region. High recombination frequency between class II subregions, *Immunogenetics,* 27, 273, 1988.
13. **van Eijk, M. J. T., Russ, I., and Lewin, H. A.,** Order of bovine DRB3, DYA and PRL determined by sperm typing, *Mammalian Genome,* 4, 113, 1993.

14. **Lindberg, P. G. and Andersson, L.,** Close association between DNA polymorphism of bovine major histocompatibility complex class I genes and serological BoLA-A specificities, *Anim. Genet.,* 19, 245, 1988.
15. **Joosten, I., Teale, A. J., van der Poel, A., and Hensen, E. J.,** Biochemical evidence of the expression of two major histocompatibility complex class I genes on bovine peripheral blood mononuclear cells, *Anim. Genet.,* 23, 113, 1992.
16. **Kemp, S. J., Tucker, E. M., and Teale, A. J.,** A bovine monoclonal antibody detecting a class I BoLA antigen, *Anim. Genet.,* 21, 153, 1990.
17. **Toye, P. G., MacHugh, N. D., Bensaid, A. M., Alberti, S., Teale, A. J., and Morrison, W. I.,** Transfection into mouse L cells of genes encoding two serologically and functionally distinct bovine class I MHC molecules from a MHC-homozygous animal: evidence for a second class I locus in cattle, *Immunology,* 70, 20, 1990.
18. **Ennis, P. D., Jackson, A. P., and Parham, P.,** Molecular cloning of bovine class I MHC cDNA, *J. Immunol.,* 141, 642, 1988.
19. **Bensaid, A., Kaushal, A., Baldwin, C. L., Clevers, H., Young, J. R., Kemp, S. J., MacHugh, N. D., Toye, P. G., and Teale, A. J.,** Identification of expressed bovine class I MHC genes at two loci and demonstration of physical linkage, *Immunogenetics,* 33, 247, 1991.
20a. **Ellis, S.A., Braem, K.A., and Morrison, W.I.,** Transmembrane and cytoplasmic domain sequences demonstrate at least two expressed bovine MHC class I loci, *Immunogenetics,* 37, 49, 1993.
20b. **Andersson, L., Böhme, J., Peterson, P. A., and Rask, L.,** Genomic hybridization of bovine class II major histocompatibility genes: 2. Polymorphism of DR genes and linkage disequilibrium in the DQ-DR region, *Anim. Genet.,* 17, 295, 1986.
21. **Andersson, L. Böhme, J., Rask, L., and Peterson, P. A.,** Genomic hybridization of bovine class II major histocompatibility genes. 1. Extensive polymorphism of DQα and DQβ genes, *Anim. Genet.,* 17, 95, 1986.
22. **Andersson, L. and Rask, L.,** Characterization of the MHC class II region in cattle. The number of DQ genes varies between haplotypes, *Immunogenetics,* 27, 110, 1988.
23. **Sigurdardóttir, S., Lundén, A., and Andersson, L.,** Restriction fragment length polymorphism of DQ and DR class II genes of the bovine MHC, *Anim. Genet.,* 19, 133, 1988.
24. **Muggli-Cockett, N. E. and Stone, R. T.,** Identification of genetic variation in the bovine major histocompatibility complex DRβ-like genes using sequenced bovine genomic probes, *Anim. Genet.,* 19, 213, 1988.
25. **Muggli-Cockett, N. E. and Stone, R. T.,** Partial nucleotide sequence of a bovine major histocompatibility class II DRB-like gene, *Anim. Genet.,* 20, 361, 1989.
26. **Groenen, M. A. M., van der Poel, J. J., Dijkhof, R. J. M., and Giphart, M. J.,** Cloning of the bovine major histocompatibility complex class II genes, *Anim. Genet.,* 20, 267, 1989.
27. **Groenen, M. A. M., van der Poel, J. J., Dijkhof, R. J. M., and Giphart, M. J.,** The nucleotide sequence of bovine MHC class II DQB and DRB genes, *Immunogenetics,* 31, 37, 1990.
28. **van der Poel, J. J., Groenen, M. A. M., Dijkhof, R. J. M., Ruyter, D., and Giphart, M. J.,** The nucleotide sequence of the bovine MHC class II alpha genes: DRA, DQA, and DYA, *Immunogenetics,* 31, 29, 1990.
29. **Stone, R. T. and Muggli-Cockett, N. E.,** Partial nucleotide sequence of a novel bovine major histocompatibility complex class II β-chain gene, BoLA-DIB, *Anim. Genet.,* 21, 353, 1990.
30. **Burke, M. G., Stone, R. T., and Muggli-Cockett, N. E.,** Nucleotide sequence and Northern analysis of a bovine major histocompatibility class II DRβ-like cDNA, *Anim. Genet.,* 22, 343, 1991.
31. **Sigurdardóttir, S., Borsch, C., Gustafsson, K., and Andersson, L.,** Cloning and sequence analysis of 14 DRB alleles of the bovine MHC by using the polymerase chain reaction, *Anim. Genet.,* 22, 199, 1991.
32. **Sigurdardóttir, S., Borsch, C., Gustafsson, K., and Andersson, L.,** Gene duplications and sequence polymorphism of bovine class II DQB genes, *Immunogenetics,* 35, 205, 1992.
33. **Xu, A., Clarke, T. J., Teutsch, M. R., Schook, L. B., and Lewin, H. A.,** Sequencing and genetic analysis of a bovine MHC DQB cDNA clone, *Anim. Genet.,* 22, 381, 1991.
34. **Muggli-Cockett, N. E. and Stone, R. T.,** Restriction fragment length polymorphisms in bovine major histocompatibility complex class II β-chain genes using bovine exon-containing hybridization probes, *Anim. Genet.,* 22, 123, 1991.

35. **Stone, R. T. and Muggli-Cockett, N. E.,** BoLA-DIB: species distribution, linkage with DOB, and Northern analysis, *Anim. Genet.,* 24, 41, 1993.
36. **Trowsdale, J., Ragoussis, J., and Campbell, R. D.,** Map of the human MHC, *Immunol. Today,* 12, 443, 1991.
37. **Davies, C. J., van der Poel, J. J., Groenen, M. A. M., and Giphart, M. J.,** Identification of really interesting new genes (RING) in the bovine lymphocyte antigen complex, *Anim. Genet.,* 23 (Suppl. 1), 35 (abstract), 1992.
38. **Powis, S. J., Deverson, E. V., Coadwell, W. J., Ciruela, A., Huskisson, N. S., Smith, H., Butcher, G. W., and Howard, J. C.,** Effect of polymorphism of an MHC-linked transporter on the peptides assembled in a class I molecule, *Nature,* 357, 211, 1992.
39. **Teutsch, M. R., Beever, J. E., Stewart, J. A., Schook, L. B., and Lewin, H. A.,** Linkage of complement factor B gene to the bovine major histocompatibility complex, *Anim. Genet.,* 20, 427, 1989.
40. **Skow, L. C., Womack, J. E., Petrash, J. M., and Millers, W. L.,** Synteny mapping of the genes for 21 steroid hydroxylase, alpha A crystallin and class I bovine leucocyte antigen in cattle, *DNA,* 7, 143, 1988.
41. **Caldwell, J., Bryan, C. F., Cumberland, P. A., and Weseli, D. F.,** Serologically detected lymphocyte antigens in Holstein cattle, *Anim. Blood Grps. Biochem. Genet.,* 8, 197, 1977.
42. **Amorena, B. and Stone, W. H.,** Serologically defined (SD) locus in cattle, *Science,* 201, 159, 1978.
43. **Spooner, R. L., Leveziel, H., Grosclaude, F., Oliver, R. A., and Vaiman, M.,** Evidence for a possible major histocompatibility complex (BLA) in cattle, *J. Immunogenet.,* 5, 335, 1978.
44. **Spooner, R. L., Oliver, R. A., Sales, D. I., McCoubrey, C. M., Millar, P., Morgan, A. G., Amorena, B., Bailey, E., Bernoco, D., Brandon, M., Bull, R. W., Caldwell, J., Cwik, S., van Dam, R. H., Dodd, J., Gahne, B., Grosclaude, F., Hall, J. G., Hines, H., Leveziel, H., Newman, M. J., Stear, M. J., Stone, W. H., and Vaiman, M.,** Analysis of alloantisera against bovine lymphocytes. Joint report of the 1st International Bovine Lymphocyte Antigen (BoLA) Workshop, *Anim. Blood Grps. Biochem. Genet.,* 10, 63, 1979.
45. **Amorena, B. and Stone, W. H.,** Bovine lymphocyte antigens (BoLA): a serologic, genetic and histocompatibility analysis, *Tissue Antigens,* 16, 212, 1980.
46. **Stear, M. J., Newman, M. J., and Nicholas, F. W.,** Two closely linked loci and one apparently independent locus code for bovine lymphocyte antigens, *Tissue Antigens,* 20, 289, 1982.
47. **Anon.,** Proceedings of the Second International Bovine Lymphocyte Antigen (BoLA) Workshop, *Anim. Blood Grps Biochem. Genet.,* 13, 33, 1982.
48. **Bull, R. W., Lewin, H. A., Wu, M. C., Peterbaugh, K., Antczak, D., Bernoco, D., Cwik, S., Dam, L., Davies, C., Dawkins, R. L., Dufty, J. H., Gerlach, J., Hines, H. C., Lazary, S., Leibold, W., Leveziel, H., Lie, Ø., Lindberg, P. G., Meggiolaro, D., Meyer, E., Oliver, R., Ross, M., Simon, M., Spooner, R. L., Stear, M. J., Teale, A. J., and Templeton, J. W.,** Joint report of the Third International Bovine Lymphocyte Antigen (BoLA) Workshop, Helsinki, Finland, 27 July 1986, *Anim. Genet.,* 20, 109, 1989.
49. **Bernoco, D., Lewin, H. A., Andersson, L., Arriens, M. A., Byrns, G., Cwik, S., Davies, C. J., Hines, H. C., Leibold, W., Lie, Ø., Meggiolar D., Oliver, R. A., Østergård, H., Spooner, R. L., Stewart-Haynes, J. A., Teale, A. J., Templeton, J. W., and Zanotti, M.,** Joint report of the Fourth International Bovine Lymphocyte Antigen (BoLA) Workshop, East Lansing, Michigan, USA, 25 August 1990, *Anim. Genet.,* 22, 477, 1991.
50. **Davies, C. J., Joosten, I., Bernoco, D., Arriens, M. A., Bester, J., Ceriotti, G., Ellis, S., Hensen, E. J., Hines, H. C., Horin, P., Kristensen, B., Lewin, H. A., Meggiolaro, D., Morgan, A. L. G., Morita, M., Nilsson, Ph. R., Oliver, R. A., Orlova, A., Østergård, H., Park, C., Schuberth, H.-J., Simon, M., Spooner, R. L., and Stewart, J. A.,** Polymorphism of bovine MHC class I genes. Joint report of the Fifth International Bovine Lymphocyte Antigen (BoLA) Workshop, Interlaken, Switzerland, 1 August 1992, *Anim. Genet.,* in press, 1993.
51. **Davies, C. J., Joosten, I., Andersson, L., Arriens, M. A., Bernoco, D., Byrns, G., Bissumbhar, B., van Eijk, M. J. T., Ellegren, H., Kristensen, B., Lewin, H. A., Morgan, A. L. G., Muggli-Cockett, N. E., Nilsson, Ph. R., Oliver, R. A., Park, C. A., van der Poel, J. J., Polli, M., Spooner, R. L., and Stewart, J. A.,** Polymorphism of bovine MHC class II genes. Joint report of the Fifth International Bovine Lymphocyte Antigen (BoLA) Workshop, Interlaken, Switzerland, 1 August 1992, *Anim. Genet.,* in press, 1993.

52. **Oliver, R. A., McCoubrey, C. M., Millar, P., Morgan, A. L. G., and Spooner, R. L.,** A genetic study of bovine lymphocyte antigens (BoLA) and their frequency in several breeds, *Immunogenetics,* 13, 127, 1981.
53. **Stear, M. J., Pokorny, T. S., Muggli, N. E., and Stone, R. T.,** Breed differences in the distribution of BoLA-A locus antigens in American cattle, *Anim. Genet.,* 19, 171, 1988.
54. **Kemp, S. J., Spooner, R. L., and Teale, A. J.,** A comparative study of major histocompatibility complex antigens in East African and European cattle breeds, *Anim. Genet.,* 19, 17, 1988.
55. **Davies, C. J. and Antczak, D. F.,** Mixed lymphocyte culture studies reveal complexity in the bovine major histocompatibility complex not detected by class I serology, *Anim. Genet.,* 22, 31, 1991.
56. **Teale, A. J., Morrison, W. I., Goddeeris, B. M., Groocock, C. M., Stagg, D. A., and Spooner, R. L.,** Bovine alloreactive cytotoxic cells generated *in vitro:* target specificity in relation to BoLA phenotype, *Immunology,* 55, 355, 1985.
57. **Spooner, R. L., Innes, E. A., Millar, P., Simpson, S. P., Webster, J., and Teale, A. J.,** Bovine alloreactive cytotoxic cells generated in vitro detect BoLA w6 subgroups, *Immunology,* 61, 85, 1987.
58. **Joosten, I., Oliver, R. A., Spooner, R. L., Williams, J. L., Hepkema, B. G., Sanders, M. F., and Hensen, E. J.,** Characterization of class I bovine lymphocyte antigens (BoLA) by one-dimensional isoelectric focusing, *Anim. Genet.,* 19, 103, 1988.
59. **Watkins, D. I., Shadduck, J. A., Stone, M. E., Lewin, H. A., and Letvin, N. L.,** Isoelectric focusing of bovine major histocompatibility complex class I molecules, *J. Immunogenet.,* 16, 233, 1989.
60. **Oliver, R. A., Brown, P., Spooner, R. L., Joosten, I., and Williams, J. L.,** The analyses of antigen and DNA polymorphism within the bovine major histocompatibility complex. 1. The class I antigens, *Anim. Genet.,* 20, 31, 1989.
61. **Viuff, B., Østergård, H., Aasted, B., and Kristensen, B.,** One-dimensional isoelectric focusing and immunoblotting of bovine major histocompatibility complex (BoLA) class I molecules and correlation with class I serology, *Anim. Genet.,* 22, 147, 1991.
62. **Usinger, W. R., Curie-Cohen, M., and Stone, W. H.,** Lymphocyte-defined loci in cattle, *Science,* 196, 1017, 1977.
63. **Bach, F. H., Widmer, M. B., Bach, M. L., and Klein, J.,** Serologically defined and lymphocyte-defined components of the major histocompatibility complex in the mouse, *J. Exp. Med.,* 136, 1430, 1972.
64. **Usinger, W. R., Curie-Cohen, M., Benforado, K., Pringnitz, D., Rowe, R., Splitter, G. A., and Stone, W. H.,** The bovine major histocompatibility complex (BoLA): close linkage of the genes controlling serologically defined antigens and mixed lymphocyte reactivity, *Immunogenetics,* 14, 423, 1981.
65. **Newman, M. J., Campion, J. E., and Stear, M. J.,** Mixed lymphocyte reactivity in cattle, *Tissue Antigens,* 20, 100, 1982.
66. **Teale, A. J. and Kemp, S. J.,** A study of BoLA class II antigens with BoT4+ T lymphocyte clones, *Anim. Genet.,* 18, 17, 1987.
67. **Glass, E. J., Oliver, R. A., Williams, J. L. W., and Millar, P.,** Alloreactive T-cell recognition of bovine major histocompatibility complex class II products defined by one-dimensional isoelectric focusing, *Anim. Genet.,* 23, 97, 1992.
68. **Teale, A. J. and Baldwin, C. L.,** Functional aspects of BoT4+ alloreactive T lymphocytes transformed by Theileria parva, *Anim. Genet.,* 19 (Suppl. 1), 31, 1988.
69. **van Rood, J. J., van Leeuwen, A., and Ploem, J. S.,** Simultaneous detection of two cell populations by two-colour fluorescence and application to the recognition of B-cell determinants, *Nature,* 262, 795, 1976.
70. **Newman, M. J., Adams, T. E., and Brandon, M. R.,** Serological and genetic identification of a bovine B lymphocyte alloantigen system, *Anim. Blood Grps. Biochem. Genet.,* 13, 123, 1982.
71. **Newman, M. J. and Stear, M. J.,** The antibody response to bovine lymphocyte alloantigens, *Vet. Immunol. Immunopathol.,* 4, 615, 1983.
72. **Mackie, J. T. and Stear, M. J.,** The definition of five B lymphocyte alloantigens closely linked to BoLA class I antigens, *Anim. Genet.,* 21, 69, 1990.
73. **Davies, C. J. and Antczak, D. F.,** Production and characterization of alloantisera specific for bovine class II major histocompatibility complex antigens, *Anim. Genet.,* 22, 417, 1991.
74. **Arriens, M. A., Hesford, F., Ruff, G., and Lazary, S.,** Production of alloantibodies against bovine B-lymphocyte antigens, *Anim. Genet.,* 22, 399, 1991.

75. **Williams, J. L., Oliver, R. A., Morgan, A. L. G., Glass, E. J., and Spooner, R. L.,** Production of alloantisera against class II bovine lymphocyte antigens (BoLA) by cross-immunization between class I matched cattle, *Anim. Genet.,* 22, 407, 1991.
76. **Joosten, I., Sanders, M. F., van der Poel, A., Williams, J. L., Hepkema, B. G., and Hensen, E. J.,** Biochemically defined polymorphism of bovine MHC class II antigens, *Immunogenetics,* 29, 213, 1989.
77. **Watkins, D. I., Shadduck, J. A., Rudd, C. E., Stone, M. E., Lewin, H. A., and Letvin, N. L.,** Isoelectric focusing of bovine major histocompatibility complex class II molecules, *Eur. J. Immunol.,* 19, 567, 1989.
78. **Bissumbhar, B., Joosten, I., Davis, W. C., and Hensen, E. J.,** Isoelectric focusing of bovine MHC DQ allelic variants, *Anim. Genet.,* 23 (Suppl. 1), 39, 1992.
79. **Andersson, L., Sigurdardóttir, S., Borsch, C., and Gustafsson, K.,** Evolution of MHC polymorphism: extensive sharing of polymorphic sequence motifs between human and bovine DRB alleles, *Immunogenetics,* 33, 188, 1991.
80. **van Eijk, M. J. T., Stewart-Haynes, J. A., and Lewin, H. A.,** Extensive polymorphism of the BoLA-DRB3 gene distinguished by PCR-RFLP, *Anim. Genet.,* 23, 483, 1992.
81. **Ellegren, H., Davies, C. J., and Andersson, L.,** Strong association between polymorphisms in an intronic microsatellite and in the coding sequence of the BoLA-DRB3 gene; implications for microsatellite stability and PCR-based DRB3 typing, *Anim. Genet.,* 24, 269, 1993.
82. **Joosten, I., Hensen, E. J., Sanders, M. F., and Andersson, L.,** Bovine MHC class II restriction fragment length polymorphism linked to expressed polymorphism, *Immunogenetics,* 31, 123, 1990.
83. **Davies, C. J., Andersson, L., Joosten, I., Mariani, P., Gasbarre, L. C., and Hensen, E. J.,** Characterization of bovine MHC class II polymorphism using three typing methods: serology, RFLP and IEF, *Eur. J. Immunogenet.,* 19, 253, 1992.
84. **Cameron, P. U., Tabarias, H. A., Pulendran, B., Robinson, W., and Dawkins, R. L.,** Conservation of the central MHC genome: PFGE mapping and RFLP analysis of complement, HSP70, and TNF genes in the goat, *Immunogenetics,* 31, 253, 1990.
85. **Grossberger, D., Hein, W., and Marcuz, A.,** Class I major histocompatibility complex cDNA clones from sheep thymus: alternate splicing could make a long cytoplasmic tail, *Immunogenetics,* 32, 77, 1990.
86. **Puri, N. K., Gogolin-Ewens, K. J., and Brandon, M. R.,** Monoclonal antibodies to sheep MHC class I and class II molecules: biochemical characterization of three class I gene products and for distinct subpopulations of class II molecules, *Vet. Immunol. Immunopathol.,* 15, 59, 1987.
87. **Joosten, I., Ruff, G., Sanders, M. F., and Hensen, E. J.,** Use of isoelectric focusing to define MHC class I polymorphism in goats, *Anim. Genet.,* 24, 47, 1993.
88. **Scott, P. C., Choi, C.-L., and Brandon, M. R.,** Genetic organization of the ovine MHC class II region, *Immunogenetics,* 25, 116, 1987.
89. **Scott, P. C., Gogolin-Ewens, K. J., Adams, T. E., and Brandon, M. R.,** Nucleotide sequence, polymorphism, and evolution of ovine MHC class II DQA genes, *Immunogenetics,* 34, 69, 1991.
90. **Scott, P. C., Maddox, J. F., Gogolin-Ewens, K. J., and Brandon, M. R.,** The nucleotide sequence and evolution of ovine MHC class II B genes: DQB and DRB, *Immunogenetics,* 34, 80, 1991.
91. **Deverson, E. V., Wright, H., Watson, S., Ballingall, K., Huskisson, N., Diamond, A. G., and Howard, J. C.,** Class II major histocompatibility complex genes of the sheep, *Anim. Genet.,* 22, 211, 1991.
92. **Mann, A. J., Abraham, L. J., Cameron, P. U., Robinson, W., Giphart, M. J., and Dawkins, R. L.,** The caprine MHC contains DYA genes, *Immunogenetics,* 37, 292, 1992.
93. **Ballingall, K. T., Wright, H., Redmond, J., Dutia, B. M., Hopkins, J., Lang, J., Deverson, E. V., Howard, J. C., Puri, N., and Haig, D.,** Expression and characterization of ovine major histocompatibility complex class II (OLA-DR) genes, *Anim. Genet.,* 23, 347, 1992.
94. **Puri, N. K. and Brandon, M. R.,** Sheep MHC class II molecules: II. Identification and characterization of four distinct subsets of sheep MHC class II molecules, *Immunology,* 62, 575, 1987.
95. **Ford, C. H. J.,** Genetic studies of sheep leukocyte antigens, *J. Immunogenet.,* 2, 31, 1975.
96. **Millot, P.,** Genetic control of lymphocyte antigens in sheep: the OLA complex and two minor loci, *Immunogenetics,* 9, 509, 1979.
97. **Stear, M. J. and Spooner, R. L.,** Lymphocyte antigens in sheep, *Anim. Blood Grps. Biochem. Genet.,* 12, 265, 1981.

98. **Cullen, P. R., Bunch, C., Brownlie, J., and Morris, P. J.,** Sheep lymphocyte antigens: a preliminary study, *Anim. Blood Grps. Biochem. Genet.,* 13, 149, 1982.
99. **Chardon, P., Kirszenbaum, M., Cullen, P. R., Geffrotin, C., Auffray, C., Strominger, J. L., Cohen, D., and Vaiman, M.,** Analysis of the sheep MHC using HLA class I, II, and C4 cDNA probes, *Immunogenetics,* 22, 349, 1985.
100. **van Dam, R. H., Borst-van Werkhoven, C., van der Donk, J. A., and Goudswaard, J.,** Histocompatibility in ruminants. The production and evaluation of alloantibodies for GLA typing in goats, *J. Immunogenet.,* 3, 237, 1976.
101. **Nesse, L. L. and Larsen, H. J.,** Lymphocyte antigens in Norwegian goats: serological and genetic studies, *Anim. Genet.,* 18, 261, 1987.
102. **Nesse, L. L. and Ruff, G.,** A comparison of lymphocyte antigens specificities in Norwegian and Swiss goats, *Anim. Genet.,* 20, 71, 1989.
103. **van Dam, R. H., van Kooten, P. J. S., van der Donk, J. A., and Goudswaard, J.,** Phenotyping by the mixed lymphocyte reaction in goats (LD typing), *Vet. Immunol. Immunopathol.,* 2, 321, 1981.
104. **van Dam, R. H., D'Amaro, J., van Kooten, P. J. S., van der Donk, J. A., and Goudswaard, J.,** The histocompatibility complex GLA in the goat, *Anim. Blood Grps. Biochem. Genet.,* 10, 121, 1979.
105. **Ruff, G.,** Investigations on the Caprine Leucocyte Antigen (CLA) System, Thesis no. 8468, 1987, Swiss Federal Institute of Technology Zurich.
106. **Eugui, M. and Emery, D. L.,** Genetically restricted cell-mediated cytotoxicity in cattle immune to Theileria parva, *Nature,* 290, 251, 1981.
107. **Goddeeris, B. M., Morrison, W. I., Teale, A. J., Bensaid, A., and Baldwin, C. L.,** Bovine cytotoxic T-cell clones specific for cells infected with the protozoan parasite *Theileria parva:* parasite strain specificity and class I major histocompatibility complex restriction, *Proc. Natl. Acad. Sci., U.S.A.,* 83, 5238, 1986.
108. **Goddeeris, B. M., Morrison, W. I., and Teale, A. J.,** Generation of bovine cytotoxic cell lines, specific for cells infected with the protozoan parasite Theileria parva and restricted by products of the major histocompatibility complex, *Eur. J. Immunol.,* 16, 1243, 1992.
109. **Preston, P. M., Brown, C. G. D., and Spooner, R. L.,** Cell-mediated cytotoxicity in Theileria annulata infection of cattle with evidence for BoLA restriction, *Clin. Exp. Immunol.,* 53, 88, 1983.
110. **Splitter, G. A., Eskra, L., and Abruzzini, A. F.,** Cloned bovine cytotoxic T cells recognize bovine herpes virus-1 in a genetically restricted, antigen-specific manner, *Immunology,* 63, 145, 1988.
111. **Rothel, J. S., Dufty, J. H., and Wood, P. R.,** Studies on the bovine major histocompatibility class I and class II antigens using homozygous typing cells and antigen-specific BoT4+ blast cells, *Anim. Genet.,* 21, 141, 1990.
112. **Glass, E. J., Oliver, R. A., and Spooner, R. L.,** Bovine T cells recognize antigen in association with MHC class II haplotypes defined by one-dimensional isoelectric focusing, *Immunology,* 72, 380, 1991.
113. **Glass, E. J., Oliver, R. A., Collen, T., Doel, T. R., Dimarchi, R., and Spooner, R. L.,** MHC class II restricted recognition of FMDV peptides by bovine T cells, *Immunology,* 74, 594, 1991.
114. **Canals, A., Gasbarre, L. C., Davies, C. J., and Zarlenga, D.,** Oesophagostomum radiatum-specific bovine T cell clones: specificity, MHC restriction and cytokine transcription, *Vet. Immunol. Immunopathol.,* in press, 1993.
115. **Lie, Ø., Solbu, H., Larsen, H. J., and Spooner, R. L.,** Possible association of antibody responses to human serum albumin and (T,G)-A—L with the bovine major histocompatibility complex (BoLA), *Vet. Immunol. Immunopathol.,* 11, 333, 1986.
116. **Glass, E. J., Oliver, R. A., and Spooner, R. L.,** Variation in T cell responses to ovalbumin in cattle: evidence for Ir gene control, *Anim. Genet.,* 21, 15, 1990.
117. **Tiwari, J. L. and Terasaki, P. I.,** HLA and Disease Associations, Springer Verlag, New York, 1985.
118. **Meredith, D., Elser, A. H., Wolf, B., Soma, L. R., Donawick, W. J., and Lazary, S.,** Equine leukocyte antigens: relationships with sarcoid tumors and laminitis in two pure breeds, *Immunogenetics,* 23, 221, 1986.
119. **Lewin, H. A. and Bernoco, D.,** Evidence for BoLA-linked resistance and susceptibility to subclinical progression of bovine leukaemia virus infection, *Anim. Genet.,* 17, 197, 1986.
120. **Lewin, H. A., Wu, M., Stewart, J. A., and Nolan, T. J.,** Association between BoLA and subclinical bovine leukemia virus infection in a herd of Holstein-Friesian cows, *Immunogenetics,* 27, 338, 1988.
121. **Dam, L. and Østergård, H.,** Investigation of associations between BoLA and bovine virus diarrhoea, *Anim. Blood Grps. Biochem. Genet.,* 16 (Suppl. 1), 88 (Abstr.), 1985.

122. **Oddgeirsson, O., Simpson, S. P., Morgan, A. L. G., Ross, D. S., and Spooner, R. L.,** Relationship between the bovine major histocompatibility complex (BoLA), erythrocyte markers and susceptibility to mastitis in Icelandic cattle, *Anim. Genet.*, 19, 11, 1988.
123. **Lundén, A., Sigurdardóttir, S., Edfors-Lilja, I., Danell, B., Rendel, J., and Andersson, L.,** The relationship between bovine major histocompatibility complex class II polymorphism and disease studied by use of bull breeding values, *Anim. Genet.*, 21, 221, 1990.
124. **Weigel, K. A., Freeman, A. E., Kehrli, M. E., Jr., Stear, M. J., and Kelley, D. H.,** Association of class I bovine lymphocyte antigen complex alleles with health and production traits in dairy cattle, *J. Dairy Sci.*, 73, 2538, 1990.
125. **Stear, M. J., Tierney, T. J., Baldock, F. C., Brown, S. C., Nicholas, F. W., and Rudder, T. H.,** Class I antigens of the bovine major histocompatibility system are weakly associated with variation in faecal worm egg counts in naturally infected cattle, *Anim. Genet.*, 19, 115, 1988.
126. **Stear, M. J., Hetzel, D. J. S., Brown, S. C., Gershwin, L. J., Mackinnon, M. J., and Nicholas, F. W.,** The relationships among ecto- and endoparasite levels, class I antigens of the bovine major histocompatibility system, immunoglobulin E levels and weight gain, *Vet. Parasitol.*, 34, 303, 1990.
127. **Stear, M. J., Spradbrow, P. B., Tierney, T. J., Nicholas, F. W., and Bellows, R. A.,** Failure to find an association between ocular squamous cell carcinoma and class I antigens of the bovine major histocompatibility system, *Anim. Genet.*, 20, 233, 1989.
128. **Østergård, H., Kristensen, B., and Andersen, S.,** Investigations in farm animals of associations between the MHC system and disease resistance and fertility, *Livestock Production Sci.*, 22, 49, 1989.
129. **Gupta, P. and Ferrer, J. F.,** Expression of Bovine Leukemia Virus genome is blocked by a nonimmunoglobulin protein in plasma from infected cattle, *Science*, 215, 405, 1982.
130. **Lewin, H. A., Nolan, T. J., and Schook, L. B.,** Altered expression of class II antigens on peripheral blood B lymphocytes from BLV-infected cows with persistent lymphocytosis, in Antigen Presenting *Cells: Diversity, Differentiation and Regulation*, Schook, L. B. and Tew, J. G., Eds., Alan R. Liss, New York, 1987, 211.
131. **Lewin, H. A.,** Disease resistance and immune response genes in cattle: strategies for their detection and evidence of their existence, *J. Dairy Sci.*, 72, 1334, 1989.
132. **Millot, P., Chatelain, J., and Cathala, F.,** Sheep major histocompatibility complex OLA: gene frequencies in two French breeds with scrapie, *Immunogenetics*, 21, 117, 1985.
133. **Millot, P., Chatelain, J., Dautheville, C., Salmon, D., and Cathala, F.,** Sheep major histocompatibility (OLA) complex: linkage between a scrapie susceptibility/resistance locus and the OLA complex in Ile-de-France sheep progenies, *Immunogenetics*, 27, 1, 1988.
134. **Ruff, G. and Lazary, S.,** Evidence for linkage between the caprine leukocyte antigen (CLA) system and susceptibility to CAE virus-induced arthritis in goats, *Immunogenetics*, 28, 303, 1988.
135. **Beever, J. E., George, P. D., Fernando, R. L., Stormont, C. J., and Lewin, H. A.,** Associations between genetic markers and growth and carcass traits in a paternal half-sib family of Angus cattle, *J. Anim. Sci.*, 68, 337, 1990.
136. **Stear, M. J., Pokorny, T. S., Muggli, N. E., and Stone, R. T.,** The relationships of birth weight, preweaning gain and postweaning gain with the bovine mojor histocompatibility system, *J. Anim. Sci.*, 67, 641, 1992.
137. **Hines, H. C., Allaire, F. R., and Michalak, M. M.,** Association of bovine lymphocyte antigens with milk fat percentage differences, *J. Dairy Sci.*, 69, 3148, 1986.
138. **Batra, T. R., Lee, A. J., Gavora, J. S., and Stear, M. J.,** Class I alleles of the bovine major histocompatibility system and their association with economic traits, *J. Dairy Sci.*, 72, 2115, 1989.
139. **Lundén, A., Andersson-Eklund, L., and Andersson, L.,** Lack of association between bovine major histocompatibility complex class II polymorphism and production traits, *J. Dairy Sci.*, 76, 843, 1993.
140. **Gogolin-Ewens, K. J., Lee, C. S., Mercer, W. R., and Brandon, M. R.,** Site-directed differences in the immune response to the fetus, *Immunology*, 66, 312, 1989.
141. **Low, B. G., Hansen, P. J., Drost, M., and Gogolin-Ewens, K. J.,** Expression of major histocompatibility complex antigens on the bovine placenta, *J. Reprod. Fert.*, 90, 235, 1990.
142. **Newman, M. J. and Hines, H. C.,** Production of foetally stimulated lymphocytotoxic antibodies by primiparous cows, *Anim. Blood Grps. Biochem. Genet.*, 10, 87, 1979.
143. **Hines, H. C. and Newman, M. J.,** Production of foetally stimulated lymphocytotoxic antibodies by multiparous cows, *Anim. Blood Grps. Biochem. Genet.*, 12, 201, 1981.

144. **Joosten, I., Sanders, M. F., and Hensen, E. J.,** Involvement of major histocompatibility complex class I compatibility between dam and calf in the aetiology of bovine retained placenta, *Anim. Genet.,* 22, 455, 1991.
145. **Joosten, I. and Hensen, E. J.,** Retained placenta: an immunological approach, *Anim. Reprod. Sci.,* 28, 451, 1992.
146. **DeMars, R. and Spies, T.,** New genes in the MHC that encode proteins for antigen processing, *Trends Cell Biol.,* 2, 81, 1992.
147. **Powis, S. J., Townsend, A. R. M., Deverson, E. V., Bastin, J., Butcher, G. W., and Howard, J. C.,** Restoration of antigen presentation to the mutant cell line RMA-S by an MHC-linked transporter, *Nature,* 354, 528, 1991.
148. **Spies, T. and DeMars, R.,** Restored expression of major histocompatibility class I molecules by gene transfer of a putative peptide transporter, *Nature,* 351, 323, 1991.
149. **Goldberg, A. L., and Rock, K. L.,** Proteolysis, proteasomes and antigen presentation, *Nature,* 357, 375, 1992.
150. **Arnold, D., Driscoll, J., Androlewicz, M., Hughes, E., Cresswell, P., and Spies, T.,** Proteasome subunits encoded in the MHC are not generally required for the processing of peptides bound by MHC class I molecules, *Nature,* 360, 171, 1992.
151. **Momburg, F., Otiz-Navarrete, V., Neefjes, J., Goulmy, E., van der Wal, Y., Spits, H., Powis, S. J., Butcher, G. W., Howard, J. C., Walden, P., and Hammerling, G. J.,** Proteasome subunits encoded by the major histocompatibility complex are not essential for antigen presentation, *Nature,* 360, 174, 1992.
152. **Neefjes, J. J. and Ploegh, H. L.,** Intracellular transport of MHC class II molecules, *Immunol. Today,* 13, 179, 1992.

Chapter 4

The T Cell Receptor

N. Ishiguro and W. R. Hein

CONTENTS

I. Introduction ... 59
II. Overview of the T Cell Receptor Complex ... 60
III. Invariant Components of Ruminant T Cell Receptors ... 61
A. CD3 γ, δ, and ε Chains ... 61
B. TCR ζ and η Chains ... 62
IV. Ruminant T Cell Receptor Heterodimers ... 63
A. TCR α Chains ... 63
B. TCR β Chains ... 64
C. TCR γ Chains ... 65
D. TCR δ Chains ... 67
V. Ligand Recognition and Signaling Pathways ... 68
VI. Conclusions and Future Research ... 70
A. Genomic Structure of the TCR Loci ... 70
B. V Region Repertoire Studies ... 70
C. γδ T Cells ... 71
D. Pathogenesis of Lymphoid Tumors ... 71
Acknowledgments ... 71
References ... 72

I. INTRODUCTION

The most important functional attribute of the immune system of vertebrates is the ability to discriminate with high precision between a virtually limitless range of self and foreign antigens and then to set in motion an appropriate effector response. Two different but related classes of molecules mediate antigen recognition by cells of the immune system. Immunoglobulins (Igs), which may either be expressed on the surface of B lymphocytes or secreted into biological fluids as a soluble molecule, are able to bind directly to native antigens and are responsible for mediating the humoral component of immunity, both at the recognition stage while still anchored in the B cell membrane or at the effector phase by inactivating or neutralizing antigens. An analogous structure on the surface of T lymphocytes, the T cell receptor (TCR), has many structural features in common with Igs but there are two principal differences between these two classes of receptor. First, the TCR is a dual recognition unit, whereby a single receptor complex binds simultaneously to a processed antigen fragment and a polymorphic gene product of the major histocompatibility complex (MHC). Secondly, the TCR is not secreted but remains anchored in the cell membrane so that antigen recognition by T cells is dependent on intimate interactions between them and antigen presenting cells. These features impose certain unique constraints on T cell-mediated immunity.

The TCR has been studied extensively in mice and humans since first being identified at a molecular level in 1984.[1-3] Over the last few years, significant progress has also been made in achieving an initial characterization of this structure in ruminants, using a combination of monoclonal antibody and molecular cloning techniques. The purpose of the present chapter is to review current knowledge concerning the structure, expression, and diversification of the TCR in cattle and sheep. The bulk of our present understanding of these features in ruminants is derived from sequence analysis of cDNA clones. In the present review, however, the inclusion of sequences has been kept to a minimum and those readers wanting to examine such data in detail should consult the relevant primary references. As a further aid, the EMBL database accession numbers for the components of ruminant TCRs that have been cloned to date are listed in an appendix at the end of this chapter.

0-8493-4952-4/94/$0.00+$.50

II. OVERVIEW OF THE T CELL RECEPTOR COMPLEX

The TCR is a molecular complex composed of a number of different glycoprotein subunits that is assembled in the cytoplasm of T cells during their ontogeny and when fully formed is expressed in the cell membrane.[4-6] Each TCR complex contains up to five noncovalently associated invariant components termed the CD3 γ, δ, and ε and TCR ζ and η chains. In addition, the TCR contains two disulphide-linked clonotypic heterodimer chains that are normally expressed in a mutually exclusive way; thus, an α chain may be paired with a β chain (the αβ TCR) or a γ chain is paired with a δ chain (the γδ TCR) (Figure 4-1).

The invariant components of the TCR perform at least two basic functions. First, in a structural sense they provide a molecular framework for the stable assembly and anchoring of the overall complex in the cell membrane. The invariant components are the first subunits of the TCR to be assembled in the endoplasmic reticulum during ontogeny, although the exact order of their assembly and the precise stoichiometry of the mature complex have not been fully resolved.[5] Several different forms have been detected experimentally in the cytoplasm of maturing T cells and these most likely represent intermediate stages in assembly.[5,6] The individual invariant subunits are encoded by single-copy genes and have characteristic extracellular transmembrane and cytoplasmic domains.

The second important function of the invariant components is their role in the transduction of signals across the cell membrane after cognitive interaction between the clonotypic heterodimer chains and their specific ligand. The cytoplasmic domains of the invariant molecules are relatively long and couple

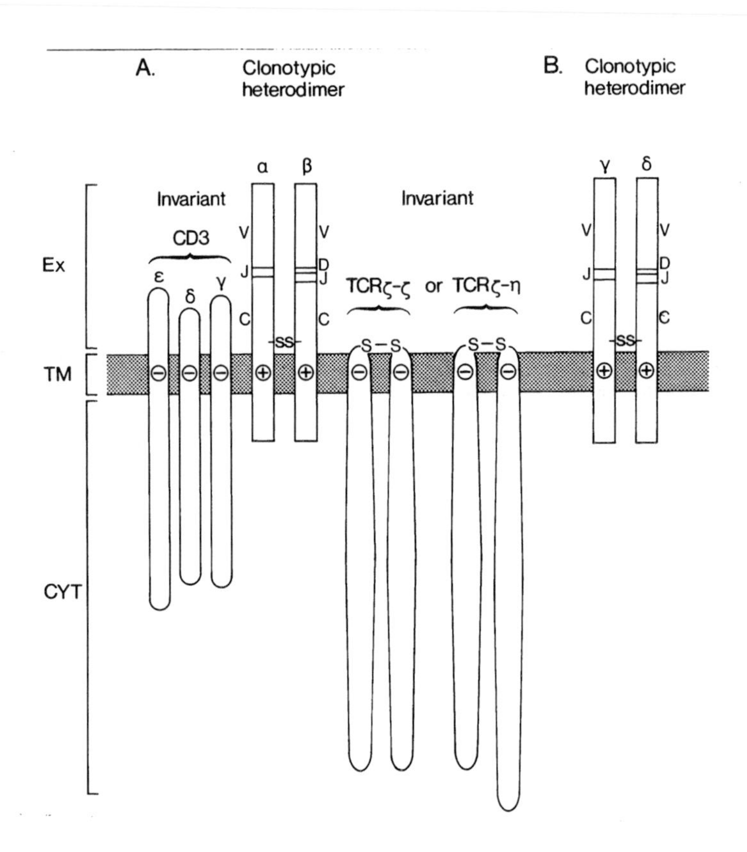

Figure 4-1 Schematic diagram showing the glycoprotein subunits that are assembled into the cell surface CD3/TCR complex. (A) The αβ TCR is composed of up to seven different subunits, with the TCR ζ and η chains being expressed as either homo- or heterodimers. The α and β clonotypic chains are assembled from variable (V), diversity (D), joining (J), and constant (C) region segments as shown. (B) An alternative pair of clonotypic heterodimers is used in the γδ TCR. The transmembrane sections of the component glycoproteins contain either a negatively or positively charged amino acid residue as indicated. (Not drawn to scale.)

receptor engagement to intracellular messenger systems. The TCR complex itself embodies at least two independent transduction units, consisting of the CD3 γδε and TCR ζ chains, respectively.[7] It is likely that differences in the biochemical pathways that are triggered after receptor occupancy will be of critical importance in determining the exact way in which a T cell responds to stimulation via the TCR.

The specificity of T cell recognition is mediated by the TCR heterodimers and has a special genetic basis. In a manner that is analogous to Ig molecules, TCR heterodimers are encoded in the germline by noncontiguous genes for variable (V), diversity (D), and joining (J) segments and functional heterodimers are assembled during T cell ontogeny by a process of gene rearrangement.[8] The diversity that is produced by the recombination process can be further accentuated by coding changes at V-D and D-J junctional regions due to exonucleolytic cleavage of nucleotides from the ends of participating gene segments and/or by the addition of non-germline encoded N or P nucleotides.[9,10] The diversity that can potentially be produced by these mechanisms is limitless and virtually ensures that the variable region of each TCR heterodimer (defined as functionally rearranged V-(D)-J segments) will have a unique structure and antigen recognition specificity (i.e., it is clonotypic). After functional rearrangement and transcription, the VDJ segments are spliced to constant region segments to form the mature TCR heterodimer mRNAs, which can then be translated into protein and assembled into the TCR complex.

III. INVARIANT COMPONENTS OF RUMINANT T CELL RECEPTORS

A. CD3 γ, δ, AND ε CHAINS

The three invariant CD3 glycoproteins are encoded by related and closely linked genes that probably arose by gene duplication and a number of structural features are common to all three chains.[4-6] The sequences of the sheep CD3 γ, δ, and ε chains have been determined by cDNA cloning[11,12] and the principal structural features are shown schematically in Figure 4-2. Partial sequences reported for the bovine homologues are very similar to the sheep.[13]

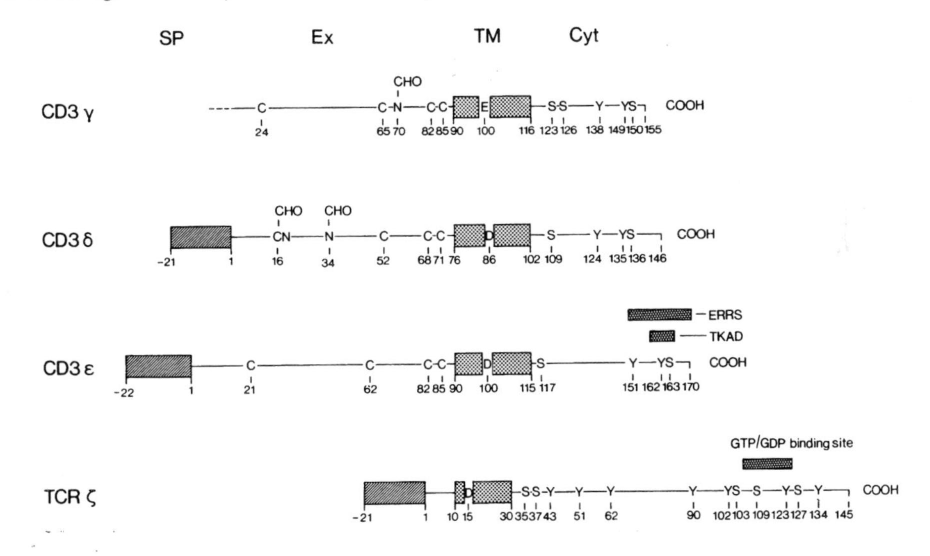

Figure 4-2 The main structural features of sheep CD3 γ,δ,ε, and TCR ζ. The boundaries between the signal peptide (SP), extracellular (Ex), transmembrane (TM), and cytoplasmic (Cyt) domains are indicated. In addition, the positions of a number of amino acid residues considered to contribute to structural or functional properties of the chains are shown. These include extracellular cysteines (C), sites of potential asparagine (N) linked glycosylation, negatively charged glutamic acid (E) or aspartic acid (D) residues in the TM domain, and serine (S) and tyrosine (Y) residues in the cytoplasmic domain which are potential sites of phosphorylation during signal transduction. Above the cytoplasmic regions, the positions in the sheep sequences that correspond to the recently identified endoplasmic reticulum retention signal (ERRS)[14] and tyrosine kinase activation domain (TKAD)[15] of CD3 ε are indicated. Note that all the invariant chains have similar motifs in this region. The proposed location of the GTP/GDP binding site[18] in the TCR ζ chain is also shown. The diagrams are based on sequences and comparative alignments reported in References 11 and 12.

Sheep CD3 γ, δ, and ε chains have a high level of overall similarity to the corresponding human and mouse sequences (70 to 80% nucleotide identity over the full coding region). However, at the protein level, different regions of the chains have been conserved to varying degrees and the sheep CD3 glycoproteins have a number of specific differences. The transmembrane and cytoplasmic domains show the highest level of amino acid identity to human and mouse (70 to 86%), while the extracellular regions have been less well conserved (46 to 59%). In comparison to human and mouse protein sequences, a five amino acid motif has been deleted from the cytoplasmic domain of sheep CD3 γ and four amino acids are missing from the extracellular region of sheep CD3 δ.[11] The extracellular region of CD3 ε shows rather more variability between different mammals, and deletions have occurred at either or both of two short stretches of sequence in the extracellular domain; in this regard, sheep CD3 ε is most similar in overall structure to the mouse glycoprotein.[12]

It is likely that the small structural differences noted above have only minor, if any, significance in terms of the assembly and functioning of the TCR complex since other amino acid motifs that have recently been shown to play important roles in these processes have been retained in the sheep sequences. For example, an endoplasmic reticulum retention signal that is involved in TCR complex assembly and a tyrosine kinase activation domain involved in signaling have been identified in the cytoplasmic region of CD3 ε;[14,15] and both domains are almost exactly conserved in the sheep protein (Figure 4-2). Further examples of the conservation of functionally important domains in other CD3 chains will undoubtedly follow. As in other species, the transmembrane regions of sheep CD3 chains contain a negatively charged amino acid residue (either aspartic acid or glutamic acid) which is considered to be an important site for interaction with the TCR heterodimer chains.[11,12] The small differences in the structure of CD3 chains apparent between species, including ruminants, may partly reflect co-evolutionary adaptations to changes in other parts of the TCR complex, such as the TCR heterodimers (see below), so that a stable overall structure is maintained.

To date, only limited studies of CD3 expression have been completed in sheep and cattle, particularly at the surface protein level, due mainly to a lack of specific monoclonal antibodies. Transcripts of the CD3 chains are detectable in the cytoplasm of sheep and bovine T cells using Northern hybridization, and the mRNAs encoding each chain have sizes similar to that reported in human and mouse T cells.[11,12] Using cross-reactive antisera, two CD3 components with molecular sizes of 21 and 23 kDa can be immunoprecipitated from surface-labeled sheep T cells, but the specific identity of these glycoproteins remains unclear.[16] The sheep CD3 γ and CD3 δ genes occur in single copy and are closely linked, so it is likely that the genomic arrangement of the CD3 gene group in ruminants is similar to that in humans and mice.[11]

B. TCR ζ AND η CHAINS

Two additional invariant components of the TCR complex, the ζ and η chains, have also been identified. The sequences of these chains are unrelated to the CD3 proteins, they are not members of the immunoglobulin superfamily of cell-surface molecules, and their genes are not linked to the CD3 group.[17] Therefore, they are properly considered as separate and distinct TCR components, although they are frequently grouped with the CD3 subunits. The two chains are assembled in the TCR complex in one of two dimeric forms, with ζ–ζ homodimers usually being more abundant than ζ–η heterodimers (see Figure 4-1). These dimeric modules are able to mediate a signal transduction pathway independently of the CD3 components and are therefore considered to play a key role in T cell activation.[7]

In contrast to the CD3 proteins, the TCR ζ chain has a very short extracellular domain, (just nine amino acids) and these have been exactly conserved between sheep and all other species characterized.[12] The transmembrane domain has also been well conserved but there are clear differences in the long cytoplasmic tail. This region of TCR ζ has been well conserved between humans, mice, and rats, but 15 unique substitutions and a two amino acid insertion have occurred in sheep.[12] Furthermore, some of these changes are located at a site that appears to be involved in GTP/GDP binding during signal transduction (Figure 4-2).[18] Therefore, in comparison to the CD3 chains which show a gradient of increasing conservation toward their C termini, the TCR ζ chain shows the opposite trend. Whether or not the accelerated rate of amino acid replacements in the cytoplasmic domain of sheep TCR ζ has any functional consequences remains an open question, although it suggests the possibility of different T cell signaling modalities in ruminants.

The TCR η chain arises as a differentially spliced product of one exon of the TCR ζ gene and, so far, has been identified at a protein level only in the mouse.[19] The TCR ζ and η chains have identical N termini, including the extracellular domain, but differ in the cytoplasmic region. The exon predicted to encode the human TCR η chain has been cloned recently.[20] To date, no information is available concerning the existence or nature of this TCR component in ruminants. However, since the predicted human and mouse TCR η sequences differ widely in the cytoplasmic region, and in view of the divergence noted above in the sheep TCR ζ chain, the structure of this component in ruminants would be of some interest.

IV. RUMINANT T CELL RECEPTOR HETERODIMERS

Four different polypeptides that have a similar overall structure are involved in the formation of the TCR heterodimer — these are termed the TCR α, β, γ, and δ chains. Each TCR chain can be subdivided into specific regions, including a hydrophobic leader (L) segment, a variable (V) segment, a diversity (D) segment in the case of β and δ chains, a joining (J) segment, and the constant (C) region.[8,21] The primary structures of a number of such regions from each of the TCR heterodimer chains of sheep or cattle have been inferred from the translated sequences of cDNA clones and the lengths of the different segments of these molecules are summarized in Table 4-1. Relevant features of each of these chains will be discussed in turn.

A. TCR α CHAINS

The analysis of multiple cDNA clones and Southern blot analysis has established that both sheep and cattle contain a single Cα gene segment, as in other animals.[22,23] The Cα region of both species consists of 140 amino acid residues and contains four potential sites for N-linked glycosylation, one more than in human, mouse, or rabbit Cα regions (see Figure 4-3). The TCR α locus of cattle and sheep appear to contain a large number of V and J segment genes and some of these have been identified as cDNA transcripts. The five bovine Vα familes identified so far as full-length sequences show 16 to 47% amino acid identity between families. Two full-length sheep Vα sequences have been described. Several different Jα regions ranging in length between 20 and 24 codons occur in the cDNA clones analyzed thus far from each animal and, in one case, the same J segment occurs in both sheep and cattle. Mature transcripts of ruminant TCR α chains are 1.5 to 1.6 kb in length.

Table 4-1 **Number of Amino Acid Residues in Different Segments of Ruminant TCR Heterodimer Chains**

	Leader	Variable	Diversity + joining	Constant
Bovine				
α	20–23	91–93	20–24	140
β	16–19	95–97	18–23	Cβ1 178
				Cβ2 178
γ	13–14	98–100	15–21	Cγ1 222[a]
				Cγ2 211
				Cγ3 194
				Cγ4 210
δ	20	93–96	23–39	155
Sheep				
α	21	90	20	140
β	18	92–94	17–21	178
γ	13–14	98–108	16–21	Cγ1 210
				Cγ2 220
				Cγ3 195
				Cγ4 207
				Cγ5 168
δ	17–20	85–97	25–43	155

[a] Transcripts of the bovine Cγ1 segment contain a stop codon in the cytoplasmic domain. The corresponding gene may therefore be a nonfunctional pseudogene.

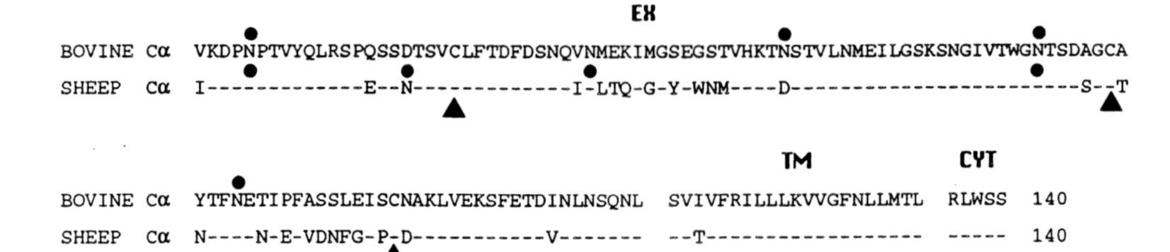

Figure 4-3 Amino acid sequences of the Cα region of cattle and sheep. The extracellular (EX), transmembrane (TM), and cytoplasmic (CYT) domains are shown above the single letter amino acid sequences, and the location of cysteine residues (▲) and potential sites for N-linked glycosylation (●) are indicated.

B. TCR β CHAINS

All other animals analyzed so far contain two germline Cβ gene segments, but the number of these genes in ruminants is not yet entirely clear. Two distinct Cβ transcripts (designated Cβ1 and Cβ2) have been identified in bovine cDNA clones.[24] However, the analysis of corresponding sheep clones revealed five different Cβ transcripts: three as full-length and two as partial sequences, and Southern blot analysis of sheep DNA suggests the possible existence of three Cβ genes.[25] The amino acid sequences of cattle and sheep Cβ regions are aligned in Figure 4-4. On the basis of sequence alone, it is not possible to decide whether the different transcripts arose from separate genes or whether they represent different alleles since all are highly similar (bovine Cβ1 and Cβ2 differ by 6 amino acids, the 3 full-length sheep Cβ sequences differ by 3–6 amino acids). It would be somewhat unusual if sheep DNA contains an additional Cβ gene segment, and thus a careful analysis of appropriate genomic clones will be necessary to ascertain whether this is the case. All of the characterized TCR β C regions have an identical size of 178 amino acids. The bovine Cβ1 chain has one, whereas bovine Cβ2 and all sheep Cβ chains have two potential sites for N-linked glycosylation.

The independently isolated cDNA clones sequenced so far from cattle and sheep allow the provisional identification of seven to nine Vβ families, although many of the sequences are not full length.[24,25] Interestingly, there are indications that some Vβ families may be used preferentially in certain lymphoid

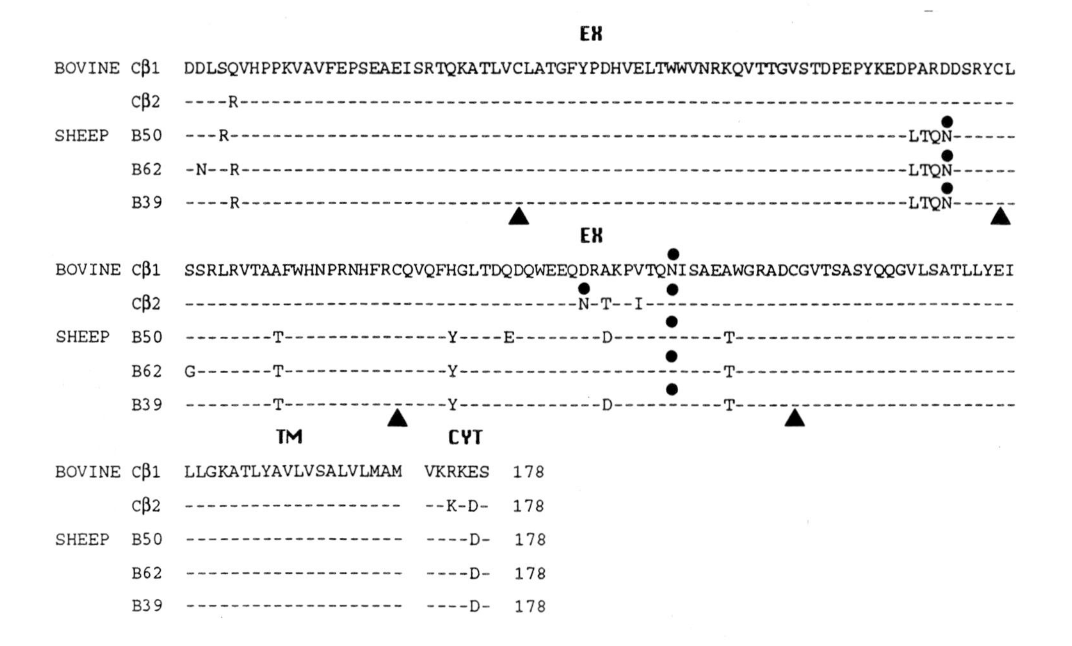

Figure 4-4 Amino acid sequences of Cβ regions of cattle and sheep; annotation as in Figure 4-3.

compartments or T cell subsets. Thus, the bovine Vβ12 gene segment, represented by clone BTB13, was isolated with a high frequency from a PBL cDNA library and one sheep Vβ family was frequently expressed by the CD4+ subset. Similar instances of preferential V gene usage have been described in rodents and humans, particularly during immune responses to specific antigens, and it is likely that future analyses in ruminants will also reveal further examples.

The junctional regions of ruminant TCR β chains are highly diverse and several Jβ segments can be identified. However, it is not yet possible to differentiate between the Dβ segments and N nucleotide additions at junctional regions of transcripts since the corresponding germline sequences of TCR β segments are not known. Transcripts of the TCR β chains occur as 1.3 and 0.9 kb mRNA, which represent functionally (V-D-J-C) or incompletely (D-J-C) rearranged forms, respectively.

C. TCR γ CHAINS

In contrast to the TCR α and TCR β chains, which show a high level of similarity to human and mouse counterparts in terms of the likely frequency and structure of gene segments, the ruminant TCR γ chains have some unusual features. First, a greater number of C region segments have been identified in sheep and cattle than in other species. So far, five different functional Cγ chains have been identified in sheep, three in cattle, and an additional Cγ pseudogene probably occurs in cattle. Second, the ruminant Cγ segments are more diverse in length and structure, and some of them represent the longest mammalian TCR C regions identified to date.[26-30]

The predicted protein sequences of ruminant Cγ chains are aligned in Figure 4-5. The lengths of the functional chains range from 168 (sheep Cγ5) to 220 (sheep Cγ2) amino acids, due to variability in the length of the connecting peptide or hinge segment between the immunoglobulin-like domain and the transmembrane domain. This region of human Cγ chains also varies in length and has been shown to be due to duplication or triplication of the short exon encoding this region.[31] It is likely that similar events have occurred in ruminants since repetitive sequence motifs corresponding approximately to the length of the human exon can be identified in the hinges of sheep Cγ chains.[26] In addition, some Cγ chains have variable numbers of repeats of a five amino acid motif (concensus sequence TTEPP) near the 5′ end of the hinge (see Figure 4-5). These motifs, which are absent from human and mouse sequences, could indicate the presence of additional Cγ exons in the ruminant genes.

Ruminant Cγ chains are also unusual in terms of the content of cysteine residues. All of them have one cysteine residue at the beginning of the hinge, which probably forms a disulfide link to the TCR δ chains, and two cysteine residues in the immunoglobulin domain, likely to be involved in intrachain disulfide bonding. In addition, some of the Cγ chains contain an extra two cysteine residues in the segment of the hinge predicted to have arisen by duplication (Figure 4-5). The transmembrane regions of all other TCR chains contain a charged lysine residue that is thought to stabilize the TCR complex by interacting with negatively charged residues in the corresponding region of the CD3 chains. However, some ruminant Cγ chains are exceptional since, in the bovine Cγ1, Cγ2, and Cγ3 chains and in the sheep Cγ2 chain, the lysine residue has been replaced by either leucine or threonine. Exactly what these unusual features mean in structural or functional terms remains unclear, but they suggest the possibility of different degrees of disulfide bonding and membrane stabilization between different isotypic forms of ruminant γδ TCRs.

The diversity of Vγ and Jγ segments has been examined by sequencing a large number of cDNA clones derived from peripheral sheep lymphocytes. Sheep express at least ten different Vγ segments that fall into six families, and six Jγ segments have also been identified.[27] Two families, Vγ2 and Vγ5, contain four and two members, respectively, and the V regions within each family show 81 to 97% nucleotide identity. All other Vγ families are thus far represented by single V region sequences and the level of nucleotide identity between families ranges from 40 to 75%. Two of the sheep Vγ regions have high levels of identity to either human or mouse Vγ sequences; at the protein level, sheep Vγ4 has 74.7% identity to mouse Vγ3, and sheep Vγ3 shares 64.7% of amino acid residues with human Vγ10. The remainder of the sheep Vγ regions appear to have no clear homologues in the other species.[27] Seven bovine Vγ sequences have been determined to date, and these show a high level of identity to the sheep Vγ2 and Vγ5 families.[30]

With only rare exceptions, the sheep TCR γ cDNA clones that have been sequenced so far show nearly invariant rearrangement patterns whereby a particular Vγ segment is always rearranged to the same Jγ segment. Furthermore, the different combinations of rearranged V-J segments are then spliced specifically to one of the five Cγ segments.[27] This pattern of predominating rearrangement and splicing events is similar to what has been reported in the mouse, and it suggests that the genomic organization of the TCR γ locus may be quite similar in these two species. In mice, the TCR γ locus consists of a number of tandem repeats

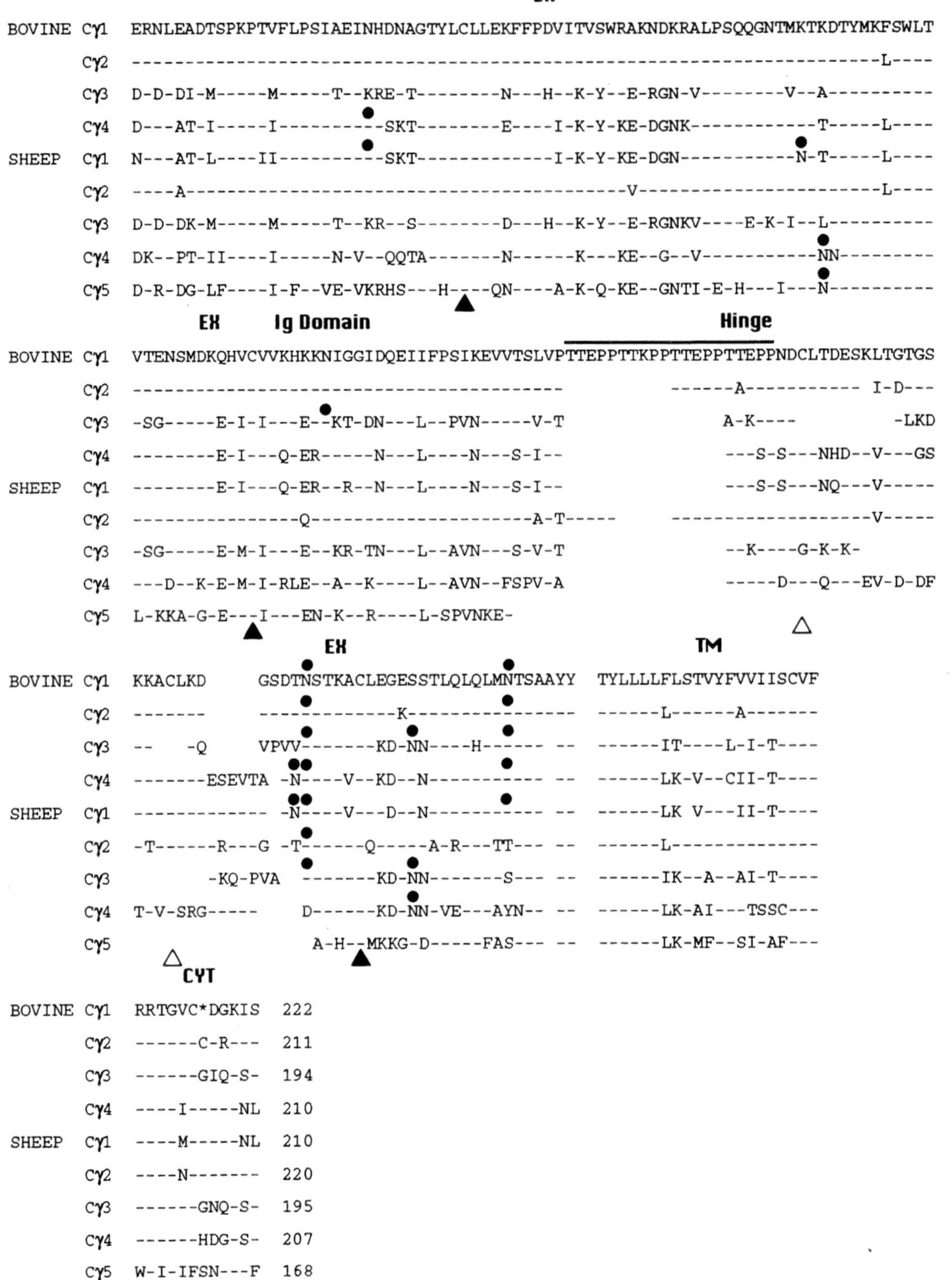

Figure 4-5 Amino acid sequences of Cγ regions of cattle and sheep. Annotation as in Figure 4-3, with some additional features. The approximate location of the Ig domain and hinge segment within the extracellular region is shown; the location of extra cysteine residues in the hinge segment of some Cγ chains is indicated and the position of repeated TTEPP motifs near the end of the hinge is overlined. Note also the concentration of potential N-linked glycosylation sites in the hinge region.

of V-J-C gene segments, and the rearrangement and splicing events are usually confined to the gene segments within one of these "transcription units".[32] It would therefore be no great surprise if genomic analyses subsequently show that the sheep TCR γ locus contains at least five such V-J-C units. If the sheep locus turns out instead to have a structure more reminiscent of the human homologue, where the Cγ genes are closely linked and lie downstream of the main Jγ and Vγ clusters,[32] then there must be a requirement for very precise control over the gene rearrangement and splicing processes.

All sheep clones sequenced so far have shown extensive coding diversity at the Vγ-Jγ junction, although these regions are relatively short, ranging from one to four amino acids between predicted V and J gene segments.[27] The coding changes appear to result from exonucleolytic cleavage of nucleotides from participating Vγ and Jγ gene segments and by the addition of N nucleotides. An additional region of increased coding variability has been identified near the N terminus of sheep Vγ chains, since the three most variable amino acid residues at positions other than the V-J junction cluster to this part of the mature protein. The presence of hypervariability at these positions may indicate a special role for these parts of the chain in ligand recognition.

D. TCR δ CHAINS

Both cattle and sheep contain a single Cδ gene that encodes a protein of 155 amino acids (Figure 4-6).[26,28] The Cδ regions of the two ruminant species are very similar in sequence to each other (85% amino acid identity) and less so to the human and mouse counterparts (~65% amino acid identity). In a manner similar to the Cγ chains, the immunoglobulin-like transmembrane and cytoplasmic domains of Cδ have been extensively conserved between species and the greatest differences are also located in the hinge region, although the degree of difference is far less. Southern blot analysis shows that the sheep Cδ gene is located within the TCRα locus;[23] so far, the locations of Jδ and Vδ elements have not been determined, but it is likely that they also occur within this locus, as in humans and mice.

The sequencing of a large number of cDNA clones containing Vδ segments expressed by peripheral lymphocytes of sheep has revealed some unusual features concerning the Vδ repertoire.[27] The 62 clones sequenced so far contain 28 distinct Vδ regions that form four families. The Vδ1 family is unusually large and contains 25 of the V regions, while the Vδ2, Vδ3, and Vδ4 families are represented by single members. The level of nucleotide identity between members of Vδ1 ranges from 79 to 97%, while the other families are highly divergent from the Vδ1 group and from each other (DNA identity 42–67%). Three different Jδ elements have been identified in sheep and, in contrast to the Vγ-Jγ junctions, there was no particular bias in terms of Vδ-Jδ combinations.[27] Although not yet analyzed as extensively, the bovine TCR δ repertoire appears to be comparable to that in sheep. So far, seven distinct full-length bovine Vδ and two Jδ sequences have been determined and these are very similar to the sheep Vδ1 family and the sheep Jδ segments, respectively.[30]

In contrast to humans and mice, therefore, sheep and cattle γδ T cells express an unusually large number of different Vδ segments. In the other two species, there is a degree of sharing of V regions between the TCR α and TCR δ loci.[33,34] However, thus far, there is no evidence to support the idea that the diversity of sheep Vδ regions can be explained in this way. Experiments using PCR amplification with various C-region specific primers indicate that the Vδ1 family is used only rarely, if at all, as part of the αβ TCR.[27] Furthermore, the different Vδ families show low levels of sequence identity to sheep and

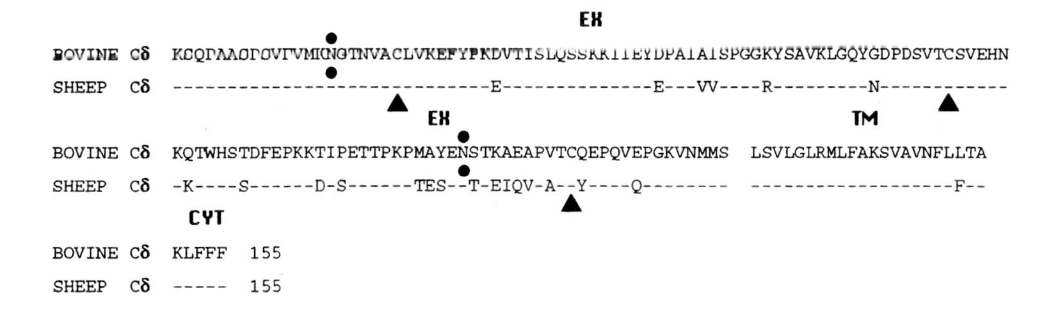

Figure 4-6 Amino acid sequences of the Cδ region of cattle and sheep. Annotation as in Figure 4-3.

bovine Vα families. The large number of Vδ segments therefore appear to have evolved as a separate gene pool that is specifically utilized by the γδ TCR.

Sheep Vδ chains show intriguing patterns of amino acid hypervariability. As in all other antigen receptor chains, extensive diversity is generated at the junctional region which forms the third complementarity determining region (CDR3). The Vδ-Dδ-Jδ junctions in sheep cDNA clones range in length from 1 to 18 amino acids. The complexity of this region varies at different developmental stages; the junctions are longer in clones derived from adult animals than from fetuses.[27] These differences probably reflect the usage of multiple Dδ elements and more extensive N nucleotide addition during T cell development in the older animals.

In addition to CDR3, two other positions of hypervariability occur in sheep Vδ chains, around residues 29–31 and 54–57, which appear to correspond to the CDR1 and CDR2 regions found in other antigen receptor V segments.[27] This implies that specific selection has occurred at these sites during evolution of sheep Vδ chains, suggesting that the γδ TCR of sheep may interact with antigen fragments attached to a presenting molecule in a more specific way than has been suggested from studies in humans and mice (see below).

V. LIGAND RECOGNITION AND SIGNALING PATHWAYS

As outlined previously, the specificity of T cell recognition is mediated by the unique variable regions that are assembled in the TCR heterodimer during ontogeny. This is shown convincingly in transgenic experiments where TCR α or β chains were introduced into cell lines; the antigen recognition specificity of the T cells changed to that encoded by the transgene (reviewed in Reference 35). However, it is wrong to suppose that the TCR is the sole molecular complex that is involved in antigen recognition since a number of accessory molecules that play supportive roles in this process have also been identified. Some of these appear to regulate adhesion events between T cells and antigen presenting cells (APCs), while others mediate signaling pathways. These processes have not been studied to any extent in cattle and sheep, but they are crucial to an overall understanding of TCR function and are therefore considered briefly here, as summarized schematically in Figure 4-7.

TCR αβ heterodimers recognize peptide fragments of antigens while they are bound within a cleft on the surface of MHC class I and II molecules.[8,35] There is a defined size of peptide that can be bound in the cleft and most MHC I bound fragments are eight to nine amino acids in length, while peptide fragments complexed to MHC II are more variable and range from 12 to 22 amino acids.[36-38] Although the interaction between the TCR and its ligand has not yet been studied directly, in a way similar to X-ray crystallography of Ig complexes, a number of mutagenesis studies of the αβ TCR give strong indications as to how this happens.[35] Mutational changes to the CDR3 regions alter the antigen specificity of the receptor, so this region is predicted to be the major site of interaction with the peptide. The CDR1 and CDR2 regions appear to interact with polymorphic determinants on the body of the MHC I or II molecules since mutation of V regions at these sites alters the MHC restriction pattern of the TCR. The dual nature of TCR-peptide-MHC interaction therefore provides a coherent molecular explanation for the phenomenon of MHC-restricted recognition of antigen by T cells (Figure 4-7).

The way in which γδ T cells recognize antigens is far less predictable. In the human and mouse γδ TCR, the combinatorial diversity generated by gene rearrangement is markedly restricted although considerable diversity is present at the CDR3 region.[32] One view therefore is that the antigen recognition potential of these cells may be somewhat limited and they could have a specialized role in defined compartments of the immune system such as mucosal surfaces.[39] Regions corresponding to CDR1 and CDR2 have not been identified in human and mouse TCR γ or δ sequences, so their presence in the sheep Vδ regions as outlined above, together with the far greater segmental diversity in sheep, is of particular significance in that it suggests that this receptor may also interact with a diverse population of antigens associated with a presenting ligand of some sort that could have more limited polymorphism. The necessity for and identity of a presenting ligand is, however, contentious since different human or mouse γδ T cell clones have shown widely variable requirements during antigen recognition *in vitro,* including a lack of defined restriction elements, restriction to either classical MHC class I or II antigens, and restriction by various nonclassical MHC molecules such as the TL group.[40] A potential role for the CD1 antigens has also been suggested. In short, the studies performed so far on antigen recognition by γδ T cells have failed to produce a satisfying concensus view and further work on these cells in ruminants may be particularly useful.

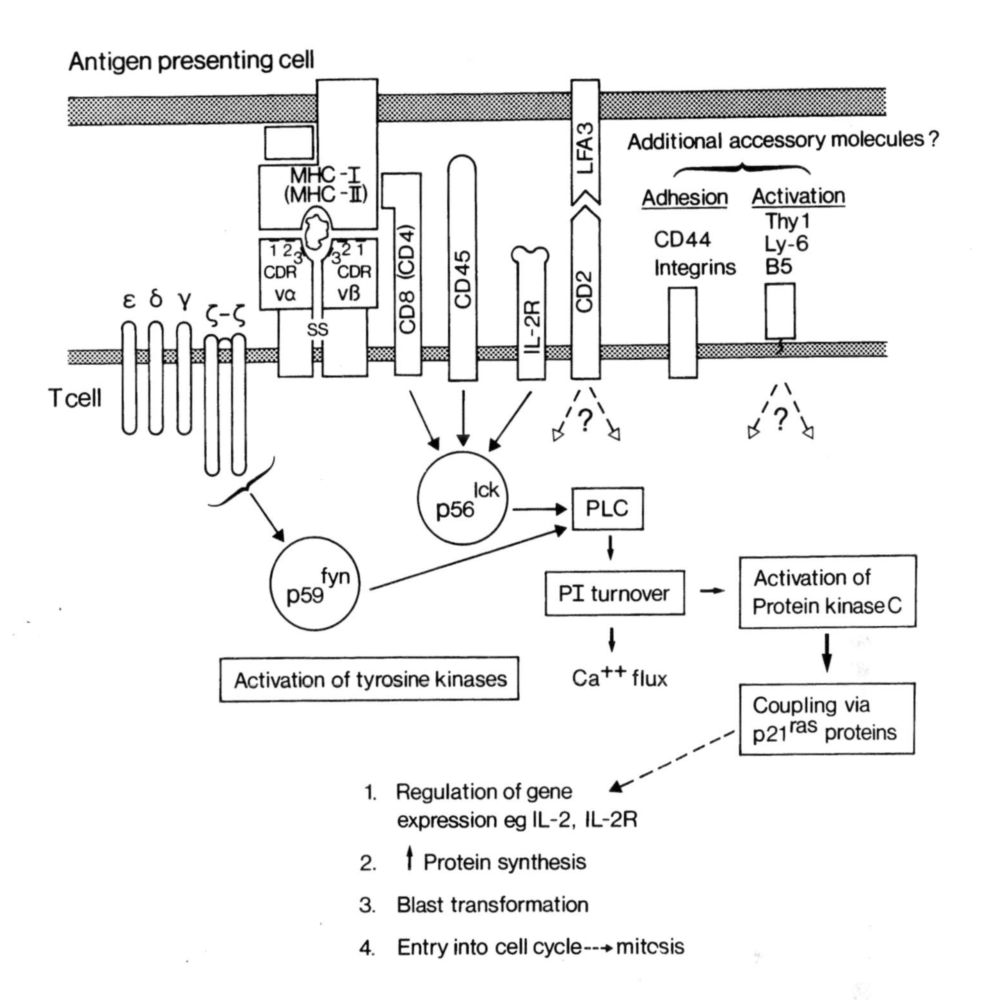

Figure 4-7 Schematic diagram showing the predicted nature of TCR-ligand interaction, some accessory molecules involved in this process, and possible signaling pathways. Many stages in the activation cascade are still poorly defined and the mechanisms which ultimately couple the more proximal events to the cell nucleus remain completely unknown. For detailed reviews and discussion of the pathways summarized in the figure, see References 4, 35, 41, and 44–47.

The affinity of the TCR for a peptide fragment bound to MHC determinants is predicted to be very low, and the requirement for close interaction between a T cell and an APC therefore introduces a need for additional stabilization of the two cells. For αβ T cells, the CD4 and CD8 molecules may be the most important accessory structures involved in this process since they have been shown to bind to defined nonpolymorphic regions of MHC class II or I molecules, respectively (reviewed in Reference 35 and 41). A range of other adhesion molecules that probably also assist have been identified, including CD2 which binds specifically to LFA3, the CD44 antigen, and a number of integrins.[42,43] Since most γδ T cells do not express either CD4 or CD8, and in ruminants they also lack CD2, other molecules may subserve a similar accessory role in this lineage, although as yet no candidates have been identified by functional criteria.

Following cognitive interaction between the TCR complex and a specific ligand, a series of biochemical pathways are activated to transduce a signal from the cell membrane to the nucleus. Some of the components involved in this activation cascade have been identified (see Figure 4-7). Two tyrosine kinases, $p59^{fyn}$ and $p56^{lck}$, appear to play pivotal roles in the most proximal part of the cascade.[41, 44-47] Stimulation via the TCR activates $p59^{fyn}$, while the CD4 and CD8 molecules are associated with $p56^{lck}$ and when cross-linked with the TCR, will transduce a signal via this tyrosine kinase pathway. Therefore, in addition to their stabilizing role, the CD4 and CD8 molecules can also modulate signaling when bound to the TCR/MHC ligand complex. The tyrosine phosphatase CD45 can also modify the activity of p56^{lck}. Once activated, the tyrosine kinases then phosphorylate a number of intracellular substrates, including phospholipase C (PLC) causing an increase in phosphoinositide (PI) breakdown. This results in a strong Ca^{++} flux and the activation of the serine/threonine kinase, protein kinase C. Signaling activity has also been associated with some other T cell structures, including CD2, Thy-1, Ly-6, and sheep T cells express a distinct cell surface molecule termed "B5" which is also capable of transducing an activation signal.[48] However, the exact way in which these molecules interact with known components of the activation cascade remains unclear. Again, the absence of CD4, CD8, and CD2 from most ruminant γδ T cells introduces the possibility that these cells may have different signaling properties.

The enzymatic pathways mentioned above appear to operate in the plane of the cell membrane.[41] The exact interrelationships between them and the precise mechanism by which activation is eventually transduced to the cell nucleus is likely to be complex and remains largely unknown, although the p21^{ras} proteins have recently been identified as important coupling components.[47] An appropriate signal is assumed to eventually be transduced to the cell nucleus, which then causes an upregulation of transcription of the genes encoding T cell lymphokines and their appropriate receptors. Additional signals may then be transduced via these newly expressed receptors, probably by utilizing the same enzymatic machinery; for example, the IL2 receptor is also associated with p56^{lck} (Figure 4-7). This second signal is essential for the clonal expansion of T cells during immune responses since stimulation via either the TCR or CD2 alone induces only a $G_0 \rightarrow G_1$ transition by T cells; transition from $G_1 \rightarrow S$, and ultimately to mitosis, only occurs after restimulation via the IL2 receptor.[47]

VI. CONCLUSIONS AND FUTURE RESEARCH

Over the last 5 years, significant progress has been made in achieving a basic characterization of the molecular structure of ruminant T cell receptors and a good deal of sequence data is now available for many gene segments at each of the TCR loci. Not unexpectedly, the overall structure and organization of this complex appears similar to the mouse and human counterparts, especially in the number and location of cysteine residues involved in disulfide bond formation and in the transmembrane and cytoplasmic regions which play an important role in signal transduction. These overall similarities notwithstanding, some interesting differences have also been revealed between ruminants and other species, chiefly in terms of the relative usage and complexity of the different TCR loci and the presence of unusual Cγ chains. To conclude this chapter, we will briefly indicate some of the follow-up work that needs to be done and identify a few research areas where the current knowledge of ruminant TCRs could be exploited.

A. GENOMIC STRUCTURE OF THE TCR LOCI

All of the information accumulated so far about the TCR complex has been derived from the analysis of cDNA clones, and there is a need to gain insights into the genomic organization of the relevant genes in ruminants. This information will shed further light on the overall evolution of the mammalian TCR and will help to resolve more specific issues such as the genetic basis for the unusual structure of ruminant Cγ segments. A detailed knowledge of the number and relative position of individual gene segments would, among other things, assist the interpretation of junctional diversity in cDNA clones and allow the planning of future experiments to study the control and pattern of gene rearrangement during T cell ontogeny. Conversely, the cDNA sequence data already available will allow DNA probes that are specific for a large number of V, D, J, and C gene segments to be prepared, thereby facilitating the analysis of genomic clones.

B. V REGION REPERTOIRE STUDIES

There are many experimental possibilities whereby the polymerase chain reaction (PCR), cDNA library construction, *in situ* hybridization, or other molecular techniques can be applied to directly measure V gene expression in different populations of T lymphocytes in sheep and cattle. Examples include the ontogeny of V gene repertoires at different stages in fetal and post-natal development[27] (see also Chapter 2); V gene usage in lymphocytes recirculating by one of the many accessible lymphatic pathways or by lymphocytes resident in a particular tissue; V gene usage by different phenotypic T cell subsets; V gene expression by T cells responding to specific antigen challenge; or any combination of these and other experimental conditions. Strategies of this sort, which seek to combine the natural experimental advantages offered by large animals with the recent developments in TCR knowledge, are certain to produce a great deal of novel data with both fundamental and applied interest.

C. $\gamma\delta$ T CELLS

The existence of an expanded repertoire of γ and δ gene segments,[27] together with the prominence of $\gamma\delta$ T cells in cattle and sheep,[49] suggests the possibility that these cells perform some special function(s) in ruminants that has largely been attenuated in the immune system of other animals. Continued characterization of these cells at molecular, and more importantly at functional, levels may therefore uncover important insights.

D. PATHOGENESIS OF LYMPHOID TUMORS

The availability of TCR probes will allow lymphoid tumors [29] and other transformed T cells to be characterized at a clonal level and this will assist the study of the pathogenesis and epidemiology of associated diseases.

ACKNOWLEDGMENTS

The Basel Institute for Immunology was founded and is supported by F. Hoffmann-La Roche and Co., Basel, Switzerland.

Appendix I EMBL Database Accession Numbers for Sequences of TCR Components

	Sheep	Cattle
Invariant components		
CD3 γ	X52994	
CD3 δ	X52993	
CD3 ε	Z12969	
TCR ζ	Z12968	
TCR chains: C-regions		
Cα	M55622	D90030
Cβ	M94181-M94183	D90139-D90140
Cγ1	Z12964	D90413
Cγ2	Z12965	D90415
Cγ3	Z12966	D90414
Cγ4	Z12967	X63680
Cγ5	Z13986	
Cδ	Z12963	D90419
TCR chains: V-regions		
Vα	M55622	D90010-D90029
Vβ		D90121-D90133
Vγ1	Z12998	
Vγ2	Z12999-Z13002	D13648
Vγ3	Z13003	
Vγ4	Z13004	
Vγ5	Z13005-Z13006	D13649-D13654
Vγ6	Z13007	
Vδ1	Z12970-Z12994	D13655-D13661
Vδ2	Z12995	
Vδ3	Z12996	
Vδ4	Z12997	

Note: The accession numbers for all ruminant TCR components included in the nucleotide databases as at November 1992 are shown.

REFERENCES

1. **Yanagi, Y., Yoshikai, Y., Leggett, K., Clark, S. P., Aleksander, I., and Mak, T. W.**, A human T cell-specific cDNA clone encodes a protein having extensive homology to immunoglobulin chains, *Nature*, 308, 145, 1984.
2. **Hedrick, S. M., Cohen, D. I., Nielsen, E. A., and Davis, M. M.**, Isolation of cDNA clones encoding T cell-specific membrane-associated proteins, *Nature*, 308, 149,1984.
3. **Hedrick, S. M., Nielsen, E. A., Kavaler, J., Cohen, D. I., and Davis, M. M.**, Sequence relationships between putative T-cell receptor polypeptides and immunoglobulins, *Nature*, 308, 153, 1984.
4. **Ashwell, J. D. and Klausner, R. D.**, Genetic and mutational analysis of the T-cell antigen receptor, *Annu. Rev. Immunol.*, 8, 139, 1990.
5. **Exley, M., Terhorst, C., and Wileman, T.**, Structure, assembly and intracellular transport of the T cell receptor for antigen, *Sem. Immunol.*, 3, 283, 1991.
6. **Frank, S. J., Engel, I., Rutledge, T. M., and Letourneur, F.**, Structure/function analysis of the invariant subunits of the T cell antigen receptor, *Sem. Immunol.*, 3, 299, 1991.
7. **Wegener, A.-M., Letourneur, F., Hoeveler, A., Brocker, T., Luton, F., and Malissen, B.**, The T cell receptor/CD3 complex is composed of at least two autonomous transduction modules, *Cell*, 68, 83, 1992.
8. **Davis, M. M. and Bjorkman, P. S.**, T-cell antigen receptor genes and T-cell recognition, *Nature*, 334, 395, 1988.
9. **Tonegawa, S.**, Somatic generation of antibody diversity, *Nature*, 302, 575, 1983.
10. **Lafaille, J. J., DeCloux, A., Bonneville, M., Takagaki, Y., and Tonegawa, S.**, Junctional sequences of T cell receptor γδ genes: implications for γδ T cell lineages and for a novel intermediate of V-(D)-J joining, *Cell*, 59, 859, 1989.
11. **Hein, W. R. and Tunnacliffe, A.**, Characterization of the CD3γ and δ invariant subunits of the sheep T-cell antigen receptor, *Eur. J. Immunol.*, 20, 1505, 1990.
12. **Hein, W. R. and Tunnacliffe, A.**, Invariant components of the sheep T-cell antigen receptor: cloning of the CD3 ε and TCR ζ chains, *Immunogenetics*, 37, 279, 1993.
13. **Clevers, H., MacHugh, N. D., Bensaid, A., Dunlap, S., Baldwin, C. L., Kaushal, A., Iams, K., Howard, C. J., and Morrison, W. I.**, Identification of a bovine surface antigen uniquely expressed on $CD4^-CD8^-$ T cell receptor γ/δ^+ T lymphocytes, *Eur. J. Immunol.*, 20, 809, 1990.
14. **Mallabiabarrena, A., Fresno, M., and Alarcon, B.**, An endoplasmic reticulum retention signal in the CD3ε chain of the T-cell receptor, *Nature*, 357, 593, 1992.
15. **Letourneur, F. and Klausner, R. D.**, Activation of T cells by a tyrosine kinase activation domain in the cytoplasmic tail of CD3ε, *Science*, 255, 79, 1992.
16. **Hein, W. R., Dudler, L., Beya, M.-F., Marcuz, A., and Grossberger, D.**, T cell receptor gene expression in sheep: differential usage of TCR1 in the periphery and thymus, *Eur. J. Immunol.*, 19, 2297, 1989.
17. **Weissman, A. M., Hou, D., Orloff, D. G., Modi, W., Seuanez, H., O'Brien, S. J., and Klausner, R. D.**, Molecular cloning and chromosomal localization of the human T-cell receptor ζ chain: distinction from the CD3 complex, *Proc. Natl., Acad. Sci. U.S.A.*, 85, 9709, 1988.
18. **Peter, M. E., Hall, C., Ruhlmann, A., Sancho, J., and Terhorst, C.**, The T-cell receptor ζ chain contains a GTP/GDP binding site, *EMBO J.*, 11, 933, 1992.
19. **Jin, Y.-J., Clayton, L. K., Howard, F. D., Koyasu, S., Sieh, M., Steinbrich, R., Tarr, G. E., and Reinherz, E. L.**, Molecular cloning of the CD3η subunit identifies a CD3ζ-related product in thymus derived cells, *Proc. Natl. Acad. Sci. U.S.A.*, 87, 3319, 1990.
20. **Jensen, J. P., Hou, D., Ramsburg, M., Taylor, A., Dean, M., and Weissman, A. M.**, Organization of the human T cell receptor ζ/η gene and its genetic linkage to the FcγRII-FcγRIII gene cluster, *J. Immunol.*, 148, 2563, 1992.
21. **Strominger, J. L.**, Developmental biology of T cell receptors, *Science*, 244, 943, 1989.
22. **Ishiguro, N., Tanaka, A., and Shinagawa, M.**, Sequence analysis of bovine T-cell receptor α chain, *Immunogenetics*, 31, 57, 1990.
23. **Hein, W. R., Marcuz, A., Fichtel, A., Dudler, L., and Grossberger, D.**, Primary structure of the sheep T-cell receptor α chain, *Immunogenetics*, 34, 39, 1991.
24. **Tanaka, A., Ishiguro, N., and Shinagawa, M.**, Sequence and diversity of bovine T-cell receptor β-chains, *Immunogenetics*, 32, 263, 1990.

25. **Grossberger, D., Marcuz, A., Fichtel, A., Dudler, L., and Hein, W. R.**, Sequence analysis of sheep T-cell receptor β chains, *Immunogenetics*, 37, 222, 1993.
26. **Hein, W. R., Dudler, L., Marcuz, A., and Grossberger, D.**, Molecular cloning of sheep T-cell receptor γ and δ chain constant regions: unusual primary structure of γ chain hinge segments, *Eur. J. Immunol.*, 20, 1795, 1990.
27. **Hein, W. R. and Dudler, L.**, Divergent evolution of T cell repertoires: extensive diversity and developmentally regulated expression of the sheep γδ T cell receptor, *EMBO J.*, 12, 715, 1993.
28. **Takeuchi, N., Ishiguro, N., and Shinagawa, M.** Molecular cloning and sequence analysis of bovine T-cell receptor γ and δ chain genes, *Immunogenetics*, 35, 89, 1992.
29. **Ishiguro, N., Matsui, T., and Shinagawa, M.**, Differentiation analysis of bovine T lymphosarcoma, *Vet. Immunol. Immunopathol.*, in press, 1993.
30. **Ishiguro, N., Aida, Y., Shinagawa, T., and Shinagawa, M.**, Molecular structures of cattle T-cell receptor gamma and delta chains predominantly expressed on peripheral blood lymphocytes, *Immunogenetics,* 38, 437, 1993.
31. **Lefranc, M.-P. and Rabbitts, T. H.**, Two tandemly organized human genes encoding the T-cell γ constant-region sequences show multiple rearrangements in different T-cell types, *Nature*, 316, 464, 1985.
32. **Raulet, D. H.**, The structure, function and molecular genetics of the γ/δ T cell receptor, *Annu. Rev. Immunol.*, 7, 175, 1989.
33. **Takihara, Y., Reimann, J., Michalopoulos, E., Ciccone, E., Moretta, L., and Mak, T. W.**, Diversity and structure of human T cell receptor δ chain genes in peripheral blood γ/δ-bearing T lymphocytes, *J. Exp. Med.*, 169, 393, 1989.
34. **Miossec, C., Faure, F., Ferradini, L., Roman-Roman, S., Jitsukawa, S., Ferrini, S., Moretta, A., Triebel, F., and Hercend, T.**, Further analysis of the T cell receptor γ/δ+ peripheral lymphocyte subset. The Vδ1 gene segment is expressed with either Cα or Cδ, *J. Exp. Med.*, 171, 1171, 1990.
35. **Jorgenson, J. L., Reay, P. A., Ehrich, E. W., and Davis, M. M.**, Molecular components of T-cell recognition, *Annu. Rev. Immunol.*, 10, 835, 1992.
36. **Schumacher, T. N., de Bruijn, M. L., Vernie, L. N., Kast, W. M., Melief, C. J., Neefjes, J. J., and Ploegh, H. L.**, Peptide selection by MHC class I molecules, *Nature,* 350, 703, 1991.
37. **Rudensky, A. Y., Preston-Hurlburt, P., Hong, S.-C., Barlow, A., and Janeway, C. A., Jr.**, Sequence analysis of peptides bound to MHC class II molecules, *Nature*, 353, 622, 1991.
38. **Rudensky, A. Y., Preston-Hurlburt, P., Al-Ramadi, B. K., Rothbard, J., and Janeway, C. A., Jr.**, Truncation variants of peptides from MHC class II molecules suggest sequence motifs, *Nature*, 359, 429, 1992.
39. **Janeway, C. A. Jr., Jones, B., and Hayday, A.**, Specificity and function of T cells bearing γδ receptors, *Immunol. Today*, 9,73, 1988.
40. **Matis, L. A. and Bluestone, J. A.**, Specificity of γδ receptor bearing T cells, *Sem. Immunol.*, 3, 75, 1991.
41. **Janeway, C. A., Jr.**, The T cell receptor as a multicomponent signalling machine: CD4/CD8 coreceptors and CD45 in T cell activation, *Annu. Rev. Immunol.*, 10, 645, 1992.
42. **Springer, T. A.**, Adhesion receptors of the immune system, *Nature*, 346, 425, 1990.
43. **Hynes, R. O.**, Integrins: versatility, modulation and signaling in cell adhesion, *Cell,* 69, 11, 1992.
44. **Keegan, A. D. and Paul, W. E.**, Multichain immune recognition receptors: similarities in structure and signaling pathways, *Immunol. Today*, 13, 63, 1992.
45. **Klausner, R. D. and Samelson, L. E.**, T cell antigen receptor activation pathways: the tyrosine kinase connection, *Cell*, 64, 875, 1991.
46. **Abraham, R. T., Karnitz, L. M., Secrist, J. P., and Leibson, P. J.**, Signal transduction through the T-cell antigen receptor, *Trends Biochem. Sci.*, 17, 434, 1992.
47. **Downward, J., Graves, J., and Cantrell, D.**, The regulation and function of $p21^{ras}$ in T cells, *Immunol. Today*, 13, 89, 1992.
48. **Hein, W. R., McClure, S., Beya, M.-F., Dudler, L., and Trnka, Z.**, A novel glycoprotein expressed on sheep T and B lymphocytes is involved in a T cell activation pathway, *J. Immunol.*, 140, 2869, 1988.
49. **Hein, W. R. and Mackay, C. R.**, Prominence of γδ T cells in the ruminant immune system, *Immunol. Today*, 12, 30, 1991.

Chapter 5

The Cytokines: Origin, Structure and Function

D. M. Haig, C. J. McInnes, P. R. Wood and H. F. Seow

CONTENTS

I. Introduction 75
II. The cDNAs and Genes 76
   A. Cloning Strategies 76
   B. The Interferons 76
      1. Alpha Interferon (IFN-α) 76
      2. Omega Interferon (IFN-ω) 76
      3. Beta Interferon (IFN-β) 78
      4. Gamma Interferon (IFN-γ) 78
   C. The Interleukins and Tumor Necrosis Factor 78
      1. Interleukin-1 (IL-1) 78
      2. Interleukin-6 (IL-6) 79
      3. Tumor Necrosis Factor (TNF-α) 79
      4. Interleukin-2 (IL-2) 79
      5. Interleukin-4 (IL-4) 80
   D. The Hemopoietins (Colony-Stimulating Factors) 80
      1. Multipotential Colony-Stimulating Factor (Multi-CSF, Interleukin-3) 82
      2. Granulocyte-Macrophage Colony-Stimulating Factor (GM-CSF) 82
      3. Granulocyte-Colony-Stimulating Factor (G-CSF) and Interleukin-7 (IL-7) 82
   E. Cytokine Genes and Control Elements 82
III. Cytokine Assays and *In Vitro* Biology 83
   A. Assays 83
   B. *In Vitro* Biology 83
IV. *In Vivo* Biology 84
   A. Lymph Studies 84
   B. Other *In Vivo* Studies 84
V. Cytokine Receptors 85
VI. Conclusions 86
Acknowledgments 86
References 86

I. INTRODUCTION

The cytokines are polypeptides that stimulate membrane receptor-mediated biochemical events in the same or other cells. They include the lymphokines, interleukins, and growth factors that regulate tissue development or repair, and immune and inflammatory responses. Following activation, cells produce cytokines locally in the tissues in the picomolar to nanomolar concentration range. The majority of cytokine mRNAs are expressed transiently at low level, are degraded rapidly, and generally the proteins have a short half-life in the circulation.[1,2] Cytokines show varying degrees of pleiotropy, stimulating responses in a variety of different target cells. Furthermore, cytokines can function singly or more often in networks, positively and negatively influencing cell responses and, in some cases, synergizing with other cytokines to amplify a particular cellular response.[2] In mouse and man, new cytokines are being discovered all the time and the list of activities for any one of them is being extended. The importance of cytokines in immunology is that they direct the quality (type) of immune response that ensues following specific recognition of antigen by cells of the immune system or are effector molecules in innate inflammatory responses following tissue injury. For example, in mouse and man following antigen challenge, mutually exclusive sets of cytokines are produced by two distinct populations of memory $CD4^+$ T cells which are referred to as TH1 and TH2.[3,4] Activated TH1 cells produce IFN-γ and lymphotoxin

0-8493-4952-4/94/$0.00+$.50

among other cytokines, whereas TH2 cells produce IL-4, IL-5, and IL-10. The TH1 cytokines are key mediators of cell-mediated immunity involving delayed-type hypersensitivity, whereas the TH2 cytokines are mediators of immune and inflammatory responses typically observed following helminth parasite infection.[4] These include enhanced levels of IgE and a local mast cell and eosinophil response in the tissues. IFN-γ can downregulate IL-4 production by TH2 cells, and IL-10 has been shown to prevent induction of TH1 cytokine production.[4,5] Cytokines direct immunoglobulin isotype selection by B cells during the immune response. IL-4 is involved in a pathway that stimulates IgE and IgG_1 production in the mouse, whereas IFN-γ is involved in IgG_{2a} isotype selection which mediates antibody-dependent cellular cytotoxicity responses.[6]

As the importance of cytokines to the development of immune and inflammatory responses has emerged from studies in mouse and man, it has become necessary to identify equivalent and novel cytokines in ruminants. This is desirable because the human and mouse cytokines show varying degrees of biological cross-reactivity in ruminant species. This is currently a rapidly expanding area of ruminant biology, and this review will be restricted to an overview of selected ruminant cytokine genes and proteins involved in immune and inflammatory responses.

II. THE cDNAs AND GENES

A. CLONING STRATEGIES

Table 5-1 lists the strategies that have been used to clone the ruminant cytokine cDNAs or genes. In general, the equivalent human DNA sequences are useful as probes for cross-hybridization or when using the PCR technique, but this approach may not be feasible if there is significant sequence divergence between the human cDNA and the ruminant cDNA of interest. This is the case with IL-3 where the DNA homology between species is very low and attempts to isolate the ovine cDNA by traditional methods using human sequences have been unsuccessful. The strategy adopted relied on the prediction that the genes for IL-3 and GM-CSF should be within 10 to 20 kb of each other in the ruminant genome, as they are in mouse and man.[40,41] By using the ovine GM-CSF cDNA to isolate GM-CSF genomic DNA, it was possible to pick up the IL-3 gene by rigorous cross-hybridization analysis with a fragmented human IL-3 cDNA.[33]

B. THE INTERFERONS

The IFNs are a group of proteins which induce an antiviral state in a number of different cell types.[42] Type I IFNs are produced predominantly by leukocytes and fibroblasts. Type II IFN (IFN-γ) is produced predominantly by activated T-cells and plays a major role in specific immune responses.[4] Type I IFNs can be separated into three distinct gene families, namely alpha (α), beta (β), and omega (ω, formerly IFN-α_{II}), with the trophoblast-derived IFN possibly constituting a fourth type I gene family. IFN-γ remains the only type II IFN identified so far.

The trophoblast IFN, for which there are at least six different genes, are expressed predominantly in the trophectoderm between days 13 and 21 of pregnancy. They have been the subject of excellent reviews[43,44] and therefore will not be covered in any more detail here.

1. Alpha Interferon (IFN-α)

IFN-α in cattle is encoded by a multigene family.[7-9] It is not known how many of the genes are functional and how many encode pseudogenes. Three functional bovine IFN-α genes have been cloned and share 95% nucleotide homology and 93% aa identity with each other and 64% aa identity with the consensus of expressed hu IFN-α genes.[7-9] The genes are intronless and code for mature proteins of 165/166 aa depending on the exact cleavage site of the signal peptide.

2. Omega Interferon (IFN-ω)

IFN-ω is closely related to the trophoblast IFNs, sharing up to 85% nucleotide homology in their coding regions. At least four genes encoding IFN-ω have been found in cattle[10] and man,[8,45] although in both cases only one functional gene has been characterized, the others being pseudogenes. An apparently functional ovine IFN-ω gene has been cloned also.[24] IFN-ω genes may not exist in all mammals as none have been found so far in mice.[10] IFN-ω genes encode proteins of 195 aa which, after cleavage of signal peptide, result in mature proteins of 172/174 aa depending on the exact site of cleavage. The functional ovine and bovine proteins share approximately 61% identity with each other, and 67 and 63%, respectively, with the human protein.

Table 5-1 **Cloning Strategies for Ruminant Cytokines**

Cytokine	RNA source	Stimulus	Cloning	Ref.	EMBL accession no.
Bovine					
IFN-α (genes)	—	—	λL47.1[a]	7,8,9	M10952-M10954, M11001
IFN-ω (genes)	—	—	λGEM-11[a]	8,10	M11002, M38190
IFN-β (genes)	—	—	λcharon 30[a]	11	
IFN-γ (cDNA)	Lymph node cells	ConA	Plasmid	12	M29867
IL-1α (cDNA)	Alveolar macrophages	LPS	λgt10	13,14	X12497, M37210
IL-1β (cDNA)	Alveolar macrophages	LPS	λgt10	14,15	M35589, M37211
IL-2 (cDNA)	Lymph node cells, lymphocytes	ConA	Plasmid	16,17	M12791, M13204
IL-4 (cDNA)	Lymph node cells	Con A	PCR	18	M84745
IL-6 (cDNA)				19	X57317
IL-7 (cDNA)				20	X64540
GM-CSF (cDNA)	BT2 T-lymphocytes	PMA PHA	λgt10	21,22	
TNF-α (cDNA)				23	
Ovine					
IFN-ω (gene)	—	—		24	X59068
IFN-γ (cDNA)	Lymph node cells	ConA	PCR	25,26	X52640
IL-1α (cDNA)	Alveolar macrophages	LPS	λgt10	27	X56754
IL-1β (cDNA)	Alveolar macrophages	LPS	λgt10, PCR	27,28,29	X56755, X54796, X56972
IL-2 (cDNA)	Lymph node cells	ConA	PCR	30,31	X53934, X55641
IL-3 (cDNA)	Lymph node cells	ConA	PCR	32	Z18291
IL-3 (gene)	—	—	Cosmid[a]	33	Z18897
IL-4 (cDNA)	Lymph node cells	PMA + A23187	PCR	34	M96845
GM-CSF (cDNA)	Lymph node cells, alveolar macrophages	ConA, LPS	PCR	35,36	X53561
TNF-α (cDNA)	Liver, alveolar macrophages	LPS	λgt11, PCR	37,38,39	X55966, X55152, X56756

Note: Abbreviations: A23187 calcium ionophore; Con-A, Concanavalin-A; LPS, lipopolysaccharide; PMA, phorbol myristate acetate; PCR, polymerase chain reaction.
[a]Genomic DNA. Spaces indicate strategy not known to reviewers (unpublished).

3. Beta Interferon (IFN-β)

IFN-β in man and mice is encoded by a single gene, but in cattle at least three functional genes have been isolated, with Southern hybridization analysis detecting as many as five.[11] Ovine IFN-β genes have not been cloned. As with human and murine IFN-β genes, those of cattle are intronless. They encode proteins of 186 aa in length, compared to 187 aa in man, the first 21 of which form the signal sequence. Overall, the mature bovine proteins share approximately 85% identity with each other and 55% identity with the human protein.[46] All three bovine proteins have two potential Asn-linked glycosylation sites compared to one in the human protein and three in the murine protein,[47] the locations of which differ in all three species. Glycosylation is not important for biological activity.[11] The three Cys residues found in the human IFN-β are also conserved in the three bovine proteins. One of these (Cys^{141}) was shown to be vital for maintaining antiviral activity.[48]

4. Gamma Interferon (IFN-γ)

Bovine and ovine IFN-γ cDNAs have been cloned.[12,25,26] Their predicted amino acid sequences, together with that of the human sequence, are shown in Figure 5-1. Human IFN-γ associates to form a dimer, with each subunit consisting of six αhelices (A-F).[51] The two pairs of helices which are predicted to be buried in the core of the dimer (C and F) are well conserved in the ruminant proteins, which therefore would have similar three-dimensional structures. Human IFN-γ, however, is not active on either bovine or ovine cells, the specificity presumably residing in the receptor binding domains. The areas of ruminant IFN-γ which differ most from the human protein (helices D and E, and the loop between helices A and B) are those predicted to be on the surface of the molecule and therefore most likely to be involved in receptor binding.

C. THE INTERLEUKINS AND TUMOR NECROSIS FACTOR

IL-1α, IL-1β, IL-6, TNF-α, and TNF-β (lymphotoxin) are among the most pleiotropic cytokines known.[52-54] They can be produced by and are potentially active upon a large number of nucleated cell types. All of them are particularly effective at initiating cytokine cascades from macrophages, fibroblasts, stromal and endothelial cells, and can be produced as a result of tissue injury or a specific immune response. IL-1α and IL-1β increase the expression of IL-2 receptors and mediate cartilage degradation, activities that have been exploited for their assay (Table 5-2). IL-6 has a role in B cell differentiation and immunoglobulin production. Both IL-1 and IL-6 induce hepatic acute phase proteins and regulate aspects of hemopoiesis. TNF-α ("macrophage derived") and TNF-β (lymphotoxin, "lymphocyte-derived") have cytotoxic/cytostatic effects on some tumor cells, exhibit anti-viral activity, and can augment the expression of MHC class I and II molecules on target cells. Ovine and/or bovine cDNAs for all of the above cytokines have been cloned.

1. Interleukin-1 (IL-1)

IL-1 is encoded by two separate genes, namely IL-1α and IL-1β.[52] The gene products share little amino acid identitity with each other (26% for human IL-1α and IL-1β; 22% for murine). However, both proteins bind to the same receptor with similar affinities and thus mediate indistinguishable biological functions.[55] The cDNAs for both ovine[27-29] and bovine[13-15] IL-1α and β have been cloned. Their predicted

```
        -23                     1         10    #  20        30        40        50        60
Ovine   MKYTSSFLALLLCVLLGFSGSYGQGPFFKEIENLKEYFNASNPDVAKGGPLFSEILKNWKEESDKKIIQSQIVSFYFKLFENL
Bovine  -----Y-------G-----------Q--R------------S------------------D----------------------
Human   -----YI--FQ--IV--SL-C-C-D-YV--A----K----GHS---DN-T--LG----------R--M-----------K-F

                  70        80 #      90       100       110       120       130       140
Ovine   KDNQVIQRSMDIIKQDMFQKFLNGSSEKLEDFKRLIQIPVDDLQIQRKAINELIKVMNDLSPKSNLRKRKRSQNLFRGRRASM
Bovine  ---------------------------------K---------------------------------------------T
Human   --D-S--K-VET--E--NV--F-SNKK-RD--EK-TNYS-T--NV-----H---Q--AE---AAKTG------M--------Q
```

Figure 5-1 The predicted amino acid sequences of ovine, bovine and human IFN-γ. Numbering is with respect to the ovine protein with the predicted NH_2-terminus numbered 1. Only the aa which differ from the consensus ovine sequence are shown. The two putative Asn-linked glycosylation sites in the ruminant proteins are indicated (#). The ruminant proteins share 96% identity with each other and approximately 63 and 47% with the human[49] and murine[50] proteins, respectively.

Table 5-2 **Assays for Ruminant Cytokines**

Cytokine	Antibodies	ELISA	Ref.	Bioassays
Bovine IL-1α; IL-1β			14,83	Thymocyte co-mitogen
Ovine IL-1α; IL-1β			84	Cartilage degradation
			27,85	Murine NOB-1/CTLL assays
Bovine TNF-α	+	+	86–89	Murine L929 or WEH1-164 or porcine
Ovine TNF-α	+	+	90	PK(15)-1512 fibroblast cytotoxicity assays
Ovine IL-3			33, 35	Bone marrow cell soft agar
Ovine GM-CSF	+	+	91	clonogenic assay
Bovine GM-CSF				
Bovine Type 1 IFNs			7,9,11	Anti-viral activity on homologous
Ovine Type 1 IFNs			92	adherent cell lines
Bovine IL-2			16	BT-2 T-lymphoblast proliferation
Bovine IFN-γ	+	+	12,93	Antiviral activity on homologous
Ovine IFN- γ	+	+	92	adherent cell lines

Note: Interspecies cytokine reactivities: bovine and ovine IL-1α and β, TNF-β, Type 1 IFNs, and IFN-α cross-react between the two species. The anti-bovine IFN-α ELISA detects ovine IFN-β and bovine IFN-α at similar levels of sensitivity. Ovine and bovine GM-CSF exhibit a small degree only of interspecies reactivity and IL-3 has not been tested. Abbreviations: see text. Details of reagents and techniques can be obtained from the references.

aa sequences, together with that of the human sequence, is shown in Figure 5-2. IL-1 is biologically active between species and critical regions should be represented in the 16% of aa which are conserved between the secreted forms of IL-1 from sheep, cattle and man. Indeed, there is a tetrapeptide (Ile-Thr-Asp-Phe) conserved near the -COOH end of all six proteins which is known to be required for the biological activity of the human protein.[58] Furthermore, the conservation of proline, leucine, and phenylalanine residues could have an important role in maintaining particular secondary and tertiary structures.

2. Interleukin-6 (IL-6)

The cDNAs for both ovine (A. Nash, Melbourne University and D. Sargan, Edinburgh University, personal communication) and bovine IL-6[19] have been cloned recently.

3. Tumor necrosis factor (TNF-α)

The bovine[23] and ovine[37-39] TNF-α cDNAs have been cloned. A comparison of their predicted aa sequences with that of the human sequence is shown in Figure 5-3. Of the three ovine TNF-α sequences reported, two have identical protein sequences, with the third lacking an amino acid in the leader sequence. Although the latter has not been shown to be biologically active, it may represent an allelic difference since Southern hybridization analysis has shown the presence of only one TNF-α gene in the ovine genome.[39]

4. Interleukin-2 (IL-2)

IL-2 is both an autocrine and paracrine growth factor produced predominantly by activated T cells and which, among other functions, stimulates CD4 and CD8 T cell proliferation and activates non-MHC-restricted killer cells.[61]

Both the bovine[16,17] and ovine[30,31] IL-2 cDNAs have been cloned. A comparison of their predicted amino acid sequences with that of the human sequence is shown in Figure 5-4. The mature IL-2 proteins are predicted to start with the dipeptide Ala-Pro, a feature common to a number of other cytokines such as IL-1β, IL-3, and GM-CSF. A three-dimensional study with human IL-2 revealed short helices, (aa residues 11-19, 33-56, 66-78, 83-101, 107-113, and 117-133; numbering with respect to the mature protein), three of which contain residues that interact with the IL-2 receptor.[64] The areas of highest

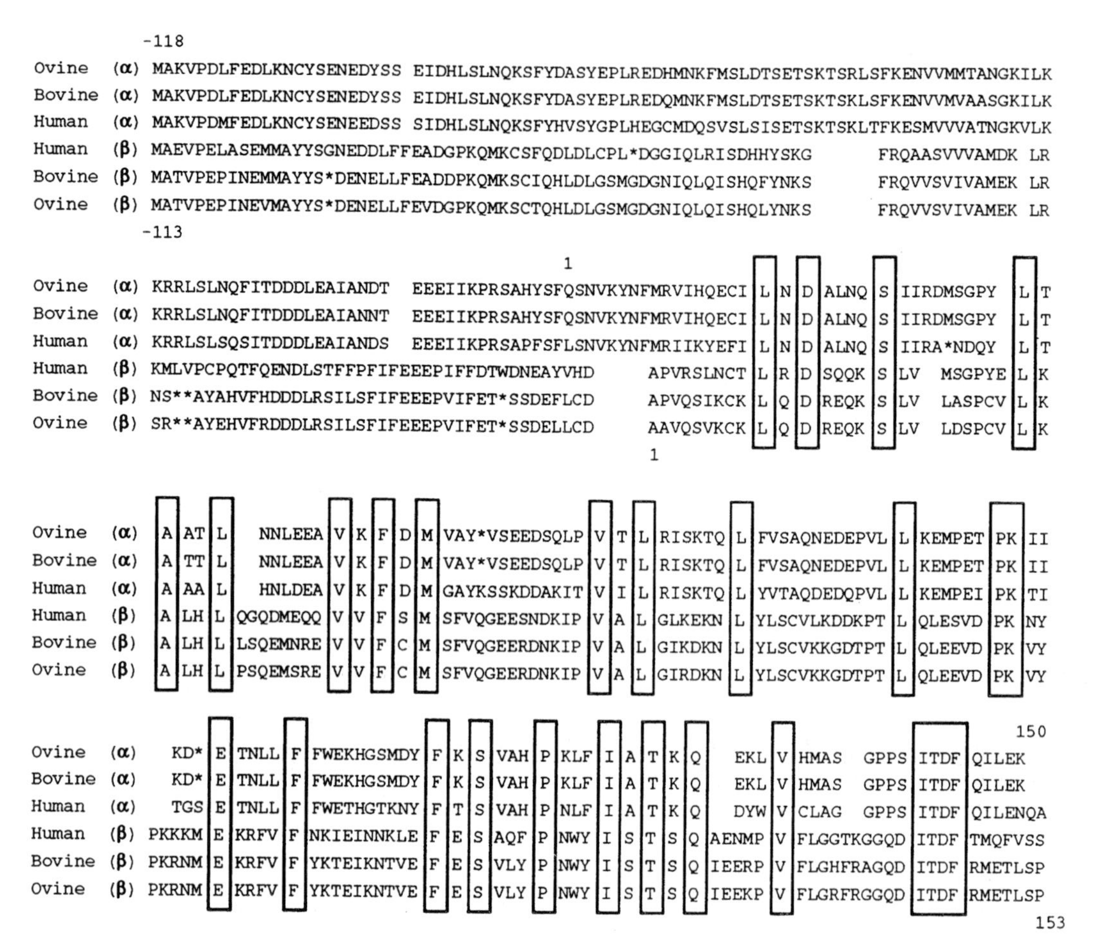

Figure 5-2 The predicted amino acid sequences of ovine, bovine, and human IL-1α and IL-1β. Numbering is with respect to the ovine proteins with the predicted NH_2 termini of the secreted proteins numbered 1. There is approximately 16% identity between all six secreted proteins, the conserved amino acids being boxed. Spaces have been introduced into the protein sequences to allow for the optimal alignment, amino acid deletions are indicated(*). The ruminant IL-1αs share 97% identity with each other and 73 and 62% identity with the human[56] and murine[57] proteins, respectively. The IL-1βs share 95% identity with each other and approximately 61 and 58% with the human and murine counterparts.[56] Bovine IL-1α and IL-1β share 23% identity, while the ovine proteins are 21% identical.

variability between the ruminant and human proteins lie between these helices and presumably represent residues that are relatively nonessential for biological activity since IL-2 is cross-reactive between species.

5. Interleukin-4 (IL-4)

IL-4 is produced by the TH2 subset of memory T cells and stimulates resting B cells to enter the cell cycle and is involved in immunoglobulin class switching to IgE and IgG_1 in mice.[65] IL-4 also affects macrophage function and is therefore involved in inflammatory responses.[66] The bovine and ovine IL-4 cDNAs have been cloned.[18,34] Their predicted amino acid sequences, together with that of the human sequence, are given in Figure 5-5.

D. THE HEMOPOIETINS (COLONY-STIMULATING FACTORS)

The hemopoietins or colony-stimulating factors (CSFs) are a family of glycoprotein cytokines that regulate the development of the blood cell lineages.[69] They include steel factor (also known as stem cell factor or kit ligand), multi-CSF (IL-3), GM-CSF, M-CSF, and G-CSF.[70] These cytokines stimulate the survival, proliferation, and differentiation of hemopoietic stem and progenitor cells, acting in coordi-

```
        -77                                                                                 1
Ovine   MSTKSMIRDVELAEEVLSNKAGGPQGSRSCWCLSLFSFLLVAGATTLFCLLHFGVIGPQREEQSPAGPSFNRPLVQTL
Bovine  ---K--------------SE----------L--------------------------------------I-S------
Human   ---------------A-PK-T-------R-LF-------I----------------------F*-RDL-LIS--A-AV

                10       #20        30        40        50        60       70
Ovine   RSSSQASNNKPVAHVVANISAPGQLRWGDSYANALMANGVELKDNQLVVPTDGLYLIYSQVLFRGHGCPSTPLFLTHT
Bovine  -------S---------D-NS------W------------K-E-------AE-------------Q---*P-PV----
Human   ----RTPSD---------PQAE---Q-LNRR----L------R-------SE-----------K-Q-----HVL----

        80        90        100       110       120       130       140       150
Ovine   ISRIAVSYQTKVNILSAIKSPCHRETLEGAEAKPWYEPIYQGGVFQLEKGDRLSAEINLPEYLDYAESGQVYFGIIAL
Bovine  --------------------------P-W------------------------------D----------------
Human   -------------L--------Q---P-------------L-----------------R-D---F------------
```

Figure 5-3 The predicted amino acid sequences of ovine, bovine and human TNF-α. Numbering is with respect to the ovine protein with the predicted NH_2-terminus numbered 1. Only those amino acids which differ from the consensus ovine sequence are shown. Amino acid deletions are indicated (*). The putative N-glycosylation site in the ovine protein is indicated (#). The ruminant TNFαs share 90% identity with each other and approximately 79% with the human[59] and 72% with the murine[60] proteins, respectively.

```
                                                                                     #
        -20       -10        1         10        20        30        40        50
Ovine   MYKIQPLSCIALTLALVANGAPTSSSTGNTMKEVKSLLLDLQLLLEKVKNPENLKLSRMHTFNFYMPKVNATELKHLK
Bovine  -----L--------------------------------------------------------D--V-----------
Human   --RM-L------S----T-S-------KK-QLQLEH------MI-NGIN-YK-P--T--L--K-----*K-------Q

          60        70        80        90        100       110       120       130
Ovine   CLLEELKLLEEVLDLAPSKNLNTREIKDSMDNIKRIVLELQGSETRFTCEYDDATVKAVEFLNKWITFCQSIYSTMT
Bovine  -------------N--------P-------------------------------N-------------------
Human   --E----P-----N--Q---FHL-P*R-LIS--NV-----K----T-M---A-E-ATI-----R--------I--L-
```

Figure 5-4 The predicted amino acid sequences of ovine, bovine, and human IL-2. Numbering is with respect to the mature ovine IL-2 protein. Only those amino acids which differ from the consensus ovine protein are shown, with amino acid deletions being indicated(*). The putative Asn-linked glycosylation site found in the ruminant proteins is indicated (#). The ruminant proteins share 97% identity with each other and approximately 65 and 50% identity with the human[62] and murine[63] IL-2s, respectively.

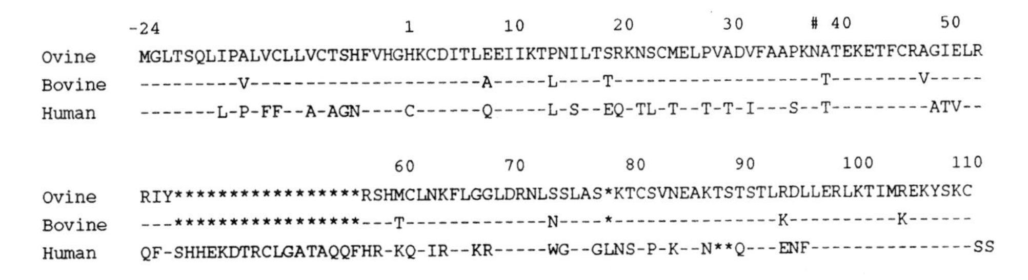

Figure 5-5 The predicted amino acid sequences of ovine, bovine, and human IL-4. Numbering is with respect to the ovine IL-4 sequence with the predicted NH_2 -terminus of the mature protein numbered 1. Only those amino acids which differ from the consensus ovine protein are shown with amino acid deletions indicated (*). The major difference between the ruminant and human proteins is the deletion of a continuous stretch of 17 aa. The murine protein lacks a continuous stretch of 7 aa in this same region. The putative Asn-linked gylcosylation site which is conserved in all IL-4 proteins is indicated (#). The ruminant proteins share 92% identity with each other and 57 and 37% with the human[67] and murine[68] proteins, respectively.

nately produced networks (e.g., IL-3 and GM-CSF) and/or sequentially. The more lineage-restricted cytokines (e.g., GM-CSF and G-CSF, respectively), can also activate mature macrophages/neutrophils and neutrophils in the tissues, respectively, and therefore contribute to immune and inflammatory responses. CD4+ T cells are a major source of IL-3 and GM-CSF, while stromal cells, fibroblasts, and endothelial cells produce GM-CSF and M-CSF upon activation.[71] Three ruminant CSFs have been cloned, namely ovine IL-3, bovine and ovine GM-CSF, and ovine G-CSF.

1. Multipotential Colony-Stimulating Factor (Multi-CSF, Interleukin-3)

Despite having broadly similar biological activities in different species, the primary sequence of IL-3 is probably the least well conserved with, for example, human IL-3 sharing only 29% identity with the murine protein.[72] The cDNA for ovine IL-3 has been cloned[32] and its predicted amino acid sequence compared with that of mouse and man is shown in Figure 5-6. The alignment shows that there are only 17 amino acids in the mature protein that are conserved in all three species (Figure 5-6). Three of these, Pro[32], Lys[107], and Leu[108] (numbering with respect to the mature ovine protein), have been identified as being putatively important in the binding of IL-3 to its receptor.[74] Deletion analysis of the murine and human proteins has shown the importance of a conserved disulfide bridge to the biological activity of the protein. Deletion mutants lacking the bridge had between 7- and 400-fold lower activities than those retaining the disulfide bridge.[75] The ovine protein contains no Cys residues and therefore has no disulfide bridges to stabilize its tertiary structure. Nevertheless, the secondary structure predicted for human IL-3 is also predicted for the ovine protein.

2. Granulocyte-Macrophage Colony-Stimulating Factor (GM-CSF)

The cDNAs for both bovine[21,22] and ovine[35,36] GM-CSF have been cloned. Their predicted amino acid sequences together with that of man, are shown in Figure 5-7. Hybrid analysis with the human sequence and murine proteins suggests that the NH_2-terminalα helix stretching from residue 15-27 is essential for high-affinity binding of GM-CSF to its receptor.[78] This region of GM-CSF shows a number of differences between the human and ruminant proteins and may account for the lack of biological cross-reactivity between these species. On the other hand, this region is perfectly conserved between the ovine and bovine proteins and may account for the fact that the recombinant ovine protein appears to be biologically active on a bovine macrophage cell line (G. Entrican, personal communication).

3. Granulocyte-Colony-Stimulating Factor (G-CSF) and Interleukin-7 (IL-7)

The ovine G-CSF has recently been cloned (P. O'Brien, CSIRO, personal communication), as has the bovine IL-7 cDNA[20] (EMBL accession no. X64540).

E. CYTOKINE GENES AND CONTROL ELEMENTS

Apart from the type I IFN genes, genomic DNA clones for ovine IL-3,[33] GM-CSF (C. McInnes, unpublished results) and IFN-γ (C. McInnes, unpublished results) have been cloned. In general, they are

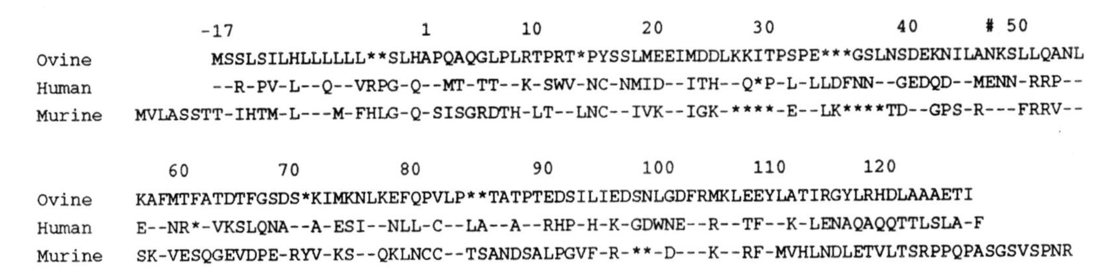
```
             -17              1        10         20         30          40     # 50
Ovine          MSSLSILHLLLLLL**SLHAPQAQGLPLRTPRT*PYSSLMEEIMDDLKKITPSPE***GSLNSDEKNILANKSLLQANL
Human          --R-PV-L--Q--VRPG-Q--MT-TT--K-SWV-NC-NMID--ITH--Q*P-L-LLDFNN--GEDQD--MENN-RRP--
Murine     MVLASSTT-IHTM-L---M-FHLG-Q-SISGRDTH-LT--LNC--IVK--IGK-****-E--LK****TD--GPS-R---FRRV--

          60        70        80          90        100        110        120
Ovine      KAFMTFATDTFGSDS*KIMKNLKEFQPVLP**TATPTEDSILIEDSNLGDFRMKLEEYLATIRGYLRHDLAAAETI
Human      E--NR*-VKSLQNA--A-ESI--NLL-C--LA--A--RHP-H-K-GDWNE--R--TF--K-LENAQAQQTTLSLA-F
Murine     SK-VESQGEVDPE-RYV-KS--QKLNCC--TSANDSALPGVF-R-**-D---K--RF-MVHLNDLETVLTSRPPQPASGSVSPNR
```

Figure 5-6 The predicted amino acid sequences of ovine, human, and murine IL-3. Numbering is with respect to the ovine protein with the predicted NH_2-terminus of the mature proteins numbered 1. The ovine sequence is given as the consensus along with those amino acids which differ in the other two species. The putative Asn-linked glycoslyation site of the ovine proteins is indicated (#). Overall, the ovine protein shares 36% identity with the human protein[72] and 30% with the murine IL-3.[73]

```
        -17             1           10          20       #  30           40          50
Ovine   MWLQNLLLLGTVVCSFSAPTRQPSPVTRPWQHVDAIKEALSLLNDSTDTAAVMDETVEVVSEMFDSQEPTCL
Bovine  --------------------P-NTA-------------------H-S--D---ND-*-----K---------
Human   ----S-------A--I---A-S---S-Q--E--N--Q--RR---L-R----E-N-----I-----L------

             60          70          80          90          100         110         120
Ovine   QTRLELYKQGLRGSLTSLTGSLTMMASHYKKHCPPTQETSCETQIITFKSFKENLKDFLFIIPFDCWEPVQK
Bovine  ----K---N--Q------M-------T--E------P----G--F-S--N---D--E------------A--
Human   ----------------K-K-P---------Q-----P----A------E----------LV----------E
```

Figure 5-7 The predicted amino acid sequences of ovine, bovine, and human GM-CSF. Numbering is with respect to the mature ovine protein. Only those amino acids which differ from the consensus ovine sequence are shown with amino acid deletions indicated(*). The potential Asn-linked glycoslyation site common to all three proteins is indicated (#). The ovine protein shares 80% identity with both the bovine and human[76] proteins and 57% identity to the murine[77] GM-CSF, whereas the bovine protein is 71% identical to the human protein and 56% identical to mouse protein.

very similar to their human counterparts in overall size and structure. For example, the ovine IL-3 gene is located approximately 10 kb from the GM-CSF gene and has five exons and four introns spanning approximately 2 to 2.5 kb of DNA. In addition, although the human and ovine IL-3 proteins have diverged considerably since their evolutionary split, the promoter regions controlling their expression are conserved. Although the ovine IL-3 gene does not possess a classical TATA box and has only a small GC-rich enhancer region, there are a number of *cis*-acting elements which are highly conserved between the species. These are: an AP-1 nuclear factor binding site; the decanucleotide CK1 element (5′-GGAAGGTTCCA-3′) which has been found in a number of cytokine genes;[79,80] the CK2 element (5′-TCAGATA-3′) found only in IL-3 and GM-CSF genes;[80] and the octanucleotide (5′-ATGAATAA-3′) which binds the OCT-1 protein and which is also found in the human IL-2 promoter.[81]

The promoter region of the ovine IFN-γ gene is also highly conserved with that of the human IFN-γ gene with at least 85% homology between the two promoters extending for over 350 bases upstream of the transcription initiation site (C. McInnes, unpublished results).

A common feature in most cytokine mRNA is the presence of AU-rich stretches in their 3′-noncoding regions, including multiple copies of an AUUUA (or UAUUU) motif. These repeats are thought to influence RNA stability.[1,2] Multiple copies of the AU-rich motif have been found in most of the ruminant cytokines that have been cloned. Furthermore, expression of recombinant ovine IL-1 and TNF-α in mammalian cells using constructs lacking the AU motifs resulted in a higher yield of protein than when constructs retaining the AU motifs were used.[27,39] The AU-rich region of the bovine IL-2 gene has also been shown to bind a non-histone protein HMG-1, the biological significance of which is not known.[82]

III. CYTOKINE ASSAYS AND *IN VITRO* BIOLOGY

Recombinant ruminant cytokines have been produced in bacterial, yeast, and mammalian expression systems. Details can be obtained from the references listed in Table 5-1.

A. ASSAYS

Table 5-2 lists the commonly used assays for the ruminant cytokines.

B. *IN VITRO* BIOLOGY

A comprehensive study of the *in vitro* biology of individual ruminant cytokines has not been performed as most of the genes have been cloned only recently. Where studied, ruminant cytokines are similar in activity to their human counterparts. For example, ovine and bovine IFN-? have antiviral activity,[12,92] anti-cell proliferative effects, and upregulate MHC class I and II molecules on macrophages.[94,95] IFN-γ is produced by activated T cells but not B cells.[92] The colony-stimulating factors IL-3 and GM-CSF in sheep show overlapping as well as distinct activities in stimulating the development of hemopoietic progenitor cells into the various blood cell lineages. Recombinant ovine IL-3 (roIL-3) stimulates early multipotential and erythroid progenitor cell development as well as eosinophil, macrophage, and mast cell development.[33] Recombinant ovine GM-CSF (roGM-CSF) stimulates mixed neutrophil/macrophage as well as single macrophage, neutrophil, and eosinophil colony development in soft agar clonogenic assays of ovine bone marrow cells.[35] Thus, roIL-3 and roGM-CSF have similar activities to their human counterparts except that ovine IL-3 is a mast cell growth factor, whereas human IL-3 may not be.[96]

Ruminant cytokine mRNAs are rapidly and transiently expressed following activation of appropriate cells. For example, ovine lymph node cells activated with concanavalin A and phorbol ester produced IFN-γ mRNA within 4 h which was undetectable by 72 h after activation.[97] IFN-γ secreted by the cells was detected at 8 h and persisted beyond 96 h following activation. Ovine alveolar macrophages stimulated with LPS were shown to express mRNA for IL-1α, IL-1β, and TNF-α differentially in time between 1 and 24 h following activation.[94] TNF-α also has effects on the ovine pituitary, increasing both growth hormone and IL-6 mRNA expression.[98]

IV. *IN VIVO* BIOLOGY

A. LYMPH STUDIES

A major advantage in studying immune responses in ruminants is the ability to cannulate afferent and efferent lymph around lymph nodes draining various tissues.[99] Important novel information on the kinetics of local *in vivo* immune responses can be obtained as well as fundamental information on lymphocyte recirculation (see Chapter 7). A study of the efferent lymph response of sheep to local challenge with ovalbumin showed that an initial predominantly $CD4^+$ T cell response within the first 3 days was associated with production of IL-2-like activity into the lymph plasma, whereas a subsequent predominantly $CD8^+$ T cell response between 3 and 6 days was associated with an increased number of cells capable of responding to IL-2.[100] In our laboratories, we are currently studying the cutaneous immune and inflammatory response of sheep challenged with orf virus (a DNA parapox virus). The virus infects via abraded skin and replicates in regenerating epidermal cells. The lesion is local and resolves at around 7 to 10 days following challenge.[101,102] RNA analysis was performed on the cells of afferent lymph draining from the skin site of sheep challenged with Orf virus, and in cells of the efferent lymph leaving the prefemoral lymph node of another animal. Within hours of virus challenge, expression of TNF-α, IL-1β, IL-2, IL-3, IFN-α and GM-CSF mRNA could be detected in the cells from afferent lymph, followed by a second response 5 to 6 days later. This latter response is thought to be due to replicating virus in the regenerating epidermis. In efferent cells, only GM-CSF mRNA was detected in the first 24 h following challenge with the major peak of IL-2, IL-3, GM-CSF, and IFN-γ mRNA expression occurring 5 days post-challenge. IFN-γ biological activity was also studied in lymph following virus challenge.[103] Within hours of challenge, IFN-γ could be detected in the plasma of efferent lymph, but not the cells. Only after 2 to 6 days following challenge did the cells from efferent lymph produce IFN-γ activity. Taken together, these observations are consistent with a rapid response to Orf virus challenge by afferent lymph memory T cells and a delayed response of efferent lymphocytes possibly reflecting recruitment of lymphocytes from the blood, and activation within the lymph node before leaving in the efferent lymph. Figure 5-8 shows *in situ* expression of ovine GM-CSF mRNA by the afferent lymph lymphocytes 6 days following Orf virus challenge. Using such techniques, fundamental questions can be asked about the function of cytokines in cell recruitment and activation events within lymph nodes. For example, it has been known for some time that following antigen challenge, there is a drop in cell output in efferent lymph known as "shutdown", which precedes a recruitment phase of lymphocytes from the blood measured as an increase in cell output in efferent lymph.[99] IFN-α has been demonstrated to induce shutdown[104] and IFN-γ must be considered as another possible factor involved in this phenomenon.[103]

The role of adjuvants and a range of nominal protein and polysaccharide antigens in stimulating a response of IFN-γ and antibody in efferent lymph has also been studied.[105] Oil adjuvants, Quil A, and dextran sulfate in the presence and absence of antigen were capable of stimulating lymph IFN-γ responses. Alhydrogel did not stimulate IFN-γ synthesis. Antigen challenge revealed no correlation between the production of IFN-γ and particular immunoglobulin isotypes in this study.

B. OTHER *IN VIVO* STUDIES

The effect of intradermally administered recombinant human or recombinant ruminant cytokines on leukocyte migration into ovine skin has been studied. Recombinant bovine IFN-γ recruited lymphocytes into skin sites, whereas recombinant human IL-1α preferentially recruited neutrophils.[106] Recombinant human TNF-α recruited lymphocytes[107] or both lymphocytes and neutrophils,[106] whereas recombinant

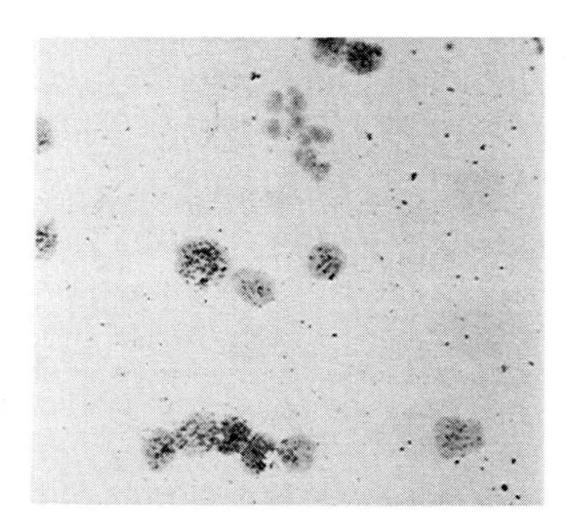

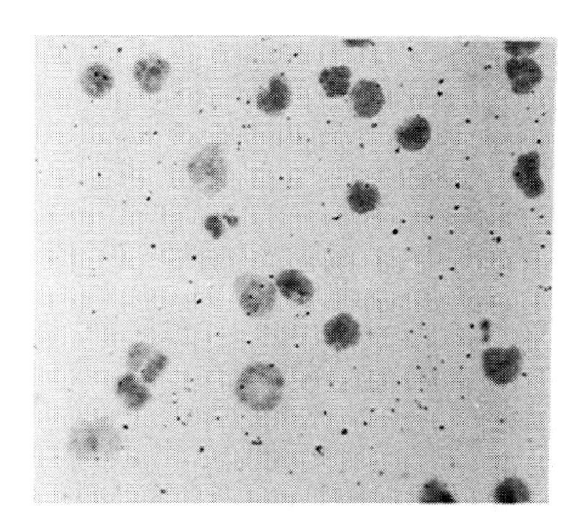

Figure 5-8 *In situ* localization of ov GM-CSF mRNA in afferent lymph cells 6 days after cutaneous challenge of sheep 1215 with Orf virus. (a) antisense GM-CSF RNA. (b) sense control GM-CSF RNA. Sense control and antisense riboprobes representing a portion of the coding region of ovine GM-CSF were labeled with [^{35}S]-dUTP by standard techniques. Lymph cell cytospots were fixed with 4% paraformaldehyde and hybridization conditions were : 2×10^6 c.p.m. probe/ slide, 42°C overnight in 50% formamide, 1 × Denhardts, 10% dextran sulfate, 100 μ*M* dithiothreitol, 300 m*M* NaCl, 20 m*M* Tris-HCl pH 7.5, 5 m*M* EDTA. For autoradiography, the slides were washed and exposed in emulsion for 10 days. Afferent lymph cells were 10% veiled dendritic cells/macrophages and 90% lymphocytes of which 65% were $CD4^+$ T cells, 17% $CD8^+$ T cells, 10% γδT cells, and 8% B cells. The large veiled dendritic cells/ macrophages did not express GM-CSF.

human IL-8 recruited predominantly neutrophils.[106] These results are similar to those obtained for other species, and it has not been demonstrated whether the individual cytokines alone are responsible for these effects or, as is more likely, they are part of a cytokine network.

TNF-α is implicated in the pathogenesis of certain chronic pulmonary infections. For example, infusion of TNF-α into sheep mimics the pathology of endotoxic shock[108] and ovine alveolar macrophages stimulated with *Corynebacterium pseudotuberculosis* or *Pasteurella haemolytica* rapidly produce TNF-?.[86]

V. CYTOKINE RECEPTORS

The activities of the cytokines are tightly regulated *in vivo*. There are several mechanisms for this, including regulation of secretion regulation of receptor expression and regulation by soluble binding proteins and inhibitors. Defining the mechanisms that regulate cytokine receptor expression is important for an overall understanding of cytokine biology. In mouse and man, receptor families have been identified. For example, the receptors for steel factor and M-CSF (c-kit and c-fms proto oncogene products, respectively) are related to the receptor for platelet-derived growth factor (PDGF) in that they are associated with a signal-tranducing tyrosine kinase which is activated when receptors are cross-linked by ligand.[109] A hemopoietin family of receptors which lack tyrosine kinase has also been described for IL-3, IL-5, IL-6, IL-7, and GM-CSF.[110] These receptors exhibit high- and low-affinity forms (first described for the IL-2 receptor[61,111]) in which an association of two separate subunits is required for high-affinity ligand binding. IL-3, GM-CSF, and IL-5 share a common receptor (β) subunit on hemopoietic cells which associates with specific αsubunits for high affinity binding.[112,113]

Bovine[114,115] and ovine[116,117] cDNAs coding for the low-affinity p55 form of the IL-2 receptor have been cloned. The predicted amino acid sequence for the ovine protein shares 94, 72, and 63% identity with the predicted bovine, human, and murine sequences, respectively.[117]

Molecular cloning in man and mouse has shown that there are two types of TNF receptor, namely TNF-R1 and TNF-R2.[118] The extracellular portions of both receptors are found naturally as soluble TNF binding proteins. Both species-specific and species-independent TNF effects on mouse cell lines have been described. Ovine TNF-α is only poorly active on murine cell lines for cytotoxicity assays, whereas it is highly active on a porcine fibroblast cell line[90] (Table 5-2). Bovine and ovine TNF-α receptors have been measured on macrophages, lymphocytes, and neutrophils using labeled ligand.[119,120] Density of receptor expression did not appear to correlate with biological response as neutrophils had the lowest receptor density but were the most responsive cell type, for example, to TNF-mediated chemotaxis.[119] IFN-γ down regulated TNF-α receptors on all cell types and IL-2 upregulated lymphocyte (but not macrophage or neutrophil) TNF-α receptor expression.

VI. CONCLUSIONS

The structure and biological function of the ruminant cytokine genes and proteins obtained so far are essentially similar to their human counterparts even though some do not share extensive protein identity with their human homologues. One example of this is IL-3. The fact that the regulatory regions (and introns) of the IL-3 gene are relatively more conserved between species than the protein coding regions may indicate that evolutionary pressure has served to ensure the expression of a cytokine the activity of which is essentially redundant. On the other hand, it may indicate that the bulk of the primary protein structure itself can be flexible, with only a few key aa being required to maintain biological activity. The availability of ruminant cytokine and cytokine receptor genes and proteins to compare to those of other species will facilitate such analyses of structure in relation to function.

The ability to measure the production and activity of cytokines will greatly enhance our understanding of the ruminant host response to a large number of biological insults. New methods for diagnosing diseases are already being applied. The specific cell-mediated immune response to *Mycobacterium bovis* in cattle can be detected by measurement of blood cell production of IFN-γ.[93] The potential of cytokines as adjuvants to direct potent specific immune responses to antigens can easily be evaluated in ruminant animals and their usefulness as therapeutic agents can be similarly determined (Chapter 14).

The molecular and cellular analysis of cytokine production and utilization in afferent and efferent lymph and within single lymph nodes will be a productive area of research in which ruminant immunology will not simply be following in the wake of rodent and human immunology. Ruminant animals are excellent species in which to study the factors affecting the development and activation of the γδT cell subsets, as large numbers of these enigmatic cells can be obtained from the blood of young animals.[121] Fundamental and novel information on the development of B cells from ovine ileal Peyer's patches can also be expected. Many more ruminant cytokine and receptor genes will be cloned over the next few years, with a consequent great leap forward in our understanding of ruminant immune and inflammatory processes.

ACKNOWLEDGMENTS

The authors are supported by funds from the Scottish Office Department of Agriculture and Fisheries and the Australian Wool Corporation. We thank Isobel Brown for typing the manuscript and the contribution of unpublished information by David Sargan of Edinburgh University and Nyree Myatt of the Moredun Research Institute.

REFERENCES

1. **Shaw, G. and Kamen, R. A.,** A conserved AU sequence from the 3′ untranslated region of GM-CSF mRNA mediates selective mRNA degradation, *Cell*, 46, 659, 1986.
2. **Miyajima, A., Miyatake, S., Schreurs, J., De Vries, J., Arai, N., Yokota, T., and Arai, K-I.,** Coordinate regulation of immune and inflammatory responses by T-cell-derived lymphokines, *FASEB J.*, 2, 2462, 1988.
3. **Mosmann, T. R., Cherwinski, H., Bond, M. W., Giedlin, M. A., and Coffman, R. L.,** Two types of murine helper T cell clone. 1. Definition according to profiles of lymphokine activities and secreted proteins, *J. Immunol.*, 136, 2348, 1986.
4. **Mosmann, T. R. and Coffman, R. L.,** Heterogeneity of cytokine secretion patterns and functions of helper T-cells, *Adv. Immunol.*, 46, 111, 1989.

5. **Fiorentino, D. F., Zlotnik, A., Vieira, P., Mosmann, T. R., Howard, M., Moore, K. W., and O'Garra, A.,** IL-10 acts on the antigen presenting cell to inhibit cytokine production by TH_1 cells, *J. Immunol.*, 146, 2444, 1991.
6. **Snapper, C. M. and Paul, W. E.,** Interferon-gamma and B cell stimulatory factor-1 reciprocally regulate immunoglobulin isotype production, *Science*, 236, 944, 1987.
7. **Velan, B., Cohen, S., Grosfeld, H., Lietner, M., and Shaferman, A.,** Bovine interferon alpha genes. Structure and expression, *J. Biol. Chem*, 260, 5498, 1985.
8. **Capon, D. J., Shepard, H. M., and Goeddel, D.V.,** Two distinct families of human and bovine interferon-alpha genes are coordinately expressed and encode functional polypeptides, *Mol. Cell Biol.*, 5, 768, 1985.
9. **Velan, B., Cohen, S., Grosfeld, H., and Shafferman, A.,** Isolation of bovine IFN-α genes and their expression in bacteria, *Meth. Enzymol.*, 119, 464,1986.
10. **Hansen, T. R., Leaman, D. W., Cross, J. C. Mathialagan, N., Bixby, J. A., and Roberts, R. M.,** The genes for the trophoblast interferons and the related interferon-α possess distinct 5′-promoter and 3′-flanking sequences, *J. Biol. Chem.*, 266, 2060, 1991.
11. **Leung, D. N., Capon, D. J., and Goeddel, D. V.,** The structure and bacterial expression of three distinct bovine interferon-β genes, *Biotechnology*, 2, 458, 1984.
12. **Cerretti, D. P., McKereghan, K., Larsen, A., Cosman, D., Gillis, S., and Baker, P. E.,** Cloning sequence and expression of bovine interferon-γ, *J. Immunol.*, 136, 4561, 1986.
13. **Leong, S. R., Flaggs, G. M., Lawman, M., and Gray, P. W.,** The nucleotide sequence for the cDNA of bovine interleukin-1 alpha, *Nucl. Acids. Res.*, 16, 9053, 1988.
14. **Maliszewski, C. R., Baker, P. E., Schoenborn, M. E., Davis, B. S., Cosman, D., Gillis, S., and Cerretti, D. P.,** Cloning sequence and expression of bovine interleukin-1α and interleukin-1β complementary DNA's, *Mol. Immunol.*, 25, 429, 1988.
15. **Leong, S. R., Flaggs, G. M., Lawman, M., and Gray, P. W.,** The nucleotide sequence for the cDNA of bovine interleukin-1 beta, *Nucl. Acids Res.*, 16, 9054, 1988.
16. **Cerretti, D. P., McKereghan, K., Larsen, A., Cantrell, M. A., Anderson, D., Gillis, S., Cosman, D., and Baker, P. E.,** Cloning sequence and expression of bovine IL-2, *Proc. Natl. Acad. Sci. U.S.A.*, 83, 3223, 1986.
17. **Reeves, R., Spies, A. G., Nissen, M. S., Buck, C. D., Weinberg, A. D., Barr, P. J., Magnuson, N. S., and Magnuson, J. A.,** Molecular cloning of a functional bovine interleukin 2 cDNA, *Proc. Natl. Acad. Sci. U.S.A.*. 83, 3228, 1986.
18. **Heusseler, V., Eichhorn, M., and Dobbelaere, D.,** Cloning of a full length cDNA encoding bovine interleukin-4 by the polymerase chain reaction, *Gene,* 114, 273, 1992.
19. **Droogmans, L., Cludtz, I., Cleuter, Y., Kettmann, R., and Burny, A.,** Nucleotide sequence of bovine interleukin-6 cDNA, *J. DNA Seq. Mapping,* 2, 411, 1992.
20. **Cludts, I., Kettman, R., Cleuter, Y., Burny, A., and Droogmans, L.,** Nucleotide sequence of bovine interleukin-7 cDNA, unpublished, EMBL Accession number x64540.
21. **Maliszewski, C. R., Schoenborn, M. A., Cerretti, D. P., Wignall, J., Picha, K.S., Cosman, D., Tushinski, R. J., Gillis, S., and Baker, P. E.,** Bovine GM-CSF molecular cloning and biological activity of the recombinant protein, *Mol. Immunol.*, 25, 843, 1988.
22. **Leong, S. R., Flaggs, G. M., Lawman, M. J. P., and Gray, P. W.,** Cloning and expression of the cDNA for bovine granulocyte-macrophage colony-stimulating factor, *Vet. Immunol. Immunopathol.*, 21, 261, 1989.
23. **Goeddel, D. B., Aggarwal, B. B., Gray, P. W., Leung, D. W., Nedwin, G. E., Palladino, M. A., Patton, J. S., Pennica, D., Shepard, H. M., Sugarman, B. J., and Wong, G. H. W.,** Tumor necrosis factors: gene structure and biological activities, *Cold Spring Harbor Symp. Quant. Biol.*, LI, 597, 1986.
24. **Whaley, A. E., Carroll, R. S., and Imakawa, K.,** Cloning and analysis of a gene encoding ovine interferon α-II, *Gene,* 106, 281, 1991.
25. **McInnes, C. J., Logan, M., Redmond, J., Entrican, G., and Baird, G. D.,** The molecular cloning of the ovine gamma-interferon cDNA using the polymerase chain reaction, *Nucl. Acids Res.*, 18, 4012, 1990.
26. **Radford, A. J., Hodgson, A. L. M., Rothel, J. S., and Wood, P. R.,** Cloning and sequencing of the ovine gamma-interferon gene, *Aust. Vet. J.*, 68, 82, 1991.

27. **Andrews, A. E., Barcham, G. J., Brandon, M. R., and Nash, A. D.,** Molecular cloning and characterisation of ovine IL-1α and IL-1β, *Immunology.*, 74, 453, 1991.
28. **Fiskerstrand, C. and Sargan, D.,** Nucleotide sequence of ovine interleukin-1 beta., *Nucl. Acids Res.*, 18, 7165, 1990.
29. **Seow, H.-F., Rothel, J. S., David, M.-J., and Wood, P. R.,** Nucleotide sequence of ovine macrophage interleukin 1-beta cDNA, *DNA Sequence,* 1, 423, 1991.
30. **Goodall, J. C., Emery, D. C., Perry, A. C. F., English, L. S., and Hall, L.,** Complementary DNA cloning of ovine interleukin 2 by PCR, *Nucl. Acids Res.*, 18, 5883, 1990.
31. **Seow, H. F., Rothel, J. S., Radford, A. J., and Wood, P.R.,** The molecular cloning of the ovine interleukin 2 gene by the polymerase chain reaction, *Nucl. Acids Res.*, 18, 7175, 1990.
32. **McInnes, C. J. and Logan, M.,** Cloning of the ovine interleukin-3 cDNA, EMBL Accession number Z18291.
33. **McInnes, C., Haig, D. M., and Logan, M.,** The cloning and expression of the gene for ovine interleukin-3 (multi-CSF) and a comparison of the *in vitro* haemopoietic activity of ovine IL-3 with GM-CSF and human M-CSF, *Exp. Hematol.*, 21, 1528, 1993.
34. **Seow, H.-F., Rothel, J. S., and Wood, P. R.,** Cloning of interleukin 4 cDNA by PCR, *Gene,* in press.
35. **McInnes, C. J., and Haig, D. M.,** Cloning and expression of a cDNA encoding ovine granulocyte-macrophage colony-stimulating factor, *Gene,* 105, 275, 1991.
36. **O'Brien, P. M., Rothel, J. S., Seow, H. F., and Wood, P. R.,** Cloning and sequencing of the complementary DNA for ovine granulocyte-macrophage colony-stimulating factor, *Immunol. Cell Biol.*, 69, 51, 1991.
37. **Young, A. J., Hay, J. B., and Chan, J. Y. C.,** Primary structure of ovine tumor necrosis factor alpha complementary DNA, *Nucl. Acids Res.*, 18, 6723, 1990.
38. **Green, I. R. and Sargan, D. R.,** Sequence of the cDNA encoding tumor necrosis factor-α: problems with cloning by inverse PCR, *Gene,* 109, 203, 1991.
39. **Nash, A. D., Barcham, G. J., Brandon, M. R., and Andrews, A. E.,** Molecular cloning, expression and characterisation of ovine TNF-α, *Immunol. Cell Biol.*, 69, 273, 1991.
40. **Barlow, D. P., Bucan, M., Lehrach, H., Hogan, B. L. M., and Gough, N. M.,** Close genetic and physical linkage between the murine haemopoietic growth factor genes GM-CSF and multi-CSF (IL-3), *EMBO J.*, 6, 617, 1987.
41. **Yang, Y.-C., Kovacic, S., Kriz, R., Wolf, S., Clark, S. C., Wellems, T. E., et al.,** The human genes for GM-CSF and IL3 are closely linked in tandem on chromosome 5, *Blood,* 71, 958, 1988.
42. **Romeo, G., Fiorucci, G., and Rossi, B.,** Interferons in cell growth and development, *Trends Genet.*, 5, 19, 1989.
43. **Roberts, R. M.,** A role for interferons in early pregnancy, *Bioessasys,* 13, 121 1991.
44. **Stewart, H. J.,** Trophoblastic factors and the maternal recognition of pregnancy in sheep and cattle, *J. Dev. Physiol.*, 14, 115, 1990.
45. **Shepard, H. M., Eaton, D., Gray, P., Maylor, S., Hollingshead, P. and Goeddel, D.,** in: *The Biology of the Interferon System* 1984, Kirchner, H. and Schellekans, H., Eds., Elsevier, Amsterdam, 1985, 147.
46. **Ohno, S. and Taniguchi, T.,** Structure of a chromosomal gene for human interferon β, *Proc. Natl. Acad. Sci. U.S.A.*, 78, 5305, 1981.
47. **Higashi, Y., Sokawa, Y., Watanabe, Y., Kauvada, Y., Ohno, S., Takaska, C., and Taniguchi, T.,** Structure and expression of a cloned cDNA for mouse interferon-β, *J. Biol. Chem.*, 258, 9522, 1983.
48. **Shepard, H. M., Leong, D. N., Stebbing, N., and Goeddel, D. V.,** A single amino acid change in IFN-β1 abolishes its antiviral activity, *Nature,* 295, 563, 1981.
49. **Gray, P. W., and Goeddel, D. V.,** Cloning and expression of murine immune interferon cDNA, *Proc. Natl. Acad. Sci. U.S.A.*, 80, 5842, 1983.
50. **Gray, P. W., and Goeddel, D. V.,** Structure of the human immune interferon gene, *Nature,* 298, 859, 1982.
51. **Ealick, S. E., Cook, W. J., Vijay-Kumar, S., Carson, M., Nagabushan, T. L., Trotta, P. P., and Bugg, C. E.,** Three-dimensional structure of recombinant human interferon-γ, *Science*, 252, 698, 1991.
52. **Dinarello, C. A.,** Interleukin-1 and its biologically related cytokines, *Adv. Immunol.*, 44, 153, 1989.
53. **Hirano, T., Akira, S., Taga, T., and Kishimoto, T.,** Biological and clinical aspects of interleukin 6, *Immunol. Today,* 11, 443, 1990.

54. **Beutler, B. and Cerami, A.,** The biology of cachetin/TNF — a primary mediator of the host response, *Annu. Rev. Immunol.*, 7, 625. 1989.
55. **Dower, S. K., Kronheim, S. R., Happ, J. P., Cantrell, M., Delley, M., Gillis, G., Henney, C. S., and Urdal, D. L.,** The cell surface receptors for interleukin-1α and interleukin-1β are identical, *Nature,* 324, 266, 1986.
56. **March, C. J., Mosley, B., Larsen, A., Cerretti, D. P., Braedt, G., Price, V., Gillis, S., Henney, C. S., Krenheim, S. R., Grabstein, K., Conlon, P. J., Happ, T. P., and Cosman, D.,** Cloning sequence and expression of two distinct human interleukin-1 complementary DNAs, *Nature,* 315, 641, 1985.
57. **Lomedico, P. T., Gubler, U., Hellman, C. P., Dukovich, M., Giri, J. G., Pan, Y.-C. E., Coller, K., Semionaw, R., Chua, A. O., and Mizel, S. B.,** Cloning and expression of murine interleukin-1 cDNA in *Escherichia coli, Nature,* 312, 458, 1984.
58. **Mosley, B., Dewer, S. K., Gillis, S., and Cosman, D.,** Determination of the minimum polypeptide lengths of the functionally active sites of human interleukins 1α and 1β, *Proc. Natl. Acad. Sci. U.S.A.*, 84, 4572, 1987.
59. **Pennica, D., Nedwin, G. E., Hayflick, J. S., Seeburg, P. H., Derynck, R., Palladino, M. A., Kohr, W. J., Aggarwal, B. B., and Goeddel, D. V.,** Human tumor necrosis factor: precursor structure, expression and homology to lymphotoxin, *Nature,* 312, 724, 1984.
60. **Pennica, D., Hayflick, J. S., Bringman, T. S., Palladino, M. A., and Goeddell, D. V.,** Cloning and expression in *E. coli* of the cDNA for murine tumour necrosis factor, *Proc. Natl. Acad. Sci. U.S.A.,* 82, 6060, 1985.
61. **Robb, R. J.,** Interleukin-2: the molecule and its function, *Immunol. Today,* 5, 203, 1984.
62. **Taniguchi, T., Matsui, H., Fujita, T., Takaoka, C., Kashima, N., Yoshimoto, R., and Hamuro, J.,** Structure and expression of a cloned cDNA for human interleukin-2, *Nature,* 302, 305, 1983.
63. **Kashima, N., Nishi-Takaoka, C., Fujita, T., Taki, S., Yamada, G., Hamuro, J., and Taniguchi, T.,** Unique structure of murine interleukin-2 as deduced from cloned cDNAs, *Nature,* 313, 402, 1985.
64. **Brandhuber, B. J., Boone, T., Kenney, W. C., and McKay, D. B.,** Three-dimensional structure of interleukin-2, *Science*, 238, 1797, 1987.
65. **Street, N. E. and Mosmann, T. R.,** IL-4 and IL-5: the role of two multifunctional cytokines and their place in the network of cytokine interactions, *Biotherapy,* 2, 347, 1990.
66. **Crawford, R., Finbloom, M., O'Hara, J., Paul, W. E., and Meltzer, M.,** B-cell stimulatory factor 1 (interleukin-4) activates macrophages for increased tumoricidal activity and expression of Ia antigens, *J. Immunol.*, 139, 135, 1988.
67. **Yokota, T., Otsuka, T., Mosmann, T., Banchereau, J., De France, T., Balanchard, D., De Vries, J. E., Lee, F., and Arai, K.,** Isolation and characterization of a human interleukin cDNA clone, homologous to mouse B-cell stimulatory factor 1, that expresses B-cell and T-cell stimulating activities, *Proc. Natl. Acad. Sci. U.S.A.,* 83, 5894, 1986.
68. **Lee, F., Yokota, T., Otsuka, T., Meyerson, P., Villaret, D., Coffman, R., Mosmann, T., Rennick, D., Roehmn, N., Smith, C., Zlotnik, A., and Arai, K.,** Isolation and characterization of a mouse interleukin cDNA clone that expresses B-cell stimulatory factor 1 activities and T-cell and mast cell stimulating activities, *Proc. Natl. Acad. Sci. U.S.A.,* 83, 2061, 1986.
69. **Metcalf, D.,** *The Haemopoietic Colony Stimulating Factors,* Elsevier, Amsterdam, 1984.
70. **Haig, D. M.,** Haemopoietic stem cells and the development of the blood cell repertoire, *J. Comp. Pathol.*, 106, 121, 1992.
71. **Bagby, G., Shaw, G., and Segal, G. M.,** Human vascular endothelial cells, granulopoiesis, and the inflammatory response, *J. Invest. Dermatol.*, 93, 48, 1989.
72. **Yang, Y.-C. and Clark, S. C.,** Molecular cloning of a primate cDNA and the human gene for interleukin 3, in *Lymphokines,* Vol. 15, Schrader, J. W., Ed., Academic Press, London, 1988, 375.
73. **Clark-Lewis, I. and Schrader, J. W.,** Molecular structure and biological activities of P cell-stimulating factor (Interleukin 3), in *Lymphokines,* Vol. 15, Schrader, J. W., Ed. Academic Press, London, 1988, 1.
74. **Lokker, N. A., Strittmatter, U., Steiner, C., Fagg, B., Graff, P., Kocher, H. P., and Zenk, G.,** Mapping the epitopes of neutralizing anti-human IL-3 monoclonal antibodies, *J. Immunol.*, 146, 893, 1991.

75. **Lokker, N. A., Zenke, G., Strittmatter, U., Fagg, B., and Movva, N. R.,** Structure activity relationship study of human interleukin-3: role of the C-terminal region for biological activity, *EMBO J.*, 10, 2125, 1991.
76. **Cantrell, M. A., Anderson, D., Cerretti, D. P., Price, V., McKerrighan, K., Tushinski, R. J., Moshizuki, D. Y., Larsen, A., Grabstein, K., Gillis, S., and Cosman, D.,** Cloning sequence and expression of a human granulocyte/macrophage colony-stimulating factor, *Proc. Natl. Acad. Sci. U.S.A.,* 83, 6250, 1985.
77. **Gough, N. M., Gough, J., Metcalf, D., Kelso, A., Grail, D., Nicola, N. A., Burgess, A., W. and Dunn, A. R.,** Molecular cloning of a cDNA encoding a murine haemopoietic growth regulator, granulocyte-macrophage colony stimulating factor, *Nature,* 309, 763, 1984.
78. **Shanafelt, A. B., Miyajima, A., Kitamura, J., and Kastelein, R. A.,** The amino-terminal helix of GM-CSF and IL-5 governs high affinity binding to their receptors, *EMBO J.*, 10, 4105, 1991.
79. **Mathey-Prevot, B., Andrews, N. C., Murphy, H. J. S., Kreissman, S. G., and Nathan, D. G.,** Positive and negative elements regulate human interleukin 3 expression, *Proc. Natl. Acad. Sci. U.S.A.,* 87, 5046, 1990.
80. **Shannon, M. F., Gamble, J. R., and Vadas, M. A.,** Nuclear proteins interacting with the promoter region of the human granulocyte/macrophage colony-stimulating factor gene, *Proc. Natl. Acad. Sci. U.S.A.,* 85, 674, 1988.
81. **Emmel, E. A., Verweij, C. L., Durand, B. B., Higgins, K. M., Lacy, E., and Crabtree, G. R.,** Cyclosporin A specifically inhibits function of nuclear proteins involved in T-cell activation, *Science,* 1617, 1989.
82. **Reeves, R., Eltan, T. S., Nissen, M. S., Lehn, D., and Johnson, K. K.,** Posttranscriptional gene regulation and specific binding of the nonhistone protein HMG-1 by the 3′ untranslated region of bovine interleukin-2 cDNA, *Proc. Natl. Acad. Sci. USA,* 84, 6531, 1987.
83. **Fiskerstrand, C., Green, I. R., Roy, D. J., and Sargan, D. R.,** Cloning and expression of ovine interleukin-1α and interleukin-1β, *Cytokine,* 4, 418, 1992.
84. **Saklatvala, J.,** Tumour necrosis factor-α stimulates resorption and inhibitis synthesis of proteoglycan in cartilage, *Nature,* 322, 547, 1989.
85. **Gearing, A. J., Bird, C. R., Bristow, A., Poole, S., and Thorp, E.,** A simple and sensitive bioassay for interleukin-1 which is unresponsive to 10^3 U/ml of interleukin-2, *J. Immunol. Meth.*, 99, 7, 1987.
86. **Ellis, J. A., Lairmore, M. D., O'Toole, D. T., and Campos, M.,** Differential induction of tumour necrosis factor alpha in ovine pulmonary alveolar macrophages following infection with *Corynebacterium pseudotuberculosis, Pasteurella haemolytica,* or *Lentiviruses, Infect. Immun.,* 59, 3254, 1991.
87. **Ellis, J. A., Godson, D., Campos, M., Babiuk, L. A., and Sileghem, M.,** Capture immunoassay for ruminant tumour necrosis factor-α: comparison with bioassay, *Vet. Immunol. Immunopathol.*, 35, 289, 1993.
88. **Adams, J. L., Semrad, S. D., and Czuprynski, C. J.,** Administration of bacterial lipopolysaccharide elicits circulating tumour necrosis factor-alpha in neontal calves, *J. Clin. Microbiol.*, 28, 998, 1990.
89. **Kenison, D. C., Elsasser, T. H., and Fayer, R.,** Radioimmunoassay for bovine tumour necrosis factor: concentrations and circulating molecular forms in bovine plasma, *J. Immunoassay,* 11, 177, 1990.
90. **Green, I. R., Fiskerstrand, C., Bertoni, G., Roy, D. J., Peterhans, E., and Sargan, D. R.,** Expression and characterisation of bioactive recombinant ovine tumour necrosis factor: some species specificity in cytotoxic responses to TNFα, *Cytokine,* in press.
91. **Haig, D. M., Brown, D., and MacKellar, A.,** Ovine haemopoiesis: the development of bone marrow-derived colony-forming cells *in vitro* in the presence of factors derived from lymphoid cells and helper T-cells, *Vet. Immunol. Immunopathol.,* 25, 125, 1990.
92. **Entrican, G., Haig, D. M., and Norval, M.,** Identification of ovine interferons: differential activities derived from fibroblast and lymphoid cells, *Vet. Immunol. Immunopathol.*, 21, 187, 1989.
93. **Rothel, J. S., Jones, S. L., Corner, L. A., Cox, J. C., and Wood, P. R.,** A sandwich immunoassay for bovine interferon-γ and its use for the detection of tuberculosis in cattle, *Res. Vet. Sci.*, 49, 46, 1990.
94. **Nash, A. D., Barcham, G. J., Andrew, A. E., and Brandon, M. R.,** Characterisation of ovine alveolar macrophages: regulation of surface antigen expression and cytokine production, *Vet. Immunol. Immunopathol.*, 31, 77, 1992.
95. **Bielefeldt-Ohmann, H., Lawman, M. J. P., and Babiuk, L. A.,** Bovine interferon: its biology and application in veterinary medicine, *Antiviral Res.,* 7, 187, 1987

96. **Valent, P., Besemer, J., Sillaber, C., Butterfield, J., Eher, R., Majdic, O., Kishi, K., Keleptko, W., Eckersberger, F., Lechner, K., and Bettelheim, P.**, Failure to detect IL-3 binding sites on human mast cells, *J. Immunol.*, 145, 3432, 1990.
97. **Entrican, G., McInnes, C. J., Rothel, J. S., and Haig, D. M.**, Kinetics of ovine interferon-gamma production: detection of mRNA and characterisation of biological activity, *Vet. Immunol. Immunopathol.*, 33, 171, 1992.
98. **Nash, A. D., Brandon, M. R., and Bello, P. A.**, Effects of tumour necrosis factor-α on growth hormone and interleukin 6 mRNA in ovine pituitary cells, *Mol. Cell Endocrinol.*, 84, R31, 1992.
99. **Hay, J. B. and Cahill, R. N.**, Lymphocyte migration patterns in sheep, in *Animal Models of Immunological Processes,* Hay J., Ed., Academic Press, London, 1982, 97.
100. **Bujdoso, R., Young, P., Hopkins, J., Allen, D., and McConnell, I.**, Non random migration of CD4 and CD8 T cells: changes in the CD4:CD8 ratio and interleukin responsiveness of efferent lymph cells following *in vivo* antigen challenge, *Eur. J. Immunol.*, 19, 1779, 1989.
101. **Yirrell, D. L., Reid, H. W., Norval, M., and Howie, S. E. M.**, Immune response of lambs to experimental infection with orf virus, *Vet. Immunol. Immunopathol.*, 22, 321, 1989.
102. **Jenkinson, D. M., Hutchison, G., Onwuka, S. K., and Reid, H. W.**, Changes in the MHC Class II+ dendritic cell population of ovine skin in response to orf virus infection, *Vet. Dermatol.*, 2, 1, 1991.
103. **Haig, D. M., Entrican, G., Yirrell, D., Deane, D., Miller, H. R. P., Norval, M., and Reid, H. W.**, Differential appearance of interferon-γ and colony stimulating activity in afferent versus efferent lymph following orf virus infection of sheep, *Vet. Dermatol.*, in press.
104. **Hein, W. R. and Supersaxo, A.**, Effect of interferon-alpha-2a on the output of recirculating lymph lymphocytes from single nodes, *Immunology.*, 64, 469, 1988.
105. **Emery, D. L., Rothel, J. S., and Wood, P. R.**, Influence of antigens and adjuvants on the production of gamma interferon and antibody by ovine lymphocytes, *Immunol. Cell. Biol.*, 68, 127, 1990.
106. **Colditz, I. G. and Watson, D. L.**, The effect of cytokines and chemotactic agonists on the migration of T-lymphocytes into skin, *Immunology*, 76, 272, 1992.
107. **Kalaaji, A. N., McCullough, K., and Hay, J. B.**, The enhancement of lymphocyte localisation in skin sites of sheep by tumour necrosis factor alpha, *Immunol. Lett.*, 23, 143, 1989.
108. **Johnson, J., Myerick, B., Jesmok, G., and Brigham, K. L.**, Human recombinant tumour necrosis factor alpha infusion mimics endotoxaemia in awake sheep, *J. Appl. Physiol.*, 66, 1448, 1989.
109. **Ullrich, A. and Schlesinger, J.**, Signal transduction by receptors with tyrosine kinase activity, *Cell,* 61, 203, 1990.
110. **Bazan, J. F.**, Structural design and molecular evolution of a cytokine receptor superfamily, *Proc. Natl. Acad. Sci. U.S.A.,* 87, 6934, 1990.
111. **Malkovsky, M. and Sondel, P. M.**, Interleukin-2 and its receptor: structure, function and therapeutic potential, *Blood Rev.,* 1, 254, 1987.
112. **Kitamura, T., Sato, N., Arai, K.-I., and Miyajima, A.**, Expression cloning of the human IL-3 receptor cDNA reveals a shared subunit for the human IL-3 and GM-CSF receptors, *Cell,* 66, 1165, 1991.
113. **Tavernier, J., Devos, R., Cornelis, S., Tuypens, T., Van der Heyden, J., Fiers, W., and Plaetnick, G.**, A human high affinity interleukin-5 receptor (IL-5R) is composed of an IL-5 specific αchain and a βchain shared with the receptor for GM-CSF, *Cell,* 66, 1175, 1991.
114. **Weinberg, A. D., Shaw, J., Paetkav, V., Bleackley, R. C., Magnuson, N. S., Reeves, R., and Magnuson, J. A.**, Cloning of cDNA for the bovine IL-2 receptor (bovine Tac Antigen), *Immunology,* 63, 603, 1988.
115. **Siess, D. C., Magnuson, N. S., and Reeves, R.**, Characterisation of the bovine receptor(s) for interleukin-2, *Immunology*, 68, 190, 1989.
116. **Verhagen, A. M., Andrews, A. E., Brandon, M. R., and Nash, A. D.**, Molecular cloning, expression and characterization of the ovine IL-2Rα chain, *Immunology*, 76, 1, 1992.
117. **Bujdoso, R., Sargan, D., Williamson, M., and McConnell, I.**, Cloning of a cDNA encoding the ovine interleukin-2 receptor 55-kDa protein, CD25, *Gene,* 113, 283, 1992.
118. **Tartaglia, L., Weber, R., Figan, I., Reynolds, C., Palladino, M., and Goeddel, D.**, The two different receptors for tumour necrosis factor mediate distinct cellular responses, *Proc. Natl. Acad. Sci. U.S.A.*, 88, 9292, 1991.

119. **Bielefeldt Ohmann, H., Campos, M., McDougall, L., Lawman, M. J. P., and Babiuk, L. A.,** Expression of tumour necrosis factor-α receptors on bovine macrophages, lymphocytes and polymorphonuclear leukocytes, internalisation of receptor-bound ligands and some functional effects, *Lymphokine Res.*, 9, 43, 1990.
120. **Winstanley, F. P.,** Detection of TNF alpha receptors on ovine leucocytes by flow cytofluorimetry, *Res. Vet. Sci.*, 53, 129, 1992.
121. **Hein, W. R. and MacKay, C. R.,** Prominence of γδT cells in the ruminant immune system, *Immunol. Today,* 12, 30, 1991.

Chapter 6

Induction of T Cell-Mediated Immune Responses in Ruminants

Declan J. McKeever

CONTENTS

I. Introduction 93
II. T Cell Recognition 94
III. Antigen Processing and Presentation 94
IV. Accessory Cells for MHC Class II-Restricted Immune Responses *In Vivo* 96
A. Afferent Lymph Veiled Cells (ALVCs) 97
B. Uptake of Antigen by ALVCs 101
C. ALVC Function *In Vivo* 102
D. Relative Efficiency of ALVCs and Other Antigen-Presenting Cell Populations 102
V. Concluding Remarks 103
Acknowledgments 103
References 104

I. INTRODUCTION

An understanding of the factors involved in the induction of primary immune responses is essential to the development of improved vaccines and immunotherapeutic measures. It is generally accepted that the generation of immunity is dependent on the induction of effective T cell responses, and this applies equally to responses effected by T cell populations and those mediated by antibody or cells of nonlymphocytic lineages. The elucidation of the basis of T cell recognition in recent years has highlighted the importance of antigen processing and presentation in the outcome of interactions between the immune system and invading pathogens. The study of antigen processing and presentation is now a major area of immunology that has attracted the attention of scientists drawn from a broad range of disciplines including cell biology, molecular genetics, and protein chemistry. Enormous progress has been made in this area in recent years. Notwithstanding the quality of these discoveries, our current perception of the events that surround the induction of immune responses is largely based on observations made in systems with limited physiological or pathological significance.

Although relatively poorly characterized in molecular terms, the induction of immune responses in ruminants has been studied closely at a physiological level. This is mainly due to the ease with which long-term lymphatic cannulation can be conducted in these species. Observations that have been made using these systems have complemented the large volume of information derived from *in vitro* studies in man and mouse to provide a more comprehensive view of the inductive requirements of primary immunity in mammalian species. Their importance in the field is however often overshadowed by the more fundamental molecular advances made in human and murine systems.

This review will attempt to highlight the progress that has been made in identifying the factors involved in the the induction of immune responses in ruminants, against the background of current knowledge of T cell function in other species. Because the induction of immune responses mediated by $CD8^+$ T cells *in vivo* is in general poorly understood, and no less so in ruminants, the discussion will focus primarily on class II MHC-restricted T cell immunity. These responses are believed to be essential amplifying components for both humoral and cellular immunity, and also provide effector function in the control of infections, both directly and through the elaboration of cytokine molecules. Specific activation of class II-restricted T cells is a feature of a number of infectious diseases in ruminants, including foot and mouth disease,[1,2] brucellosis,[3,4] babesiosis,[5] and East Coast Fever of cattle.[6] Immunological aspects of some of these diseases are discussed elsewhere in this volume.

0-8493-4952-4/94/$0.00+$.50

II. T CELL RECOGNITION

The repertoire of T cell receptors (TCRs) that is available to the immune system is strongly influenced by the interaction of immature T cells with self major histocompatibility (MHC) antigens on thymic epithelial cells. Through a combination of positive and negative selection processes in the thymus,[7,8] a cohort of cells is generated whose receptors bind self-MHC with a weak affinity; high-affinity binding sufficient to initiate activation occurs only when the MHC molecule is altered by the presence of a foreign antigenic determinant. The specificity of the T cell receptor is believed to result from structural characteristics similar to those that give rise to the binding sites of the immunoglobulin (Ig) molecule. Whereas the entire recognition site of the Ig molecule is dedicated to binding antigen, distinct hypervariable regions on the outer domains of the TCR α and β chains are believed to interact with the MHC molecule and its associated foreign antigenic determinant.[9,10] The MHC-specific component of this interaction is the basis of the MHC restriction of T cell responses that confounded immunologists for many years; T cell populations respond to the antigens for which they are specific only in the context of the MHC molecule on which they were educated.

Two major subsets of T cells differ in the MHC product that restricts their responses, and these are distinguished by the expression of accessory molecules that interact with the MHC molecule and presumably stabilize the TCR-MHC-antigen complex. Class I MHC-restricted T cells express the heterodimeric CD8 antigen, which is known to bind the $\alpha3$ domain of the class I MHC molecule.[11,12] Those T cells restricted by class II MHC molecules express the CD4 antigen, which has been shown to bind the $\beta2$ domain of the class II MHC molecule.[13,14] These subpopulations are also broadly distinguished by their functional characteristics; $CD4^+$ T cells have been associated with helper/inducer immune function, while the $CD8^+$ subset is largely responsible for cytotoxic T cell (CTL) activity.[7] It has recently been established that two subsets of $CD4^+$ T cells, referred to as TH1 and TH2, can be distinguished on the basis of function and cytokine secretion.[15,16]

III. ANTIGEN PROCESSING AND PRESENTATION

It has been known for some time that specialized antigen presenting cells (APCs) are required for the induction of MHC class II-restricted responses.[17] Inhibition studies with mild fixatives and lysosomotropic reagents such as chloroquine have indicated that extracellular antigens are taken up by APCs and processed in an acid compartment of the cell.[18-21] The basis of this processing event has been shown to be the fragmentation of the antigen,[22,23] with immunogenic peptides generated by this process binding to class II MHC molecules[24,25] (Figure 6-1). Processing appears to occur in the early endosomal compartment of the APC and its molecular basis has been extensively studied.[26-30] A number of observations have implicated the aspartic protease cathepsin D and the cysteine protease cathepsin B in these events. Both enzymes are required in cell-free systems for the processing and presentation of ovalbumin for recognition by murine T cell clones,[26] although the activity of cathepsin B alone appears sufficient to process the antigen conalbumin for T cell recognition.[28] That other proteases may also be involved is indicated by the observation that different protease inhibitors have distinct effects on the recognition of a given antigen by T cell clones.[27] These results are consistent with several proteases being involved in antigen processing for class II MHC-restricted T cell responses, with different immunogenic peptides requiring different sets of proteases. Perhaps surprisingly, the molecular aspects of antigen processing by bovine APCs have been well characterized. In an analysis of the processing of sperm whale myoglobin by intact bovine pulmonary macrophages, Van Noort et al.[30] observed that almost all cleavages of the antigen during endosomal processing within these cells were consistent with the combined activities of cathepsin D and cathepsin B. Initial cleavage of the molecule was attributable to cathepsin D, and it was observed in addition that each of the T cell epitopes that have been defined for the molecule occurred at the extreme amino terminal of a cathepsin D fragment. Cathepsin B was seen to have strong carboxypeptidase activity on cathepsin D-released peptides, suggesting that this enzyme is responsible for trimming initial fragments to the size that is characteristic of immunogenic peptides.

The interaction between processed peptides and class II MHC molecules is also believed to occur in the endosomal compartment, although the exact location of this event has not been determined. There is strong evidence[31-33] that peptides bind to newly synthesized MHC molecules rather than those derived from the cell surface. Nascent class II molecules in the endoplasmic reticulum associate with a nonpolymorphic glycoprotein known as the invariant chain,[34] which has signals that target the complex

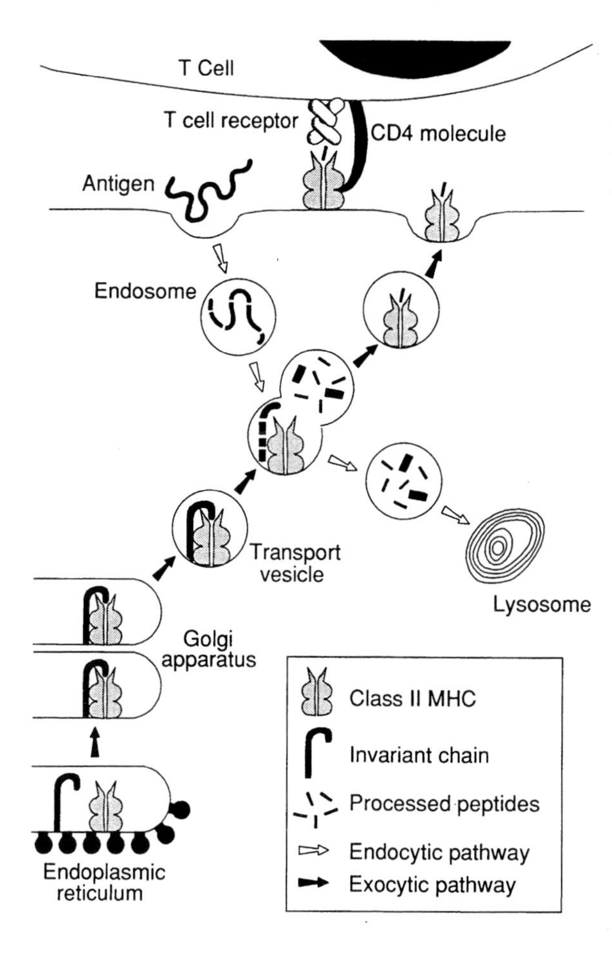

Figure 6-1 Schematic representation of the current perception of the processes involved in class II MHC-restricted antigen presentation. Exogenous antigens are taken up by antigen presenting cells and degraded to peptide fragments within the endosomal compartment of the cell. Nascent class II MHC molecules are targeted to this compartment by the invariant chain, which there becomes degraded, allowing processed peptides to bind to the MHC molecule. The peptide-MHC complex is then transported to the cell surface for recognition by the antigen receptor of specific T cells. Binding of the MHC-peptide complex by the T cell receptor is stabilized by the CD4 accessory molecule.

to the endosomal compartment[35,36] (Figure 6-1). The invariant chain has also been shown to prevent the binding of endogenous peptides to MHC class II molecules,[37,38] which presumably ensures that the peptide binding site is preserved for interactions with processed exogenous peptides. The invariant chain later dissociates from the class II molecule, probably through the action of proteases.[39] A recent study has shown that class II molecules intersect the endocytic pathway in a mildly acidic late endosomal compartment, by which time they are no longer associated with the invariant chain[33] (Figure 6-1). It is at some point between this location and the cell surface that processed peptides bind to the MHC molecules (Figure 6-1). There is strong evidence that this process requires an acid pH,[40,41] and it has been reported that, in the absence of bound peptide, class II molecules are actually unstable at neutral pH.[42]

The precise nature of peptide binding to MHC molecules has been firmly established in recent years. Solution of the crystal structure of the human class I MHC molecule HLA-A2[43] revealed a structure with a groove at its outer extremity, formed by two α-helices superimposed on an array of β-pleated sheets. Electron-dense material within this groove was assumed to represent antigenic peptides. Subsequent observations have confirmed this assumption,[44,45] and more recently, the conformation of peptides within the binding groove has been accurately defined.[46,47] Predictive models of class II MHC molecular structure have suggested that they bind processed peptides in a similar manner,[48] and this is consistent with preliminary crystallographic analysis.[49] However, peptides eluted from immunopurified class II MHC molecules are generally longer than those obtained from class I MHC, and it has been suggested that structural differences may allow class II MHC-associated peptides to extend beyond the extremeties of the cleft.[50]

Although engagement of the TCR by the peptide-MHC complex and appropriate accessory molecules appears sufficient to induce lymphokine release by T cells, certain co-stimulator factors are required for the initiation of cell division. For the TH2 subset of murine CD4$^+$ T cells, interleukin 1 (IL-1) expressed on the APC surface has been shown to fulfill this role.[51,52] It is likely that cytokines are also important in the initiation of ruminant T cell responses. A recent study has indicated that infection of bovine macrophages with *Brucella abortus* induces surface expression of IL-1.[4] Infected macrophages efficiently present the bacterial antigens to immune T cells.[3] Downregulation of membrane expression of IL-1 is observed in the presence of the cytokine γ-interferon, and this is associated with considerable reduction in specific T cell proliferation.[4] These observations suggest that post-recognition signals between APCs and antigen-specific T cells are probably similar in ruminants and mice.

IV. ACCESSORY CELLS FOR MHC CLASS II-RESTRICTED IMMUNE RESPONSES *IN VIVO*

A number of cell types, including macrophages, dendritic cells, epidermal Langerhans cells,[53] B cells,[54] and T cells,[55] have been shown in mouse and man to provide antigen presenting function for class II MHC-restricted T cell responses *in vitro*. The significance of these observations in the induction of immunity *in vivo* is not clear, and there is much evidence that dendritic cells constitute the chief accessory cell population for primary T cell responses.[51,56-65] Initiation of immune responses is believed to occur in the lymph node that drains the site of antigenic challenge,[66] and induction of primary $CD4^+$ T cell responses within lymph nodes has been shown to be dependent on the accessory function of interdigitating cells (IDCs), a dendritic cell population found in the lymph node paracortex[67] (Figure 6-2). The activity

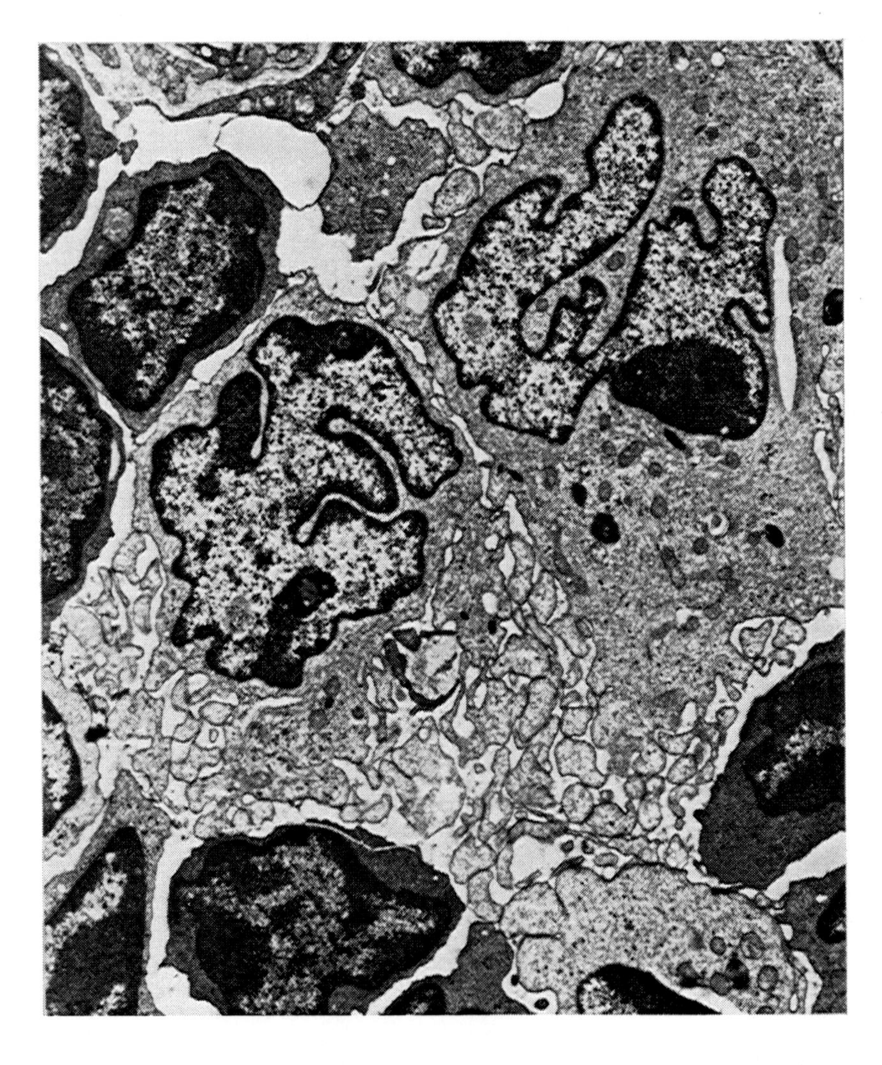

Figure 6-2 Electron micrograph showing two interdigitating cells within the paracortex of a bovine lymph node. (Courtesy of Dr. W. I. Morrison, × 14,500)

of this population is therefore fundamental to the generation of primary immunity. There is strong evidence that IDCs are derived from dendritic cells in afferent lymph known as veiled cells.[68-70] Observations in a number of species have shown that epidermal Langerhans cells, afferent lymph veiled cells (ALVCs), and IDCs are phenotypically similar, and have suggested that, together with dendritic cells in the dermis, these cell types represent different migratory stages of the same population.[71-74] These observations have been consolidated in cattle by distribution analysis of a number of leukocyte surface antigens, including integrin molecules and the CD1 specificity, and two partially characterized bovine leukocyte antigens defined by monoclonal antibodies (MAb) ILA24 and ILA53.[75] The β2 integrins are a family of leukocyte adhesion molecules that consist of a constant β chain (CD18) associated with one of the α chains, CD11a, CD11b, or CD11c. These molecules participate in adhesion-dependent functions of immune cells, including the binding of lymphocytes to target cells and their trafficking between lymphoid compartments.[76] The CD1 specificity encompasses a family of molecules expressed variously on cortical thymocytes, skin dendritic cells, and a subpopulation of B lymphocytes.[77] The CD1b molecule is present on cortical thymocytes and skin dendritic cells but is not expressed by epidermal Langerhans cells (Figure 6-3a).[77] The bovine homologue of CD1b is defined by MAb TH97a.[75] The MAb ILA53 defines a 220-kDa antigen termed WC6 that is expressed by bovine dendritic cells and at a lower level by both T and B lymphocytes; despite its similarity in molecular weight to CD45, sequential immunoprecipitation studies have shown that it is distinct from CD45.[78] MAb ILA24 recognizes a heterodimeric molecule (p110/75) expressed by bovine cells of the granulocyte-macrophage lineage.[79]

The distribution of these antigens among dendritic cells of skin, afferent lymph, and lymph nodes is summarized in Table 6-1. The CD11a antigen is expressed by all of these populations, as well as by blood monocytes and follicular dendritic cells (FDCs) within lymph node germinal centers. However, CD11c, WC6, and p110/75 are expressed together only on Langerhans cells, ALVCs, dermal dendritic cells, and IDCs, and so collectively appear to characterize this lineage (Figures 6-3b and 6-4). None of these markers are expressed by FDCs, an observation that supports evidence from other systems[80] that FDCs are not bone marrow-derived dendritic cells, but rather originate from mesenchymal tissue. In addition, unlike ALVCs, bovine FDCs do express CD11b. ALVCs are also phenotypically distinct from blood monocytes (Table 6-1); the latter do not express CD1 or the specificity defined by ILA53, and stain positively for the presence of CD11b.

A. AFFERENT LYMPH VEILED CELLS (ALVCs)

Because of their likely importance in the generation of primary immunity, considerable effort has been focused on the phenotypic and functional characterization of ALVCs. Much of this has been accomplished in ruminant systems. By cannulation of efferent lymphatic ducts after removal of the lymph node, afferent lymph can be collected from cattle and sheep with relative ease.[72,75,81-83] Anastomosis of the interrupted lymphatic circulation allows direct flow of afferent lymph into efferent vessels, and appreciable volumes of the fluid can be collected from cannulated animals over extended periods. Veiled cells constitute 15 to 20% of the cell output collected from cattle[75,81] and 5 to 20% of that derived from sheep.[83] When observed under phase contrast microscopy, ALVCs have a characteristic veiled or frilly appearance; when treated with Romanowski stains, they appear as large cells with abundant pale agranular cytoplasm and a nuclear morphology that ranges from oval to lobulated.

Ultrastructurally, the most striking features of ALVCs are a highly convoluted plasma membrane, the presence of numerous endocytic vesicles and intermediate filaments, and a poorly developed lysosomal apparatus as compared with monocytes and macrophages (Figures 6-5A and B). A small minority can be seen to contain Birbeck granules (Figure 6-5B), which are a characteristic feature of epidermal Langerhans cells. This suggests that Langerhans cells constitute a small proportion of bovine ALVCs, the remainder presumably being derived from dermal dendritic cells and dendritic cells from other tissues drained by the lymphatic. A number of additional observations have confirmed that ALVCs are phenotypically heterogeneous. Only a subpopulation of bovine ALVCs express the CD1b specificity,[75] and expression is not uniform on positive cells. Similarly, the CD1 specificity is not expressed on all ALVCs of the sheep, and experiments in this species have indicated that its expression is enhanced after *in vivo* exposure to antigen.[82] Bovine ALVCs are, in addition, heterogeneous for the expression of CD11a and p110/75,[75] the determinant defined by MAb ILA24. Expression of these antigens by ALVCs is almost mutually exclusive, with only a minor population exhibiting the p110/75^{+}/CD11a^{+} phenotype.[75] The phenotypic characteristics of ALVC derived from cattle and sheep are summarized in Table 6-2.

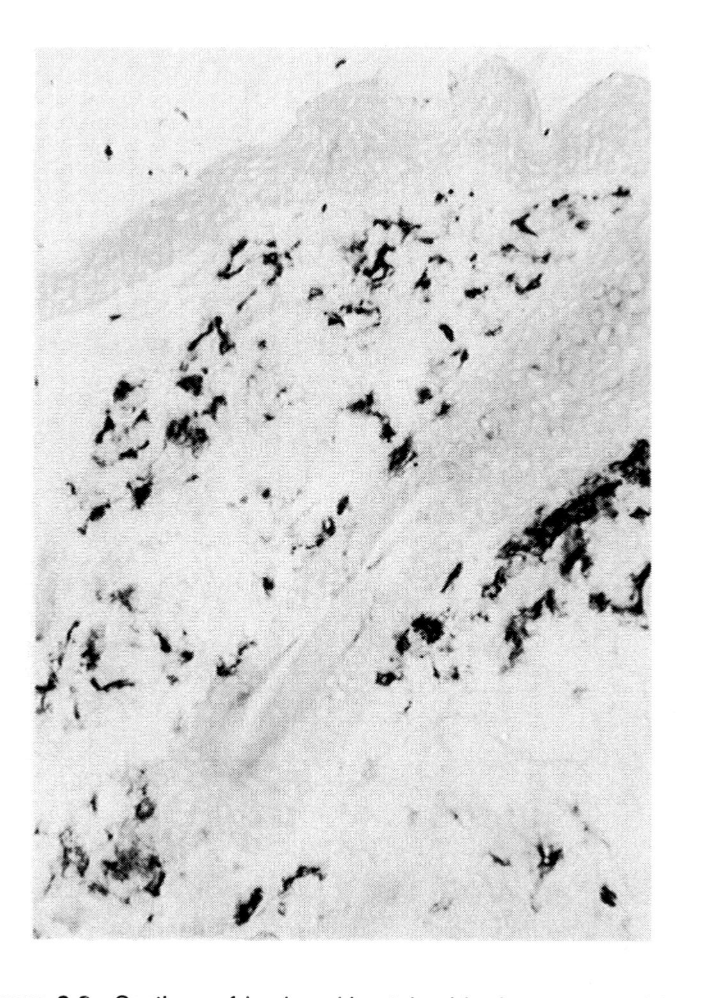

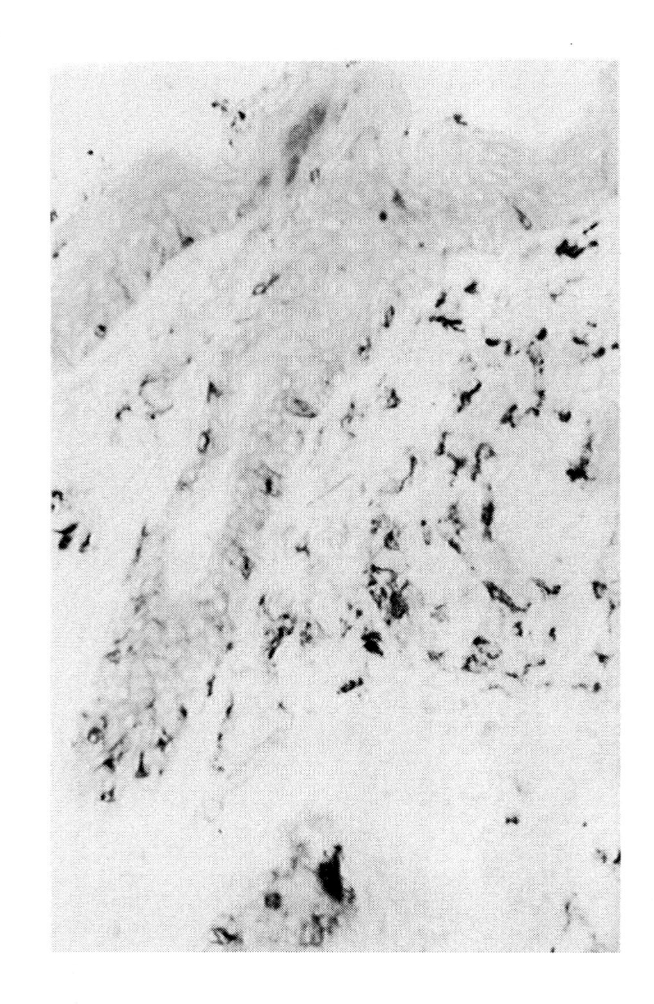

Figure 6-3 Sections of bovine skin stained by immunoperoxidase with MAb specific for (A) CD1b and (B) CD11c. Cells expressing CD1b are confined mainly to the dermis, whereas cells expressing CD11c are present in both the dermis and epi-dermis. (Courtesy of Dr. W. I. Morrison, $\times$ 200).

Table 6-1 **Phenotypic Relationships of Bovine Dendritic Cells and Monocytes**

Tissue	Cell type	Specificity[a]					
		CD11a	CD11b	CD11c	WC6	p110/75	CD1b
Skin	Langerhans cell	+	–	+	+	+	–
	Dermal dendritic cell	+	+	+	+	+	+
Afferent lymph	ALVCs	+/–	–	+	+	+/–	+/–
Lymph node	Interdigitating cell	+	+	+	+	+	+
	Follicular dendritic cell	+	+	–	–	–	–
Blood	Monocytes	+	+	+	–	+	–

[a]In the case of afferent lymph, the CD11a, p110/75, and CD1b specificities are expressed only on a subpopulation of veiled cells; this is indicated as +/–. Expression of the antigens by cells in skin and lymph nodes was evaluated by immunohistology, and therefore, when positive cells were detected it was not possible to determine whether they represented all or a proportion of the dendritic cell population.

The capacity of ALVCs to present soluble antigens to T cells has been well documented cattle and in sheep,[75,81,82] and heterogeneity within the population is also reflected in this function. Cell sorting experiments with bovine ALVCs using MAb ILA24 have revealed that bovine $CD4^+$ T cell clones specific for trypanosomal variable surface glycoprotein (VSG) respond to the antigen in the presence of $p110/75^+$ ALVCs, but not when it is presented by the $p110/75^-$ population.[75] Similar experiments with a CD1-specific MAb revealed no association of this specificity with antigen presenting function. Sorted $ILA24^+$ and $ILA24^-$ ALVCs are, however, equally efficient in the induction of allogeneic mixed leukocyte reactions (author's unpublished observations). Whether functional heterogeneity among ALVCs arises from the presence of

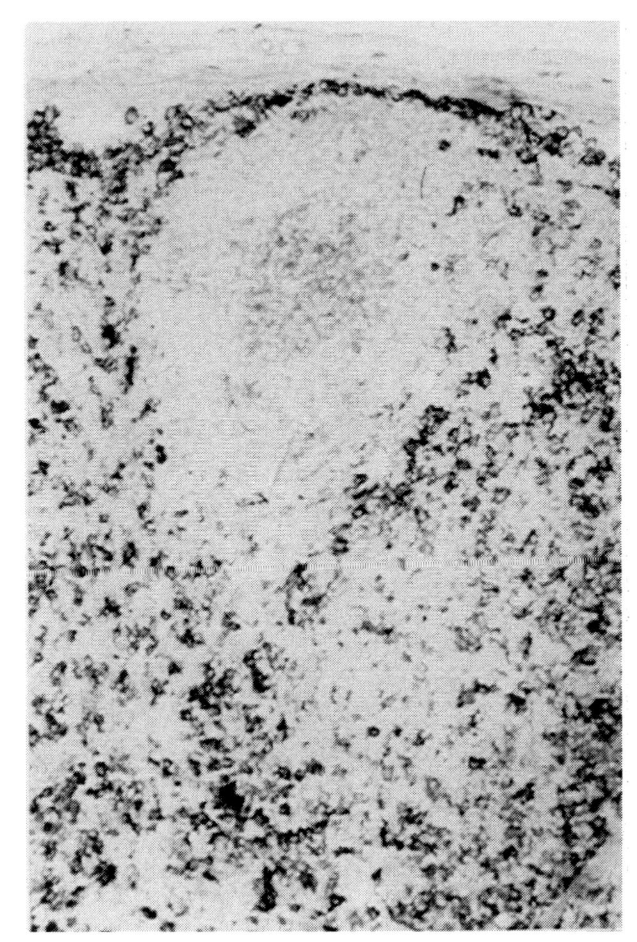

Figure 6-4 Section of bovine lymph node stained with a MAb specific for CD11c. Numerous large positive cells are present in the T-dependent paracortical areas, whereas the B-dependent area is devoid of positive cells. (Courtesy of Dr. W. I. Morrison, × 125).

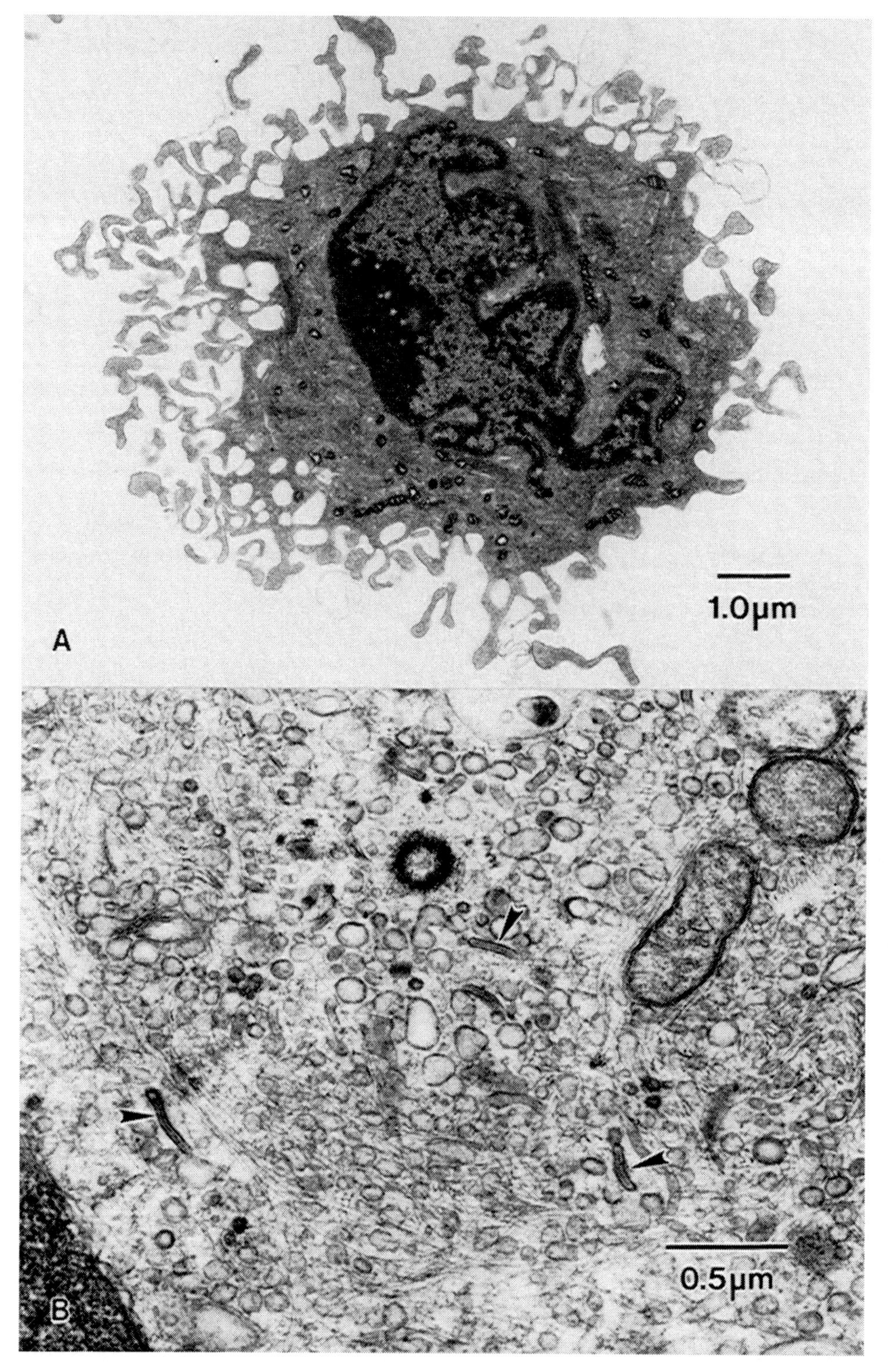

Figure 6-5 Panel A: electron micrograph of a veiled cell purified from bovine afferent lymph. Panel B: veiled cell pictured at higher magnification showing Birbeck granules (arrows) and numerous endocytic vesicles and intermediate filaments.

Table 6-2 **Phenotypic Characteristics of Ruminant ALVCs**

Specificity	Cattle		Sheep	
	Expression	Ref.	Expression	Ref.
Class I MHC	+	75	+	72, 83
Class II MHC	+	75	+	72, 83, 105
CD1	+/–	75, 106	+/–	72,83
CD2	–[a]		+	72
CD11a	+/–	75		
CD11b	–	75		
CD11c	+	75		
CD45			+	83
p110/75	+/–	75		
LFA3			+	72
sIg	+/–	86	+	72
FcR	–	86	+	89

[a]D. J. McKeever, and N. D. MacHugh, unpublished results.

distinct populations or merely reflects different states of maturity is not clear. Recent observations in the mouse indicate that, after a period in culture, epidermal Langerhans cells lose the capacity to present soluble antigens but remain effective in stimulating allogeneic T cells.[53,84] It is speculated[53] that when Langerhans cells take up antigen in the epidermis, they process it rapidly, and by upregulating their synthesis of class II MHC molecules, efficiently express peptide-MHC complexes on their surface. In the process of maturation to dendritic cells in the lymph node, they may downregulate class II MHC synthesis and processing mechanisms, but efficiently present existing MHC-peptide complexes. There are indications that such a situation may apply to ALVCs. Lobulated nuclear morphology is a characteristic of more differentiated cells and is seen in IDCs within lymph nodes of cattle and other species[84] (Figure 6-2). The majority of p110/75$^-$ ALVCs have this appearance, while cells with more immature oval to reniform nuclei dominate the p110/75$^+$ subpopulation.[75] This suggests that p110/75$^+$ ALVCs may be relatively immature, and in line with the proposal of Romani et al.,[53] more efficient in the processing of native antigens for recognition by T cells. Alternatively, p110/75$^-$ ALVCs may provide accessory function for a population of T lymphocytes that is not represented among the clones that were tested.

A number of studies have examined the mechanisms of antigen presentation by ALVCs. In cattle and sheep, ALVCs rapidly assimilate soluble antigens *in vivo* and *in vitro* and process them for recognition by immune T cells.[72,86] This capacity has been shown in bovine ALVCs to be consistent with processing mechanisms within the endocytic pathway; antigen presentation is inhibited by the presence of chloroquine and light fixation before but not after exposure to antigen. Class II MHC molecules are strongly expressed by ALVCs derived from both cattle and sheep,[75,82] and the abrogation of antigen presentation in the presence of MAb specific for these elements[86] confirms that ALVCs function is class II MHC restricted. Interestingly, studies in the sheep have revealed an increase in expression of class II MHC by ALVCs in lymph draining a secondary antigenic challenge.[82] Although not seen following primary exposure to antigen, this upregulation is associated with markedly enhanced antigen presenting function. This may indicate a regulatory influence for modulation of class II MHC expression in ovine immune function.[82]

B. UPTAKE OF ANTIGEN BY ALVCs

Ovine ALVCs bear surface IgM and IgG,[87] and phagocytose immune complexes *in vivo*.[87] Bujdoso et al.[72] considered that antibody passively acquired by ALVCs may act as a receptor for antigen and therefore be of considerable importance to antigen processing. Between 60 and 70% of ALVCs derived from sheep express Fc receptors, and uptake of antigens by these cells *in vivo* and *in vitro* is more efficient in the presence of specific antibody.[89] In kinetic studies, maximal labeling of ALVCs with FITC-ovalbumin was observed in the presence of specific antibody after 5 min *in vitro* and 20 to 40 min *in vivo*. *In vitro* labeling was equally efficient at 4 as at 37°C, suggesting that these results may reflect binding rather than uptake. Although surface IgM is occasionally observed on bovine ALVCs, it has not been possible to demonstrate the presence of Fc receptors on these cells in rossetting assays (author's unpublished observations) or by the use of FITC-labeled immune complexes.[86] We have observed that bovine ALVCs assimilate large quantities of radiolabeled antigen *in vitro* in the absence of immune serum.[75] In addition, bovine ALVCs from immunized animals present antigen efficiently in the absence of specific antibody, and their ability to present antigen is not enhanced by the addition of immune serum.[75,86] The exact mechanism whereby ALVCs take up antigens is therefore not clear. A number of observations have indicated that ALVCs and dendritic cells from a variety of species are only poorly phagocytic *in vitro*.[90,91] However, Drexhage and others[92] reported that porcine ALVCs can engulf large volumes of extracellular fluid by sweeping movements of their long cytoplasmic processes. It has been observed that while ALVCs derived from sheep and rat are relatively inefficient in the uptake of latex beads, they can assimilate large quantities of viral particles.[90] Based on these observations, it is likely that antigen uptake by ALVCs is largely the result of fluid phase pinocytosis. This is supported by the observation that after incubation at 37°C for 1 h in the presence of horseradish peroxidase, a pinocytic tracer, bovine ALVCs can be seen to distribute the marker throughout their endosomal compartment (P. Webster, personal communication). In this regard, it has been demonstrated that murine dendritic cells, although poorly phagocytic,[93] are efficient in the process of fluid phase pinocytosis and in the trafficking of material through the endosomal compartment.[94]

C. ALVC FUNCTION *IN VIVO*

A prominent role for ALVCs in the induction of immune responses in ruminants is indicated by their capacity to assimilate and process soluble antigens inoculated *in vivo*. Ovine ALVCs collected 24 h after intradermal inoculation of soluble antigen in the area drained by a cannulated lymphatic have been shown to present it efficiently to specific T cells.[72] Kinetic studies in cattle have indicated that uptake of antigens by ALVCs *in vivo* is very rapid. Maximal presenting function is observed in ALVCs collected within 30 min of the inoculation of antigen, and is no longer detectable after 48 h.[86] When compared with observations of specialized APC in other species, these kinetics suggest that ALVCs are extremely effective accessory cells; splenic dendritic cells derived from mice have been reported to require an overnight exposure to antigen for optimal presenting function.[58] Recent studies in this laboratory have provided strong evidence that antigen-loaded ALVCs present in afferent lymph are responsible for the induction of primary T cell responses in the draining lymph node.[86] An overnight collection of bovine afferent lymph can yield up to 10^8 purified ALVCs,[75] so that large numbers can be pulsed with antigen *in vitro* and returned to the donor or inoculated in other recipient cattle. In a number of experiments that involved intradermal inoculation of 10^7 autologous or MHC-identical antigen-pulsed ALVCs in naive cattle, specific T cell responses were detected as soon as 1 week after inoculation, using both ovalbumin and trypanosomal VSG.

D. RELATIVE EFFICIENCY OF ALVCs AND OTHER APC POPULATIONS

Other cell populations in cattle that have been shown to present soluble antigens to T cells in a specific manner include peripheral blood monocytes[79,95] and activated B cells.[96] We have observed, in addition, that bovine activated T cells, which express class II MHC, can present exogenous antigens, although less efficiently than other APC (author's unpublished observations). Although the precise significance of these populations to the induction of primary and secondary responses *in vivo* is not clear, observations in human and murine systems have suggested that antigen-specific B cells may be involved in the induction of secondary T cell responses. Human antigen-specific B lymphocytes present antigen to T cells approximately 10^4 times more efficiently than nonspecific B cells.[97] This superiority is abrogated by the addition of anti-immunoglobulin antibodies, indicating that surface immunoglobulin is responsible for antigen uptake. These characteristics have also been reported for murine B cells,[98] and it has also been observed in this species that resting as well as activated B cells are effective in antigen presentation.[99] These observations suggest that circulating immune B lymphocytes constitute a pool of APC that is capable of rapid amplification of T cell responses on secondary exposure to antigen. Although B cells from immunized cattle have been shown to present soluble antigen to immune T cells after activation with bacterial lipopolysaccharide, resting bovine B cells do not perform this function.[96] The role of surface immunoglobulin in these processes has not been examined in cattle, but the efficiency of antigen presentation by bovine B cells is comparable to that of blood monocytes, suggesting that similar mechanisms of uptake and processing are employed by both populations.[96]

In contrast, bovine ALVCs have been shown to be 10 to 100 times more efficient than blood monocytes in antigen presenting function. This superiority can be observed in allogeneic[75] and autologous mixed lymphocyte reactions (author's unpublished observations) and in responses of specific T cells to soluble antigens.[75] Since both cell populations assimilate antigens effectively,[75] the latter observation does not appear to be related to antigen uptake. Expression of class II MHC antigens is markedly higher in bovine ALVCs than in purified monocyte populations,[75] and this factor is likely to contribute to their superior presenting function. The potency of murine dendritic cells as APCs has been attributed to their ability to cluster with responding lymphocytes.[56,100,101] In particular, dendritic cells are capable of forming clusters with lymphocytes in the absence of antigen, and this characteristic may be responsible for their ability to induce primary T cell responses.[56] There is evidence from studies in mice that the CD11a antigen is involved in stabilizing antigen-specific clusters,[102] presumably through interaction with its ligand ICAM-1.[103] Initial antigen-independent clustering between murine dendritic cells and T lymphocytes is not, however, inhibited by CD11a-specific mAb,[102] suggesting that other molecules are involved in these interactions. Antigen-independent clustering is also a feature of interactions between bovine ALVCs and immune lymphocytes.[75] That reactivity to antigen resides in these clusters is illustrated by the fact that antigen-specific T cell clones can readily be generated from them.[75,86] In contrast, blood monocytes form clusters with immune lymphocytes only in the presence of antigen, and these clusters are smaller than those observed with ALVCs.[75] It seems likely that the superiority of ALVCs over monocytes in antigen presenting function is related to their capacity to cluster with lymphocytes in the absence of antigen, although other factors may be involved. Cluster formation *in vitro* has also been described between ovine ALVCs and allogeneic efferent lymph lymphocytes[72] and appears necessary for the generation of allogeneic mixed lymphocyte reactions. The surface molecules participating in the formation of these clusters have not been defined.

V. CONCLUDING REMARKS

The bulk of the observations outlined in this review highlight the importance of dendritic cells, and in particular ALVCs, in the induction of $CD4^+$ T cell responses in ruminants. Clear evidence is now available for involvement of these cells in the uptake of antigens in the periphery, its carriage in afferent lymph to the draining lymph node, and in the stimulation of specific T cell responses within the node. It is also clear that ALVCs are highly adapted for antigen presenting function. However, available functional data on ALVCs are derived from studies of cells collected from the periphery and relate only to soluble antigens. Whether ALVCs in other locations behave similarly, and whether these cells play a major role in the induction of immunity to intracellular pathogens, remains to be established. In this regard, dendritic cells have been shown to be potent stimulators of primary cytotoxic T cell responses in both alloreactive[59] and virus[61] systems.

The importance of ALVCs in the generation of primary immunity raises the possibility of vaccine delivery strategies based on the targeting of antigens to them. This approach has been used successfully in the induction of murine immune responses to avidin when conjugated to the dendritic cell-specific MAb 33D1.[104] ALVCs are phenotypically distinct from blood monocytes and dendritic cells in the skin and lymph node,[75] and we have recently identified an MAb that appears specific for ALVCs (N. D. MacHugh, personal communication). In the light of the observations presented in this review, this reagent may provide the basis for an effective antigen delivery system for cattle.

ACKNOWLEDGMENTS

The author is grateful to Dr. M. Shaw for the provision of electron micrographs, and to Dr. W. I. Morrison for photomicrographs.

REFERENCES

1. **Collen, T. and Doel, T. R.,** Heterotypic recognition of foot-and-mouth disease virus by cattle lymphocytes, *J. Gen. Virol.*, 71, 309, 1991.
2. **van Lierop, M. J., van Maanen, K., Meloen, R. H., Rutten, V.P., de Jong, M. A., and Hensen, E. J.,** Proliferative lymphocyte responses to foot-and-mouth disease virus and three FMDV peptides after vaccination or immunisation with these peptides in cattle, *Immunology*, 75, 406, 1992.
3. **Splitter, G. A. and Everlith, K. M.,** Collaboration of bovine T lymphocytes and macrophages in T-lymphocyte response to *Brucella abortus*, *Infect. Immun.*, 51, 776, 1986.
4. **Splitter, G. A. and Everlith, K. M.,** *Brucella abortus* regulates bovine macrophage-T-cell interaction by major histocompatibility complex class II and interleukin-1 expression, *Infect. Immun.*, 57, 1151, 1989.
5. **Brown, W. C. and Logan, K. S.,** *Babesia bovis*: bovine helper T cell lines reactive with soluble and membrane antigens of merozoites, *Exp. Parasitol.*, 74, 188, 1992.
6. **Baldwin, C. L., Goddeeris, B. M., and Morrison, W. I.,** Bovine helper T cell clones specific for lymphocytes infected with *Theileria parva* (Muguga), *Parasite Immunol.*, 9, 499, 1987.
7. **Von Boehmer, H.,** The developmental biology of T lymphocytes, *Annu. Rev. Immunol.*, 6, 309, 1988.
8. **Nikolic-Zugic, J.,** Phenotypic and functional stages in the intrathymic development of $\alpha\beta$ T cells, *Immunol. Today*, 12, 65, 1991.
9. **Davis, M. M. and Bjorkmann, P. J.,** T-cell receptor genes and $\alpha\beta$ T-cell recognition, *Nature*, 334, 395, 1988.
10. **Matis, L. A.,** The molecular basis of T-cell specificity, *Annu. Rev. Immunol.*, 8, 65, 1990.
11. **Potter, T. A., Rajan, T. V., Dick, R. F., II, and Bluestone, J. A.,** Substitution at residue 227 of H-2 class I molecules abrogates recognition by CD8-dependent but not CD8-independent cytotoxic T lymphocytes, *Nature*, 337, 73, 1989.
12. **Salter, R. D., Benjamin, R. J., Wesley, P. K., Buxton, S. E., Garrett, T. P. J., Clayberger, C., Krensky, A. M., Norment, A. M., Littman, D. R., and Parham, P.,** A binding site for the T-cell co-receptor CD8 on the alpha-3 Domain of HLA-A2, *Nature*, 345, 41, 1990.
13. **Konig, R., Huang, L.-Y., and Germain, R. N.,** MHC class II interaction with CD4 mediated by a region analagous to the MHC class I binding site for CD8, *Nature*, 356, 796, 1992.
14. **Cammarota, G., Scheirle, A., Takacs, B., Doran, D. M., Knorr, R., Bannwarth, W., Guardiola, J., and Sinigaglia, F.,** Identification of a CD4 binding site on the β2 domain of HLA-DR molecules, *Nature*, 356, 799, 1992.
15. **Mosmann, T. R., Cherwinski, H., Bond, M. W., Giedlin, M. A., and Coffman, R. L.,** Two types of murine helper T cell clone. I. Definition according to profiles of lymphokine activities and secreted proteins, *J. Immunol.*, 136, 2348, 1986.
16. **Mosmann, T. R. and Coffman, R. L.,** TH1 and TH2 cells: different patterns of lymphokine secretion lead to different functional properties, *Annu. Rev. Immunol.*, 7, 145, 1989.
17. **Chesnut, R. W. and Grey, H. M.,** Antigen presenting cells and mechanisms of antigen presentation, *Crit. Rev. Immunol.*, 5, 263, 1985.
18. **Zeigler, K. and Unanue, E.,** Decrease in macrophage antigen catabolism caused by ammonia and chloroquine is associated with inhibition of antigen presentation, *Proc. Natl. Acad. Sci. U.S.A.*, 79, 175, 1982.
19. **Zeigler, K. and Unanue, E.,** Identification of a macrophage antigen-processing event required for I region-restricted antigen presentation to T lymphocytes, *J. Immunol.*, 127, 1869, 1981.
20. **Chesnut, R., Colon, S., and Grey, H.,** Requirements for the processing of antigens by antigen-presenting B cells. I. Functional comparison of B cell tumours and macrophages, *J. Immunol.*, 129, 2382, 1982.
21. **Grey, H., Colon, S., and Chesnut, R.,** Requirements for the processing of antigens by antigen presenting B cells. II. Biochemical comparison of the fate of antigen in B-cell tumours and macrophages, *J. Immunol.*, 129, 2389, 1982.
22. **Shimonkevitz, R., Kappler, J., Marrack, P., and Grey, H. M.,** Antigen recognition by H-2-restricted T cells. I. Cell-free antigen processing, *J. Exp. Med.*, 158, 303, 1983.
23. **Shimonkevitz, R., Colon, S., Kappler, J. W., Marrack, P., and Grey, H. M.,** Antigen recognition by H-2-restricted T cells. II. A tryptic ovalbumin peptide that substitutes for processed antigen, *J. Immunol.*, 133, 2067, 1984.

24. **Babbitt, D. P., Allen, P. M., Matsueda, G., Haber, E., and Unanue, E. R.,** Binding of immunogenic peptides to Ia histocompatibility molecules, *Nature*, 317, 359, 1985.
25. **Buus, S., Colon, S., Smith, C., Freed, J. H., Miles, C., and Grey, H. M.,** Interaction between a "processed" ovalbumin peptide and Ia molecules, *Proc. Natl. Acad. Sci. U.S.A.*, 83, 3968, 1986.
26. **Diment, S.,** Different roles for thiol and aspartyl proteases in antigen presentation of ovalbumin, *J. Immunol.*, 145, 417, 1990.
27. **Puri, J. and Factorovich, Y.,** Selective inhibition of antigen presentation to cloned T cells by protease inhibitors, *J. Immunol.*, 141, 3313, 1988.
28. **Gradehant, G. and Ruede, E.,** The endo/lysosomal protease cathepsin B is able to process conalbumin fragments for presentation to T cells, *Immunology*, 74, 393, 1991.
29. **Van Noort, J. M. and Van der Drift, A. C. M.,** The selectivity of cathepsin D suggests an involvement of the enzyme in the generation of T-cell epitopes, *J. Biol. Chem.*, 264, 14159, 1989.
30. **Van Noort, J. M., Boon, J., Van der Drift, A. C. M., Wagenaar, J. P. A., Boots, A. M. H., and Boog, C. J.,** Antigen processing by endosomal proteases determines which sites of sperm-whale myoglobin are eventually recognised by T cells, *Eur. J. Immunol.*, 21, 1989, 1991.
31. **Neefjes, J. J., Stollorz, V., Peters, P. J., Geuze, H. J., and Ploegh, H. L.,** The biosynthetic pathway of MHC class-II but not class-I molecules intersects the endocytic route, *Cell*, 61, 171, 1990.
32. **Davidson, H. W., Reid, P. A., Lanzavecchia, A., and Watts, C.,** Processed antigen binds to newly synthesised MHC class II molecules in antigen-specific B lymphocytes, *Cell*, 67, 105, 1991.
33. **Peters, P., Neefjes, J. J., Oorschot, V., Ploegh, H. L., and Geuze, H. J.,** Segregation of MHC class II molecules from class I molecules in the Golgi complex for transport to lysosomal compartments, *Nature*, 349, 669, 1991.
34. **Kvist, S., Wiman, K., Claesson, L., Peterson, P. A., and Dobberstein, B.,** Membrane insertion and oligomeric assembly of HLA-DR histocompatibility antigens, *Cell*, 29, 61, 1982.
35. **Lotteau, V., Teyton, L., Peleraux, A., Nilsson, T., Karisson, L., Schmid, S. L., Quaranta, V., and Peterson, P. A.,** Intracellular transport of class II MHC molecules directed by invariant chain, *Nature*, 348, 600, 1990.
36. **Bakke, O. and Dobberstein, B.,** MHC class II invariant chain contains a sorting signal for endocytic compartments, *Cell*, 63, 707, 1990.
37. **Roche, P. A. and Cresswell, P.,** Invariant chain association with HLA-DR molecules inhibits immunogenic peptide binding, *Nature*, 345, 615, 1990.
38. **Teyton, L., O'Sullivan, D., Dickson, P., Lotteau, V., Sette, A., Fink, P., and Peterson, P.,** Invariant chain distinguishes between the exogenous and endogenous antigen presentation pathways, *Nature*, 348, 39, 1990.
39. **Nowell, J. and Quaranta, V.,** Chloroquine affects biosynthesis of Ia molecules by inhibiting dissociation of invariant (gamma) chains from alpha-beta dimers in B cells, *J. Exp. Med.*, 162, 1371, 1985.
40. **Jensen, P. E.,** Regulation of antigen presentation by acidic pH, *J. Exp. Med.*, 171, 1779, 1990.
41. **Jensen, P. E.,** Enhanced binding of peptide antigen to purified class II major histocompatibility glycoproteins at acidic pH, *J. Exp. Med.*, 174, 1111, 1991.
42. **Sadegh-Nasseri, S. and Germain, R. N.,** A role for peptide in determining MHC class II structure, *Nature*, 353, 167, 1991.
43. **Bjorkman, P. J., Saper, M. A., Strominger, J. L., and Wiley, D. C.,** Structure of the human class I histocompatibility antigen, HLA-A2, *Nature*, 329, 506, 1987.
44. **Saper, M. A., Bjorkman, P. J., and Wiley, D. C.,** Refined structure of the human histocompatibility antigen HLA-A2 at 2.6 Å resolution, *J. Mol. Biol.*, 219, 277, 1991.
45. **Garret, T. P. J., Saper, M. A., Bjorkman, P. J., Strominger, J. L., and Wiley, D. C.,** Specificity pockets for the side chains of peptide antigens in HLA-Aw68, *Nature*, 342, 692, 1989.
46. **Guo, H., Jardetzky, T. S., Garret, T. P. J., Lane, W. S., Strominger, J. L., and Wiley, D. C.,** Different length peptides bind to HLA-Aw68 similarly at their ends but bulge out in the middle, *Nature*, 360, 364, 1992.
47. **Silver, M. L., Guo, H., Strominger, J. L., and Wiley, D. C.,** Atomic structure of a human MHC molecule presenting an influenza virus peptide, *Nature*, 360, 367, 1992.
48. **Brown, J. H., Jardetzky, T., Saper, M. A., Samraoui, B., Bjorkman, P. J., and Weley, D. C.,** A hypothetical model of the foreign antigen binding side of class II histocompatibility molecules, *Nature*, 332, 845, 1988.

49. **Brown, J. H., Jardetzky, T., Gorga, J. C., Stern, L. J., Strominger, J. L., and Wiley, D. C.,** Crystal structure of the class II histocompatibility antigen HLA-DR1, *J. Cell. Biochem.*, Suppl. 17C, 66, 1993.
50. **Rudensky, A., Preston-Hurlburt, P., Hong, S.-C., Barlow, A., and Janeway, C. A.,** Sequence analysis of peptides bound to MHC class II molecules, *Nature*, 353, 622, 1991.
51. **Inaba, K. and Steinman, R. M.,** Resting and sensitized T lymphocytes exhibit distinct stimulatory (antigen-presenting cell) requirements for growth and lymphokine release, *J. Exp. Med.*, 160, 1717, 1984.
52. **Weaver, C. T. and Unanue, E. R.,** The costimulatory function of antigen-presenting cells, *Immunol. Today*, 11, 49, 1990.
53. **Romani, N., Koide, S., Crowley, M., Witmer-Pack, M., Livingstone, A. M., Fathman, C. G., Inaba, K., and Steinman, R. M.,** Presentation of exogenous protein antigens by dendritic cells to T cell clones. Intact protein is presented best by immature, epidermal Langerhans cells, *J. Exp. Med.*, 169, 1169, 1989.
54. **Chesnut, R. W. and Grey, H. M.,** Studies on the capacity of B cells to serve as antigen-presenting cells, *J. Immunol.*, 126, 1075, 1981.
55. **Lanzavecchia, A., Roosnek, E., Gregory, T., Berman, P., and Abrignani, S.,** T cells can present antigens such as HIV gp120 targeted to their own surface molecules, *Nature*, 334, 530, 1988.
56. **Inaba, K. and Steinman, R. M.,** Accessory cell-T lymphocyte interactions. Antigen-dependent and -independent clustering, *J. Exp. Med.*, 163, 247, 1986.
57. **Steinman, R. M. and Witmer, M. D.,** Lymphoid dendritic cells are potent stimulators of the primary mixed leukocyte reaction in mice, *Proc. Natl. Acad. Sci. U.S.A.*, 75, 5132, 1978.
58. **Inaba, K., Metlay, J. P., Crowley, M. T., and Steinman, R. M.,** Dendritic cells pulsed with protein antigens *in vitro* can prime antigen specific, MHC-restricted T cells *in situ, J. Exp. Med.*, 172, 631, 1990.
59. **Inaba, K., Young, J. W., and Steinman, R. M.,** Direct activation of CD8+ cytotoxic T lymphocytes by dendritic cells, *J. Exp. Med.*, 166, 182, 1987.
60. **Young, J. W. and Steinman, R. M.,** Dendritic cells stimulate primary human cytolytic lymphocyte responses in the absence of CD4+ helper T cells, *J. Exp. Med.*, 171, 1315, 1990.
61. **Macetonia, S. E., Taylor, P. M., Knight, S. C., and Askonas, B. A.,** Primary stimulation by dendritic cells induces antiviral proliferative and cytotoxic T cell responses *in vitro, J. Exp. Med.*, 169, 1255, 1989.
62. **Langhoff, E. and Steinman, R. M.,** Clonal expansion of human T lymphocytes initiated by dendritic cells, *J. Exp. Med.*, 169, 315, 1989.
63. **Metlay, J. P., Pure, E., and Steinman, R. M.,** Distinct features of dendritic cells and anti-Ig activated B cells as stimulators of the primary mixed leukocyte reaction, *J. Exp. Med.*, 169, 239, 1989.
64. **Inaba, K., Steinman, R. M., Van Voorhis, W. C., and Muramatsu, S.,** Dendritic cells are critical accessory cells for thymus-dependent antibody responses in mouse and man, *Proc. Natl. Acad. Sci. U.S.A.*, 80, 6014, 1983.
65. **Knight, S. C., Mertin, J., Stackpoole, A., and Clarke, J.,** Induction of immune responses *in vivo* with small numbers of veiled (dendritic) cells, *Proc. Natl. Acad. Sci. U.S.A.*, 80, 6032, 1983.
66. **Morris, B.,** The cells of lymph and their role in immunological reactions, *Handbuch der Allg. Pathol.*, 3(6), 405, 1972.
67. **Breel, M., Mebius, R., and Kraal, G.,** The interaction of interdigitating cells and lymphocytes *in vitro* and *in vivo*, *Immunology*, 63, 331, 1988.
68. **Kelly, R. H., Balfour, B. M., Armstrong, J. A., and Griffiths, S.,** Functional anatomy of lymph nodes. II. Peripheral lymph-borne mononuclear cells, *Anat. Rec.*, 190, 5, 1978.
69. **Fossum, S.,** Lymph-borne dendritic leukocytes do not recirculate, but enter the lymph node paracortex to become interdigitating cells, *Scand. J. Immunol.*, 27, 97, 1988.
70. **Kraal, G., Breel, M., Janse, M., and Bruin, G.,** Langerhans cells, veiled cells, and interdigitating cells in the mouse recognized by a monoclonal antibody, *J. Exp. Med.*, 163, 981, 1986.
71. **Balfour, B. M., Drexhage, H. A., Kamperdijk, E. W. A., and Hoefsmit, E. A.,** Antigen presenting cells, including Langerhans cells, veiled cells and interdigitating cells, in *Microenvironments in Haemopoietic and Lymphoid Differentiation*, Pitman, London, 1981, 281.
72. **Bujdoso, R., Hopkins, J., Dutia, B., Young, P., and McConnell, I.,** Characterisation of sheep afferent lymph veiled cells and their role in antigen carriage, *J. Exp. Med.*, 170, 1285, 1989.

73. **Silberberg-Sinakin, I., Thorbecke, G. J., Baer, R. L., Rosenthal, S. A., and Berezowsky, V.,** Antigen bearing Langerhans' cells in the skin, dermal lymphatics and in lymph nodes, *Cell. Immunol.*, 25, 137, 1976.
74. **Hoefsmit, E. C. M., Duijvestign, A. M., and Kamperdijk, E. W.,** Relation between Langerhans cells, veiled cells and interdigitating cells, *Immunobiology*, 161, 255, 1982.
75. **McKeever, D. J., MacHugh, N. D., Goddeeris, B. M., Awino, E., and Morrison, W. I.,** Bovine afferent lymph veiled cells differ from blood monocytes in phenotype and are superior in accessory function, *J. Immunol.*, 147, 3703, 1991.
76. **Hemler, M. E.,** Adhesive protein receptors on haemopoietic cells, *Immunol. Today*, 9, 109, 1988
77. **Cattoretti, G., Berti, E., Mancuso, A., and D'Amato, L.,** An MHC class I related family of antigens with widespread distribution on resting and activated cells, in *Leukocyte Typing III*, McMichael, A. J., Ed., Oxford University Press, New York, 1987, 89.
78. **Parsons, K. R., Bembridge, G., Sopp, P., and Howard, C. J.,** Studies of monoclonal antibodies identifying two novel bovine lymphocyte antigen differentiation clusters: workshop clusters (WC) 6 and 7, *Vet. Immunol. Immunopathol.*, in press.
79. **Ellis, J. A., Davis, W. C., MacHugh, N. D., Emery, D. L., Kaushal, A., and Morrison, W. I.,** Differentiation antigens on bovine mononuclear phagocytes identified by monoclonal antibodies, *Vet. Immunol. Immunopathol.*, 19, 325, 1988.
80. **Rademakers, L. H. P. M., van Wychen, D., and de Weger, R. A.,** Immunohistochemical localisation of intermediate filament and contractile proteins in the tonsil with reference to follicular dendritic cells, in *Lymphatic Tissues and In Vivo Immune Responses*, Imhof, B. A., Berrih-Aknin, S., and Ezine, S., Eds., Marcel Dekker, New York, 1990, 721.
81. **Emery, D. L., MacHugh, N. D., and Ellis, J. A.,** The properties and functional activity of non-lymphoid cells from bovine afferent lymph, *Immunology*, 62, 177, 1987.
82. **Hopkins, J., Dutia, B. M., Bujdoso, R., and McConnell, I.,** *In vivo* modulation of CD1 and MHC class II expression by sheep afferent lymph dendritic cells. Comparison of primary and secondary immune responses, *J. Exp. Med.*, 170, 1303, 1989.
83. **Hein, W. R., McClure, S. J., and Miyasaka, M.,** Cellular composition of peripheral lymph and skin of sheep defined by monoclonal antibodies, *Int. Arch. Allergy Appl. Immunol.*, 84, 241, 1987.
84. **Streilein, J. W. and Grammar, S. F.,** *In vitro* evidence that Langerhans cells can adopt two functionally distinct forms capable of antigen presentation to T lymphocytes, *J. Immunol.*, 143, 3925, 1989.
85. **Morrison, W. I., Lalor, P. A., Christensen, A. K., and Webster, P.,** Cellular constituents and structural organization of the bovine thymus and lymph node. In: *The Ruminant Immune System in Health and Disease*, Morrison, W. I., Ed., Cambridge University Press, Cambridge, 1986, 220.
86. **McKeever, D. J., Awino, E., and Morrison, W. I.,** Afferent lymph veiled cells prime CD4+ T cell responses *in vivo*, *Eur. J. Immunol.*, 22, 3057, 1992.
87. **Miller, H. R. P. and Adams, E. P.,** Reassortment of lymphocytes in lymph from normal and allografted sheep, *Am. J. Pathol.*, 87, 59, 1977.
88. **Hall, J. G. and Robertson, D.,** Phagocytosis *in vivo* of immune complexes by dendritic cells in the lymph of sheep, *Int. Arch. Allergy Appl. Immunol.*, 73, 155, 1984.
89. **Harkiss, G. D., Hopkins, J., and McConnell, I.,** Uptake of antigen by afferent lymph veiled cells mediated by antibody, *Eur. J. Immunol.*, 20, 2367, 1990.
90. **Barfoot, R., Denham, S., Gyure, L. A., Hall, J. G., Hobbs, S. M., and Jackson, L. E.,** Some properties of dendritic macrophages from peripheral lymph, *Immunology*, 68, 233, 1989.
91. **Rhodes, J. M., Balfour, B. M., Blom, J., and Agger, R.,** Comparison of antigen uptake by peritoneal macrophages and veiled cells from the thoracic duct using isotope-, FITC-, or gold-labelled antigen, *Immunology*, 68, 403, 1989.
92. **Drexhage, H. A., Mulink, H., de Groot, J., Clarke, J., and Balfour, B. M.,** A study of cells present in peripheral lymph of pigs with special reference to a type of cell resembling the Langerhans cell, *Cell Tissue Res.*, 202, 407, 1979.
93. **Steinman, R. M. and Nussenzweig, M. C.,** Dendritic cells: features and functions, *Immunol. Rev.*, 53, 127, 1980.
94. **Levine, T. P. and Chain, B. M.,** Endocytosis by antigen presenting cells: dendritic cells are as endocytically active as other antigen presenting cells, *Proc. Natl. Acad. Sci. U.S.A.*, 89, 8342, 1992.

95. **Glass, E. J. and Spooner, R. L.,** Requirements for MHC class II positive accessory cells in an antigen-specific bovine T cell response, *Res. Vet. Sci.*, 46, 196, 1989.
96. **Lutje, V. and Black, S. J.,** Cellular interactions regulating the *in vitro* response of bovine lymphocytes to ovalbumin, *Vet. Immunol. Immunopathol.*, 28, 275, 1991.
97. **Lanzavecchia, A.,** Antigen-specific interaction between T and B cells, *Nature*, 314, 537, 1985.
98. **Casten, L. A. and Pierce, S. K.,** Receptor mediated B cell antigen processing: increased antigenicity of a globular protein covalently coupled to antibodies specific for B cell surface structures, *J. Immunol.*, 140, 404, 1988.
99. **Jelachich, M. L., Lakey, E. K., Casten, L. A., and Pierce, S. K.,** Antigen presentation is a function of all B cell subpopulations separated on the basis of size, *Eur. J. Immunol.*, 16, 411, 1986.
100. **Nussenzweig, M. C. and Steinman, R. M.,** Contribution of dendritic cells to stimulation of the murine syngeneic mixed leukocyte reaction, *J. Exp. Med.*, 151, 1196, 1980.
101. **Flechner, E., Freudenthal, P., Kaplan, G., and Steinman, R. M.,** Antigen-specific T lymphocytes efficiently cluster with dendritic cells in the human primary mixed lymphocyte reaction, *Cell. Immunol.*, 111, 183, 1988.
102. **Inaba, K. and Steinman, R. M.,** Monoclonal antibodies to LFA-1 and to CD4 inhibit the mixed leukocyte reaction after the antigen-independent clustering of dendritic cells and T lymphocytes, *J. Exp. Med.*, 165, 1403, 1987.
103. **Makgoba, M. W., Sanders, M. E., Luce, G. E. G., Dustin, M. L., Springer, T. A., Clark, E. A., Mannoni, P., and Shaw, S.,** ICAM-1 is the ligand for LFA-dependent adhesion of B, T, and myeloid cells, *Nature*, 331, 86, 1988.
104. **Carayanniotis, G., Skea, D. L., Luscher, M. A., and Barber, B. H.,** Adjuvant-independent immunization by immunotargetting antigens to MHC and non MHC determinants *in vivo*, *Mol. Immunol.*, 28:3, 261, 1991.
105. **Hopkins, J., Dutia, B. M., and McConnell, I.,** Monoclonal antibodies to sheep lymphocytes. I. Identification of MHC class II molecules on lymphoid tissue and changes in the level of class II expression on lymph-borne cells following antigen stimulation *in vivo*, *Immunology*, 59, 433, 1986.
106. **MacHugh, N. D., Bensaid, A., Davis, W. C., Howard, C. J., Parsons, K. R., Jones, B., and Kaushal, A.,** Characterisation of a bovine thymic differentiation antigen analogous to CD1 in the human, *Scand. J. Immunol.*, 27, 541, 1988.
107. **Mackay, C., Maddox, J., Goglin-Ewens, K., and Brandon, M.,** Characterisation of two sheep lymphocyte differentiation antigens, SBU-T1 and SBU-T6, *Immunology*, 55, 729, 1985.

Chapter 7

Lymphocyte Recirculation and Homing

W. G. Kimpton, E. A. Washington, and R. N. P. Cahill

CONTENTS

I. Introduction 109
I. Lymphocyte Migration Streams 110
A. The Sheep Model 110
B. Major Pathways of Lymphocyte Migration 112
III. *In Vivo* Homing of Lymphocytes to Peripheral Lymph Nodes 113
A. Basic Physiology of Lymphocyte Traffic through Lymph Nodes 113
B. L-selectin is the Peripheral Lymph Node Homing Receptor 114
C. Differential *In Vivo* Homing of CD4, CD8, and γδ T Cells to Lymph Nodes 114
D. Altered Homing of αβ and γδ T Cells to Peripheral Lymph Nodes During an Immune Response 115
E. Naive and Memory T Cells in Ruminants 116
IV. *In Vivo* Homing of Lymphocytes to Skin and Peripheral Tissues 116
V. *In Vivo* Homing of Lymphocytes to Gut 118
A. Homing of T Cells to the Gut 118
B. *In Vivo* Homing of Lymphocytes to the Gut in the Fetus 118
C. Migration of Memory or Activated Cells 119
VI. Homing of γδ T Cells in Fetal and Post-Natal Animals 120
VII. Conclusions 120
References 121

I. INTRODUCTION

There has been a considerable increase in the understanding of many aspects of lymphocyte recirculation in recent years and it is now possible to provide a molecular basis for some of the receptor-ligand interactions which enable lymphocytes to bind to vascular endothelium. The binding of lymphocytes to endothelial cells is the crucial first step in enabling lymphocytes to leave the blood and enter the tissues and it is, therefore, critical in determining the tissue destination or homing behavior of lymphocytes. Lymphocyte-endothelial cell binding has been studied extensively by many groups over the last 10 years. These studies have been largely, but not exclusively, based on the *in vitro* binding of mouse or human lymphocytes to tissue sections or cultured endothelial cells.[1-3] It has become clear that endothelial cells express adhesion molecules on their surface which bind to complimentary ligands called *homing receptors* expressed on the surface of circulating lymphocytes, and it is now possible to construct credible, but not necessarily correct or complete, descriptions of how T cells bind to endothelium in at least some tissues.[2,4]

The *in vitro* binding characteristics of lymphocytes to endothelium have been extremely useful in identifying cell surface molecules which might play a role in lymphocyte migration; however, definitive evidence for the physiological relevance of a putative adhesion molecule or homing receptor requires experiments in intact animals. The only decisive test of the homing properties of lymphocytes centers on demonstrating their ability to home repeatedly to specific tissues such as skin or gut *in vivo*. The most extensive studies on the physiology of lymphocyte recirculation and homing *in vivo* have been performed in sheep, and this chapter will deal largely with work performed in this species.

0-8493-4952-4/94/$0.00+$.50

II. LYMPHOCYTE MIGRATION STREAMS

A. THE SHEEP MODEL

Lymphocytes in blood and lymph form part of a large pool of lymphocytes which are continuously recirculating between the various lymphoid organs and other tissues of the body, traveling from the bloodstream into lymphoid organs, then to the collecting efferent lymphatics, and back into the bloodstream to repeat the cycle. The great advantage of the sheep in studying the process of recirculation lies in the fact that chronic lymphatic fistula can be established surgically so that all the cells and lymph draining whole organs, or portions thereof, can be collected and quantified. In this way, the exact number and phenotype of cells passing through any individual organ can be measured directly and compared with the number and phenotype of cells which are being offered for extraction in the blood supplying the individual organ or tissue.

The sheep was developed as a model for studying lymphocyte recirculation by the late Bede Morris, and the traffic of cells through normal and variously stimulated tissues and organs has been extensively documented.[5-7] The last 10 years have seen a considerable improvement in the precision with which lymphocyte homing can be studied in ruminants due to the production of a large number of monoclonal antibodies which identify T cell subsets, adhesion molecules, homing receptors, and sheep immunoglobulins. Other chapters in this book deal extensively with leukocyte markers in ruminants, but Table 7-1 shows the adhesion molecules which have been used in recirculation studies in ruminants.

One of the central aims of studying the tissue-selective homing patterns of lymphocytes is to try and understand just how a series of interactions between circulating lymphocytes and blood vascular endothelium are orchestrated to bring together an array of phenotypically and functionally distinct subpopulations of lymphocytes into very different assemblies from one tissue to another. The underlying assumption is that the phenomenon of lymphocyte recirculation represents an economy of scale in the immune system whereby different cohorts of lymphocytes of appropriate antigen specificities and developmental potential are concentrated in various tissues where they may encounter antigen and fulfill their effector functions. It has been suggested that lymphocyte recruitment in antigen-stimulated lymph nodes provides a means whereby antigen presented on accessory cells could be screened by millions of different lymphocytes each with a different antigen specificity, and that the immune system might be better served by directing memory or recently activated T cells back to sites where they might meet previously encountered antigens in skin and mucous membranes.

If only memory or recently activated T cells home to skin, for example, where and how did they acquire their skin-homing properties? The answer would presumably be that some time after T cells are exported from the thymus and commence to recirculate as naive cells, they make contact with their cognate antigen in lymph nodes, resulting in the generation of progeny which proceed to recirculate through skin rather than lymph nodes. Clearly, we need to know more about the maturation of T cells after their export from the thymus to answer this question. What, for example, are the homing properties of T cells recently exported from the thymus and, in particular, what changes occur in their homing patterns after export either as a consequence of activation by antigen, or indeed independently from any effects of antigen? The sheep fetus provides a unique way of approaching these questions in an *in vivo* model because it allows the homing patterns of T cells to be studied in a situation where their naive status is absolute in an environment where there is no foreign antigen and no memory or recently activated T cells.

The sheep placenta sequesters the developing fetal lamb from foreign antigen so that it remains immunologically virgin until the first exposure to antigen, which normally occurs after birth.[8,9] A number of factors, including the length of gestation (150 days), availability of reagents, and surgical accessibility, have made the sheep fetus an excellent model for studying the development of the mammalian immune system and lymphocyte recirculation (reviewed in Reference 9). Lymphocytes start to recirculate from blood to lymph in the sheep fetus at least as early as 75 days of gestation and the size of the pool of recirculating lymphocytes increases exponentially throughout *in utero* life as the fetus grows and new cells are added to the blood, lymph, and lymphoid tissues.[8] The size and composition of the pool has been examined by collecting cells from the fetus in experiments involving chronic catheterization of blood vessels and lymphatics in fetal lambs *in utero*.[8-11] The pool increases from around $5–6 \times 10^8$ cells at 90 to 95 days gestation to about 10^{10} cells at term. The predominant cell circulating in the fetus is the T cell, and all the major T cell subsets present in the adult are circulating in the fetus.[8,9,11-15] Immunological

Table 7-1 **Molecules Used in Lymphocyte Recirculation Studies in Ruminants**

Molecular family		Possible role in lymphocyte recirculation
Selectin	L-selectin	Peripheral lymph node homing receptor; involved in αβ and γδ T cell binding to lymph node HEV; expressed on most γδ T cells regardless of tissue source; expressed on all αβ T cells in peripheral lymph node efferent lymph and about half αβ T cells in skin afferent lymph.
Addressin	PNAd	Ligand for L-selectin; expressed on peripheral lymph node endothelium. In sheep, PNAd is also found on Peyer's patch endothelium.
Integrin	LFA-1 (CD11a/CD18)	Expressed on all T cells; binds to ICAM-1,2 on endothelial cells; may provide a "secondary" adhesion bond that strengthens the lymphocyte-endothelial interaction after initial binding via homing receptors.
	VLA-4 (α4/β1) (CD49d/CD29)	Involved in homing to sites of chronic inflammation. In sheep α4 is upregulated on most memory T cells regardless of tissue source. The β1 component is high on $CD45R^-$ T cells from skin and peripheral lymph nodes but low on those from Peyer's patches, mesenteric nodes, and gut lymph.
Ig superfamily	VCAM-1	Ligand for VLA-4; expressed on inflamed endothelium in skin and peripheral nodes in sheep but not on gut vascular endothelium.
	CD2	Is expressed on most αβ T cells. Is upregulated on activated T cells but unlikely to be involved in T cell homing.
	LFA-3 (CD58)	Ligand for CD2. Expressed on all T cells.
Others	CD44	Proteoglycan cartilage-link protein; expressed at a lower level by γδ TCR than αβ TCR T cells in sheep. Binds to hyaluronate.
	T19 (WC1)	Expressed on a subset of γδ T cells in ruminants. Function is unknown but it may have a role in the homing of γδ T cells.
	$CD45R^+$	High molecular weight isoform of leukocyte common antigen; it is expressed on all B cells, and on naive αβ T cells, but not memory αβ T cells in the blood of immunized sheep and cattle.

studies have shown essentially the same pattern of development for the thymus, spleen, lymph nodes, and Peyer's patches in the human and ovine fetus, and the human infant and lamb have reached a similar stage of immunological development at birth. This is in contrast to rats and mice where the lymphoid organs are poorly developed at birth; thus the sheep fetus also serves as a useful model for immunological development in the human fetus.[15]

B. MAJOR PATHWAYS OF LYMPHOCYTE MIGRATION

A scheme summarizing the major pathways of lymphocyte migration in fetal and post-natal animals which will be discussed in this chapter is illustrated in Figure 7-1. Newly formed lymphocytes are exported from the primary lymphoid organs into the peripheral pool of lymphocytes. In sheep, the ileal Peyer's patch is the main source of virgin or naive B cells which enter the bloodstream via the ileal lymph node and intestinal lymph, after which they migrate randomly between the gut-associated lymphoid tissues and peripheral lymph nodes.[16] The thymus exports naive T cells at a rate of around 1% of total thymocytes per day, and the majority of thymic emigrants enter the bloodstream directly via thymic veins, although a proportion enter the blood via thymic lymph.[17-19] Lymphocytes in the bloodstream consist of a very heterogeneous pool of cells, the majority of which are in transit and only remain in the blood for about 30 min.[20] Depending on the age of the animal, the entire blood vascular pool of lymphocytes only represents some 1 to 2% of the total body lymphocyte population.[20] The spleen is a major site of lymphocyte recirculation and is continuously exchanging lymphocytes with the blood, with an average spleen transit time of only a few hours. The homing of lymphocytes to the spleen has been studied more extensively in pigs[21] and rats[22] than in ruminants, and it appears to be mediated by a different set of molecular interactions from those governing lymphocyte homing to other tissues.[3,21]

As well as migrating from the blood to the spleen, lymphocytes are continuously leaving the blood and entering secondary lymphoid tissues such as peripheral and mesenteric lymph nodes, Peyer's patches, and gut. This process is controlled by specialized vascular endothelial cells which line post-capillary venules at traffic sites in lymph nodes and gut. These venules have a characteristic cuboidal-shaped endothelium in

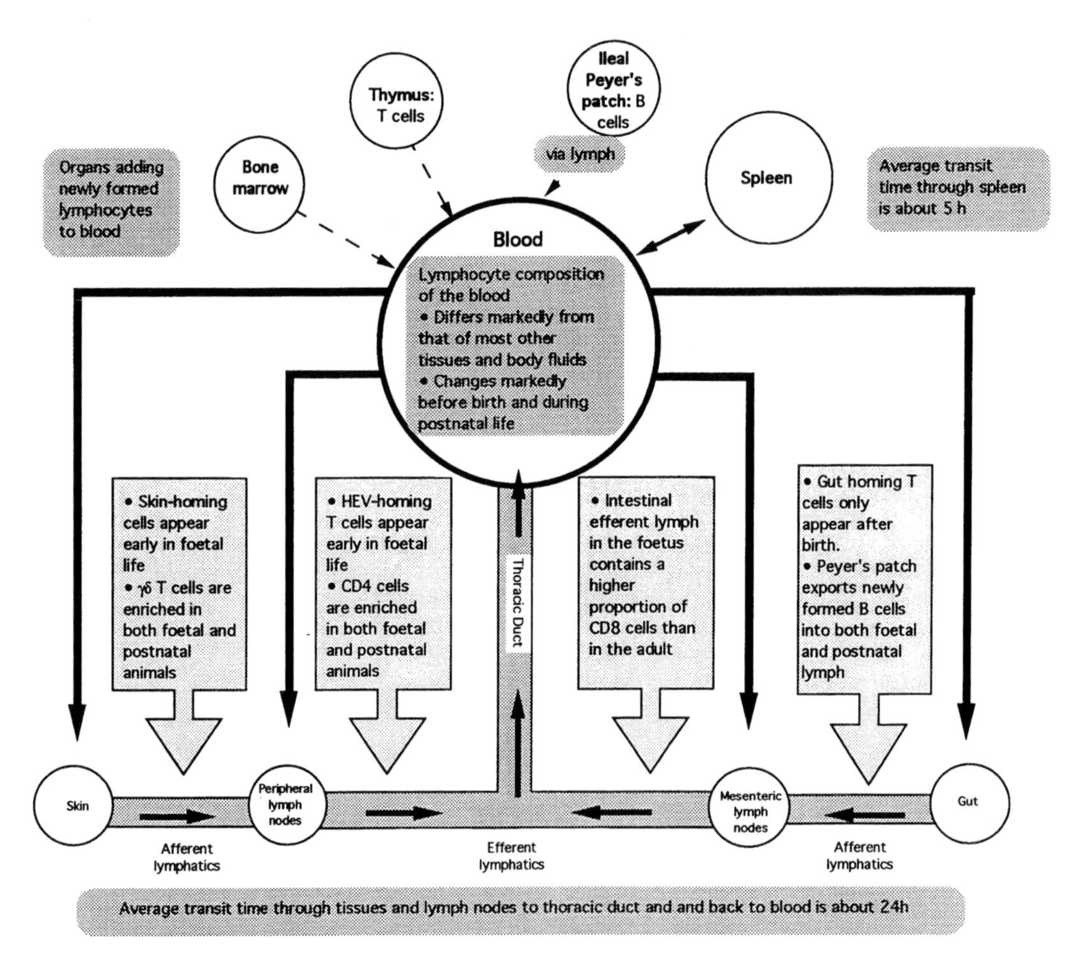

Figure 7-1 A scheme summarizing the major pathways of lymphocyte migration in fetal and post-natal animals.

rodents and humans and have become universally known as high endothelial venules (HEV). Although post-capillary venules do not have this high endothelial morphology in sheep (except in early fetal lymph nodes[23] and in adjuvant-induced granulomas[24]), we will nonetheless refer to them as HEV.

III. *IN VIVO* HOMING OF LYMPHOCYTES TO PERIPHERAL LYMPH NODES

A. BASIC PHYSIOLOGY OF LYMPHOCYTE TRAFFIC THROUGH LYMPH NODES

Much of our understanding of the basic physiology of lymphocyte traffic through lymph nodes has come from the isolated lymph node model in sheep which was developed by Morris and colleagues (reviewed in References 5 and 6). Lymphocytes can enter lymph nodes in one of only two ways. They can enter either directly from the blood via HEV in the lymph node itself or they can enter via afferent lymph draining the various organs and tissues such as skin. Regardless of their route of entry, in most species (the pig is a notable exception[25]), once lymphocytes have entered lymph nodes, they can then only leave via efferent lymph. A scheme showing some of the major features of lymphocyte homing to lymph nodes and skin which will be discussed in this chapter is shown in Figure 7-2. The classic experiments of Hall and Morris showed that in a resting lymph node unstimulated by antigen, over 90% of lymphocytes enter a lymph node via HEV, less than 10% enter via afferent lymph, and that less than 2% of lymphocytes in efferent lymph have been generated by cell division within the node itself.[26] Antigen has a profound effect on the traffic of lymphocytes through a lymph node. There is a transient fall in lymphocyte output from a lymph node in the first few hs after challenge with antigen, which is called *shutdown.*[27] The extent of shutdown varies with the antigen used, being most profound with viral antigens, and it occurs at the same time as antigen causes a massive recruitment of lymphocytes into the lymph node.[28] The hallmark of a lymph node undergoing an immune response is a massive increase in the number of lymphocytes entering the node via HEV and trafficking through into efferent lymph. This lymphocyte recruitment is associated with an increase in blood flow to the stimulated lymph node and the release of a variety of cytokines into efferent lymph.[29, 30] It is also associated with a period during which cells with specific antigen reactivity to the stimulating antigen cannot be detected in efferent

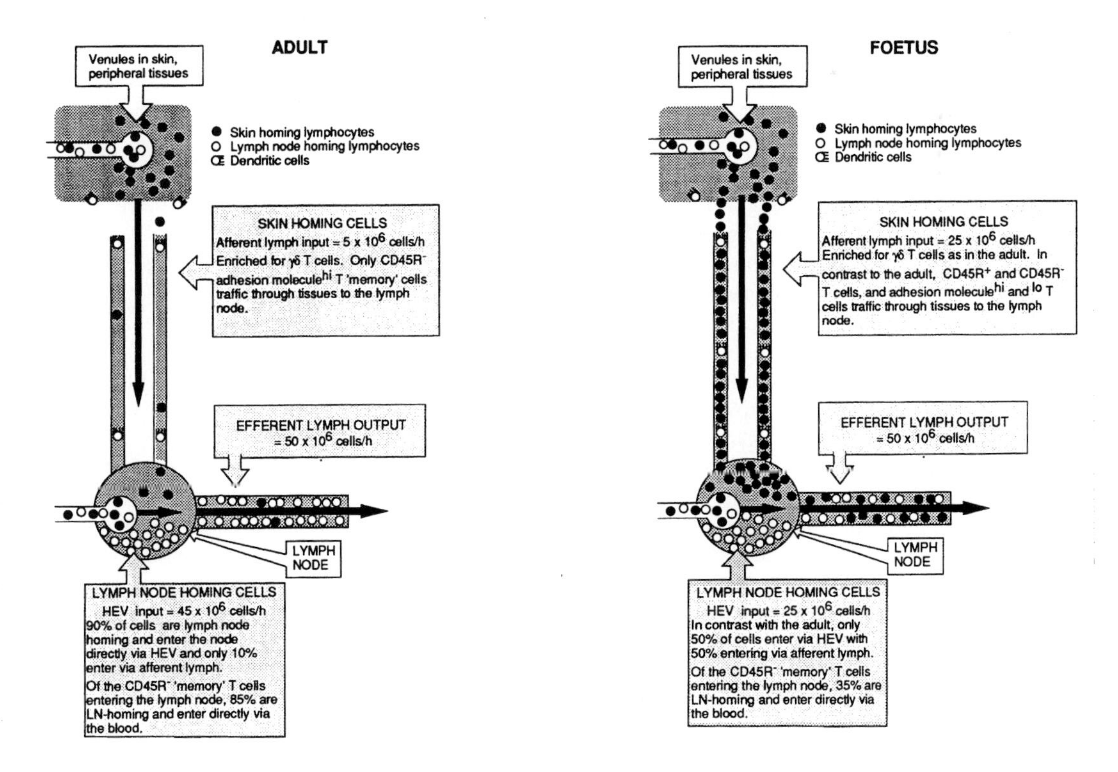

Figure 7-2 Scheme showing the preferred routes of entry of T cells to skin and lymph nodes in fetal and adult sheep.

lymph, and it is followed by the appearance of large numbers of newly formed blast cells which have been generated within the lymph node in response to antigen.[31-33] It has been shown that the cells leaving a lymph node in the efferent lymph during this period are responsible for the amplification and dissemination of the immune response, so that an intact lymphatic system is mandatory for spread of memory and effector cells throughout the body.[34]

B. L-SELECTIN IS THE PERIPHERAL LYMPH NODE HOMING RECEPTOR

The best documented homing receptor on lymphocytes is L-selectin (Mel-14, LAM-1, LECAM-1) which has been well conserved throughout mammalian evolution[35] and has been described in sheep[36] and cattle.[37] Some of the adhesion molecules which are thought to be important in lymphocyte-endothelial interactions and which have been examined in ruminants are shown in Figure 7-3. L-selectin is the prototypic homing receptor and is expressed on many T cells before they leave the thymus and recent thymic emigrants home well to lymph nodes.[38] The endothelial adhesion molecule which interacts with L-selectin is an oligosaccharide ligand expressed on HEV called the *peripheral lymph node addressin* (PNAd) and is recognized by MAb MECA-79 in mice.[3] MECA-79 has also been shown to react with sheep lymph node HEV quite intensely.[36] The majority of efferent lymph cells draining peripheral lymph nodes are L-selectin$^+$, as would be expected for lymph node homing lymphocytes.[39] They also express the β_2 integrin LFA-1 which recognizes ICAM-1 and 2 and probably other ligands on endothelium, and they express LFA-3, CD2, and CD44.[36, 39] It has been suggested that the receptor-ligand interactions mediated by these molecules are important as "secondary" adhesion molecules which are involved in strengthening the adhesion of L-selectin$^+$ cells binding to HEV.[1-4] In this context, it is important to consider the time scale of the events involved in lymphocyte homing to a particular tissue. Homing has been observed *in situ* in living animals and the relative time scale of the events involved has been quantified directly. The migration of lymphocytes to Peyer's patches in rats *in vivo* was reported as seconds for adhesion to endothelium, min for transposition across the endothelium, and hs for retention in tissue.[40] Any changes in the expression of "secondary" adhesion molecules triggered by L-selectin ligation with PNAd would have to occur within min to be important in homing and, in our view, changes in the expression of these molecules may well prove to be more important in a variety of transient cell-to-cell interactions and in cell locomotion to discrete T and B cell domains in lymph nodes after lymphocytes have negotiated HEV.

C. DIFFERENTIAL *IN VIVO* HOMING OF CD4, CD8, AND γδ T CELLS TO LYMPH NODES

Although L-selectin is on all cells entering lymph nodes and is clearly important in mediating lymphocyte entry into peripheral lymph nodes, the fact that lymphocyte subsets are present in very different proportions in blood and efferent lymph draining peripheral lymph nodes suggests that additional subset-linked selection mechanisms may be operating at HEV. CD4$^+$ T cells are present in efferent lymph in a higher proportion than blood, while γδ T cells are enriched in blood compared with efferent lymph.[41,42] Analysis of lymphocyte subset delivery to and output from prescapular lymph nodes suggested that, under steady-state conditions, CD4$^+$ cells were being extracted at a faster rate by HEV (about 1 in 2) than CD8$^+$, γδT19$^+$, and B cells (1 in 4 to 1 in 5).[42] Further experiments tested this possibility directly by labeling lymphocytes *in vitro* with fluorochrome and examining their migration *in vivo* through prescapular lymph nodes into lymph.[43] The results indicated that γδT19$^+$ cells were in fact extracted at the same rate as CD4$^+$ cells and the slower extraction rate for these cells in the steady-state studies was thought to be due to a population of γδT19$^+$ cells which remained in the blood and did not recirculate. These observations on the non-random migratory characteristics of CD4$^+$, CD8$^+$, and γδT19$^+$ lymphocyte subsets, however, are not easily explained by invoking solely tissue-specific lymphocyte-endothelial cell recognition mechanisms. One difficulty with a single specificity hypothesis in which only tissue-specific adhesion molecules are involved is that every vascular bed in which lymphocyte subsets are extracted at different rates would require its own specific population of lymphocytes which could only migrate through that particular bed, and whose extraction rates relative to each other could not be changed by any alteration in the expression of adhesion molecules on vascular endothelium. In these experiments, lymphocyte subsets of a single tissue specificity were taken from efferent lymph and returned into blood, whereupon they were extracted by the lymph nodes of origin in quite different proportions from those which existed when they were infused into the blood.[43] It was suggested that there may be two distinct families of adhesion molecules which could act jointly to regulate the homing of lymphocytes to

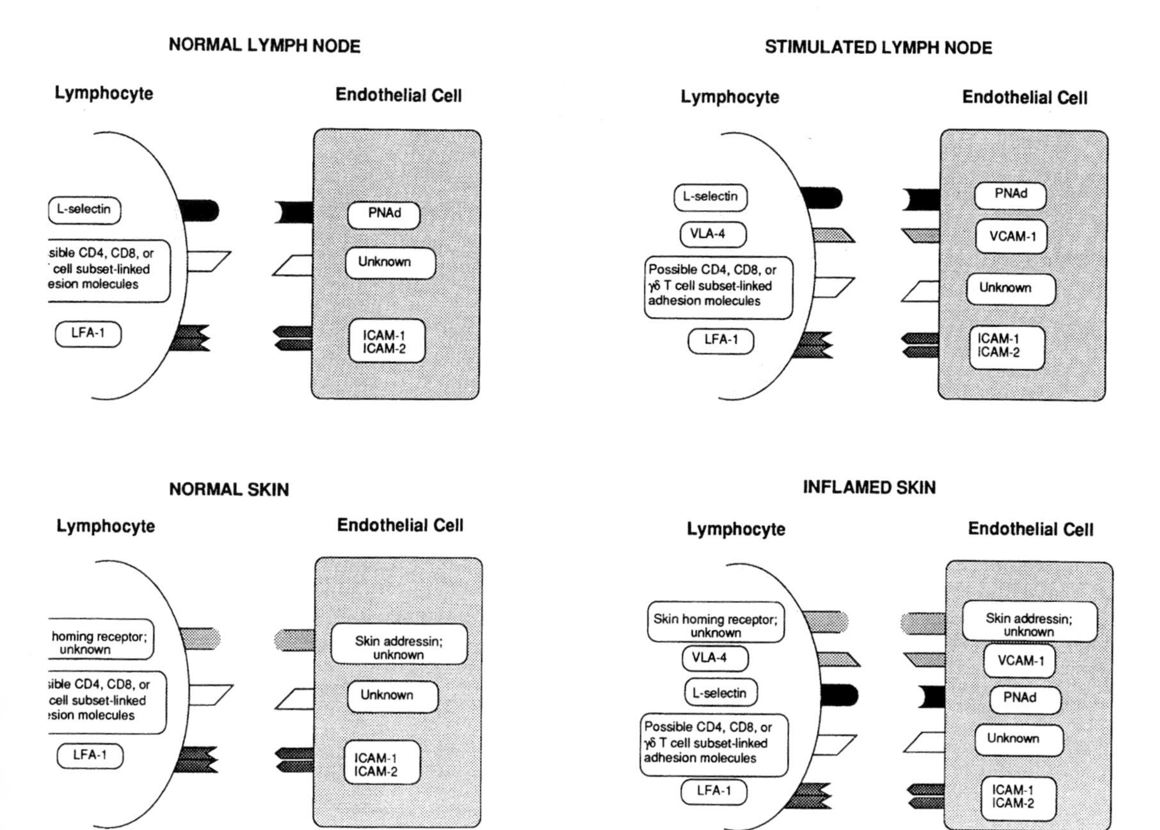

Figure 7-3 Scheme showing possible lymphocyte-endothelial interactions which may regulate the homing of T cells to normal and inflamed tissues. Direct evidence that these interactions are critical in the *in vivo* homing of T cells in ruminants is still lacking, but there is unequivocal experimental evidence for the role of L-selectin and VLA-4 in T cell homing in mice.[79,80]

peripheral lymph nodes. In this model, tissue-specific lymphocyte-endothelial interactions would enable lymphocytes to home to lymph nodes or skin, for example, and subset-specific interactions would determine the rate of extraction of lymphocyte subsets of a given tissue specificity.[41-45] The interaction of L-selectin with PNAd is certainly critical in the tissue-specific homing of cells to HEV, but experiments in which the *in vivo* migration of $CD4^+$ and $CD8^+$ cells through four different tissue compartments was examined suggested that the enrichment of $CD4^+$ cells over $CD8^+$ cells which occurs at HEV was operating independently from the tissue-specific homing of lymphocyte subsets.[45]

This also appears to be the case in fetal sheep, where all lymph node homing cells are L-selectin$^+$; but, although there are differences in the preferred routes of entry into lymph nodes, the extraction of T cells and B cells by lymph nodes in the fetus is remarkably similar to that in adult animals. Thus, $CD4^+$ cells are also the major T cell subset homing to peripheral lymph nodes in the fetus and are enriched in fetal efferent lymph compared with blood. They also appear to be extracted by HEV in fetal nodes at a faster rate than other T cells and B cells in the same way as they are in adult lymph nodes.[13] Peripheral lymph node-homing $CD8^+$ cells are present in the same proportions in efferent lymph draining peripheral lymph nodes in both fetus and adult, but there are fewer B cells in the fetus, presumably reflecting the much smaller numbers of circulating B cells (<5%) in the fetal pool.[8,9,11-14]

D. ALTERED HOMING OF αβ AND γδ T CELLS TO PERIPHERAL LYMPH NODES DURING AN IMMUNE RESPONSE

The distribution of T cell subsets in efferent lymph draining a single lymph node changes considerably after stimulation with antigen. The altered ratios of T cell subsets seen during lymphocyte recruitment suggested that altered extraction of T cell subsets by HEV may be induced by antigen.[46-48] A more recent study, using PPD as the stimulating antigen in BCG primed animals, found striking changes in the efferent lymph draining popliteal lymph nodes stimulated with PPD.[39] The output of CD4+, CD8+, $\gamma\delta$ T cells, and B cells all increased markedly, but the outputs peaked at differing times, raising the possibility that HEV in antigen-stimulated nodes were extracting T cell subsets at different rates at different times during the immune response. Of particular interest in this study were the changes in the output of cells expressing $\alpha_4\beta_1$ integrin (VLA-4) which is one of the principle adhesion molecules for extravasation of lymphocytes into inflammatory lesions, where it binds to VCAM-1 expressed on inflamed endothelium.[2] T cells migrating through lymph nodes prior to challenge with antigen were predominantly of the naive phenotype and were β_1^- and L-selectin+ (see below). During the shutdown phase immediately following antigen administration, the percentage of L-selectin+ cells fell while that of β_1^+ cells rose. The large increase in cell numbers in efferent lymph during the immune response was due to CD4+β_1^+ memory T cells which were L-selectin-. Immunohistological examination of HEV showed no changes in the expression of PNAd; but unlike HEV in resting lymph nodes, HEV following challenge with antigen were induced to express VCAM-1.[39] The authors suggested that memory T cells were being recruited into an antigen-stimulated lymph node via a completely different mechanism from the L-selectin/PNAd interaction used by naive T cells (and presumably from that used by memory T cells to enter unchallenged nodes), and they suggested that VLA-4/VCAM-1 interactions were used to recruit memory T cells into the challenged node.[39] The authors pointed out that they could not exclude the loss of L-selectin and the expression of VLA-4 after T cells had entered the lymph node, although they considered this was unlikely. In this context, it should be noted that in fetal sheep which contain only naive T cells, lymph nodes show the same phenomenon of lymphocyte recruitment after stimulation with antigen as do lymph nodes in post-natal animals.

E. NAIVE AND MEMORY T CELLS IN RUMINANTS

Naive and memory T cells in ruminants have been distinguished by their expression of various adhesion molecules (particularly of the integrin family) and by their expression of CD45R isoforms.[49,50] Ovine memory T cells were CD2^{hi}, LFA-1^{hi}, LFA-3^{hi}, and CD45RA-; and when peripheral blood lymphocytes from immunized animals were sorted into CD45RA+ and CD45RA- subsets, the proliferative response to the recall antigen resided in the CD45RA- or memory subset.[49] In these studies, the ovine homologue of CD45RA was identified using MAb 73B, which usually immunoprecipitates a single molecule with a molecular weight of M_r 220 kDa.[49] The terminology used in this review to distinguish T cells expressing different isoforms of CD45R is the following: CD45R+ (naive) cells are those cells expressing the high molecular weight p220 isoform, and CD45R- (memory) cells are those cells expressing the low molecular weight p180 isoform. When literature is referred to which uses other terminology (e.g., CD45RA+/- cells), this usage is retained. In sheep, L-selectin is expressed on a large proportion of naive T cells but fewer memory T cells are L-selectin+, raising the possibility that memory T cells show different tissue specificity from naive cells.[36,49] However, using similar proliferative assays in cattle, T cells could not be divided into naive and memory cells on the basis of L-selectin expression.[37] Although some recent studies have shown that memory T cells in rats can convert to the CD45R+ or naive phenotype, indicating that the division of naive and memory T subsets based on CD45R expression is not absolute,[51] they have nonetheless been regarded as functionally and phenotypically distinct subsets of T cells.[36] More recently, memory T cells in sheep have been functionally distinguished by their expression of b_1 integrin, the expression of which on sheep and human T cells signifies either a memory or recently activated/effector T cell.[36] When efferent lymph from 6- to 12- month-old sheep was examined, 10 to 20% of T cells were found to express b_1 integrin as opposed to 25 to 30% of b_1 expressing T cells in efferent lymph from 2 to 3-year-old animals, suggesting that the majority of T cells in efferent lymph draining resting unstimulated lymph nodes were naive T cells and that their numbers were declining with age.[36]

IV. *IN VIVO* HOMING OF LYMPHOCYTES TO SKIN AND PERIPHERAL TISSUES

In all mammals there is a continuous migration of lymphocytes between blood, tissues, and lymph and, while lymphoid tissues are the preferred site of emigration of lymphocytes from the blood, a small but substantial number of lymphocytes are present in afferent lymph draining all tissues. Skin-homing cells can be collected directly in sheep by cannulating afferent lymphatics (reviewed in Reference 7). The cellular composition of afferent lymph draining skin and peripheral tissues, and that of efferent lymph draining lymph nodes, is quite different,[41,52,53] since afferent lymph has tenfold less lymphocytes, fewer B cells, and it contains 5 to 10% veiled or dendritic cells.[24,54] Afferent dendritic cells constitutively express MHC class II, CD1, and the cell adhesion molecule LFA-3, together with the ligand for this molecule CD2.[55] When activated in secondary immune responses *in vivo*, afferent lymph dendritic cells show an enhanced ability to present antigen which correlates directly with increased expression of MHC class II.[56] Lymphocytes have also been shown to recirculate preferentially to lymph draining sites of chronic inflammation in skin.[7] Skin-homing T cells in afferent lymph also display phenotypic differences from lymph node homing cells which enter via HEV. Afferent lymph contains higher proportions of MHC class II$^+$ T cells and T cells bearing $\gamma\delta$ antigen receptors than efferent lymph,[41] and the traffic of these $\gamma\delta$ T cells appears to be altered after administration of antigen.[48, 52]

It has been suggested that naive and memory T cells recirculate via different pathways because T cells in popliteal afferent lymph of adult sheep exclusively express a memory phenotype in that they are CD45RA$^-$, β_1 integrin^{hi}, and express high levels of other adhesion molecules, namely CD2, LFA-1, LFA-3, and CD44.[39,49] In this model, naive T cells only recirculate through peripheral lymph nodes, while memory T cells home preferentially to skin and peripheral tissues where they may have an increased likelihood of reacting with previously encountered antigens. The effect of antigen on the traffic of memory T cells into afferent lymph has been examined in order to see whether antigen changed the type of cells present in afferent lymph. Following subcutaneous challenge with PPD, all afferent T cells continued to express β_1 integrin, and 1 day after challenge skin vascular endothelium was stained intensely with VCAM-1.[39] This was in contrast to the situation with afferent lymph draining a granuloma induced with Freund's complete adjuvant where about 10% of afferent T cells were β_1 integrin$^-$ and where, in addition to the presence of VCAM-1 on skin vascular endothelium, PNAd was also expressed on the endothelium,[39] suggesting an additional recruitment of naive T cells in chronic inflammation.

The input of cells into lymph nodes via HEV and efferent lymph has also been studied in the fetus in the total absence of any memory or activated T cells and compared with adult sheep (see Figure 7-2). In these experiments in order to collect all the cells migrating through skin before entering afferent lymph, a two-step procedure was used (which had been employed previously to collect large numbers of dendritic cells in afferent lymph from adult sheep).[55] After surgical removal of lymph nodes (prescapular in the fetus and popliteal in the adult), the afferent lymphatics re-anastomose with the former efferent duct to form a pseudoafferent lymphatic duct which is then cannulated at a second operation, thus enabling the total afferent input to the extirpated lymph node to be measured directly. Since cells can only enter lymph nodes via afferent lymphatics or HEV, and since the efferent cell output represents the total cell traffic through the lymph node, the total HEV input was obtained in this study by simply subtracting the total afferent input from the efferent cell output of lymph nodes in age-matched fetuses or, in the case of adult animals, the contralateral lymph node. The major difference in the entry of cells into lymph nodes before and after birth was that much larger numbers of skin-homing T cells entered lymph nodes via afferent lymph in the fetus (50%) than was the case in adult animals (<10%) as shown in Figure 7-2. Both skin homing T cells and lymph node homing T cells in the fetus were CD2^{lo}, LFA-3^{lo}, CD44^{lo}, and L-selectin$^+$, in contrast to skin-homing T cells in the adult which were CD2^{hi}, LFA-3^{hi}, CD44^{hi}, and L-selectin$^{+/-}$, so that it is not mandatory for T cells to be activated to adhesion molecule ^{hi} status in order to home to skin. The selective entry of $\gamma\delta$ T cells into skin and afferent lymph seen in adult sheep also occurred in the fetus. Although all skin-homing $\gamma\delta$ T cells in the fetus were CD45R$^-$ (as is true of adult $\gamma\delta$ T cells), a substantial minority (30 to 40%) of skin-homing $\alpha\beta$ T cells in the fetus were CD45R$^+$ in contrast to the adult where all skin-homing T cells were CD45R$^-$.[57] The capacity of both $\alpha\beta$ and $\gamma\delta$ T cells to home to skin or HEV is in fact acquired inside the thymus before T cells are exported to the periphery and commence their tissue-specific migration (unpublished results).

The capacity of cells to home to skin and peripheral tissues is established during fetal life in which instance, rather than being directed by any immunological stimulus, recirculation through skin and afferent lymph is a physiological property of immunologically naive T cells. While it is true that all T cells in afferent lymph in adult sheep express a so-called *memory phenotype*, there are such remarkable similarities in the migration of cells to fetal and adult skin; our view at present is that the critical molecules directing the homing of T cells to skin have yet to be identified since the tissue-specific homing of $\gamma\delta$ and $\alpha\beta$ T cells to fetal skin cannot be related to the presence or absence in lymph of memory or activated T cells, nor can it be related to the expression of any known homing or adhesion molecules.

It should be said, however, that although neither CD45R isoform, L-selectin, or adhesion molecule expression can be used to distinguish naive T cells in the fetus, the situation with respect to the expression of these surface molecules on T cells and their pathways of recirculation may be very different in adult animals. In adult animals where the thymus has long since involuted, one is examining T cells or their progeny which must have left the thymus many years previously, after which time their recirculation pathways and phenotypes may have changed considerably under the influence of antigen. Nonetheless, whatever effects antigen may prove to have on changing patterns of tissue-selective homing of T cells throughout post-natal life, it is apparent from these fetal studies that two major pathways of recirculation of skin-homing and lymph node-homing T cells are established in the complete absence of antigen.

V. *IN VIVO* HOMING OF LYMPHOCYTES TO GUT

A. HOMING OF T CELLS TO THE GUT

The first description of tissue-specific homing of small lymphocytes was reported in sheep when two independent studies showed preferential homing of intestinal lymph lymphocytes back to the intestines.[58,59] One study used unseparated lymphocytes and speculated that surface Ig may play a role in the preferential homing of B cells to the gut.[58] The other study used nylon-wool prepared T cells and put forward the proposition that the pool of peripheral T cells could be broadly divided into two populations — an intestinal and a peripheral lymph node pool of T cells.[59] Since that time, numerous studies in mice and humans have confirmed the existence of gut-homing and peripheral lymph node-homing T cells.[3]

The tissue-selective migration of T cell subsets to the gut has also been studied. Recent experiments in which the gut vs. lymph node homing properties of FITC-labeled T cell subsets was examined have suggested that, while $CD4^+$ cells discriminate between gut and peripheral lymph nodes since they homed preferentially to their tissue of origin; $CD8^+$cells from peripheral lymph nodes, on the other hand, did not discriminate between these two different tissues and homed equally well to gut or lymph nodes. This migration pattern was explained, at least partly, by the expression of L-selectin on the migrating cells. Thus, while virtually all $CD8^+$ cells were L-selectin$^+$ regardless of their tissue of origin, there were far fewer L-selectin$^+$ gut-homing $CD4^+$ cells. The observation that $CD8^+$ cells homed equally well to gut or peripheral lymph nodes could be explained if they had dual specificity and expressed both lymph node- and gut-homing receptors.[60]

B. *IN VIVO* HOMING OF LYMPHOCYTES TO THE GUT IN THE FETUS

There are major differences in the circulation of cells through the gut and mesenteric lymph nodes between the fetus and animals after birth.[9-11,61] The output of cells from intestinal efferent lymph draining the small intestine and associated mesenteric lymph nodes is low in the fetus, compared with lambs after birth when there is a huge increase in the mass of mesenteric lymph nodes and in the traffic of cells through the gut. There are also differences in the distribution of lymphocyte subsets appearing in intestinal efferent lymph compared with those in peripheral lymph node efferent lymph in the fetus. Efferent lymph draining the ileal lymph node, which receives lymph from almost the entire ileum (containing one continuous Peyer's patch), has been compared with efferent lymph draining the prescapular lymph node in fetuses aged around 130 days gestation. The percentage of $CD8^+$ cells circulating in ileal lymph is much higher than in peripheral lymph node efferent lymph; in fact, it is much higher than in any other tissues or organs of the fetus thus far examined. It is also much higher than that found in ileal lymph of adult sheep, where the proportions of $CD4^+$ and $CD8^+$ lymphocytes in ileal and peripheral lymph node efferent lymph are similar.[14] It is not known at present whether this high concentration of $CD8^+$ cells in ileal lymph is due to tissue-selective homing of $CD8^+$ cells to fetal ileum, whether they are being locally produced as part of some post-thymic expansion, or whether changes in the $CD4^+$ population alter the CD4/CD8 ratio. The proportion of B cells in fetal ileal lymph is also far higher than in any other lymph compartment in the fetus, although in this case it is likely (although not proven) that these cells are newly formed B cells which are being exported from the ileal Peyer's patch, which is the primary source of B cells in the sheep.[16]

Studies on intestinal efferent lymph in the fetus were unable to show preferential homing of fetal intestinal T cells back into intestinal lymph, although allogeneic lymphocytes collected from the intestinal lymph of post-natal animals and infused intravenously into fetal recipients did home preferentially back into fetal intestinal lymph.[61] The gut-homing population of T cells thus only appears after birth, and may or may not be related to the appearance of antigen-primed T cells following the introduction of environmental antigens into the gut. It should be noted that fetal gut vascular endothelial cells obviously do express tissue-specific adhesion molecules since post-natal T cells home preferentially to fetal gut in the same way they do in post-natal animals. In fact, the mucosal addressin recognized by MAb MECA-367 in mice is widely expressed on the vascular endothelium of the intestines at fetal day 14, before lymphocytes appear; and thus, expression seems, at least in part, to be ontogenetically determined.[3] The fact that gut-homing T cells only appear after birth and exposure to antigen, when there is a rapid increase in the weight of the gut-associated lymphoid tissue and the circulation of cells through the gut, suggests a possible role for antigen in the generation of gut-homing (memory?) T cells. The observations that post-natal gut-homing T cells migrate perfectly well to fetal gut, taken together with the fact that antigen-stimulated fetal lymph nodes show the same increased traffic of T cells from blood into efferent lymph as do adult lymph nodes,[62] and that fetal T cells display the same adhesion molecules and lymph node homing receptor as do adult T cells, strongly suggest that recirculating T lymphocytes in the fetus use the same basic lymphocyte-endothelial mechanisms to extravasate from the blood as they do in the adult. Future experiments, particularly on the effects of introducing antigen into discrete tissue beds in the fetus, should prove useful in further defining the role of antigen in the generation of lymphocytes with particular tissue-specific homing properties. It is also possible that there is a qualitative change in the export of T cells from the thymus after birth which results in the export of cells whose gut-homing properties are imprinted intrathymically. Further experiments, particularly on the homing properties of recent thymic emigrants, are necessary to explore these possibilities which highlight the current lack of understanding of post-thymic maturation of T cells, and the role of antigen in the generation of tissue-selective homing T cells.

C. MIGRATION OF MEMORY OR ACTIVATED CELLS

A number of lymphocyte surface molecules which appear on T cells as a consequence of activation with antigen have been proposed as gut-homing receptors. The activation-dependent integrins $\alpha_4\beta_1$, $\alpha_4\beta_p$, and $\alpha_4\beta_7$ have been implicated in lymphocyte homing in mice.[3] A recent study in sheep has compared memory cells circulating through gut and skin in sheep.[36] Only memory T cells (defined as CD2^{hi}, CD45R$^-$) homed to skin, whereas both memory and naive T cells (defined as CD2^{lo}, CD45R$^+$) circulated through gut. This study also found that when T cells from intestinal afferent lymph were labeled with fluorochrome *in vitro*, and their reappearance in gut afferent lymph was examined, only memory T cells homed back to the gut, whereas naive T cells homed to peripheral lymph nodes. Naive T cells were virtually all L-selectin$^+$, whereas only about 30% of memory T cells expressed L-selectin. Gut-homing memory T cells homed poorly to skin and expressed very low levels of α_6 and β_1 integrins, in contrast to skin-homing memory T cells which expressed high levels of these integrins. Both skin- and gut-homing T cells expressed high levels of α_4 integrin.[36] The expression of a number of endothelial cell adhesion molecules was also examined in this study, including VCAM-1, PNAd, and α_6 integrin; however, none of these molecules seemed likely candidates as ligands for gut-homing T cells. The strong expression of PNAd on Peyer's patch HEV presumably reflects the higher concentration of L-selectin$^+$ T cells in the interfollicular areas of Peyer's patches of young lambs compared with the gut of older animals.[36]

The regulatory mechanisms that generate lymphocyte subsets with tissue-selective homing properties after stimulation with antigen in sites such as gut and lymph nodes are unknown, but are presumably related to factors in the local microenvironment. In this context, it is worth noting that lymph-borne blasts generated in the gut-associated lymphoid tissues home to the small gut, whereas those from peripheral lymph nodes do not,[63,64] and major differences in the homing of lymphoblasts and lymphocytes from a variety of sources have recently been described in the pig.[65] The mucosal addressin is thought to have a role in the homing of at least some blasts to the gut in mice.[3]

Although the preferential *in vivo* homing of gut or lymph node T cells back to their tissue of origin, and the differential binding of lymphocytes to tissue sections of gut and peripheral lymph nodes are enduring and highly reproducible phenomena, the molecular basis of lymphocyte homing to the gut remains unresolved. The lymphocyte homing receptor for mucosal addressin in the mouse is still unidentified and no gut equivalent to the primary, "activation-independent" lymph node-homing receptor, L-selectin has been found, although a number of "activation-dependent" lymphocyte surface molecules linked with memory subsets have been implicated in the homing of T cells to the gut. Indeed, notwithstanding the considerable advances made in recent years — particularly in identifying the molecular basis of receptor-ligand interactions that may regulate lymphocyte homing — there remain surprising gaps in our understanding of the basic *in vivo* physiology of lymphocyte traffic through lymph nodes. For example, a recent study has found that although rat thoracic duct lymphocytes show preferential binding differences in *in vitro* binding assays, no tissue-specific migration between the Peyer's patch and peripheral lymph nodes could be demonstrated *in vivo.*[66] The origin of the lymphocytes in efferent intestinal lymph is also still uncertain,[67] since there appear to be around twice the number of lymphocytes entering the mesenteric lymph node via blood and afferent lymph than are leaving in efferent lymph.

Current physiological data cannot explain this apparent imbalance between cell input to, and cell output from, the mesenteric lymph node which, except for the spleen, is the largest single site of lymphocyte traffic in the whole animal.

VI. HOMING OF γδ T CELLS IN FETAL AND POST-NATAL ANIMALS

Sheep, along with other ruminants, and chickens have a higher proportion of γδ T cells in the peripheral T cell pool than do mice and rats.[68] Although αβ T cells constitute the majority of T cells in the periphery in sheep, both sheep and cattle show a high proportion of γδ T cells in the blood which is age related.[69] The percentage of γδ T cells in blood lymphocytes of sheep increases from around 15 to 20% at birth to 40 to 50% in the early months after birth, and then declines gradually with age to around 5 to 10% in animals 5 to 8 years of age.[68] This high concentration of γδ T cells in the blood does not occur uniformly throughout other tissues in sheep, and it may be due to differences in the homing properties of blood-borne γδ T cells at different ages. It has been shown that fetal thymectomy abolishes the rise in blood-borne γδ T cells which normally occurs after birth in sheep;[70] and recently it has been shown that the rate of export of γδ T cells from the thymus is greatly increased after birth,[19] so that in the absence of the thymus, γδ T cells do not expand their numbers in the periphery, and it appears that the increase in their numbers after birth is due solely to a large increase in their rate of export from the thymus.

The ability to examine γδ T cells throughout a long gestational period and to measure quantitatively the *in vivo* traffic of γδ T cells through a wide range of tissues, together with the high proportion of peripheral T cells expressing the γδ TcR, is proving ruminants to be an extremely useful model to study the ontogeny of γδ T cells and their unusual tissue distribution and homing properties. The value of examining γδ T cells in "γδ high" species such as ruminants has been emphasized by an important new study in sheep on the ontogeny of the γδ TcR[71] which shows that a much greater degree of γδ TcR complexity and diversity occurs in sheep than in humans or mice and that the probable ligand-mediated selection of γδ T cells during ontogeny is tightly regulated by the thymus (see Chapters 2 and 4). In ruminants (as in all other species studied), γδ T cells are present in high concentrations in association with epithelia in such diverse sites as skin, gut, tongue, trachea, bladder, and uterus.[68,72-75] Skin-homing γδ T cells have been studied in sheep by collecting cells from afferent lymph which has a higher concentration of γδ T cells than lymph draining either peripheral lymph nodes or mesenteric lymph nodes.[41,52] Afferent lymph draining skin in the fetus is similarly enriched in γδ T cells,[57] and curiously γδ T cells in both fetal and adult animals are mostly L-selectin^{hi} (unpublished results) so that the expression of L-selectin is no guarantee that a cell will home to lymph nodes rather than skin.

A possible candidate as a homing receptor for γδ T cells does exist in ruminants; it was first identified in sheep and called T19,[73,76] and later found in cattle and called WC1.[75] There are no experiments to date which bear directly on a role for T19 in the homing of γδ T cells, but this molecule clearly defines two populations of γδ T cells which appear at different times during ontogeny.[68,73] A possible role for the T19/WC1 molecules in tissue-specific homing of γδT19^{+} T cells has been raised by recent experiments in sheep and cattle. The first experiments were done in cattle where the bovine WC1 gene was cloned and evidence suggested a family of WC1 genes existed.[77] More recent studies in sheep suggest that a family of between 20 and 200 genes may code for the T19 family, and it has been suggested that the existence of a large repertoire of variants within the T19 family could be linked to a family of diverse but structurally related homing ligands for various tissue-seeking populations of γδ T cells.[78]

VII. CONCLUSIONS

In this chapter we have outlined some of the classic *in vivo* studies on lymphocyte recirculation in sheep and reviewed recent progress in understanding the physiological basis of lymphocyte recirculation and homing. We have concentrated mostly on studies in sheep which provide an ideal, if not unique, model for identifying *in vivo* tissue-selective migration streams of lymphocytes and for experimentally manipulating lymphocyte recirculation through specific tissues and organs.

A considerable array of putative homing receptors and adhesion molecules have been identified in recent years (largely from *in vitro* binding studies), but the physiological relevance of many of these molecules in tissue-specific targeting remains uncertain. While *in vitro* studies will no doubt continue to be useful in identifying molecules which might play a role in T cell migration, there is a pressing need for more *in vivo* physiological studies :

(1) To test the functional relevance of any candidate molecules which may be involved in tissue-specific lymphocyte targeting. This is an important area for future investigation to which the sheep is particularly well suited because populations of biologically purified T cells which have homed to individual tissues, such as skin or gut for example, can be collected directly by cannulating regional lymphatics draining discrete tissue beds. Candidate homing receptor molecules on T cells which have targeted individual tissues *in vivo* can then be blocked, for example with monoclonal antibodies, and the capacity of these cells to home back to their tissue of origin after reinfusion into an animal can then be examined with great precision. The adhesive interactions which control the entry of lymphocytes into individual tissues and organs *in vivo* are essential for normal immune responses. On the other hand, the presence of abnormal tissue-specific homing T cell circuits may form the basis of tissue-selective inflammatory disease. Future studies involving experimental manipulation of T cell circulation through normal and inflamed tissues *in vivo* will be a mandatory first step in the development of any therapeutic strategies based on controlling the entry of T cells into diseased tissues.

(2) To identify the extent to which unique homing mechanisms exist in tissues other than those for lymph nodes, gut, and skin. For example, additional T cell homing circuits may exist which target lymphocytes to tissues such as lungs, synovium, and pregnant uterus.

(3) To determine the role of the thymus in the generation of tissue-selective T cells. Despite the presence of unique populations of T cells whose migration appears to be highly tissue restricted, it is surprising how little is known about the regulatory mechanisms which generate T cell subsets with unique tissue-specific homing properties. It is remarkable that although it has been known for over 30 years that the thymus is responsible for the formation of the peripheral T cell pool, so little is known about the role of the thymus in the induction of tissue- and organ-specific homing of T cells. An important question which needs to be addressed in future studies is whether or not the thymus is responsible for the induction of a repertoire of different tissue-selective homing specificities on developing T cells inside the thymus, which when exported to the periphery form the basis for a series of tissue-specific migration streams of naive T cells in young animals.

(4) To determine the role of antigen in tissue-selective processes. Although the stability of a particular tissue-specific homing phenotype for any given cell is unknown, memory or antigen-primed T cells appear to display altered homing behavior. The effect of antigenic stimulation, or a lack thereof, and the effect of particular tissue microenvironments on the homing phenotype is also unknown, and more physiological studies in intact animals are needed to address these questions. Future experiments in fetal sheep will be very important in defining the role of antigen in the generation of tissue-specific pathways of recirculation because it is an *in vivo* model which allows the homing patterns of T cells to be studied in a situation where their naive status is absolute in an environment where there is no foreign antigen and no memory or recently activated T cells.

REFERENCES

1. **Kishimoto, T. K., Larson, R. S., Corbi, A. L., Dustin, M. L., Staunton, D. E., and Springer, T. A.,** The leukocyte integrins, *Adv. Immunol.*, 46, 149, 1989.
2. **Dustin, M. L. and Springer, T. A.,** Role of lymphocyte adhesion receptors in transient interactions and cell locomotion, *Annu. Rev. Immunol.*, 9, 27, 1991.
3. **Picker, L. J. and Butcher, E. C.,** Physiological and molecular mechanisms of lymphocyte homing, *Annu. Rev. Immunol.*, 10, 561, 1992.
4. **Butcher, E. C.,** Leukocyte-endothelial cell recognition: three (or more) steps to specificity and diversity, *Cell*, 67, 1033, 1991.
5. **Trnka, Z. and Cahill, R. N. P.,** Aspects of the immune response in single lymph nodes, *Monogr. Allergy*, 16, 245, 1980.
6. **Miyasaka, M. and Trnka, Z.,** Lymphocyte migration and differentiation in a large-animal model: the sheep, *Immunol. Rev.*, 91, 87, 1986.
7. **Abernethy, N. J. and Hay, J. B.,** Lymphocyte migration through skin and skin lesions, in *Migration and Homing of Lymphoid Cells*, Husband, A. J., Ed., CRC Press, Boca Raton, FL, 1988, 113.
8. **Pearson, L. D., Simpson-Morgan, M. W., and Morris, B.,** Lymphopoiesis and lymphocyte recirculation in the sheep fetus, *J. Exp. Med.*, 143, 167, 1976.

9. **Cahill, R. N. P. and Trnka, Z.,** Growth and development of recirculating lymphocytes in the sheep fetus, in *Essays on the Anatomy and Physiology of Lymphoid Tissues*, Trnka, Z. and Cahill, R. N. P., Eds., S. Karger, Basel, 1980, 38.
10. **Cahill, R. N. P., Poskitt, D. C., Hay, J. B., Heron, I., and Trnka, Z.,** The migration of lymphocytes in the fetal lamb, *Eur. J. Immunol.*, 9, 251, 1979.
11. **Cahill, R. N. P., Heron, I., Poskitt, D. C., and Trnka, Z.,** Lymphocyte recirculation in the sheep fetus, in *Blood Cells and Vessel Walls: Functional Interactions (Ciba Found. Symp. 71)*, Excerpta Medica, 1980, 145.
12. **Cahill, R. N. P., Poskitt, D. C., Heron, I., and Trnka, Z.,** Collection of lymph from single lymph nodes and the intestines of fetal lambs *in utero*, *Int. Arch. Allergy Appl. Immunol.*, 59, 117, 1979.
13. **Kimpton, W. G., Washington, E. A., and Cahill, R. N. P.,** Recirculation of lymphocyte subsets ($CD5^+$, $CD4^+$, $CD8^+$, $T19^+$, and B cells) through fetal lymph nodes, *Immunology*, 68, 575, 1989.
14. **Kimpton, W. G., Washington, E. A., and Cahill, R. N. P.,** Non-random migration of $CD4^+$, $CD8^+$ and $\gamma\delta^+T19^+$, and B cells between blood and lymph draining ileal and prescapular lymph nodes in the sheep fetus, *Int. Immunol.*, 2, 937, 1990.
15. **Kimpton, W. G., Washington, E. A., and Cahill, R. N. P.,** The development of the immune system in the fetus, in *Oxford Textbook of Fetal Physiology*, Thorburn, G. D. and Harding, R., Eds., Oxford University Press, Oxford, 1994, in press.
16. **Reynolds, J. D., Kennedy, L., Peppard, J., and Pabst, R.,** Ileal Peyer's patch emigrants are predominantly B cells and travel to all lymphoid tissues in sheep, *Eur. J. Immunol.*, 21, 283, 1991.
17. **Scollay, R., Butcher, E., and Weissman, I.,** Thymus cell migration: quantitative studies on the rate of migration of cells from the thymus to the periphery in mice, *Eur. J. Immunol.*, 10, 210, 1980.
18. **Miyasaka, M., Pabst, R., Dudler, L., Cooper, M., and Yamaguchi, K.,** Characterization of lymphatic and venous emigrants from the thymus, *Thymus*, 16, 29, 1990.
19. **Witherden, D. A.,** Emigration of T cells from the Foetal and Post-natal Thymus: A Comparison Between $\alpha\beta$ and $\gamma\delta$ Subsets, Ph.D. thesis, The University of Melbourne, 1993.
20. **Westermann, J. and Pabst, R.,** Lymphocyte subsets in the blood: a diagnostic window on the lymphoid system, *Immunol. Today*, 11, 406, 1990.
21. **Pabst, R. and Binns, R. M.,** Heterogeneity of lymphocyte homing physiology: several mechanisms operate in the control of migration to lymphoid and non-lymphoid organs *in vivo*, *Immunol. Rev.*, 108, 83, 1989.
22. **Willführ, K. H., Westermann, J., and Pabst, R.,** Absolute numbers of lymphocyte subsets migrating through the compartments of the normal and transplanted rat spleen, *Eur. J. Immunol.*, 20, 903, 1990.
23. **Morris, B. and Salami, M. A.,** The blood and lymphatic capillaries of lymph nodes in the sheep fetus and their involvement in cell traffic, *Lymphology*, 20, 244, 1987.
24. **Smith, J. B., McIntosh, G. H., and Morris, B.,** The migration of cells through chronically inflammed tissues, *J. Pathol.*, 100, 21, 1970.
25. **Binns, R. M. and Pabst, R.,** Lymphoid cell migration and homing in the young pig: alternative mechanisms in action, in *Migration and Homing of Lymphoid Cells*, Husband, A. J., Ed., CRC Press, Boca Raton, FL, 1988, 137.
26. **Hall, J. G. and Morris, B.,** The origin of cells in the efferent lymph from a single lymph node, *J. Exp. Med.*, 121, 901, 1965.
27. **Hall, J. G. and Morris, B.,** The immediate effect of antigens on the cell output of a lymph node, *Br. J. Exp. Pathol.*, 46, 450, 1965.
28. **Cahill, R. N. P., Frost, H., and Trnka, Z.,** The effects of antigen on the migration of recirculating lymphocytes through single lymph nodes, *J. Exp. Med.*, 143, 870, 1976.
29. **Hay, J. B. and Hobbs, B. B.,** The flow of blood to lymph nodes and its relation to lymphocyte traffic and the immune response, *J. Exp. Med.*, 145, 31, 1977.
30. **Bujdoso, R., Young, P., Hopkins, J., and McConnell, I.,** IL-2-like activity in lymph fluid following *in vivo* antigen challenge, *Immunology.*, 69, 45, 1990.
31. **Hay, J. B., Cahill, R. N. P., and Trnka, Z.,** The kinetics of antigen-reactive cells during lymphocyte recruitment, *Cell. Immunol.*, 10, 145, 1974.
32. **Cahill, R. N. P., Hay, J. B., Frost, H., and Trnka, Z.,** Changes in lymphocyte circulation after administration of antigen, *Haematologica*, 8, 321, 1974.

33. **Cahill, R. N. P., Frost, H., Hay, J. B., Lafleur, L., and Trnka, Z.,** Changes in mixed lymphocyte culture-reactive lymphocytes following alloimmunization of single lymph nodes in sheep, *Transplantation*, 27, 102, 1979.
34. **Hall, J. G., Morris, B., Moreno, G. D., and Bessis, M.,** The ultrastructure and function of the cells in lymph following antigenic stimulation, *J. Exp. Med.*, 125, 91, 1967.
35. **Spertini, O., Kansas, G. S., Reimann, K. A., MacKay, C. R., and Tedder, T. F.,** Function and evolutionary conservation of distinct epitopes on the leukocyte adhesion molecule-1 (TQ-1, Leu-8) that regulate leukocyte migration, *J. Immunol.*, 147, 942, 1991.
36. **Mackay, C. R., Marston, W. L., Dudler, L., Spertini, O., Tedder, T. F., and Hein, W. R.,** Tissue-specific migration pathways by phenotypically distinct subpopulations of memory T cells, *Eur. J. Immunol.*, 22, 887, 1992.
37. **Howard, C. J., Sopp, P., and Parsons, K. R.,** L-selectin expression differentiates T cells isolated from different lymphoid tissues in cattle but does not correlate with memory, *Immunology*, 77, 228, 1992.
38. **Reichert, R. A., Gallatin, W. M., Butcher, E. C., and Weissman, I. L.,** A homing receptor-bearing cortical thymocyte subset: implications for thymus cell migration and the nature of cortisone-resistant thymocytes, *Cell*, 38, 89, 1984.
39. **Mackay, C. R., Marston, W., and Dudler, L.,** Altered patterns of T cell migration through lymph nodes and skin following antigen challenge, *Eur. J. Immunol.*, 22, 2205, 1992.
40. **Ottaway, C. A.,** Dynamic aspects of lymphoid cell migration, in *Migration and Homing of Lymphoid Cells*, Husband, A. J., Ed., CRC Press, Boca Raton, FL, 1988, 167.
41. **Mackay, C. R., Kimpton, W. G., Brandon, M. R., and Cahill, R. N.,** Lymphocyte subsets show marked differences in their distribution between blood and the afferent and efferent lymph of peripheral lymph nodes, *J. Exp. Med.*, 167, 1755, 1988.
42. **Washington, E. A., Kimpton, W. G., and Cahill, R. N. P.,** $CD4^+$ lymphocytes are extracted from blood by peripheral lymph nodes at different rates from other T cell subsets and B cells, *Eur. J. Immunol.*, 18, 2093, 1988.
43. **Witherden, D. A., Kimpton, W. G., Washington, E. A., and Cahill, R. N. P.,** Non-random migration of $CD4^+$, $CD8^+$ and $\gamma\delta^+T19^+$ lymphocytes through peripheral lymph nodes, *Immunology*, 70, 235, 1990.
44. **Kimpton, W. G., Washington, E. A., and Cahill, R. N. P.,** Recirculation of lymphocyte subsets ($CD5^+$, $CD4^+$, $CD8^+$, SBU-$T19^+$, and B cells) through gut and peripheral lymph nodes, *Immunology*, 66, 69, 1989.
45. **Abernethy, N. J., Hay, J. B., Kimpton, W. G., Washington, E. A.,** and Cahill, R. N. P., Non-random recirculation of small $CD4^+$ and $CD8^+$ T lymphocytes in sheep: evidence for lymphocyte subset specific lymphocyte-endothelial cell recognition, *Int. Immunol.*, 2, 231, 1990.
46. **McClure, S. J. and Hein, W. R.,** Functional characteristics of 197^+ $CD4^-$ $CD8^-$ sheep T lymphocytes: expansion and differentiation of peripheral T cells, *Immunol. Cell Biol.*, 67, 223, 1989.
47. **Bujdoso, R., Young, P., Hopkins, J., Allen, D., and McConnell, I.,** Non-random migration of CD4 and CD8 T cells: changes in the CD4:CD8 ratio and interleukin-2 responsiveness of efferent lymph cells following *in vivo* antigen challenge, *Eur. J. Immunol.*, 19, 1779, 1989.
48. **Kimpton, W. G., Washington, E. A., and Cahill, R. N. P.,** Non-random migration of $CD4^+$, $CD8^+$ and $\gamma\delta^+T19^+$ lymphocyte subsets following *in vivo* stimulation with antigen, *Cell. Immunol.*, 130, 236, 1990.
49. **Mackay, C. R., Marston, W. L., and Dudler, L.,** Naive and memory T cells show distinct pathways of lymphocyte recirculation, *J. Exp. Med.*, 171, 801, 1990.
50. **Howard, C. J., Sopp, P., Parsons, K. R., McKeever, D. J., Taracha, E. L. N., Jones, B. V., Machugh, N. D., and Morrison, W. I.,** Distinction of naive and memory BoCD4 lymphocytes in calves with a monoclonal antibody, CC76, to a restricted determinant of the bovine leukocyte-common antigen, CD45, *Eur. J. Immunol.*, 21, 2219, 1991.
51. **Bell, E. B. and Sparshott, S. M.,** Interconversion of CD45R subsets of CD4 T cells *in vivo*, *Nature,* 348, 163, 1990.
52. **Hein, W. R., McClure, S. J., and Miyasaka, M.,** Cellular composition of peripheral lymph and skin of sheep defined by monoclonal antibodies, *Int. Arch. Allergy Appl. Immunol.*, 84, 241, 1987.
53. **McKeever, D. J., MacHugh, N. D., Goddeeris, B. M., Awino, E., and Morrison, W. I.,** Bovine afferent lymph veiled cells differ from blood monocytes in phenotype and accessory function, *J. Immunol.*, 147, 3703, 1991.

54. **Barfoot, R., Denham, S., Gyure, L. A., Hall, J. G., Hobbs, S. M., and Jackson, L. E.,** Some properties of dendritic macrophages from peripheral lymph, *Immunology*, 68, 233, 1989.
55. **Bujdoso, R., Hopkins, J., Dutia, B. M., Young, P., and McConnell, I.,** Characterization of sheep afferent lymph dendritic cells and their role in antigen carriage, *J. Exp. Med.*, 170, 1285, 1989.
56. **Hopkins, J., Dutia, B. M., Bujdoso, R., and McConnell, I.,** *In vivo* modulation of CD1 and MHC class II expression by sheep afferent lymph dendritic cells, *J. Exp. Med.*, 170, 1303, 1989.
57. **Kimpton, W. G., Washington, E. A., and Cahill, R. N. P.,** Noval homing of αβ and γδ T cells to skin and lymph nodes in the fetus, submitted, 1994.
58. **Scollay, R., Hopkins, J., and Hall, J. G.,** Possible role of surface Ig in non-random recirculation of small lymphocytes, *Nature*, 260, 528, 1976.
59. **Cahill, R. N. P., Poskitt, D. C., Frost, H., and Trnka, Z.,** Two distinct pools of recirculating T lymphocytes: migratory characteristics of nodal and and intestinal T lymphocytes, *J. Exp. Med.*, 145, 420, 1977.
60. **Katerelos, M., Washington, E. A., Cahill, R. N. P., and Kimpton, W. G.,** Tissue specific recirculation of $CD4^+$, $CD8^+$ and $\gamma\delta^+$ T cells through peripheral and intestinal lymphoid tissues of sheep, submitted, 1994.
61. **Kimpton, W. G. and Cahill, R. N. P.,** Circulation of autologous and allogeneic lymphocytes in lambs before and after birth, in *Immunology of the Sheep*, Morris, B. and Miyasaka, M., Eds., Editiones Roche, Basle, 1985, 306.
62. **Hugh, A. R., Travella, W., Simpson-Morgan, M. W., and Morris, B.,** The lymph-borne response of fetal lamb lymph nodes to challenge with *Brucella abortus in utero*, *Aust. J. Exp. Biol. Med. Sci.*, 63, 381, 1985.
63. **Hall, J. G., Hopkins, J., and Orlans, E.,** Studies on the lymphocytes of sheep. III Destination of lymph-borne immunoblasts in relation to their tissues of origin, *Eur. J. Immunol.*, 7, 30, 1977.
64. **Hall, J. G.,** An essay on lymphocyte circulation and the gut, *Monogr. Allergy*, 16, 100, 1980.
65. **Binns, R. M., Licence, S. T., and Pabst, R.,** Homing of blood, splenic, and lung emigrant lymphoblasts: comparison with the behaviour of lymphocytes from these sources, *Int. Immunol.*, 4, 1011, 1992.
66. **Westermann, J., Blaschke, V., Zimmermann, G., Hirschfeld, U., and Pabst, R.,** Random entry of circulating lymphocyte subsets into peripheral lymph nodes and Peyer's patches: no evidence *in vivo* of a tissue-specific migration of B and T lymphocytes at the level of high endothelial venules, *Eur. J. Immunol.*, 22, 2219, 1992.
67. **Chin, G. W. and Cahill, R. N. P.,** The appearance of fluorescein-labelled lymphocytes in lymph following in vitro labelling: the route of lymphocyte recirculation through mesenteric lymph nodes, *Immunology*, 52, 341, 1984.
68. **Hein, W. R. and Mackay, C. R.,** Prominence of γδ T cells in the ruminant immune system, *Immunol. Today*, 12, 30, 1991.
69. **Washington, E. A., Kimpton, W. G., and Cahill, R. N. P.,** Changes in the distribution of αβ and γδ T cells in blood and in lymph nodes from fetal and postnatal lambs, *Dev. Comp. Immunol.*, 16, 493, 1992.
70. **Hein, W. R., Dudler, L., and Morris, B.,** Differential peripheral expansion and *in vivo* antigen reactivity of α/β and γ/δ T cells emigrating from the early fetal lamb thymus, *Eur. J. Immunol.*, 20, 1805, 1990.
71. **Hein, W. R. and Dudler, L.,** Divergent evolution of T cell repertoires: extensive diversity and developmentally regulated expression of the sheep γδ T cell receptor, *EMBO J.*, 12, 715, 1993.
72. **Mackay, C. R. and Hein, W. R.,** A large proportion of bovine T cells express the γδ T cell receptor and show a distinct tissue distribution and surface phenotype, *Int. Immunol.*, 1, 540, 1989.
73. **Mackay, C. R., Beya, M.-F., and Matzinger, P.,** γ/δ T cells express a unique surface molecule appearing late during thymic development, *Eur. J. Immunol.*, 19, 1477, 1989.
74. **McClure, S. J., Hein, W. R., Yamaguchi, K., Dudler, L., Beya, M., and Miyasaka, M.,** Ontogeny, morphology and tissue distribution of a unique subset of CD4-CD8- sheep T lymphocytes, *Immunol. Cell Biol.*, 67, 215, 1989.
75. **Clevers, H., Machugh, N. D., Bensaid, A., Dunlap, S., Baldwin, C. L., Kaushal, A., Iams, K., Howard, C. J., and Morrison, W. I.,** Identification of a bovine surface antigen uniquely expressed on $CD4^-$ $CD8^-$ T cell receptor γ/δ^+ T lymphocytes, *Eur. J. Immunol.*, 20, 809, 1990.

76. **Mackay, C. R., Maddox, J. F., and Brandon, M. R.,** Three distinct subpopulations of sheep T lymphocytes, *Eur. J. Immunol.*, 16, 19, 1986.
77. **Wijngaard, P. L. J., Metzellar, M. J., MacHugh, N. D., Morrison, W. I., and Clevers, H. C.,** Molecular characterization of the WC1 antigen expressed specifically on bovine $CD4^- CD8^- \gamma\delta$ T lymphocytes, *J. Immunol.*, 149, 3273, 1992.
78. **Walker, I. D., Glew, M. D., O'Keeffe, M. A., Metcalf, S. A., Clevers, H. C., Wijngaard, P. L. J., Adams, T. E., and Hein, W. R.,** The T19 gene family in sheep: isolation and charcterisation of multiple T19 genomic recombinant clones and preliminary enumeration of the T19-like genes, submitted, 1994.
79. **Gallatin, W. M., Weissman, I. L., and Butcher, E. C.,** A cell surface molecule involved in organ-specific homing of lymphocytes, *Nature*, 304, 30, 1983.
80. **Yednock, T. A., Cannon, C., Fritz, L. C., Sanchez-Madrid, F., Steinman, L., and Karin, N.,** Prevention of experimental autoimmune encephalomyelitis by antibodies against $\alpha4\beta1$ integrin, *Nature*, 356, 63, 1992.

Chapter 8

Neutrophils and Killer Cells

James A. Roth

CONTENTS

I. Introduction ... 127
II. Neutrophils ... 128
  A. Morphology, Physiology, and Biochemistry ... 128
    1. Structural Characteristics ... 128
    2. Microbicidal Mechanisms and Cytotoxicity ... 128
      a. Nonoxidative Microbicidal Mechanisms ... 128
      b. Oxidative Microbicidal Mechanisms ... 130
      c. Neutrophil-Mediated Cytotoxicity ... 130
    3. Stimulants of Neutrophil Activity ... 131
      a. Chemotaxis ... 131
      b. Ingestion, Oxidative Metabolism, and Degranulation ... 131
  B. Deficits in Neutrophil Function ... 131
    1. Age-Related Changes in Neutrophil Function ... 131
    2. Genetic Variation in Neutrophil Function ... 132
    3. Genetic Deficit in Neutrophil Function ... 132
    4. Suppression of Neutrophil Function ... 132
      a. Infectious Agents ... 132
      b. Hormones and Pharmaceuticals ... 133
      c. Nutritional Factors ... 133
  C. Enhancement of Neutrophil Function ... 133
III. Killer Cells ... 134
  A. Natural Killer Cells ... 134
  B. K Cells ... 135
  C. Lymphokine-Activated Killer Cells ... 135
IV. Summary ... 136
References ... 136

I. INTRODUCTION

Neutrophils are important components of both native and acquired immunity in ruminants. They are capable of responding rapidly to many infectious agents in the absence of a humoral and cell-mediated immune response and, therefore, play a major role in first line defense against infectious agents and, perhaps, neoplastic cells. Neutrophils are a major component of the inflammatory response to microbial invasion or tissue damage.

Numerous factors have been shown to inhibit the neutrophil response to inflammatory stimuli and to inhibit neutrophil function. These factors are associated with an increased susceptibility to infection. The ability of neutrophils to control infectious agents is enhanced in the presence of a humoral or cell-mediated immune response. Antibody to a specific pathogen may enhance the ability of neutrophils to control that pathogen through opsonization, complement fixation, and antibody-dependent cell-mediated cytotoxicity (ADCC). The cell-mediated immune response to a specific antigen enhances the functional activity of neutrophils so that they should be more effective in controlling infectious agents to which they respond.

Killer cells are lymphoid cells that are capable of non-major histocompatibility complex (MHC)-restricted cytotoxic activity for target cells. They are functionally defined in man into three types: natural killer (NK) cells, killer (K) cells, and lymphokine-activated killer (LAK) cells. These are not discrete cell types, as a single cell may be capable of all three functional activities. The NK cells are functionally defined as lymphoid cells that mediate spontaneous non-MHC restricted cytotoxicity (in the absence of

0-8493-4952-4/94/$0.00+$.50

antibody). Most of the cells in man with NK activity are of the large granular lymphocyte (LGL) phenotype. The K cells are defined as lymphoid cells (non-T, non-B) that have Fc receptors allowing them to bind and kill antibody-coated target cells; cell types such as macrophages, neutrophils, and eosinophils can perform this function. In man, many LGLs are capable of both NK and K activity. The LAK cells are functionally defined as lymphoid cells that mediate non-MHC-restricted cytotoxicity after activation with interleukin-2 (IL-2) or other cytokines. In man, NK cells are the predominant precursor cells for LAK cells.[1,2] The NK, K, and LAK cells are thought to be important in resistance to some viral infections, parasitic infections, and neoplasia. In ruminants, cells with NK and K activity against tumor cells have been difficult to detect; however, cells with NK activity against virally infected cells have been reported. Cells with LAK activity against tumor and virus-infected targets after cytokine activation have also been reported.

The aim of this paper is not to provide a comprehensive review of neutrophil and killer cell physiology and function, but rather to focus on the characteristics of these cells in ruminants (as compared to humans) and their role in susceptibility to, and prevention of, infectious diseases.

II. NEUTROPHILS

Bacterial infections of the respiratory tract, reproductive tract, and mammary gland are major causes of disease loss in ruminants. Because neutrophils are central in protection from bacterial infection, they have been studied extensively in ruminants. Inhibition of neutrophil function has been found to be associated with predisposition to respiratory disease and mastitis and to facilitate the survival of facultative intracellular pathogens. Treatment of animals with biologic response modifiers has been shown to enhance neutrophil function and increase resistance to bacterial diseases under experimental conditions.

Ruminant neutrophils can easily be obtained in large numbers simply by centrifuging anticoagulated whole blood, discarding the plasma and buffy coat layers, then lysing the erythrocytes with distilled water.[3] This has allowed bovine neutrophils to be used as a model system for biochemical studies of plasma membranes, biochemical characterization of the superoxide anion generating system, and biochemical characterization of granule contents.

A. MORPHOLOGY, PHYSIOLOGY, AND BIOCHEMISTRY

Most aspects of ruminant neutrophil structure and function are similar to those reported for humans and other species. Excellent references are available on neutrophil structure and function in humans.[4-7] Differences observed between neutrophil structure and function in ruminants and humans will be emphasized here.

1. Structural Characteristics

A distinguishing structural characteristic of ruminant neutrophils is the presence of a unique third granule population. Ruminant neutrophils have peroxidase-positive granules which are formed first, and peroxidase-negative granules of similar size which are formed later in development. These two granule types have the characteristics of the azurophil or primary granules and the specific or secondary granules, respectively, as observed in other species. Ruminant neutrophils also contain large peroxidase-negative granules which are formed after the primary granules and before the specific granules during granulopoiesis.[8] These granules are markedly larger than azurophil and specific granules, and have a sharply defined membrane and a uniformly low electron density. These granules are notable for containing potent oxygen-independent antimicrobial peptides (discussed below). They also contain lactoferrin but do not contain peroxidase, lysosomal hydrolases, or myeloperoxidase. Other structural aspects of ruminant neutrophils have not been observed to be markedly different from human neutrophils.[9]

2. Microbicidal Mechanisms and Cytotoxicity

Microbicidal mechanisms can be divided into those requiring oxygen and those not requiring oxygen for activity (Table 8-1). Bovine neutrophils are also capable of killing eukaryotic cells through antibody-dependent and -independent cell-mediated cytotoxicity. The molecular basis for this cytotoxic activity is not clearly understood and may involve a variety of mechanisms.

a. Nonoxidative Microbicidal Mechanisms

The microbicidal mechanisms which do not require oxygen are largely dependent upon the activity of substances found in the cytoplasmic granules (Table 8-1). There are marked differences in the granule

Table 8-1 **Granule Contents and Products of Oxidative Metabolism Contributing to the Inhibition of Microbial Growth and/or Destruction of Microbes by Ruminant Neutrophils**

Granule contents		Products of oxidative metabolism
Enzymes	**Binding proteins**	
Proteases	Lactoferrin	Superoxide anion
Elastase	Vitamin B_{12} binding protein	Singlet oxygen
Myeloperoxidase		Hydrogen peroxide
Catalase	**Cationic bactericidal**	Hydroxyl radical
Glutathione peroxidase	**proteins**	Hypochlorous acid
Glutathione reductase		Aldehydes
Acid phosphatase	Bactenecins	Chloramines
Alkaline phosphatase	Indolicidin	
Phosphodiesterase	Bactericidal permeability	
Arylsulfatase	inducing protein	
β-glucuronidase,		
α and β galactosidases,		
α and β mannosidases		
N-acetylglucosaminidase		
N-acetylgalactosaminidase-		
α-L-fucosidase		
α-Glucosidase		

contents and in the nonoxidative bactericidal mechanisms between human and ruminant neutrophils. Ruminant neutrophils have lower concentrations than human neutrophils of many granule enzymes, including β-glucuronidase, β-galactosidase, myeloperoxidase, elastase, catalase, and lysozyme.[10-13] They have higher concentrations of alkaline phosphatase, lactoferrin, vitamin B_{12} binding protein, glutathione peroxidase, and glutathione reductase.[11,12]

Human neutrophils contain several antimicrobial proteins which are associated with the azurophil granules.[5,6,14,15] These are lysozyme, cathepsin G, azurocidin (CAP 37), bactericidal-permeability-inducing protein (BPI/CAP 57), and small peptides called "defensins" (at least four).[6] Lysozyme, which is abundant in human neutrophils, has been shown to be absent in ruminant neutrophils.[10,11] Cathepsin G and azurocidin show strong homology with elastase.[6] Attempts to identify and quantify cathepsin G and azurocidin in ruminant neutrophils have apparently not been reported. BPI is a cationic protein (approximately 57 kDa) found in human and rabbit neutrophils. It shares homology with rabbit lipopolysaccharide binding protein. The antimicrobial activity of BPI is confined to the NH_2-terminal, 25-kDa domain of BPI. That domain is cationic and amphipathic and is released by the activity of elastase on BPI.[6] Bovine neutrophils have been shown to have a gene with 75% homology to the BPI gene in human neutrophils.[16]

Defensins are small cationic peptides containing 29 to 34 amino acid residues and three intramolecular disulfide bonds. The disulfide bonds give them a cyclic structure. Defensins are produced as pre-proproteins of 93 to 95 residues and are anchored in cell membranes. They are eventually processed to the 29 to 34 amino acid residue proteins found in azurophil granules.[6,14,15] Defensins have apparently not yet been isolated from ruminant neutrophils; however, bovine neutrophils have been shown to contain potent bactericidal cationic peptides that are distinct from defensins and are found in the large peroxidase-negative granules that are unique to bovine neutrophils. These have been named bactenecins[17-22] and indolicidin.[23]

Bactenecins, of which three have been purified and sequenced, are highly cationic polypeptides. The first to be sequenced contained 12 amino acids and exhibited bactericidal activity against both *Escherichia coli* and *Staphylococcus aureus*.[17] The next two to be identified were Bac5 made up of 42 amino acid residues, and Bac7 consisting of 59 amino acid residues. Both of these contained greater than 45% proline and greater than 23% arginine.[19] They have been shown to be bactericidal for *Salmonella typhimurium, Klebsiella pneumoniae*, and *E. coli*, but not for *Proteus vulgaris, S. aureus*, or *Streptococcus agalactiae*.[18] Bactenecins are synthesized in the bone marrow as pre-probactenecins; they are processed to probactenecins and stored in the large granules of immature myeloid cells.[20] The probactenecins are not bactericidal and must be cleaved by serine proteases to become biologically active.[20] Serine proteases are found in the azurophil granules and probactenecins are found in the large granules of ruminant neutrophils. Therefore, in order for biologically active bactenecins to be formed, both the azurophil granules and large granules must release their contents into the phagosome or to the exterior of the cell.[20,22] The mechanism of action for Bac5 and Bac7 involves binding to the outer membrane of Gram-negative bacteria, followed by rapid translocation to the inner membrane where they impair the electron transport chain as well as some membrane activities dependent on the production of energy. In addition, Bac7 inactivates human herpes simplex virus types 1 and 2.[20,21]

Indolicidin is a unique antimicrobial peptide, consisting of 13 amino acids, isolated from bovine neutrophils.[23] It is unusual in that five of the amino acids are tryptophan residues, three are proline, and the two carboxyl terminal residues are arginine. The terminal arginine is carboxamidated. The mole percent of tryptophan in indolicidin is the highest observed among known protein sequences. It has been shown to be bactericidal for *S. aureus* and *E. coli.*

b. Oxidative Microbicidal Mechanisms

The oxidative microbicidal mechanisms are dependent on activity of a membrane-associated oxidase enzyme complex which converts oxygen to superoxide anion. The superoxide anion undergoes further reactions to form hydrogen peroxide, hydroxyl radical, and singlet oxygen.[4] These oxygen-containing molecules are formed inside the phagosome and at the external surface of the cell, and have varying degrees of bactericidal activity. Myeloperoxidase released from azurophil granules will catalyze a reaction between hydrogen peroxide and halide ions (iodide, bromide, and chloride), resulting in the halogenation and oxidation of bacterial surface components and the generation of hypochlorous acid, aldehydes, and chloramines.[4] The myeloperoxidase-hydrogen peroxide-halide system is one of the most potent bactericidal mechanisms available to the neutrophil. Many factors associated with increased susceptibility to infection in cattle have been shown to inhibit neutrophil oxidative metabolism and/or activity of the MPO-H_2O_2-halide system (discussed below).

The oxidase enzyme complex has been studied extensively in bovine neutrophils. Superoxide anion is rapidly produced by a membrane-bound oxidase consisting of an NADPH-dependent flavoprotein and a low-potential B-type cytochrome. The enzyme catalyzes the one-electron transfer from NADPH to O_2. Activation of this enzyme occurs very rapidly after cell surface receptor binding with opsonized particles or stimulation with phorbol myristate acetate (PMA). Activation of the oxidase enzyme requires the translocation of multiple cytosolic proteins to the membrane.[24-27] In the bovine, maximal oxidase activation results in the selective translocation of cytosolic proteins of approximately 65, 53, 45, and 17 kDa to the membrane.[24] A 63-kDa protein which behaves as an oxidase activating factor in bovine neutrophils has been purified and characterized.[28] Antibodies against this protein recognize a similar sized protein in human neutrophils which is missing in neutrophils from patients with chronic granulomatous disease, who are known to lack a 67-kDa oxidase activating factor.

Activation of the bovine oxidase enzyme has been studied in a cell-free system consisting of purified neutrophil membranes, GTP γ S, arachidonic acid, and neutrophil cytosol (a cell homogenate supernatant). The bovine oxidase system behaves very similarly to the human neutrophil and B cell oxidase systems. In fact, cytosol from human neutrophils or B cells can substitute for bovine neutrophil cytosol in cell-free oxidase activation systems.[29,30] The generation of superoxide anion by bovine neutrophils is not inhibited by glycolytic (NaF) or cytoskeletal (colchicine, cytochalasin B, and prostaglandin E_1) inhibitors.[31]

c. Neutrophil-Mediated Cytotoxicity

As in other species, bovine neutrophils are cytotoxic for eukaryotic cells coated with antibody. This phenomenon is called antibody-dependent cell-mediated cytotoxicity (ADCC) and is thought to be important in resistance to many viral infections and, perhaps, elimination of tumor cells and in graft rejection. In addition to ADCC, bovine neutrophils have been shown to be active in antibody-independent cell-mediated cytotoxicity (AICC).[32-34] This AICC activity has been observed using xenogeneic erythrocytes as target cells. The AICC activity is increased by pretreatment of neutrophils with either lymphokine produced by antigen-stimulated bovine mononuclear cells, recombinant bovine interferon-γ (rbo-IFN-γ) (*in vitro* or *in vivo*), or recombinant bovine tumor necrosis factor-α (rbo-TNF-α). Bovine neutrophil ADCC activity is also increased by pretreatment with lymphokine or rbo-IFN-γ (*in vitro* or *in vivo*).[32-36] The enhancement of AICC activity requires a 2-h preincubation time, mRNA and protein synthesis by the neutrophil, and neutrophil arachidonic acid metabolism, whereas enhancement of ADCC activity by lymphokine and rbo-IFN-γ treatment does not require any of these activities.[32,34] The method of target cell recognition by neutrophils in AICC is not known. However, the AICC activity is to some extent selective in that the activated neutrophils have been shown to be cytotoxic for erythrocytes from chickens, turkeys, and humans, but not for bovine erythrocytes from genetically different individuals.[32,36] The mechanism of cytotoxicity has not been characterized. The enhancement of neutrophil AICC activity by cytokines may be somewhat unique to the bovine since a similar phenomenon was not observed in human neutrophils treated with recombinant human interferon γ (Steinbeck and Roth, unpublished observations).

3. Stimulants of Neutrophil Activity

a. Chemotaxis

There are marked differences between bovine and human neutrophils in their response to potential chemotactic factors. Bovine neutrophils are not chemotactically attracted to several factors that are known to be chemotactic for human neutrophils. These include n-formylmethionylleucylphenylalanine (FMLP) and four related peptides, culture filtrates from at least seven different bacteria including *E. coli,* and platelet activating factor.[37-39] Sheep neutrophils also fail to respond to the formyl peptides.[40] Bovine neutrophils are chemotactically attracted to the following factors *in vitro:* zymosan or LPS-activated bovine serum, partially purified bovine C5a, leukotriene B_4 (LTB_4), 5-hydroxyeicosatetraenoic acid (5-HETE), 15-HETE, lipoxin B_4, prostaglandin $F_{2\alpha}$, 17 β estradiol, *Pasteurella haemolytica* culture fluids, estrone, and supernatants from tissue suspensions of placental cotyledons and uterine wall tissue.[37-39,41-45] LTB_4, LTB_5, and 5-HETE were also shown to attract neutrophils *in vivo.*[46] Even though casein and *E. coli* culture filtrates are not chemotactic for bovine neutrophils *in vitro,*[39] they do induce a shape change in bovine neutrophils (one of the first steps in the neutrophil chemotactic response).[47]

b. Ingestion, Oxidative Metabolism, and Degranulation

IgM and IgG_2, but not IgG_1, serve as effective opsonins for bovine and sheep neutrophils. Fc receptors for IgM and IgG_2, but not IgG_1, have been found on bovine neutrophils.[48-50] Various stimuli have been used *in vitro* to induce the burst of oxidative metabolism and degranulation in bovine neutrophils. These stimuli may act differentially in neutrophil activation. For example, PMA will stimulate oxidative metabolism but not elastase release or increases in cytoplasmic calcium concentration in bovine neutrophils.[13] In vitro treatment with either concanavalin A (Con A) or calcium ionophore in the presence of low levels of cytochalasin B will cause an increase in cytoplasmic calcium concentration and elastase release without stimulating oxidative metabolism in bovine neutrophils.[13] Recombinant human C5a will also stimulate an increase in cytoplasmic calcium concentration in bovine neutrophils.[51] Zymosan particles opsonized with fresh bovine serum will stimulate an increase in cytoplasmic calcium concentration, elastase release, and oxidative metabolism.[13] Exposure to low concentrations of *P. haemolytica* leukotoxin will also stimulate both oxidative metabolism and release of specific granules.[52]

There are important differences in the responses of bovine and human neutrophils to several stimuli. Bovine neutrophils have Fc receptors for IgM, whereas human neutrophils do not.[50] Human neutrophils have receptors for, and respond to, FMLP, whereas bovine neutrophils do not.[13,37-39] PMA stimulates both elastase release and oxidative metabolism in human neutrophils, whereas it stimulates oxidative metabolism but not elastase release in bovine neutrophils. Cytochalasin B is commonly used as a pretreatment for human neutrophils *in vitro* to cause them to release granule contents extracellularly upon stimulation. An important difference between bovine and human neutrophils is that in the bovine, neutrophils derived from many animals will respond to cytochalasin B with an increase in cytoplasmic calcium concentration and release of elastase. Therefore, for neutrophils from most cattle, it is not possible to use cytochalasin B pretreatment to cause extracellular release of granule contents upon secondary stimulation as is commonly done for *in vitro* studies on human neutrophils.[13] Other investigators have been able to demonstrate that calcium ionophore and platelet activating factor induce oxidative metabolism in bovine neutrophils.[53]

B. DEFICITS IN NEUTROPHIL FUNCTION

1. Age-Related Changes in Neutrophil Function

Neonatal and young calves have an increased susceptibility to infectious disease which correlates with deficits in neutrophil function. Neutrophils from neonatal calves produce less superoxide anion in response to PMA than neutrophils from either fetal calves (210 to 220 days gestation) or adult cattle.[54,55] The deficit in superoxide anion generation is not observed when opsonized zymosan is used as a stimulus.[56] Neutrophils from neonatal calves also have decreased Fc receptor expression, decreased capping of Con A binding sites and decreased myeloperoxidase content as compared to neutrophils from adult cattle.[57] Colostrum consumption has been shown to increase neutrophil oxidative metabolism.[58] Neutrophils from neonatal calves have increased concentration of alkaline phosphatase and increased chemotactic responsiveness to zymosan-activated plasma and C5a.[45] These differences between neonatal and adult neutrophil function are apparently not related to intracellular calcium mobilization or protein kinase C activity. Neonatal and adult bovine neutrophils have similar changes in intracellular calcium concentration in response to recombinant human C5a.[51] In addition, there is no significant difference in protein kinase C activity or in the ability to translocate protein kinase C from the cytosol to the membrane in response to PMA between neutrophils from neonatal and adult cattle.[59]

Calves up to several weeks of age continue to have differences in neutrophil function as compared to adult cattle. Deficits in superoxide anion generation, iodination, ADCC, chemotaxis, random migration, and myeloperoxidase content were found in calves up to 4 months of age as compared to adult cattle. Neutrophils from these young calves have similar changes in intracellular calcium concentration in response to opsonized zymosan as neutrophils from adult cattle. They also respond similarly to *in vitro* treatment with rbo-IFN-γ.[60,61] In contrast, neutrophils from young calves were seen to have increased ability to ingest *S. aureus* up to 4 to 5 weeks of age and increased AICC activity up to 4 months of age as compared to neutrophils from adult cattle.[60,61]

2. Genetic Variation in Neutrophil Function

Neutrophil function in cattle has been shown to be influenced by both the genetic background of the sire and by the class I bovine lymphocyte antigen (BoLA) complex alleles present in the animal.[62,63] Blood neutrophil function, as well as other parameters of immune function, were evaluated on five occasions in 98 lactating Holstein cows over a 60-day period. The cows were from two genetic lines that had been selected for up to seven generations for high milk production. The results of the neutrophil function testing were correlated with the presence of specific alleles at the bovine MHC class I locus and with the sire progeny of the animals. Significant influences of both the BoLA type and sire progeny group on several aspects of neutrophil function were found.[62,63]

3. Genetic Deficit in Neutrophil Function

Cattle have been reported to have an autosomal recessive syndrome very similar to leukocyte adhesion deficiency (LAD) described in children and dogs.[64-69] LAD is caused by a defect in the family of leukocyte integrins (Mac-1, LFA-1, p150,95). Leukocyte integrins are glycoproteins expressed on leukocyte membranes which are essential for neutrophil adherence to endothelial cells and emigration from the bloodstream. One of the salient features of this syndrome is a marked neutrophilia (often exceeding 100,000 per μl) due to the fact that the neutrophils are not able to adhere efficiently to endothelial cells and leave the bloodstream. The disease syndrome in cattle is characterized by recurrent pneumonia, ulcerative and granulomatous stomatitis, enteritis with bacterial overgrowth, periodontitis, delayed wound healing, persistent neutrophilia, and death at an early age. This has been reported only in Holstein calves, and the pedigrees of all affected calves can be traced to a common sire which was used extensively in artificial insemination. The genetic defect in bovine LAD has been identified as a single amino acid substitution in a 26 amino acid sequence that shares 100% homology with human and murine CD18 protein sequences.[67] CD18 is the common β subunit found in all three members of the leukocyte integrin family. Immunofluorescent analysis for surface expression of CD18 on neutrophils can be used to identify homozygous clinically affected animals and heterozygous clinically normal carriers of the genetic defect.[65]

4. Suppression of Neutrophil Function

Many factors have been shown to interfere with neutrophil function in ruminants; these include viral infection, bacterial virulence factors, stress, hormones, drugs, and nutritional factors. In many cases, the alterations in function have been associated with increased susceptibility to infectious disease.

a. Infectious Agents

Infection with respiratory viruses has been shown to be an important predisposing factor in bacterial pneumonia of cattle. In addition, infection with these same viruses has been shown to suppress neutrophil function as well as other aspects of the immune system. Bovine viral diarrhea (BVD) virus infection has been shown to be capable of suppressing neutrophil iodination, *S. aureus* ingestion, and ADCC. Suppression of neutrophil function may last as long as 3 weeks after initial infection.[70-72] Cattle that are persistently infected with BVD virus after *in utero* infection have decreased random migration, *S. aureus* ingestion, superoxide anion generation, iodination, AICC, and cytoplasmic calcium flux in response to stimuli.[73] Infection of cattle with parainfluenza 3 virus has been shown to result in decreased oxidative metabolism and iodination activity in neutrophils.[74] Infection with bovine herpesvirus 1 has been shown to result in reduced neutrophil random migration but enhanced neutrophil *S. aureus* ingestion.[74]

Infection with bovine immunodeficiency-like virus (BIV, a bovine lentivirus) is associated with a significant decrease in neutrophil ADCC and iodination activities. These mild decreases in neutrophil function are detectable by 4 to 5 months after infection and, as yet, have not been associated with increased susceptibility to infectious disease.[75]

Virulence factors from several important bacterial pathogens of cattle have been shown to interfere directly with neutrophil function. *Pasteurella haemolytica* has a leukotoxin which is cytotoxic for bovine neutrophils as well as other leukocytes.[76] At sub-cytolytic concentrations, the leukotoxin impairs bovine neutrophil function. Capsular material and lipopolysaccharide from *P. haemolytica* has also been shown to impair bovine neutrophil function.[77,78] *Moraxella bovis* also has a cytotoxin which impairs neutrophil function.[79-82] *Brucella abortus* and *Haemophilus somnus* are two facultative intracellular pathogens that have been shown to be resistant to killing by bovine neutrophils.[83] Both of these pathogens have been shown to have RNA components associated with their surface. These components inhibit degranulation of neutrophil primary granules and may be important virulence factors for the intracellular survival of these bacteria.[84-88]

Infection with coccidia has been shown to suppress neutrophil function in cattle and to be associated with increased susceptibility to bacterial pneumonia. Both subclinical and clinical infections with *Eimeria* spp. were found to be associated with decreased neutrophil oxidative metabolism and iodination activity.[89]

b. Hormones and Pharmaceuticals

Physical and psychological distress are known to be immunosuppressive and to predispose to infection. Stress-induced suppression of neutrophil function is probably mediated by hormones released in response to stress. Hormones that may play a role include cortisol, catecholamines, endogenous opiates, and perhaps others.[90] Elevated cortisol levels are known to suppress neutrophil function in cattle.[91-93]

Dexamethasone is a potent glucocorticoid capable of strongly suppressing neutrophil function as well as other aspects of immune function in ruminants.[92,94-96] Dexamethasone is also known to markedly increase susceptibility to bacterial, viral, and protozoal infections.[89,92,97] Pharmacologic doses of dexamethasone alter nearly every parameter of neutrophil function. The alteration of neutrophil function by dexamethasone and other glucocorticoids is mediated by at least two mechanisms. Dexamethasone treatment of cattle inhibits the production of lipoxygenase products of arachidonic acid metabolism in neutrophils. This is responsible for a decrease in neutrophil oxidative metabolism and iodination activity.[98] Suppression of neutrophil ADCC activity and the enhancement of random migration observed after dexamethasone treatment in cattle is apparently mediated through glucocorticoid effects on monocytes which subsequently release factors altering neutrophil function.[99]

Administration of pharmacologic dosages of progesterone to bovine steers has also been shown to inhibit neutrophil iodination and enhance neutrophil random migration.[100] This correlates with alterations of neutrophil function during the normal estrous cycle in cows. When progesterone is high during the estrous cycle, neutrophil oxidative metabolism, iodination, and ADCC are depressed and random migration is enhanced.[101] Certain antibiotics, particularly chloramphenicol, have also been shown to inhibit bovine neutrophil function.[102-104]

c. Nutritional Factors

Nutritional deficiencies have been associated with increased incidence of infectious diseases and decreased neutrophil function in ruminants. Nutritional deficiencies that have been shown to decrease neutrophil function in ruminants include deficiencies in copper,[105,106] cobalt,[107] selenium,[108] molybdenum, thiamine, and sulfur.[106]

C. ENHANCEMENT OF NEUTROPHIL FUNCTION

Bovine neutrophil function has been shown to be enhanced by treatment of cattle with a compound (avridine) that is capable of inducing interferon as well as other cytokines.[109,110] The enhanced neutrophil activity after *in vivo* treatment with a cytokine inducer can be mimicked by *in vitro* treatment with supernatants from antigen-stimulated mononuclear cells.[32] The exposure to mononuclear cell supernatant altered nearly every neutrophil function measured, including inhibition of random migration, enhancement of adherence to plastic, enhancement of *S. aureus* ingestion, enhancement of oxidative metabolism, enhancement of ADCC, and induction of AICC (as previously discussed). The only parameter of neutrophil function measured that was not enhanced was iodination. Many cytokines that are likely to be present in an antigen-stimulated mononuclear cell supernatant have been shown to be capable of directly enhancing bovine neutrophil function (discussed below). Some neutrophil activities can be enhanced almost immediately upon contact with cytokine, while others require preincubation of 1 to 2 h to allow for neutrophil protein synthesis. These cytokines would be released during a cell-mediated immune response. Therefore, neutrophils may be important effector cells contributing to enhanced immunity to disease in animals with cell-mediated immunity. For example, rbo-IFN-γ-activated bovine neutrophils have been shown to kill *B. abortus* more efficiently *in vitro* (although the effect was small).[111] Rbo-IFN-γ treatment of dexamethasone-immunosuppressed calves enhanced their resistance to an *H. somnus* challenge.[97]

Rbo-IFN-γ has been shown to be capable of altering most of the same neutrophil functions as a crude mononuclear cell supernatant. In general, IFN-γ, as well as avridine (an interferon inducer) have had greater activity on the function of neutrophils from immunosuppressed than from control animals.[35,73,97,109] Enhancement of different neutrophil functions by rbo-IFN-γ requires different metabolic events within the neutrophil. For example, enhancement of the AINC response requires both arachidonic acid metabolism and protein synthesis; inhibition of random migration requires only protein synthesis and enhancement of ADCC requires neither.[34]

Bovine neutrophils have been shown to have enhanced functional activity in response to several recombinant cytokines *in vitro* and/or *in vivo* including: rbo-IFN-γ,[33-35,73,111-113] rbo-IFN-α,[112,114] rbo-IL-1β,[115] rbo-TNF-α,[36] rbo-g-CSF,[55,116] and rbo-gm-CSF.[117] Bovine neutrophils have also been shown to respond to recombinant cytokines from other species, including recombinant human (rhu) IL-1α,[118] rhu TNF-α,[118] rhu IL-1β,[115] rhu IL-2,[119] and recombinant murine IL-1α.[115] Bovine neutrophils not only respond to IL-1, but also have been shown to synthesize and release IL-1.[120]

In addition to recombinant cytokines, other compounds administered to cattle have been shown to enhance bovine neutrophil function. Dihydroheptaprenol administered to cattle has been shown to increase the number of neutrophils in the peripheral blood, and enhance neutrophil oxidative metabolism, *S. aureus* ingestion, and bactericidal activity.[121,122] High dosages of ascorbic acid administered to cattle tended to reverse or prevent some of the immunosuppressive effects of dexamethasone on bovine neutrophil function.[123]

III. KILLER CELLS

Killer cells are a family of lymphoid cells described in man that are capable of non-MHC restricted cytotoxic activity. They are functionally defined by the three types to be discussed here: NK, K, and LAK cells. The activities of these cell types in the bovine have previously been reviewed.[124,125]

A. NATURAL KILLER CELLS

Natural killer cells are functionally defined as lymphoid cells that mediate natural (spontaneous) non-MHC-restricted cytotoxicity. In humans, NK cells do not express on their surface the complete CD3 complex or any of the four chains of the T cell receptor, but do have surface expression of the low-affinity receptor for the Fc component of IgG (FCRIII or CD16) and the CD56 molecule. NK cells are capable of mediating cytotoxicity even in the absence of MHC class I or MHC class II molecules on the target cell surface. In humans, most of the NK cells have the large granular lymphocyte morphology. However, some resting T cells and activated cytotoxic T lymphocytes also have the large granular lymphocyte morphology. In humans, NK cells are capable of using the low-affinity Fc receptor for IgG to bind to IgG on target cells and mediate ADCC. When mediating ADCC, these cells are classified as killer (K) cells.[2]

Bovine peripheral blood lymphocytes are inefficient in mediating lysis of target cells typically used to evaluate NK cell function in other species. Most investigators have found only minimal lysis of tumor cells (K562, YAC-1, HL60S, and HL60R) by nonactivated bovine peripheral blood lymphocytes.[126-129] Bovine peripheral blood lymphocytes will bind to the tumor cell targets but will not efficiently lyse them unless activated by IL-2.[128] One group of investigators was able to detect lysis of YAC-1 cells by bovine peripheral blood lymphocytes.[130] It is possible that this discrepancy may have been due to *in vivo* activation of NK cells since other investigators have shown that unpurified bovine lymphokines are capable of inducing LAK activity for YAC-1 cells by bovine lymph node lymphocytes.[131]

In contrast to the difficulty in demonstrating bovine NK activity against tumor cell targets, bovine NK activity has been shown for various virus-infected cells, including cell lines infected with PI3 virus,[132] bovine leukemia virus,[129] and bovine herpesvirus 1 (BHV-1).[133,134] Bovine NK activity against BHV-1-infected cells requires target cell expression of BHV-1 major cell surface glycoproteins.[135] Specifically, cell surface expression of BHV-1 glycoproteins GI and GIV, but not GIII, makes the target cells susceptible to bovine NK activity.[136] This indicates that bovine NK cells may be capable of mediating antigen-specific cell lysis.

Bovine cells with NK activity against virus-infected target cells have been shown to be large, low-density mononuclear cells that do not express MHC class II molecules.[134] Identification of granules in bovine NK cells similar to those observed in human NK cells has been difficult, with no conclusive results.[125] Goddeeris et al.[137] have cloned two populations of lymphocytes that are capable of non-MHC-restricted cell killing. These cells did not express the CD4, CD5, CD6, or WC1 (γ/δ T cell marker) surface antigens. One of the clones did express the CD2 antigen and the other clone expressed homodimeric CD8. Neither clone was found to contain mRNA for the γ or δ chains of CD3 or the α chain of the T cell receptor. They did have mRNA for the δ chain of the T cell receptor and a truncated form of the β chain for the T cell receptor. These two clones exhibited moderate levels of cytotoxic activity on autologous lymphoblasts and on autologous and allogeneic lymphoblasts infected with *Theileria parva*.[137] Since these cell lines were maintained as actively proliferating cells in culture, they may represent either NK or LAK cells.

Natural killer cells are believed to be important in assisting to control viral infections and some parasitic infections. Suppression of NK cell activity may decrease the animal's resistance to these conditions. *In vitro* exposure of bovine peripheral blood mononuclear cells to dexamethasone, betamethasone, flunixin, and phenylbutazone has been shown to decrease NK cytotoxic activity to BHV-1-infected target cells. The levels of flunixin, dexamethasone, and betamethasone that inhibited NK activity *in vitro* are levels that are likely to be attainable *in vivo.*[138] Decreases in NK or LAK activity after *in vivo* treatment with pharmaceuticals or exposure to infectious agents has apparently not been reported in ruminants.

B. K CELLS

The K cells are functionally defined as lymphoid cells (non-T, non-B) that have Fc receptors which enable them to bind to antibody-coated target cells and destroy them through an ADCC mechanism. Other nonlymphoid cells such as neutrophils, macrophages, and eosinophils are also capable of destroying antibody-coated cells through ADCC. In man, many LGLs are capable of both NK and K activity.

Campos and Rossi were unable to demonstrate ADCC activity using purified peripheral blood lymphocytes.[128,139] After activation with IL-2, the lymphocytes were capable of mediating ADCC and are then considered to be LAK cells.[128] This author was unable to find any published evidence of ADCC activity by nonactivated ruminant peripheral blood lymphocytes. Therefore, the existence of K cell activity in ruminants has apparently not yet been proven.

C. LYMPHOKINE-ACTIVATED KILLER CELLS

Lymphokine-activated killer (LAK) cells are defined as lymphocytes that mediate non-MHC-restricted cytotoxicity against a broad array of targets following *in vivo* or *in vitro* activation with IL-2 alone or in combination with other cytokines. In humans, NK cells are the predominant precursors of LAK cells; however, some T cells may also serve as LAK precursors.[1]

Ruminant peripheral blood lymphoid cells have been shown to be capable of LAK activity after incubation with a variety of cytokines. A bovine lymphokine preparation containing IL-2 induced LAK activity in bovine peripheral blood mononuclear leukocytes for K562 and YAC-1, but not HSB2 target cells.[140] Bovine LAK cells induced by incubation of lymph node lymphocytes with lymphokine-containing media were cytotoxic for allogeneic fibroblasts, a human lung carcinoma cell line (A549) and a bovine neurofibrosarcoma cell line (LMS).[131] Recombinant human IL-2 induced LAK activity *in vivo* and *in vitro* in sheep peripheral blood lymphocytes toward K562 cells and sheep pulmonary microvascular endothelial cells. This suggests that LAK cells might mediate the increased pulmonary vascular permeability observed following recombinant human IL-2 infusion in sheep.[141] Incubation of bovine peripheral blood mononuclear leukocytes with rhuIL-2, rhuIL-4, rbo-IFN-α, and rbo-IFN-γ induced LAK activity for both

K562 and YAC-1 target cells. Purified hu-IFN-β and rhuIL-1 did not induce LAK activity in bovine leukocytes.[126]

Bovine peripheral blood mononuclear leukocytes were found to be capable of binding to tumor cell targets (K562, HL60S, and HL60R), but would not lyse the target cells unless they were activated by IL-2. This activity was dose dependent and required the continuous presence of IL-2; Twelve to eighteen hours of incubation with IL-2 was sufficient to result in activation. The peripheral blood mononuclear leukocytes also did not mediate ADCC unless they were activated by IL-2.[128] Cold target competition experiments demonstrated that the cell population responsible for ADCC after IL-2 activation was the same as that responsible for direct lysis of tumor cells.[140] The LAK cells induced from bovine peripheral blood mononuclear leukocytes were found to be nonadherent (to nylon wool and sephadex G-10), non-monocyte, non B-cells that were $CD2^+$, $CD4^-$, and $CD8^-$.[128]

IV. SUMMARY

Neutrophils and the killer family of cells have been extensively studied in ruminants. Because they are readily available in large numbers, bovine neutrophils have been used in a number of laboratories as the preferred model for the study of neutrophil biology. Bovine neutrophils have also been extensively studied for their important role in resistance to infectious diseases in cattle. Decreased neutrophil function due to immaturity, parturition, stress, hormonal changes, viral infection, genetic defects, parasitic infection, or the action of bacterial virulence factors has been associated with increased susceptibility to disease. The relatively recent findings that neutrophil function can be enhanced through the action of various cytokines indicates that activated neutrophils may be important effector cells in the cell-mediated immune response. Enhancement of neutrophil function through the action of cytokines has shown promise for increasing resistance to infectious diseases in cattle.

The killer family of lymphocytes, as well as neutrophils, may be cytotoxic for virally infected cells, tumor cells, or foreign cells. There are a number of differences in the cytotoxic activity of these cells between humans and ruminants. Bovine NK cells may bind to tumor target cells, but have very little cytotoxic activity unless they are activated by IL-2. Similarly, bovine lymphocytes are not capable of mediating ADCC activity unless activated by IL-2 or, perhaps, other cytokines. In humans, both of these activities can be demonstrated without apparent previous cytokine activation. Once activated by cytokines, these cells, in both humans and ruminants, are considered to be LAK cells. Neutrophils in ruminants, but not in humans, have been shown to be active in antibody-independent cell-mediated cytotoxicity. Neutrophils from most normal cattle have low levels of AICC activity which is markedly enhanced after *in vitro* or *in vivo* treatment with INF-γ or certain other cytokines. Therefore, neutrophils and the killer family of cells in ruminants have been shown to be responsive to the presence of cytokines and are active participants in the cell-mediated immune response.

REFERENCES

1. **Herberman, R. B.,** Lymphokine-activated killer (LAK) cells, in *Encyclopedia of Immunology*, Roitt, I. M. and Delves, P. J., Eds., Academic Press, San Diego, 1992, 1013.
2. **Trinchieri, G.,** Natural killer (NK) cells, in *Encyclopedia of Immunology*, Roitt, I. M. and Delves, P. J., Eds., Academic Press, San Diego, 1992, 1136.
3. **Carlson, G. P. and Kaneko, J. J.,** Isolation of leukocytes from bovine peripheral blood, *Proc. Soc. Exp. Biol. Med.*, 142, 853, 1973.
4. **Densen, P. and Mandell, G. L.,** Granulocytic phagocytes, in *Principles and Practice of Infectious Diseases*, Mandell, G. L., Douglas, R. G., and Bennett, J. E., Eds., Churchill Livingstone, New York, 1990, 81.
5. **Lehrer, R. I. and Ganz, T.,** Antimicrobial polypeptides of human neutrophils, *Blood*, 76, 2169, 1990.
6. **Spitznagel, J. K.,** Antibiotic proteins of human neutrophils, *J. Clin. Invest.*, 86, 1381, 1990.
7. **Clark, R. A.,** The human neutrophil respiratory burst oxidase, *J. Infect. Dis.*, 161, 1140, 1990.
8. **Baggiolini, M., Horisberger, U., Gennaro, R., and Dewald, B.,** Identification of three types of granules in neutrophils of ruminants, *Lab. Invest.*, 52, 151, 1985.
9. **Bertram, T. A.,** Neutrophilic leukocyte structure and function in domestic animals, *Adv. Vet. Sci. Comp. Med.*, 30, 91, 1985.

10. **Healy, P. J.,** Lysosomal hydrolase activity in leucocytes from cattle, sheep, goats, horses and pigs, *Res. Vet. Sci.*, 33, 275, 1982.
11. **Rausch, P. G. and Moore, T. G.,** Granule enzymes of polymorphonuclear neutrophils: a phylogenetic comparison, *Blood*, 46, 913, 1975.
12. **Gennaro, R., Schneider, C., DeNicola, G., Cian, F., and Romeo, D.,** Biochemical properties of bovine granulocytes, *Proc. Soc. Exp. Biol. Med.*, 157, 342, 1978.
13. **Brown, G. B. and Roth, J. A.,** Comparison of the response of bovine and human neutrophils to various stimuli, *Vet. Immunol. Immunopathol.*, 28, 201, 1991.
14. **Boman, H. G.,** Antibacterial peptides: key components needed in immunity, *Cell*, 65, 205, 1991.
15. **Lehrer, R. I., Ganz, T., and Selsted, M. E.,** Defensins: endogenous antibiotic peptides of animal cells, *Cell*, 64, 229, 1991.
16. **Leong, S. R. and Camerato, T.,** Nucleotide sequence of the bovine bactericidal permeability increasing protein (BPI), *Nucl. Acids Res.*, 18, 3052, 1990.
17. **Romeo, D., Skerlavaj, B., Bolognesi, M., and Gennaro, R.,** Structure and bactericidal activity of an antibiotic dodecapeptide purified from bovine neutrophils, *J. Biol. Chem.*, 263, 9573, 1988.
18. **Gennaro, R., Skerlavaj, B., and Romeo, D.,** Purification, composition, and activity of two bactenecins, antibacterial peptides of bovine neutrophils, *Infect. Immun.*, 57, 3142, 1989.
19. **Frank, R. W., Gennaro, R., Schneider, K., Przybylski, M., and Romeo, D.,** Amino acid sequences of two proline-rich bactenecins, *J. Biol. Chem.*, 265, 18871, 1990.
20. **Zanetti, M., Litteri, L., Gennaro, R., Horstmann, H., and Romeo, D.,** Bactenecins, defense polypeptides of bovine neutrophils, are generated from precursor molecules stored in the large granules, *J. Cell Biol.*, 111, 1363, 1990.
21. **Skerlavaj, B., Romeo, D., and Gennaro, R.,** Rapid membrane permeabilization and inhibition of vital functions of Gram-negative bacteria by bactenecins, *Infect. Immun.*, 58, 3724, 1990.
22. **Zanetti, M., Litteri, L., Griffiths, G., Gennaro, R., and Romeo, D.,** Stimulus-induced maturation of probactenecins, precursors of neutrophil antimicrobial polypeptides, *J. Immunol.*, 146, 4295, 1991.
23. **Selsted, M. E., Novotny, M. J., Morris, W. L., Tang, Y. Q., Smith, W., and Cullor, J. S.,** Indolicidin, a novel bactericidal tridecapeptide amide from neutrophils, *J. Biol. Chem.*, 267, 4292, 1992.
24. **Doussiere, J., Pilloud, M. C., and Vignais, P. V.,** Cytosolic factors in bovine neutrophil oxidase activation, *Biochemistry*, 29, 2225, 1990.
25. **Doussiere, J., Pilloud, M. C., and Vignais, P. V.,** Activation of bovine neutrophil oxidase in a cell free system; GTP-dependent formation of a complex between a cytosolic factor and a membrane protein, *Biochem. Biophys. Res. Commun.*, 152, 993, 1988.
26. **Pilloud, M. C., Doussiere, J., and Vignais, P. V.,** The superoxide generating oxidase activation of bovine neutrophils; evidence for synergism of multiple cytosolic factors in a cell-free system, *FEBS Lett.*, 257, 167, 1989.
27. **Ligeti, E., Tardif, M., and Vignais, P. V.,** Activation of superoxide generating oxidase of bovine neutrophils in a cell-free system; interaction of a cytosolic factor with the plasma membrane and control by G nucleotides, *Biochemistry*, 28, 7116, 1989.
28. **Pilloud-Dagher, M. C. and Vignais, P. V.,** Purification and characterization of an oxidase activating factor of 63 kilodaltons from bovine neutrophils, *Biochemistry*, 30, 2753, 1991.
29. **Ligeti, E., Doussiere, J., and Vignais, P. V.,** Activation of the superoxide-generating oxidase in plasma membrane from bovine polymorphonuclear neutrophils by arachidonic acid, a cytosolic factor of protein nature, and nonhydrolyzable analogues of GTP, *Biochemistry*, 27, 193, 1988.
30. **Cohen-Tanugi, L., Morel, F., Pilloud-Dagher, M. C., Seigneurin, J. M., Francois, P., Bost, M., and Vignais, P. V.,** Activation of superoxide-generating oxidase in an heterologous cell-free system derived from Epstein-Barr-virus-transformed human B lymphocytes and bovine neutrophils, *Eur. J. Biochem.*, 202, 649, 1991.
31. **Silva, I. D. and Jain, N. C.,** Effects of glycolytic and cytoskeletal inhibitors on phagocytic and nitroblue tetrazolium reductive activities of bovine neutrophils, *Am. J. Vet. Res.*, 50, 1175, 1989.
32. **Lukacs, K., Roth, J. A., and Kaeberle, M. L.,** Activation of neutrophils by antigen-induced lymphokine with emphasis on antibody-independent cytotoxicity, *J. Leukoc. Biol.*, 38, 557, 1985.
33. **Steinbeck, M. J., Roth, J. A., and Kaeberle, M. L.,** Activation of bovine neutrophils by bovine recombinant γ interferon, *Cell. Immunol.*, 98, 137, 1986.

34. **Steinbeck, M. J., Webb, S. A., and Roth, J. A.,** Role for arachidonic acid metabolism and protein synthesis in recombinant bovine γ interferon induced activation of bovine neutrophils, *J. Leukoc. Biol.*, 46, 450, 1989.
35. **Roth, J. A. and Frank, D. E.,** Recombinant bovine interferon γ as an immunomodulator in dexamethasone-treated and nontreated cattle, *J. Interferon Res.*, 9, 143, 1989.
36. **Chiang, Y.-W., Murata, H., and Roth, J. A.,** Activation of bovine neutrophils by recombinant bovine tumor necrosis factor-α, *Vet. Immunol. Immunopathol.*, 29, 329, 1991.
37. **Gray, G. D., Knight, K. A., Nelson, R. D., and Herron, M. J.,** Chemotactic requirements of bovine leukocytes, *Am. J. Vet. Res.*, 43, 757, 1982.
38. **Carroll, E. J., Mueller, R., and Panico, L.,** Chemotactic factors for bovine leukocytes, *Am. J. Vet. Res.*, 43, 1661, 1982.
39. **Craven, N.,** Chemotactic factors for bovine neutrophils in relation to mastitis, *Comp. Immunol. Microbiol. Infect. Dis.*, 9, 29, 1986.
40. **Chenoweth, D. E., Lane, T. A., Rowe, J. G., and Hugli, T. E.,** Quantitative comparisons of neutrophil chemotaxis in four animal species, *Clin. Immunol. Immunopathol.*, 15, 525, 1980.
41. **Brunner, C. J., George, A. L., III, and Hsu, C. L.,** Chemotactic response of bovine neutrophils to *Pasteurella haemolytica* culture fluid, *Vet. Immunol. Immunopathol.*, 21, 279, 1989.
42. **Hoedemaker, M., Lund, L. A., and Wagner, W. C.,** Function of neutrophils and chemoattractant properties of fetal placental tissue during the last month of pregnancy in cows, *Am. J. Vet. Res.*, 53, 1524, 1992.
43. **Hoedemaker, M., Lund, L. A., Weston, P. G., and Wagner, W. C.,** Influence of conditioned media from bovine cotyledon tissue cultures on function of bovine neutrophils, *Am. J. Vet. Res.*, 53, 1530, 1992.
44. **Hoedemaker, M., Lund, L. A., and Wagner, W. C.,** Influence of arachidonic acid metabolites and steroids on function of bovine polymorphonuclear neutrophils, *Am. J. Vet. Res.*, 53, 1534, 1992.
45. **Zwahlen, R. D. and Roth, D. R.,** Chemotactic competence of neutrophils from neonatal calves; functional comparison with neutrophils from adult cattle, *Inflammation*, 14, 109, 1990.
46. **Heidel, J. R., Taylor, S. M., Laegreid, W. W., Silflow, R. M., Liggitt, H. D., and Leid, R. W.,** In vivo chemotaxis of bovine neutrophils induced by 5-lipoxygenase metabolites of arachidonic and eicosapentaenoic acid, *Am. J. Pathol.*, 134, 671, 1989.
47. **Forsell, J. H., Kateley, J. R., and Smith, C. W.,** Bovine neutrophils treated with chemotactic agents: morphologic changes, *Am. J. Vet. Res.*, 46, 1971, 1985.
48. **Mukkur, T. K. S. and Inman, F. P.,** Relative importance of salmonella-specific antibody isotypes in phagocytosis of *Salmonella typhimurium* by ovine mammary neutrophils, *Res. Vet. Sci.*, 46, 153, 1989.
49. **Howard, C. J., Taylor, G., and Brownlie, J.,** Surface receptors for immunoglobulin on bovine polymorphonuclear neutrophils and macrophages, *Res. Vet. Sci.*, 29, 128, 1980.
50. **Grewal, A. S., Rouse, B. T., and Babiuk, L. A.,** Characterization of surface receptors on bovine leukocytes, *Int. Arch. Allergy Appl. Immunol.*, 56, 289, 1978.
51. **Doré, M., Slauson, D. O., Suyemoto, M. M., and Neilsen, N. R.,** Calcium mobilization in C5a-stimulated adult and newborn bovine neutrophils, *Inflammation*, 14, 71, 1990.
52. **Czuprynski, C. J., Noel, E. J., Ortiz-Carranza, O., and Srikumaran, S.,** Activation of bovine neutrophils by partially purified *Pasteurella haemolytica* leukotoxin, *Infect. Immun.*, 59, 3126, 1991.
53. **Freiburghaus, J., Jörg, A., and Müller, T.,** Luminol-dependent chemiluminescence in bovine eosinophils and neutrophils: differential increase of intracellular and extracellular chemiluminescence induced by soluble stimulants, *J. Biolumin. Chemilumin.*, 6, 115, 1991.
54. **Clifford, C. B., Slauson, D. O., Neilsen, N. R., Suyemoto, M. M., Zwahlen, R. D., and Schlafer, D. H.,** Ontogeny of inflammatory cell responsiveness: superoxide anion generation by phorbol ester-stimulated fetal, neonatal, and adult bovine neutrophils, *Inflammation*, 13, 221, 1989.
55. **Doré, M., Slauson, D. O., and Neilsen, N. R.,** Decreased respiratory burst activity in neonatal bovine neutrophils stimulated by protein kinase C agonists, *Am. J. Vet. Res.*, 52, 375, 1991.
56. **Holden, W., Slauson, D. O., Zwahlen, R. D., Suyemoto, M. M., Doré, M., and Neilsen, N. R.,** Alterations in complement-induced shape change and stimulus-specific superoxide anion generation by neonatal calf neutrophils, *Inflammation*, 13, 607, 1989.
57. **Zwahlen, R. D., Wyder-Walther, M., and Roth, D. R.,** Fc receptor expression, Concanavalin A capping, and enzyme content of bovine neonatal neutrophils: a comparative study with adult cattle, *J. Leukoc. Biol.*, 51, 264, 1992.

58. **Lombardo, P. S., Todhunter, D. A., Scholz, R. W., and Eberhart, R. J.,** Effect of colostrum ingestion on indices of neutrophil phagocytosis and metabolism in newborn calves, *Am. J. Vet. Res.*, 40, 362, 1979.
59. **Doré, M., Neilsen, N. R., and Slauson, D. O.,** Protein kinase-C activity in phorbol myristate acetate-stimulated neutrophils from newborn and adult cattle, *Am. J. Vet. Res.*, 53, 1679, 1992.
60. **Hauser, M. A., Koob, M. D., and Roth, J. A.,** Variation of neutrophil function with age in calves, *Am. J. Vet. Res.*, 47, 152, 1986.
61. **Lee, C. C. and Roth, J. A.,** Differences in neutrophil function in young and mature cattle and their response to IFN-γ, *Comp. Haematol. Int.*, in press, 1992.
62. **Kehrli, M. E., Jr., Weigel, K. A., Freeman, A. E., Thurston, J. R., and Kelley, D. H.,** Bovine sire effects on daughters' *in vitro* blood neutrophil functions, lymphocyte blastogenesis, serum complement and conglutinin levels, *Vet. Immunol. Immunopathol.*, 27, 303, 1991.
63. **Weigel, K. A., Kehrli, M. E., Jr., Freeman, A. E., Thurston, J. R., Stear, M. J., and Kelley, D. H.,** Association of class I bovine lymphocyte antigen complex alleles with *in vitro* blood neutrophil functions, lymphocyte blastogenesis, serum complement and conglutinin levels in dairy cattle, *Vet. Immunol. Immunopathol.*, 27, 321, 1991.
64. **Hagemoser, W. A., Roth, J. A., Lofstedt, J., and Fagerland, J. A.,** Granulocytopathy in a Holstein heifer, *J. Am. Vet. Med. Assoc.*, 183, 1093, 1983.
65. **Kehrli, M. E., Jr., Schmalstieg, F. C., Anderson, D. C., van der Maaten, M. J., Hughes, B. J., Ackermann, M. R., Wilhelmsen, C. L., Brown, G. B., Stevens, M. G., and Whetstone, C. A.,** Molecular definition of the bovine granulocytopathy syndrome: identification of deficiency of the Mac-1 (CD11b/CD18) glycoprotein, *Am. J. Vet. Res.*, 51, 1826, 1990.
66. **Kehrli, M. E., Jr., Ackermann, M. R., Shuster, D. E., van der Maaten, M. J., Schmalstieg, F. C., Anderson, D. C., and Hughes, B. J.,** Bovine leukocyte adhesion deficiency β_2 integrin deficiency in young Holstein cattle, *Am. J. Pathol.*, 140, 1489, 1992.
67. **Shuster, D. E., Bosworth, B. T., and Kehrli, M. E., Jr.,** Sequence of the bovine CD18-encoding cDNA: comparison with the human and murine glycoproteins, *Gene*, 114, 267, 1992.
68. **Nagahata, H., Noda, H., Takahashi, K., Kurosawa, T., and Sonoda, M.,** Bovine granulocytopathy syndrome: neutrophil dysfunction in Holstein Friesian calves, *J. Vet. Med. Assoc.*, 34, 445, 1987.
69. **Takahashi, K., Miyagawa, K., Abe, S., Kurosawa, T., Sonoda, M., Nakade, T., Nagahata, H., Noda, H., Chihaya, Y., and Isogai, E.,** Bovine granulocytopathy syndrome of Holstein-Friesian calves and heifers, *Jpn. J. Vet. Sci.*, 49, 733, 1987.
70. **Roth, J. A., Kaeberle, M. L., and Griffith, R. W.,** Effects of bovine viral diarrhea virus infection on bovine polymorphonuclear leukocyte function, *Am. J. Vet. Res.*, 42, 244, 1981.
71. **Roth, J. A. and Kaeberle, M. L.,** Suppression of neutrophil and lymphocyte function induced by a vaccinal strain of bovine viral diarrhea virus with and without the concurrent administration of ACTH, *Am. J. Vet. Res.*, 44, 2366, 1983.
72. **Bolin, S.R., Roth, J.A., Uhlenhopp, E.K., and Pohlenz, J.F.,** Immunologic and virologic findings from a case of chronic bovine viral diarrhea, *J. Am. Vet. Med. Assoc.*, 191, 425, 1987.
73. **Brown, G. B., Bolin, S. R., Frank, D. E., and Roth, J. A.,** Defective function of leukocytes from cattle persistently infected with bovine viral diarrhea virus and the influence of recombinant cytokines, *Am. J. Vet. Res.*, 52, 381, 1991.
74. **Briggs, R. E., Kehrli, M., and Frank, G. H.,** Effects of infection with parainfluenza-3 virus and infectious bovine rhinotracheitis virus on neutrophil functions in calves, *Am. J. Vet. Res.*, 49, 682, 1988.
75. **Flaming, K., van der Maaten, M., Whetstone, C., Carpenter, S., Frank, D., and Roth, J.,** The effect of bovine immunodeficiency-like virus on immune function in experimentally-infected cattle, *Vet. Immunol. Immunopathol.*, 36, 91, 1993.
76. **O'Brien, J. K. and Duffus, W. P. H.,** *Pasteurella haemolytica* cytotoxin: relative susceptibility of bovine leucocytes, *Vet. Microbiol.*, 13, 321, 1987.
77. **Czuprynski, C. J., Noel, E. J., and Adlam, C.,** Modulation of bovine neutrophil antibacterial activities by *Pasteurella haemolytica* A1 purified capsular polysaccharide, *Microb. Pathog.*, 6, 133, 1989.
78. **Paulsen, D. B., Confer, A. W., Clinkenbeard, K. D., and Mosier, D. A.,** *Pasteurella haemolytica* lipopolysaccharide-induced arachidonic acid release from and neutrophil adherence to bovine pulmonary artery endothelial cells, *Am. J. Vet. Res.*, 51, 1635, 1990.

79. **Thomas, C. B., van Ess, P., Wolfgram, L. J., Riebe, J., Sharp, P., and Schultz, R. D.,** Adherence to bovine neutrophils and suppression of neutrophil chemiluminescence by *Mycoplasma bovis*, *Vet. Immunol. Immunopathol.*, 27, 365, 1991.
80. **Kagonyera, G. M., George, L., and Miller, M.,** Effects of *Moraxella bovis* and culture filtrates on ^{51}Cr-labeled bovine neutrophils, *Am. J. Vet. Res.*, 50, 18, 1989.
81. **Kagonyera, G. M., George, L. W., and Munn, R.,** Cytopathic effects of *Moraxella bovis* on cultured bovine neutrophils and corneal epithelial cells, *Am. J. Vet. Res.*, 50, 10, 1989.
82. **Hoien-Dalen, P. S., Rosenbusch, R. F., and Roth, J. A.,** Comparative characterization of the leukocytic and hemolytic activity of *Moraxella bovis*, *Am. J. Vet. Res.*, 51, 191, 1990.
83. **Czuprynski, C. J. and Hamilton, H. L.,** Bovine neutrophils ingest but do not kill *Haemophilus somnus* in vitro, *Infect. Immun.*, 50, 431, 1985.
84. **Canning, P. C., Roth, J. A., Tabatabai, L. B., and Deyoe, B. L.,** Isolation of components of *Brucella abortus* responsible for inhibition of function in bovine neutrophils, *J. Infect. Dis.*, 152, 913, 1985.
85. **Hubbard, R. D., Kaeberle, M. L., Roth, J. A., and Chiang, Y. W.,** *Haemophilus somnus*-induced interference with bovine neutrophil functions, *Vet. Microbiol.*, 12, 77, 1986.
86. **Bertram, T. A., Canning, P. C., and Roth, J. A.,** Preferential inhibition of primary granule release from bovine neutrophils by an extract from *Brucella abortus*, *Infect. Immun.*, 52, 285, 1986.
87. **Chiang, Y.-W., Kaeberle, M. L., and Roth, J. A.,** Identification of the suppressive components in *Haemophilus somnus* fractions which inhibit bovine polymorphonuclear leukocyte function, *Infect. Immun.*, 52, 792, 1986.
88. **Canning, P. C., Roth, J. A., and Deyoe, B. L.,** Release of 5′-guanosine monophosphate and adenine by *Brucella abortus* and their role in the intracellular survival of the bacteria, *J. Infect. Dis.*, 154, 464, 1986.
89. **Roth, J. A., Jarvinen, J. A., Frank, D. E., and Fox, J. E.,** Alteration of neutrophil function associated with coccidiosis in cattle: influence of decoquinate and dexamethasone, *Am. J. Vet. Res.*, 50, 1250, 1989.
90. **Kelley, K. W.,** Cross-talk between the immune and endocrine systems, *J. Anim. Sci.*, 66, 2095, 1988.
91. **Roth, J. A., Kaeberle, M. L., and Hsu, W. H.,** Effects of ACTH administration on bovine polymorphonuclear leukocyte function and lymphocyte blastogenesis, *Am. J. Vet. Res.*, 43, 412, 1982.
92. **Roth, J. A. and Kaeberle, M. L.,** Effects of glucocorticoids on the bovine immune system, *J. Am. Vet. Med. Assoc.*, 180, 894, 1982.
93. **Murata, H. and Hirose, H.,** Suppression of bovine neutrophil function by sera from cortisol-treated calves, *Br. Vet. J.*, 147, 63, 1991.
94. **Roth, J. A. and Kaeberle, M. L.,** Effects of in vivo dexamethasone administration on in vitro bovine polymorphonuclear leukocyte function, *Infect. Immun.*, 33, 434, 1981.
95. **Roth, J. A. and Kaeberle, M. L.,** Effect of levamisole on lymphocyte blastogenesis and neutrophil function in dexamethasone-treated cattle, *Am. J. Vet. Res.*, 45, 1781, 1984.
96. **Kaeberle, M. L. and Roth, J. A.,** Effects of thiabendazole on dexamethasone-induced suppression of lymphocyte and neutrophil function in cattle, *Immunopharmacology*, 8, 129, 1984.
97. **Chiang, Y.-W., Roth, J. A., and Andrews, J. J.,** Influence of recombinant bovine interferon γ and dexamethasone on pneumonia attributable to *Haemophilus somnus* in calves, *Am. J. Vet. Res.*, 51, 759, 1990.
98. **Webb, D. S. A. and Roth, J. A.,** Relationship of glucocorticoid suppression of arachidonic acid metabolism to alteration of neutrophil function, *J. Leukoc. Biol.*, 41, 156, 1987.
99. **Frank, D. E. and Roth, J. A.,** Factors secreted by untreated and hydrocortisone-treated monocytes that modulate neutrophil function, *J. Leukoc. Biol.*, 40, 693, 1986.
100. **Roth, J. A., Kaeberle, M. L., and Hsu, W. H.,** Effect of estradiol and progesterone on lymphocyte and neutrophil function in steers, *Infect. Immun.*, 35, 997, 1982.
101. **Roth, J. A., Kaeberle, M. L., Appell, L. H., and Nachreiner, R. F.,** Association of increased estradiol and progesterone blood values with altered bovine polymorphonuclear leukocyte function, *Am. J. Vet. Res.*, 44, 247, 1983.
102. **Paape, M. J., Miller, R. H., and Ziv, G.,** Effects of florfenicol, chloramphenicol, and thiamphenicol on phagocytosis, chemiluminescence, and morphology of bovine polymorphonuclear neutrophil leukocytes, *J. Dairy Sci.*, 73, 1734, 1990.
103. **Nickerson, S. C., Paape, M. J., and Dulin, A. M.,** Effect of antibiotics and vehicles on bovine mammary polymorphonuclear leukocyte morphologic features, viability, and phagocytic activity in vitro, *Am. J. Vet. Res.*, 46, 2259, 1985.

104. **Ziv, G., Paape, M. J., and Dulin, A. M.,** Influence of antibiotics and intramammary antibiotic products on phagocytosis of *Staphylococcus aureus* by bovine leukocytes, *Am. J. Vet. Res.*, 44, 385, 1983.
105. **Jones, D. G. and Suttle, N. F.,** Some effects of copper deficiency on leucocyte function in sheep and cattle, *Res. Vet. Sci.*, 31, 151, 1981.
106. **Olkowski, A. A., Gooneratne, S. R., and Christensen, D. A.,** Effects of diets of high sulphur content and varied concentrations of copper, molybdenum and thiamine on *in vitro* phagocytic and candidacidal activity of neutrophils in sheep, *Res. Vet. Sci.*, 48, 82, 1990.
107. **MacPherson, A., Gray, D., Mitchell, G. B. B., and Taylor, C. N.,** Ostertagia infection and neutrophil function in cobalt-deficient and cobalt-supplemented cattle, *Br. Vet. J.*, 143, 348, 1987.
108. **Aziz, E. and Klesius, P. H.,** Effect of selenium deficiency on caprine polymorphonuclear leukocyte production of leukotriene B_4 and its neutrophil chemotactic activity, *Am. J. Vet. Res.*, 47, 426, 1986.
109. **Roth, J. A. and Kaeberle, M. L.,** Enhancement of lymphocyte blastogenesis and neutrophil function by avridine in dexamethasone-treated and nontreated cattle, *Am. J. Vet. Res.*, 46, 53, 1985.
110. **Woodard, L. F., Jasman, R. L., Farrington, D. O., and Jensen, K. E.,** Enhanced antibody-dependent bactericidal activity of neutrophils from calves treated with a lipid amine immunopotentiator, *Am. J. Vet. Res.*, 44, 389, 1983.
111. **Canning, P. C. and Roth, J. A.,** Effects of *in vitro* and *in vivo* administration of recombinant bovine interferon γ on bovine neutrophil response to *Brucella abortus*, *Vet. Immunol. Immunopathol.*, 20, 119, 1989.
112. **Bielefeldt Ohmann, H. and Babiuk, L. A.,** Alteration of some leukocyte functions following *in vivo* and *in vitro* exposure to recombinant bovine α- and γ-interferon, *J. Interferon Res.*, 6, 123, 1986.
113. **Sample, A. K. and Czuprynski, C. J.,** Recombinant bovine interferon-γ, but not interferon-α, potentiates bovine neutrophil oxidative responses in vitro, *Vet. Immunol. Immunopathol.*, 25, 23, 1990.
114. **Bielefeldt Ohmann, H. and Babiuk, L. A.,** Effect of bovine recombinant α-1 interferon on inflammatory responses of bovine phagocytes, *J. Interferon Res.*, 4, 249, 1984.
115. **Lederer, J. A. and Czuprynski, C. J.,** Characterization and identification of interleukin 1 receptors on bovine neutrophils, *J. Leukoc. Biol.*, 51, 586, 1992.
116. **Kehrli, M. E., Jr., Cullor, J. S., and Nickerson, S. C.,** Immunobiology of hematopoietic colony-stimulating factors: potential application to disease prevention in the bovine, *J. Dairy Sci.*, 74, 4399, 1991.
117. **Reddy, P. G., McVey, D. S., Chengappa, M. M., Blecha, F., Minocha, H. C., and Baker, P. E.,** Bovine recombinant granulocyte-macrophage colony-stimulating factor enhancement of bovine neutrophil functions in vitro, *Am. J. Vet. Res.*, 51, 1395, 1990.
118. **Sample, A. K. and Czuprynski, C. J.,** Priming and stimulation of bovine neutrophils by recombinant human interleukin-1 α and tumor necrosis factor α, *J. Leukoc. Biol.*, 49, 107, 1991.
119. **Roth, J. A., Abruzzini, A. F., and Frank, D. E.,** Influence of recombinant human interleukin-2 administration on lymphocyte and neutrophil function in clinically normal and dexamethasone-treated cattle, *Am. J. Vet. Res.*, 51, 546, 1990.
120. **Canning, P. C. and Neill, J. D.,** Isolation and characterization of interleukin-1 from bovine polymorphonuclear leukocytes, *J. Leukoc. Biol.*, 45, 21, 1989.
121. **Yoneyama, O., Osame, S., Kimura, M., Araki, S., and Ichijo, S.,** Enhancement of neutrophil function by dihydroheptaprenol in adult cows, *Jpn. J. Vet. Sci.*, 51, 1283, 1989.
122. **Yoneyama, O., Osame, S., Ichijo, S., Kimura, M., Araki, S., Suzuki, M., and Imamura, E.,** Effects of dihydroheptaprenol on neutrophil functions in calves, *Br. Vet. J.*, 145, 531, 1989.
123. **Roth, J. A. and Kaeberle, M. L.,** In vivo effect of ascorbic acid on neutrophil function in healthy and dexamethasone-treated cattle, *Am. J. Vet. Res.*, 46, 2434, 1985.
124. **Lawman, M. P., Gauntlett, D., and Gallery, F.,** Antibody-dependent cellular cytotoxicity and natural killer cell activity in cattle: mechanisms of recovery from infectious diseases, in *The Ruminant Immune System in Health and Disease*, Morrison, W. I., Ed., Cambridge University Press, Cambridge, 1986, 346.
125. **Cook, C. G. and Splitter, G. A.,** Comparison of bovine mononuclear cells with other species for cytolytic activity against virally-infected cells, *Vet. Immunol. Immunopathol.*, 20, 239, 1989.
126. **Jensen, J. and Schultz, R. D.,** Bovine natural cell mediated cytotoxicity (NCMC): activation by cytokines, *Vet. Immunol. Immunopathol.*, 24, 113, 1990.
127. **Campos, M., Bielefeldt Ohmann, H., Rapin, N., and Babiuk, L. A.,** Demonstration of the *in vitro* antiviral properties of bovine lymphokine-activated killer (LAK) cells, *Viral Immunol.*, 4, 259, 1991.

128. **Campos, M., Rossi, C. R., Bielefeldt Ohmann, H., Beskorwayne, T., Rapin, N., and Babiuk, L. A.,** Characterization and activation requirements of bovine lymphocytes acquiring cytotoxic activity after interleukin-2 treatment, *Vet. Immunol. Immunopathol.*, 32, 205, 1992.
129. **Yamamoto, S., Onuma, M., Kodama, H., Koyama, H., Mikami, T., and Izawa, H.,** Existence of cytotoxic activity against BLV-transformed cells in lymphocytes from normal cattle and sheep, *Vet. Immunol. Immunopathol.*, 8, 63, 1985.
130. **Bielefeldt Ohmann, H., Davis, W. C., and Babiuk, L. A.,** Functional and phenotypic characteristics of bovine natural cytotoxic cells, *Immunobiology*, 169, 503, 1985.
131. **Cook, C. G. and Splitter, G. A.,** Lytic function of bovine lymphokine-activated killer cells from a normal and a malignant catarrhal fever virus-infected animal, *Vet. Immunol. Immunopathol.*, 19, 105, 1988.
132. **Campos, M., Rossi, C. R., and Lawman, M. J. P.,** Natural cell-mediated cytotoxicity of bovine mononuclear cells against virus-infected cells, *Infect. Immun.*, 36, 1054, 1982.
133. **Chung, S. and Rossi, C. R.,** Natural cell-mediated cytotoxicity to cells infected with infectious bovine rhinotracheitis virus, *Vet. Immunol. Immunopathol.*, 14, 45, 1987.
134. **Cook, C. G. and Splitter, G. A.,** Characterization of bovine mononuclear cell populations with natural cytolytic activity against bovine herpes virus 1-infected cells, *Cell. Immunol.*, 120, 240, 1989.
135. **Cook, C. G., Letchworth, G. J., and Splitter, G. A.,** Bovine naturally cytolytic cell activation against bovine herpes virus type 1-infected cells does not require late viral glycoproteins, *Immunology*, 66, 565, 1989.
136. **Palmer, L. D., Leary, T. P., Wilson, D. M., and Splitter, G. A.,** Bovine natural killer-like cell responses against cell lines expressing recombinant bovine herpes virus type 1 glycoproteins, *J. Immunol.*, 145, 1009, 1990.
137. **Goddeeris, B. M., Dunlap, S., Bensaid, A., MacHugh, N. D., and Morrison, W. I.,** Cell surface phenotype of two cloned populations of bovine lymphocytes displaying non-specific cytotoxic activity, *Vet. Immunol. Immunopathol.*, 27, 195, 1991.
138. **O'Brien, M. A. and Duffus, W. P. H.,** The effects of dexamethasone, betamethasone, flunixin and phenylbutazone on bovine natural-killer-cell cytotoxicity, *J. Vet. Pharmacol. Ther.*, 13, 292, 1990.
139. **Campos, M. and Rossi, C. R.,** Inability to detect a K cell in bovine peripheral blood leukocytes, *Vet. Immunol. Immunopathol.*, 8, 351, 1985.
140. **Campos, M. and Rossi, C. R.,** Cytotoxicity of bovine lymphocytes after treatment with lymphokines, *Am. J. Vet. Res.*, 47, 1524, 1986.
141. **Duke, S. S., King, L. S., Jones, M. R., Newman, J. H., Brigham, K. L., and Forbes, J. T.,** Human recombinant interleukin 2-activated sheep lymphocytes lyse sheep pulmonary microvascular endothelial cells, *Cell. Immunol.*, 122, 188, 1989.

Chapter 9

East Coast Fever (*Theileria Parva*): Cell-Mediated Immunity and Protection

Bruno M. Goddeeris, W. Ivan Morrison, Evans L. Taracha, and Declan J. McKeever

CONTENTS

I. The Parasite and the Disease 143
II. Vaccination and Parasite Strains 144
III. The Parasite Genome and Antigens 145
IV. Mechanisms of Protective Immunity 145
V. T Cell-Mediated Immunity 146
A. Induction of *Theileria Parva*-Specific Proliferation 146
B. CD4+ T Cells 146
C. CD8+ T Cells 147
1. *Theileria*-Specific CTL 148
2. Relationship of Parasite Strain Specificity of CTL with Cross-Protection 148
3. Influence of the MHC Phenotype on Strain Specificity of the CTL Response 149
4. Influence of the Immunizing Parasite Strain on MHC Restriction and Strain Specificities of the CTL Response 150
5. Cellular and Chemical Basis of the Antigenic Specificity of CTL 150
D. Influence of Cytokines on *Theileria*-Infected Cells 151
VI. Epilogue 152
References 153

I. THE PARASITE AND THE DISEASE

Theileria parva is a protozoan parasite of the class of *Sporozoea* to which malaria parasites also belong. It causes an acute lymphoproliferative disease in cattle known as East Coast fever, which is endemic in a large area of eastern, central, and southern Africa (reviewed by Irvin and Morrison[1]). The presence of the disease closely follows the distribution of the tick vector *Rhipicephalus appendiculatus*. In susceptible cattle exposed to infection with *T. parva*, morbidity and mortality are approximately 100 and 95%, respectively.

When an infected tick commences feeding on a new host, parasites present in the salivary gland undergo sporogony, resulting in the formation of sporozoites. At 3 days after tick attachment, the sporozoites are injected with saliva into the mammalian host. In the bovine host, sporozoites can invade lymphocytes in a matter of minutes by receptor-mediated endocytosis.[2] This may involve lymphocytes in the inflammatory reaction at the site of tick attachment or may occur in the regional lymph node following migration in afferent lymph. Binding of sporozoites to, and infection of host cells can be blocked specifically by anti-MHC class I or anti-β_2 microglobulin monoclonal antibodies (MAb), suggesting that class I molecules are involved in the entry process.[3] During the subsequent 12 to 24 h, the host cell membrane surrounding the parasite is destroyed so that the parasite comes to lie free in the cytoplasm. The intracellular parasite starts to undergo nuclear division, resulting after 1 to 2 days in the appearance of typical multinucleate bodies, named *schizonts*. At this time, the host cell undergoes blast transformation and starts to proliferate; the schizont divides synchronously with the host cell so that there is clonal expansion of the cell population initially infected.[4] As the infection progresses, some schizonts undergo merogony, giving rise to merozoites which, upon release, enter erythrocytes and develop into piroplasms. Continuation of the life cycle is ensured by ingestion of infected erythrocytes by ticks.

Schizont-infected cells are responsible for the pathogenicity of the infection (reviewed by Irvin and Morrison[1]). These infected cells behave essentially like tumor cells: they multiply in an uncontrolled manner in the lymphoid tissues and invade a wide variety of tissues, including the respiratory and

0-8493-4952-4/94/$0.00+$.50

gastrointestinal tracts. In the later stages of the disease, there is widespread lymphocytolysis which affects both parasitized and nonparasitized cells. This results in a profound depletion of lymphocytes, both in the solid lymphoid tissues and in the recirculating pool. In the terminal stages of the disease, severe pulmonary edema develops, causing respiratory distress with death 2 to 4 weeks after tick attachment.

The unique relationship of the parasite with the host lymphocyte allows the establishment of continuously growing infected cell lines by infecting normal lymphocytes *in vitro* with sporozoites.[5] This culture system has proved invaluable in defining the nature of the target cells for infection and in studying T cell responses against the parasite. By the use of MAb specific for subpopulations of bovine lymphocytes and accessory cells, it has been shown that B cells, both major subpopulations of αβT cells ($CD4^+$ and $CD8^+$), and γδT cells can be infected *in vitro* with the parasite and transform into permanently growing cell lines.[6,7] Also, *in vitro* studies have shown that afferent lymph veiled cells can be invaded by sporozoites which develop subsequently into early schizonts, but it has not been possible to obtain transformed cell lines of these cells.[9] However, the majority of infected lymphocytes detected *in vivo* are T cells. A striking feature during infection is the appearance of $CD4^+/CD8^+$-infected lymphocytes; since $CD4^+$ T cells often acquire CD8 after *in vitro* infection, these double-positive cells *in vivo* are thought to originate from $CD4^+$ cells.[8]

II. VACCINATION AND PARASITE STRAINS

It has been recognized for many years that those cattle that fortuitously recover from East Coast fever are immune to subsequent challenge. Already at the beginning of this century, Theiler[10] and Spreull[11] were able to immunize cattle (against homologous challenge) with *T. parva*, by inoculating animals with schizont-infected cells obtained from spleens and lymph nodes of animals dying of East Coast fever. Later, with the possibility of *in vitro* infection of lymphocytes, *T. parva*-infected cell lines were used.[12] An important finding was that, while 10^8 or more allogeneic infected cells were needed to induce protection against challenge, such protection could be achieved with as few as 10^2 autologous infected cells.[13] A number of findings indicated that in the case of allogeneic cells, schizonts have to transfer into cells of the recipient animal for successful immunization. Indeed, it was possible to block induction of immunity with allogeneic infected cells by prior induction of anti-schizont antibodies with killed parasites in the recipient cattle.[14] The main practical disadvantage in using schizont-infected cells is the large number of infected allogeneic cells required for inducing immunity: approximately 100 ml of cultured cell suspension is required to provide 10^8 cells. Another drawback in the use of allogeneic infected cells is the possible outcome of fatal theileriosis when allogeneic infected cells are more or less histocompatible with the recipient animal.

Immunization protocols based on infection of animals with sporozoites and treatment with theileristatic or theilericidal drugs have also been recognized for many years.[15] With the advent of slow-release formulations of oxytetracycline in the early 1970s, the method became refined to a single dose of long-acting tetracycline given at the same time as the infective stabilate of sporozoites (homogenate of partially fed infected ticks).[16] However, immunization with one stock of the parasite did not provide protection against all other stocks.[16,17] Nevertheless, parasite strain heterogeneity with respect to protection appeared to be limited as the use of mixtures of two to three stocks for immunization provided protection against challenge with a number of stocks from different geographical locations.

Although cross-protection studies are the best way to identify the immunologically important strains of the parasite, they are expensive and do not always give conclusive results. Indeed, when breakthrough infections occur, they often do so only in a proportion of animals and usually do not occur reciprocally between strains. These findings may be partly due to the use of uncloned parasite stocks, some of which appear to be heterogeneous as indicated by laboratory findings (see below). Nonetheless, even when cloned parasites are used, not all animals immunized with a particular clone (strain) of the parasite react similarly on cross-challenge with another clone (strain). In view of the practical significance of strain heterogeneity, in the last few years considerable effort has been devoted to developing laboratory techniques for identifying parasite strains, principally by production of parasite-specific MAb and DNA probes. A series of MAb, raised against the schizont stage were found to react with parasites of some stocks but not others, and in some instances reacted with only some of the cloned parasites derived from a particular stock.[18,19] These MAb were shown to be specific for the same antigen, termed the *polymorphic immunodominant molecule* (PIM, see later). In addition to the polymorphic antibody determinants detected on PIM, differences between and within stocks in the M_r of the molecule were observed by testing MAb specific for conserved epitopes, in Western blots.[20,21] Differences in the protein spot patterns of schizont proteins separated by two-dimensional gel electrophoresis have also been detected between and within strains of *T. parva,* although the precise nature of the variable proteins has not been defined.[22]

Restriction fragment length polymorphism (RFLP) of parasite DNA has been detected in two ways. First, pulsed field gel electrophoresis of parasite DNA digested with rare cutter restriction enzymes has revealed distinctive RFLP patterns for different parasite stocks.[23] Second, RFLP between and within stocks has been detected using cloned homologous DNA probes representing repetitive sequences in the *T. parva* genome.[24-27]

The inter- and intra-stock differences detected by these techniques, either singly or in combination, highlight the potential difficulties in interpreting the results of cross-protection experiments with *T. parva* and emphasize the need for cloned parasite populations for such studies. Consequently, it has not yet been possible to determine whether or not a relationship exists between the detectable protein and DNA polymorphisms and protection between strains.

III. THE PARASITE GENOME AND ANTIGENS

The parasite genome in the mammalian host is haploid, consists of four choromosomes, and is estimated at 10^7 bp.[28] Sexual recombination probably happens during development in the tick; zygotes are formed in the gut by fusion of gametes derived from male and female piroplasms and reduction to the haploid state is believed to occur during sporogony in the tick salivary gland.[29] Sexual recombination in the tick between different strains could play an important role in generating new haploid genomes (i.e., new strains) of the parasite. As East Coast fever is lethal for the majority of cattle, the African buffalo (*Syncerus caffer*) which is a healthy carrier of different strains of the parasite is probably an ideal brewing pot for delivering strain mixtures of piroplasms to the tick.

For the last 10 years, a major effort has been directed toward identification and characterization of parasite antigens. Some of these antigens appear to be restricted in expression to the sporozoite stage, while others are expressed on the sporozoite and the schizont. Recently, much attention has focused on a sporozoite-restricted surface antigen, the 67-kDa antigen or p67, as polyclonal and monoclonal antibodies specific for this antigen neutralize the infectivity of sporozoites *in vitro*.[30,31] The gene encoding the p67 antigen has been cloned,[32] and recombinant protein has been shown to protect the majority of immunized animals against challenge with live sporozoites[33] (see later). Additional sporozoite antigens were identified by screening a gt11 expression library of genomic DNA fragments with a polyclonal bovine antiserum which neutralized sporozoite infectivity.[34,35] Three distinct gt11 recombinant clones were identified and their products used in the antibody-select method to affinity-purify the specific antibodies from bovine anti-sporozoite serum. When tested in Western blots of sporozoite antigens, these affinity-purified antibodies identified three groups of antigens: one group contained 104-105, 90, 85, and 35-kDa proteins and was found to react with the microneme/rhoptry complexes of the sporozoite; another group included antigens of 69, 67, 52, 47, and 43 kDa which were related to p67 described above; and a third group containing an apparently distinct pair of antigens of 47 and 43 kDa. There is evidence to suggest that the related antigens within each group probably arose through protein breakdown or transcription of alternatively spliced RNA.

Efforts at defining antigens of the schizont stage of the parasite have been less successful. The majority of recovered bovine sera show dominant activity against one protein in Western blots of schizont antigens. Moreover, most of the MAb raised against schizonts are specific for this antigen. The antigen is polymorphic between strains, as detected both by reactivity with some of the MAb and differences in M_r in SDS-PAGE, which ranged from 117 to 62 kDa.[20,21] Because of these properties, the antigen has been named the PIM antigen. Differences in M_r for PIM of cloned infected cell lines suggest that at least four different parasite types (strains) are present in the original Marikebuni stock, while the Muguga stock appears to be of one parasite type.[21] The antigen has also been detected on the surface of sporozoites by *in vitro* neutralization of sporozoites with specific antibodies and immunoelectronmicroscopy.[21] The gene has recently been isolated and sequenced and awaits further analysis (P. Toye, personal communication).

IV. MECHANISMS OF PROTECTIVE IMMUNITY

There is evidence that protective immunity against East Coast fever is operative against two stages of the infection: the invading sporozoite and the schizont-infected cell.

Sporozoite-neutralizing antibodies have been detected in the blood of hyperimmunized animals, and MAb with neutralizing activity have been raised against sporozoites.[30,31] Recent data indicate that recombinant p67 antigen can induce protection in cattle.[33] However, only a proportion of animals was protected and the levels of sporozoite-neutralizing antibody in serum did not closely correlate with protection; some animals with high neutralizing antibody titers were not protected (T. Musoke, personal communication).

Animals that have recovered from primary infections or been immunized by infection and treatment have very low levels of sporozoite-neutralizing antibodies. A number of observations strongly indicate that in such animals, immunity is directed against the schizont stage. First, following sporozoite challenge of animals immunized by infection and treatment, infection develops to the schizont stage before it is eliminated. Second, such immunized animals are immune to challenge with doses of autologous schizont-infected cells, which in naive animals result in lethal infections. Third, animals immunized by inoculation with schizont-infected cells are immune against sporozoite challenge.

The biological properties of the parasite, together with a number of experimental observations, would indicate that this protective immunity is cell mediated rather than antibody mediated. Despite the fact that schizont-specific antibodies are abundant in serum of immune animals, the mode of replication of the parasite dictates that the schizonts are inaccessible to such antibodies. Moreover, the failure of immune sera to induce complement-mediated lysis or antibody-dependent cell-mediated cytotoxicity of parasitised cells indicate that infected cells do not express parasite antigens recognizable by antibodies on their surface.[36] Indeed, cattle inoculated with serum or concentrated γ-globulins from immune cattle, remained susceptible to challenge.[37] Also, cattle that were immunized with heat-killed schizont-infected cells or semipurified preparations of schizont antigens remained fully susceptible to challenge, although they produced antibody titers comparable to those accompanying immunization with the live parasites.[14,38] Conversely, protection could be conferred to chimeric twin animals by adoptive transfer of thoracic duct lymphocytes.[39] A role for cytotoxic T lymphocytes (CTL) in immunity was suggested by their detection in the peripheral blood concomitantly with the elimination of the schizont-infected cells from the local draining lymph node in animals undergoing primary or secondary infections.[40,41] More formal evidence that $CD8^+$ CTL play a key role in immunity has come from recent studies which have shown that protection can be conferred by adoptive transfer of cells highly enriched for $CD8^+$ lymphocytes expressing potent cytotoxic activity.[42,43]

V. T CELL-MEDIATED IMMUNITY

A. INDUCTION OF *THEILERIA PARVA*-SPECIFIC PROLIFERATION

Theileria parva-infected cells induce proliferative responses in autologous mononuclear cells of the peripheral blood (PBMC) of naive as well as immune animals; this reaction has been named the autologous *Theileria* mixed leukocyte reaction (MLR).[44] The proliferative response is independent of the presence of monocytes and indeed is generally of greater magnitude when monocytes are absent.[45] The detection of the response in both naive and immune animals suggests that at least part of the proliferation is not specific for *Theileria parva* antigens. It has been shown that some infected cell lines secrete a growth factor with IL-2-like (IL = interleukin) activity;[46] however, the unconcentrated supernatants of infected cell lines are generally not mitogenic. This nonspecific *in vitro* response may be the counterpart of a marked T cell blast response observed *in vivo* in the regional lymph node during the early stages of *T. parva* infection.[1]

Two obervations indicate that in immune cattle, part of the autologous *Theileria* MLR is parasitespecific. First, parasitized cells fixed with glutaraldehyde do not induce proliferative responses in monocyte-depleted PBMC of naive animals, but are stimulatory for lymphocytes of immune animals.[45] Second, in the autologous *Theileria* MLR, parasite-specific MHC-restricted cytotoxicity could be generated from immune but not naive animals.[45] Repeated stimulation on a weekly basis of the viable cells of the autologous *Theileria* MLR with autologous infected cells resulted in enhanced T cell proliferation and increased levels of *Theileria*-specific cytotoxicity.[47] However, after five or more stimulations, cytotoxicity sometimes became nonspecific.

B. $CD4^+$ T CELLS

T cell clones with the $CD4^+/CD8^-$ phenotype specific for parasitised cells have been derived from restimulated autologous *Theileria* MLR.[48] As anticipated from studies in other species, the clones were genetically restricted, produced T cell growth factors (probably IL-2), and proliferated in response to autologous schizont-infected cells in the absence of exogenous T cell growth factors. Antigen-specific proliferation of the clones was inhibited by MAb specific for MHC class II, indicating that they were MHC class II restricted. Some clones were specific for the immunizing strain of the parasite, while others recognized several different strains. These findings have demonstrated that CD4+ T cells that recognize *Theileria* antigen in association with MHC class II on the surface of the infected cells are generated in immune cattle.

More recently, efforts have focused on the identification of *Theileria* antigens recognized by CD4 cells. Screening for antigens presented by class II MHC molecules is more straightforward than for those presented by class I: while antigens presented by class I molecules have to be delivered within the cytosol of the presenting cell either by endogenous synthesis (endogenous antigens) or by artificial means, class II molecules can present exogenous antigens after uptake and processing by antigen presenting cells (see chapter 6). When homogenates prepared from schizont-infected cells were separated by differential centrifugation into a high-speed supernate (HSS) and a membrane fraction (with PIM), both fractions stimulated proliferation of *T. parva*-specific CD4+ T cell lines.[49] This study had been performed with two *Theileria*-specific cell lines from different animals: one cell line appeared to respond to both fractions, while the other responded only to the HSS. Helper T cell clones, which were specific for the HSS, were produced from the latter cell line and used for identifying the stimulatory parasite antigens.[50] The HSS was purified by anion exchange chromatography and hydroxyapatite chromatography and subsequently fractionated by HPLC. Three major peaks of activity with M_r of 45, 12, and 4.2 kDa, as defined from the elution profile, were detected. By metabolic labeling of the schizont-infected cells before homogenization and preparation of the HSS, the highest *Theileria*-specific antigenic activity in relation to total protein content of the separated fractions could readily be related to a 9- to 10-kDa protein as defined by SDS-PAGE.[51] However, as the same CD4+ clone also responded to an array of antigens with higher M_r, it would appear that the 9- to 10-kDa antigen is a breakdown product. This is supported by another publication of the same research group wherein proliferation of the same clones correlated in a dose-dependent manner with a 24-kDa antigen (SDS-PAGE).[52] The antigen appears to be proteinaceous as antigenicity is abrogated by treatment with papain.

Recently, a limiting dilution assay has been optimized to determine the precursor frequency of *T. parva*-specific CD4+ T cells in the peripheral blood (J. P. Scheerlinck, T. Vandeputte, E. Taracha, D. J. McKeever, and B. M. Goddeeris, unpublished data). This assay is based on the induction of parasite-specific proliferation by purified schizonts or *T. parva*-infected autologous cells; the latter were fixed to avoid induction of nonspecific proliferation. Only slightly higher precursor frequencies were obtained when *T. parva*-infected cells with or without purified schizonts, instead of purified schizonts alone, were used as antigen. The test could thus be used to screen parasite antigens, either biochemically fractionated or in expression libraries, for helper T cell epitopes on infected cells. The assay was specific for CD4+ T cells as all response was contained within the purified CD4+ cells, while no response occurred in the CD4- cell population. In a preliminary study, specific precursors were detected in four out of six immune animals examined; precursor frequencies ranged from about 1/5000 for an animal that had been challenged on several occasions, to about 1/20,000 for animals that had not been challenged or been challenged only once after immunization; the remaining two animals had a precursor frequency below the limit of detection of the assay (1/100,000).

C. CD8+ T CELLS

The first substantial evidence that cell-mediated cytotoxicity is generated against schizont-infected cells was presented by Pearson et al.[44], who showed that irradiated infected cells derived from cell lines were capable of inducing *in vitro* (i.e., in an autologous *Theileria* MLR) cytotoxic cells in PBMC from immune but not naive animals. However, cytotoxicity was not *Theileria* specific in that it was also operative against uninfected autologous blasts.[53] There is evidence that this MHC-restricted *Theileria*-nonspecific killing was due to the use of cultured cell lines for immunization with subsequent induction of responses to culture-associated antigens. Indeed, by using an *in vitro* antigen presentation system consisting of peripheral blood monocytes pulsed with membrane antigen from autologous *Theileria*-infected cells, it was shown that PBMC of animals immunized with cultured cell lines exhibited strong proliferative responses with generation of genetically restricted cytotoxic cells, while PBMC from naive animals or animals immunized with sporozoites did not (T. Tenywa, W. I. Morrison and D. L. Emery, unpublished data). However, this initial observation triggered strong interest in the potential role of CTL in protection against East Coast fever. Because of difficulties in interpreting cytotoxicity results from studies wherein allogeneic and autologous cell lines were used for immunization, no further reference will be made to such studies.

1. *Theileria*-Specific CTL

The demonstration that *Theileria*-specific cytotoxicity appeared in PBMC of animals undergoing immunization by infection with sporozoites and treatment with tetracyclines, and that this response coincided with the remission of the immunizing infection suggested a role for CTL in protection against East Coast fever.[40,41] Moreover, the observation of a degree of parasite stock specificity in the response strengthened the notion that such responses may have a protective role, as heterogeneity in parasite strains had been demonstrated in cross-immunization trials.[54]

In animals undergoing immunization by infection with sporozoites and simultaneous treatment, *Theileria*-specific cytotoxic activity could be detected for only a few days in the 3rd week after immunization, and immune animals undergoing homologous challenge displayed cytotoxic activity for only a few days during the 2nd week after challenge (between days 8 and 11).[40,41] However, following the active phase of the response, the memory cytotoxic T cell precursors (CTLp) can be reactivated by stimulation *in vitro* with autologous infected cells in an autologous *Theileria* MLR. The advent of markers (MAb) for bovine lymphocyte subpopulations (see Chapter 1), and reagents (polyclonal and MAb) for phenotyping cattle for their class I major histocompatibility complex (MHC) antigens (bovine lymphocyte antigens, BoLA) (see chapter 3), enabled the effector cells and their MHC restriction specificities to be defined. Both in the *in vivo* response and following *in vitro* stimulation of PBMC, it was shown in numerous animals that the *Theileria*-specific cytotoxic cells killed autologous parasitized lymphoblasts but not uninfected lymphoblasts and that they were MHC class I restricted.[47,55] Class I restriction was demonstrated by testing cytotoxicity on panels of MHC-matched (for one or both BoLA-A locus alleles) and mismatched infected targets, and by the ability of anti-class I MHC but not anti-class II MHC antibodies to inhibit the cytotoxicity. Moreover, using the fluorescence-activated cell sorter, all cytotoxic activity was shown to reside within the $CD8^+$ population of PBMC.[42,43,47] On subsequent cloning of the cytotoxic cells, it was confirmed that they were $CD2^+/CD8^+/CD4^-$, were T. parva-specific and were restricted by products of one of the two MHC class I haplotypes of the animal.[56,57] Thus, by analogy with findings in other animal species, the *Theileria*-specific CTL probably recognize parasite-derived peptides in association with class I MHC molecules on the surface of the infected cell.[58]

A notable feature of the *Theileria*-specific cytotoxic responses in MHC heterozygous cattle was the bias of the response to one of the MHC haplotypes.[55] Moreover, with the different class I specificities of cattle analyzed, responses restricted by some specificities (or haplotypes) consistently predominated over responses restricted by others. Thus, there appears to be a hierarchy in dominance among the class I MHC molecules which restrict the response. This is perhaps not surprising as the polymorphic nature of the antigen presenting class I molecules will result in some having a higher affinity for peptide antigens than others, and thus be better at inducing a CTL response, as has been confirmed with antigen presenting class II molecules for $CD4^+$ helper T cells.[59] Differences in T cell receptor repertoire selection by different MHC molecules during ontogeny of T cells[60] may also influence the capacity to respond to particular parasite antigenic epitopes.

Using a limiting dilution assay, *Theileria*-specific CTLp frequencies in peripheral blood of immune animals were estimated to range between about 1:2000 and 1:15,000, whereas no CTLp were detected in naive animals to a cell input of 10^5 per well.[42,43,61] By collecting from an immune animal the efferent lymph of the lymph node draining the challenge site, it was possible to compare the kinetics of appearance of CTLp in the lymph with that in the peripheral blood.[42,43] A rise in *Theileria*-specific CTLp was first detected in the lymph between days 5 and 7 after challenge with peak levels of 1:30 to 1:40 attained between days 9 and 11. A rise in CTLp in peripheral blood was first detected between days 8 and 9 and reached peak levels of 1:500 to 1:650 around day 11. Moreover, comparison of CTLp frequencies in $CD8^+$ and $CD8^-$ cell populations and in $CD8^+$ small cells and $CD8^+$ large cells (blasts) demonstrated that most, if not all, of the *Theileria*-specific cytotoxic activity was within the $CD8^+$ population and a large majority was within the blasting population. These results confirm previous assumptions that priming and reactivation of the cytotoxic T cell response occurs in the lymph node draining the site of infection. The high frequency of CTLp in efferent lymph has permitted the adoptive transfer of this activity from immune to naive monozygotic twin calves. Two calves that received 1.0×10^{10} and 8.5×10^9 cells enriched for $CD8^+$ T cells 6 to 8 days after challenge with *T. parva* rapidly eliminated the parasite.[42,43]

2. Relationship of Parasite Strain Specificity of CTL with Cross-Protection

If CTL are indeed important in protective immunity, one would expect that differences in the capacity of parasite stocks to cross-protect would be reflected in the strain specificities of the CTL responses. Most of the detailed analyses of the strain specificity of *T. parva*-specific CTL have focused on two parasite stocks, namely Muguga and Marikebuni. Cross-protection experiments with these two stocks indicate that while Marikebuni protects against Muguga, a proportion of Muguga-immunized animals is susceptible to challenge with Marikebuni. However, as already indicated, a major difficulty in interpreting results from cross-immunization trials with stocks is that some stocks are heterogeneous mixtures of strains. This is certainly the case for Marikebuni, from which at least five different strains, as defined with MAb and DNA probes, have been identified.[21,62]

In general, Muguga-immunized animals frequently generated strain-specific CTL, i.e. T cells that killed targets infected with Muguga but not those infected with Marikebuni; whereas, in the few instances that cytotoxic responses were monitored in Marikebuni-immunized animals, they were cross-reactive.[47,55-57] Studies with CTL clones derived from Muguga-immunized animals have demonstrated Muguga-specific and cross-reactive clones from different animals.[62] It is of particular note that when cross-reactive clones from Muguga-immunized animals were tested on different cloned cell lines infected with Marikebuni, they often killed only some of the cloned cell lines.[62] Differences in the Marikebuni parasites in these cell lines were detected with parasite-specific MAb and DNA probes.

To overcome the difficulties of working with parasite stocks, methods have recently been developed for obtaining populations of sporozoites from cloned parasites; they involve pick-up of infection by ticks from animals infected with doubly cloned parasitized cell lines.

A study to examine the relationship of strain specificity of CTL responses with cross-protection has been conducted using the Muguga stock of the parasite and a cloned population of parasites (3219) derived from the Marikebuni stock. The failure to detect heterogeneity within the Muguga stock with parasite-specific MAb and CTL clones — together with the finding that animals immunized with cloned cell lines infected with Muguga were protected against challenge with the Muguga stock — suggest that this stock is immunologically homogeneous. Some 17 cattle were immunized with Muguga and five with Marikebuni 3219, and the frequencies of CTLp capable of killing Muguga-infected and Marikebuni-infected target cells were determined.[42,63] Use of the limiting dilution assay allowed quantitative measurement of CTL responses directly from the animals, thus avoiding any bias in the response as a result of repeated stimulation of the T cells *in vitro*. CTLp frequencies ranged from 1:2071 to 1:14,200; 6 of the 17 Muguga-immunized animals and three of the five Marikebuni 3219-immunised animals had similar frequencies of CTLp for Muguga-infected and Marikebuni-infected cells; in the remaining animals, only CTLp reactive with the immunizing parasite were detected. Following challenge of the animals with a lethal dose of the heterologous parasite, the nine animals with cross-reactive CTLp displayed only mild reactions with transient low parasitosis in the regional lymph node, while the remaining 13 animals with strain-specific CTLp exhibited moderate to severe reactions (high fever, prolonged parasitosis) and two of them died. This correlation between strain specificity of CTL and the reactions to heterologous challenge in individual animals provides further compelling evidence that CTL play an important role in mediating immunity. The results also demonstrate that both the Muguga and Marikebuni 3219 parasite populations contain strain-specific T cell epitopes and that in a proportion of immunized animals the CTL response is directed preferentially to these epitopes despite the presence of epitopes common to both parasites. The finding that, in contrast to the parent Marikebuni stock, Marikebuni 3219 induced strain-specific CTL and immunity in some animals suggests that the broad protective capacity of the parent stock is due to the complex strain composition of this stock.

In further experiments, it was shown that Muguga-immunized animals that had strain-specific CTLp developed cross-reactive CTLp following recovery from challenge with Marikebuni 3219. Since the CTLp were activated in the limiting dilution assay only with Muguga-infected cells, it is clear that the cross-reactive epitopes are presented in an immunogenic form on the surface of Muguga-infected cells from these animals. Thus, the failure of these animals to generate a cross-reactive CTL response is not due to the absence of T cells capable of responding to the epitopes; rather, the findings suggest that there may be some form of antigenic competition resulting in the response in individual animals being predominantly directed toward a limited number of antigenic epitopes.

3. Influence of the MHC Phenotype on Strain Specificity of the CTL Response

A striking feature of the results reported above is the variation in strain specificity of the CTL response between individual animals. The potential role of MHC phenotype in influencing strain specificity has been examined in two ways. The first involved comparing the strain specificity of CTL clones originating from the same animal but having different MHC restriction specificities. While CTL responses of individual animals are often directed predominantly toward one of the class I MHC molecules,[55] it has been possible in some animals to generate sets of CTL clones restricted by two different class I specificities. Such clones provide the opportunity to examine the influence of MHC in isolation of other variables. Clones from three animals have been analyzed; two of these had been immunized with the Muguga and the other with Marikebuni 3219.[62-64] The MHC restriction specificities of the clones were defined using panels of target cells of defined MHC phenotype and by testing the capacity of class I allele-specific MAb to inhibit cytotoxicity. In each of the three animals, the two sets of clones displayed different parasite strain specificities when tested on target cells infected with different cloned populations of the Marikebuni stock. These results indicate that different antigenic epitopes are being presented by different class I molecules in the same animal. This is perhaps not surprising given recent findings in human and murine systems that each class I molecule binds peptides with a characteristic motif of anchor residues.[65,66] Thus, there is clearly the potential for variation in MHC phenotype to result in differences in strain specificity between individual animals.

The second way in which this question has been addressed is to examine the strain specificity of the CTL response in a series of animals in which the CTL are restricted by the same MHC molecule. Because the expression of two polymorphic class I loci has only recently been demonstrated in cattle,[67] definitive information relating strain specificity to MHC restriction in a reasonable number of animals is only available for one MHC haplotype. Within the group of cattle used in the cross-protection experiments described above, eight animals immunized with Muguga expressed the class I A and B locus specificities w10 and KN104, respectively, on one haplotype.[42] In six of these animals, the CTLp were restricted entirely by products of this haplotype, and in five of the six animals the CTLp were Muguga specific. In four of the animals, including the one which displayed cross-reactive CTLp, the response was shown to be restricted by the KN104 B locus product. These data indicate that while in the majority of animals KN104-restricted CTL responses to Muguga are directed to strain-specific epitopes, there is not an absolute correlation between the restricting MHC molecule and strain specificity. Such variation may relate to differences in the T cell receptor repertoire of the animals or to other host polymorphisms that affect the efficacy with which particular epitopes are presented on infected cells.

4. Influence of Immunizing Parasite Strain on MHC Restriction and Strain Specificities of the CTL Response

The issues of whether particular host genotypes are generally associated with strain-specific CTL responses induced by different parasites, and whether the same or different MHC molecules tend to be involved in restricting the responses, are important in considering immunization strategies. Access to MHC-identical animals has allowed these questions to be addressed. The responses of two sets of identical twin calves (produced by embryo splitting) and a pair of class I-identical animals have been examined following immunization of one animal of each pair with Muguga and the other with Marikebuni 3219.[42] All six animals had the w10, KN104 class I haplotype. The three Muguga-immunized animals exhibited a Muguga-specific CTL response restricted entirely by KN104. In two of the animals immunized with Marikebuni 3219, the response was also strain specific but was restricted by w10, while the third animal showed a KN104-restricted cross-reactive CTL response. The latter result is particularly intriguing since it demonstrates that genetically identical animals, presumably with similar T cell receptor repertoires, can respond to a strain-specific or a conserved epitope presented on the same class I molecule, depending on the immunizing parasite. The findings also show that the immunizing parasite can influence which class I molecules are the dominant restricting elements. From these results it is also clear that Muguga-infected cells express simultaneously strain-specific and conserved epitopes in association with the same MHC specificity on their cell membranes.

5. Cellular and Chemical Basis of the Antigenic Specificity of CTL

A number of conclusions can be drawn from the currently available information on the *Theileria*-specific CTL response. First, the response in individual animals is focused on a limited number of antigenic epitopes presented by one or two class I molecules. Second, the antigenic epitopes to which a response is generated vary from one animal to another and, depending on whether they are conserved or variable between strains, may result in a cross-reactive or parasite-specific response. Third, animals fail to respond to some antigenic epitopes despite the presence of T cells of the appropriate specificity and the demonstration that the epitopes are presented on the surface of infected cells from the particular animals.

Important questions remain to be answered. What are the factors that determine the immunodominance of particular epitopes? Why do potentially immunogenic epitopes in some circumstances fail to stimulate a detectable CTL response?

One factor that is likely to be important in determining immunodominance is the quantity of antigenic peptide bound to MHC molecules on the cell surface. It is well established in other systems that different peptides vary in their ability to bind to a particular MHC molecule, and that the capacity of a given peptide to bind to different MHC molecules also varies.[59,68] Recent findings have shown that there are characteristic anchor residues which determine the binding of peptides to each MHC molecule.[65,66] Variation in other peptide residues will influence the peptide binding affinities. Moreover, the quantities of particular peptides generated within the cytosol of infected cells and the efficacy with which they are transported into the endoplasmic reticulum, where they associate with class I, are also likely to be important contributory factors. Within this scheme, it is easy to envisage that particular peptide-MHC combinations will be more immunogenic than others and that these will vary between animals, depending largely on MHC phenotype. The absence of specifically reactive T cells will in some instances result in a failure to respond to "dominant" epitopes. It is well documented that such "holes in the repertoire" can occur because of similarity of the foreign peptide-MHC determinant to a self peptide-MHC combination, for which reactive T cells have been eliminated or rendered anergic during ontogeny.[69,70]

The absence of responses to particular epitopes in animals that are known to be capable of responding to the epitopes is more difficult to explain, but may also relate to lower abundance of the MHC-associated peptide in relation to other immunogenic peptides. Thus, it is possible that the initial activation of T cells recognizing the immunodominant antigenic specificities, followed by killing of the antigen presenting (parasitized) cells, precludes the induction of responses against the relatively less dominant epitopes. Studies to elucidate the involvement of these various phenomena in determining the antigenic specificity of the CTL response must await identification of the target antigens.

D. INFLUENCE OF CYTOKINES ON *THEILERIA*-INFECTED CELLS

While the potent cytotoxic activity of $CD8^+$ parasite-specific T cells suggests that killing of infected cells is an important means of controlling infection, other effector mechanisms involving direct or indirect effects of cytokines on infected cells may also play an important role. The potential involvement of mechanisms other than cell killing in control of infection is further emphasized by recent observations that some individual animals (particularly young animals) exhibit little or no CTL responses and have very few parasite-specific CTLp following immunization. However, information on the effects of cytokines on growth and regulation of parasitized cells is fragmentary.

Brown et al.[46] have shown that *Theileria*-specific $CD4^+$ T cells produce a T cell growth factor with the properties of IL-2. Although human rIL-2 significantly inhibits the establishment of infected cell lines,[71] IL-2 has no obvious inhibitory effect on the growth of *Theileria*-infected cell lines. On the contrary, bovine[72] and human rIL-2[71,73] have been reported to enhance the growth of infected B and T cells, but not T cells. It has been suggested that an autocrine mechanism in which IL-2 is involved contributes to the continuous proliferation of infected cell lines as IL-2-specific antibodies can reduce proliferation of infected cell lines.[74,75] Indeed, a T cell growth factor with properties similar to IL-2 has been isolated from culture supernates of infected cell lines,[46,73] and low levels of IL-2 mRNA expression have been detected in infected cell lines.[75] Transcripts were detected in infected lines of T or B cell origin and disappeared following removal of the parasite by drug treatment. Expression of Tac antigen (the low-affinity component of the IL-2 receptor) on the cell membrane and the presence of Tac mRNA have been demonstrated in parasitized T and B cell lines, supporting the view that infected cells are capable of responding to IL-2.[76]

It has been shown that many infected cell lines themselves produce IFN-γ.[77] Bovine rIFN-γ has no influence on the growth of established cell lines, although different research groups reported positive or negative effects on the establishment of infected cell lines.[71,72]

Two independent studies have provided evidence that rTNF-α (human and bovine) has a negative effect on the *in vitro* establishment of infected cell lines.[71,72] However, TNF-α had no negative influence on the growth of established cell lines; on the contrary, in one study it (bovine recombinant) consistently enhanced the proliferation of schizont-infected cells.[71]

Preston et al.[71] have demonstrated that human rIFN-α reduces the *in vitro* establishment of schizont-infected cell lines although it does not affect their growth once established. This accords with data from *in vivo* experiments where oral treatment of cattle with low doses of IFN-α (natural human) before and after infection with sporozoites of *T. parva* protected the animals against development of clinical theileriosis.[78] Another cytokine produced by activated macrophages, IL-1, has been shown to behave similarly to IFN-α *in vitro* in that it (human recombinant) inhibits the establishment of infected cell lines but has no effect on the growth of established cell lines.[71]

The results of *in vitro* studies with individual cytokines are sometimes difficult to interpret in relation to protective immunity because the same cytokine may have different effects depending on concentration, or it may require to act in concert with other cytokines to exert an effect. Moreover, when dealing with complex mixtures of cytokines, the detection of biological activity may be masked by the presence of cytokines with opposing activities. There is a need for definitive information on the profile of cytokines that are produced by both T cells and accessory cells *in vivo* during the acquisition of immunity. Currently, a panel of bovine cytokine cDNA probes, recombinant proteins, and specific monoclonal antibodies are being generated for this purpose.

VI. EPILOGUE

Taken together the experimental data on immunization against *T. parva* indicate that acquisition of immunity to the parasite is dependent on controlling the number of parasitized lymphocytes and, that if the numbers of infected cells exceed a certain threshold, the immune system is overwhelmed. Control of the infection can be brought about in two ways. First, immune responses against sporozoite-specific antigens can limit establishment of infection. Such responses have been induced with a recombinant sporozoite protein. They involve production of neutralizing antibodies which are believed to act by blocking infection of host cells. The second means of controlling the infection is by T cell-mediated immune responses against parasitized lymphocytes. The appearance of $CD8^+$ CTL concomitant with the disappearance of schizont-infected cells during immunization or challenge, and the observed correlation of parasite strain specificity of CTL responses with the outcome of challenge experiments with heterologous strains of *T. parva*, clearly indicate that $CD8^+$ CTL can control the expansion of parasitized cells. The importance of $CD8^+$ cells in protection has also been confirmed in adoptive transfer experiments.

The role that $CD4^+$ T cell responses play in immunity is less well defined. Almost certainly, they are required to provide help in the generation of anti-sporozoite antibody responses. Parasite-specific $CD4^+$ T cells reactive with infected lymphocytes have also been identified. *Theileria*-infected cells generally express high levels of both class I and class II MHC molecules. Hence, $CD4^+$ cells can interact directly with parasitized cells and thus may be brought into close proximity to the specific $CD8^+$ T cells. This may allow the $CD4^+$ cells to provide help in the generation of CTL responses. Recent experiments with purified $CD4^+$ and $CD8^+$ T cells have shown that both naive and immune $CD8^+$ T cells require help from antigen-specific $CD4^+$ T cells (Tarada and McKeever, unpublished data). If this is indeed the case, priming of $CD4^+$ cells with conserved parasite antigens in animals that mount a strain-specific CTL response may allow them to generate a more rapid CTL response upon heterologous challenge and thus enhance their chances of surviving the infection. Such a role for $CD4^+$ cells in relation to antibody responses has been proposed for infections with *Babesia bovis* in situations where animals with strain-specific antibody show immunity to challenge with a heterologous parasite strain.[79] The possibility that $CD4^+$ T cells can act as effector cells by secretion of cytokines that either act directly on parasitised cells or are responsible for recruiting other cell types such as macrophages that exert an anti-parasite effect also needs to be explored.

The available data indicate that schizont-encoded antigens recognized by T cells, in particular those recognized by $CD8^+$ CTL, are candidates for the development of subunit vaccines. The identification of $CD8^+$ T cell target antigens has thus far proved problematical because of the difficulties in using T cells to screen for endogenously processed antigens and the size of the parasite genome. However, recent technological advances that allow isolation of antigenic peptides directly from MHC molecules now offer a way forward in this area. With regard to $CD4^+$ cells, a limiting dilution proliferative assay has been established (J. P. Scheerlinck, T. van de Putte, E. Taracha, D. J. McKeever, and B. M. Goddeeris) which should facilitate screening of parasite antigens, either biochemically fractionated or in expression libraries, in order to identify the antigens recognized by T cells specific for sporozoite or schizont antigens.

REFERENCES

1. **Irvin, A. D. and Morrison, W. I.,** Immunopathology, immunology and immunoprophylaxis of *Theileria* infections, in *Immune Responses in Parasitic Infections: Immunology, Immunopathology and Immunoprophylaxis, Vol 3,* Soulsby E. J. L., Ed., CRC Press, Boca Raton, FL, 1987, 223.
2. **Fawcett, D. W., Doxsey, S., Stagg, D. A., and Young, A. S.,** The entry of sporozoites of *Theileria parva* into bovine lymphocytes in vitro. Electron microscopic observations, *Eur. J. Cell Biol.*, 27, 10, 1982.
3. **Shaw, M. K., Tilney, L. G., and Musoke, A. J.,** The entry of *Theileria parva* sporozoites into bovine lymphocytes: evidence for MHC class I involvement, *J. Cell Biol.*, 113, 87, 1991.
4. **Hulliger, L., Wilde, J. K. H., Brown, C. G. D., and Turner, L.,** Mode of multiplication of *Theileria* in cultures of bovine lymphocytic cells, *Nature*, 203, 728, 1964.
5. **Brown C. G. D., Stagg, D. A., Purnell, R. E., Kanhai, G. K., and Payne, R. C.,** Infection and transformation of bovine lymphoid cells *in vitro* by infective particles of *Theileria parva*, *Nature*, 245, 101, 1973.
6. **Baldwin, C. L., Black, S. J., Brown, W. C., Conrad, P. A., Goddeeris, B. M., Kinuthia, S. W., Lalor, P. A., MacHugh, N. D., Morrison, W. I., Morzaria, S. P., Naessens, J., and Newson, J.,** Bovine T cells, B cells, and null cells are transformed by the protozoan parasite *Theileria parva*, *Infect. Immun.*, 56, 462, 1988.
7. **Morrison, W. I., Goddeeris, B. M., Brown, W. C., Baldwin, C. L., and Teale, A.J.,** *Theileria parva* in cattle: characterization of infected lymphocytes and the immune responses they provoke, *Vet. Immunol. Immunopathol.*, 20, 213, 1989.
8. **Emery, D. L., MacHugh, N. D., and Morrison W. I.,** *Theileria parva* (Muguga) infects bovine T-lymphocytes *in vivo* and induces coexpression of BoT4 and BoT8, *Parasite Immunol.*, 10, 379, 1988.
9. **Shaw, M., Tilney, L., and McKeever, D. J.,** Tick salivary gland extract and interleukin-2 stimulation enhance susceptibility of lymphocytes to infection by *Theileria parva* sporozoites, *Infect. Immun.*, 61, 1486, 1993.
10. **Theiler, A.,** The artificial transmission of East Coast fever, in Report of the Government Veterinary Bacteriologist for the Year 1909-1910, Theiler A., Ed., The Government Printing and Stationery Office, Pretoria, 1911, 7.
11. **Spreull, J.,** East Coast fever inoculation in the Transkeian territories, South Africa, *J. Comp. Pathol. Ther.*, 27, 299, 1914.
12. **Brown, C. G. D., Cunningham, M. P., Joyner, L. P., Purnell, R. E., Branagan, D., Corry, G. L., and Bailey, K. P.,** *Theileria parva*: significance of leukocytes for infecting cattle, *Exp. Parasitol.*, 45, 55, 1978.
13. **Emery, D. L., Morrison, W. I., Büscher, G., and Nelson, R. T.,** Generation of cell-mediated cytotoxicity to *Theileria parva* (East Coast fever) after inoculation of cattle with parasitized lymphoblasts, *J. Immunol.*, 128, 195, 1982.
14. **Emery, D. L., Morrison, W. I., Nelson, R. T., and Murray, M.,** The induction of cell-mediated immunity in cattle inoculated with cell lines parasitized with *Theileria parva*, in *Current Topics in Veterinary Medicine and Animal Science 14: Advances in the Control of Theileriosis*, Irvin, A.D., Cunningham, M.P., and Young, A.S., Eds., Martinus Nijhoff, The Hague, 1981, 295.
15. **Neitz, W. O.,** Aureomycin in *Theileria parva* infection, *Nature*, 171, 34, 1953.
16. **Radley, D. E., Brown, C. G. D., Cunningham, M. P., Kimber, C. D., Musisi, F. L., Payne, R. C., Purnell, R. E., Stagg, S. M., and Young, A. S.,** East Coast fever. 3. Chemoprophylactic immunization of cattle using oxytetracycline and a combination of Theilerial strains, *Vet. Parasitol.*, 1, 51, 1975.
17. **Irvin, A. D., Dobbelaere, D. A. E., Mwamachi, D. M., Minami, T., Spooner, P. R., and Ocama, J. G. R.,** Immunisation against East Coast fever: correlation between monoclonal antibody profiles of *Theileria parva* stocks and cross-immunity in vivo, *Res. Vet. Sci.*, 35, 341, 1983.
18. **Pinder, M. and Hewett, R. S.,** Monoclonal antibodies detect antigenic diversity in *Theileria parva* parasites, *J. Immunol.*, 124, 1000, 1980.
19. **Minami, T., Spooner, P. R., Irvin, A. D., Ocama, J. G. R., Dobbelaere, D. A. E., and Fujinaga, T.,** Characterization of stocks of *Theileria parva* by monoclonal antibodies, *Res. Vet. Sci.*, 35, 334, 1983.

20. **Shapiro S. Z., Fujisaki, K., Morzaria, S. P., Webster, P., Fujinaga, T., Spooner, P. R., and Irvin, A. D.,** A life-cycle stage-specific antigen of *Theileria parva* recognised by anti-macroschizont monoclonal antibodies, *Parasitology*, 94, 29, 1987.
21. **Toye, P. G., Goddeeris, B. M., Iams, K., Musoke, A. J., and Morrison, W. I.,** Characterization of a polymorphic immunodominant molecule in sporozoites and schizonts of *Theileria parva, Parasite Immunol.*, 13, 49, 1991.
22. **Sugimoto, C., Conrad, P. A., Mutharia, L., Dolan, T. T., Brown, W. C., Goddeeris, B. M., and Pearson, T. W.,** Phenotypic characterization of *Theileria parva* schizonts by two-dimensional gel electrophoresis, *Parasitol. Res.*, 76, 1, 1989.
23. **Morzaria, S. P., Spooner, P. R., Bishop, R. P., Musoke, A. J., and Young, J. R.,** SfiI and NotI polymorphisms in *Theileria* stocks detected by pulsed field gel electrophoresis, *Mol. Biochem. Parasitol.*, 40, 203, 1990.
24. **Conrad, P. A., Iams, K., Brown, W. C., Sohanpal, B., and Ole-Moi Yoi, O.,** DNA probes detect genomic diversity in *Theileria parva* stocks, *Mol. Biochem. Parasitol.*, 25, 213, 1987.
25. **Conrad, P. A., Baldwin, C. L., Brown, W. C., Sohanpal, B., Dolan, T. T., Goddeeris, B. M., DeMartini, J. C., and Ole-Moi Yoi, O. K.,** Infection of bovine T cell clones with genotypically distinct *Theileria parva* parasites and analysis of their cell surface phenotype, *Parasitology*, 99, 205, 1989.
26. **Allsopp, B. A., and Allsopp, M. T. E. P.,** *Theileria parva* genomic DNA studies reveal intra-specific sequence diversity, *Mol. Biochem. Parasitol.*, 28, 77, 1988.
27. **Allsopp, B., Carrington, M., Baylis, H., Sohal, S., Dolan, T., and Iams, K.,** Improved characterization of *Theileria parva* isolates using the polymerase chain reaction and oligonucleotide probes, *Mol. Biochem. Parasitol.*, 35, 137, 1989.
28. **Morzaria, S. P. and Young, J. R.,** Restriction mapping of the genome of the protozoan parasite *Theileria parva, Proc. Natl. Acad. Sci. U.S.A.*, 89, 5241, 1992.
29. **Morzaria, S., Young, J., Bishop, R., Young, A., Dolan, T., and Mehlhorn, H.,** Evidence for a sexual cycle in *Theileria parva*, in *Annual Scientific Report 1991 ILRAD*, 1992, 13–14.
30. **Dobbelaere, D. A. E., Spooner, P. R., Barry, W. C., and Irvin, A. D.,** Monoclonal antibody neutralizes the sporozoite stage of different *Theileria parva* stocks, *Parasite Immunol.*, 6, 361, 1984.
31. **Musoke, A. J., Nantulya, V. M., Rurangirwa, F. R., and Buscher, G.,** Evidence for a common protective antigenic determinant on sporozoites of several *Theileria parva* strains, *Immunology*, 52, 231, 1984.
32. **Nene, V., Iams, K. P., Gobright, E., and Musoke, A. J.,** Characterisation of the gene encoding a candidate vaccine antigen of *Theileria parva* sporozoites, *Mol. Biochem. Parasitol.*, 51, 17, 1992.
33. **Musoke, A., Morzaria, S., Nkonge, C., Jones, E., and Nene, V.,** A recombinant sporozoite surface antigen of *Theileria parva* induces protection in cattle, *Proc. Natl. Acad. Sci. U.S.A.*, 89, 514, 1992.
34. **Iams, K. P., Hall, R., Webster, P., and Musoke, A. J.,** Identification of gt11 clones encoding the major antigenic determinants expressed by *Theileria parva* sporozoites, *Infect. Immun.*, 58, 1828, 1990.
35. **Iams, K. P., Young, J. R., Nene, V., Desai, J., Webster, P., Ole-MoiYoi, O. K., and Musoke, A. J.,** Characterisation of the gene encoding a 104-kilodalton microneme-rhoptry protein of *Theileria parva, Mol. Biochem. Parasitol.*, 39, 47, 1990.
36. **Creemers, P.,** Lack of reactivity of sera from *Theileria parva*-infected and recovered cattle against membrane antigens of *Theileria parva* transformed cell lines, *Vet. Immunol. Immunopathol.*, 3, 427, 1982.
37. **Muhammed, S. I., Lauerman, L. H., and Johnson, L. W.,** Effect of humoral antibodies on the course of *Theileria parva* infection (East Coast fever) in cattle, *Am. J. Vet. Res.*, 36, 399, 1975.
38. **Wagner, G. G., Duffus, W. P. H., and Burridge, M. J.,** The specific immunoglobulin response in cattle immunized with isolated *Theileria parva* antigens, *Parasitology*, 69, 43, 1974.
39. **Emery, D. L.,** Adoptive transfer of immunity to infection with *Theileria parva* (East Coast fever) between cattle twins, *Res. Vet. Sci.*, 30, 364, 1981.
40. **Eugui, E. M. and Emery, D. L.,** Genetically restricted cell-mediated cytotoxicity in cattle immune to *Theileria parva, Nature*, 290, 251, 1981.
41. **Emery, D. L., Eugui, E. M., Nelson, R. T., and Tenywa, T.,** Cell-mediated immune responses to *Theileria parva* (East Coast fever) during immunization and lethal infections in cattle, *Immunology*, 43, 323, 1981.

42. **Taracha, E. L. N.,** Investigation of Cytotoxic T-Lymphocyte Responses of Cattle to Theileria parva by Limiting Dilution Analyses, Ph.D. thesis, Brunel University, Uxbridge, U.K., 1991.
43. **McKeever, D. L., Taracha, E. L. N., Innes, E. L., MacHugh, N. D., Awino, E., Goddeeris, B. M., and Morrison, W. I.,** Adoptive transfer of immunity to *Theileria parva* in the CD8+ fraction of responding efferent lymph, Proc. Natl. Acad. Sci. U.S.A., in press.
44. **Pearson, T. W., Lundin, L. B., Dolan, T. T., and Stagg, D. A.,** Cell-mediated immunity to *Theileria*-transformed cell lines, *Nature*, 281, 678, 1979.
45. **Goddeeris, B. M. and Morrison, W. I.,** The bovine autologous *Theileria* mixed leucocyte reaction: influence of monocytes and phenotype of the parasitized stimulator cell on proliferation and parasite specificity, *Immunology*, 60, 63, 1987.
46. **Brown, W. C. and Logan, K. S.,** Bovine T-cell clones infected with *Theileria parva* produce a factor with IL 2-like activity, *Parasite Immunol.*, 8, 189, 1986.
47. **Goddeeris, B. M., Morrison, W. I., and Teale, A. J.,** Generation of bovine cytotoxic cell lines, specific for cells infected with the protozoan parasite *Theileria parva* and restricted by products of the major histocompatibility complex, *Eur. J. Immunol.*, 16, 1243, 1986.
48. **Baldwin, C. L., Goddeeris, B. M., and Morrison, W. I.,** Bovine helper T-cell clones specific for lymphocytes infected with *Theileria parva* (Muguga), *Parasite Immunol.*, 9, 499, 1987.
49. **Brown, W. C., Sugimoto, C., Conrad, P. A., and Grab, D. J.,** Differential response of bovine T-cell lines to membrane and soluble antigens of *Theileria parva* schizont-infected cells, *Parasite Immunol.*, 11, 567, 1989.
50. **Brown, W. C., Lonsdale-Eccles, J. D., DeMartini, J. C., and Grab, D.J.,** Recognition of soluble *Theileria parva* antigen by bovine helper T cell clones: characterization and partial purification of the antigen, *J. Immunol.*, 144, 271, 1990.
51. **Baldwin, C. L., Iams, K. P., Brown, W. C., and Grab, D. J.,** *Theileria parva*: CD4+ helper and cytotoxic T-cell clones react with a schizont-derived antigen associated with the surface of *Theileria parva*-infected lymphocytes, *Exp. Parasitol.*, 75, 19, 1992.
52. **Grab, D. J., Baldwin, C. L., Brown, W. C., Innes, E. A., Lonsdale-Eccles, J. D., and Verjee, Y.,** Immune CD4+ T cells specific for *Theileria parva*-infected lymphocytes recognize a 24-kilodalton protein, *Infect. Immun.*, 60, 3892, 1992.
53. **Pearson, T. W., Hewett, R. S., Roelants, G. E., Stagg, D. A., and Dolan, T. T.,** Studies on the induction and specificity of cytotoxicity to *Theileria*-transformed cell lines, *J. Immunol.*, 128, 2509, 1982.
54. **Eugui, E. M., Emery, D. L., Büscher, G., and Khaukha, G.,** Specific and non-specific cellular immune responses to *Theileria parva* in cattle, in *Current Topics in Veterinary Medicine and Animal Science 14: Advances in the Control of Theileriosis*, Irvin, A.D., Cunningham, M.P., and Young, A.S., Eds., Martinus Nijhoff, The Hague, 1981, 289.
55. **Morrison, W. I., Goddeeris, B. M., Teale, A. J., Groocock, C. M., Kemp, S. J., and Stagg, D. A.,** Cytotoxic T cells elicited in cattle challenged with *Theileria parva* (Muguga): evidence for restriction by class I MHC determinants and parasite strain specificity, *Parasite Immunol.*, 9, 563, 1987.
56. **Goddeeris, B. M., Morrison, W. I., Teale, A. J., Bensaid, A., and Baldwin, C. L.,** Bovine cytotoxic T-cell clones specific for cells infected with the protozoan parasite *Theileria parva*: parasite strain specificity and class I major histocompatibility complex restriction, *Proc. Natl. Acad. Sci. U.S.A.*, 83, 5238, 1986.
57. **Morrison, W. I., Goddeeris, B. M., and Teale, A. J.,** Bovine cytotoxic T-cell clones which recognize lymphoblasts infected with two antigenically different stocks of the protozoan parasite *Theileria parva*, *Eur. J. Immunol.*, 17, 1703, 1987.
58. **Zinkernagel, R. M., and Doherty, P. C.,** MHC-restricted cytotoxic T cells: studies on the biological role of polymorphic major transplantation antigens determining T-cell restriction-specificity, function and responsiveness, *Adv. Immunol.*, 27, 51, 1979.
59. **Buus, S., Sette, A., Colon, S. M., Miles, C., and Grey, H. M.,** The relation between major histocompatibility complex (MHC) restriction and the capacity of Ia to bind immunogenic peptides, *Science*, 235, 1353, 1987.
60. **Blackman, M. A., Kappler, J. W., and Marrack, P.,** T-cell specificity and repertoire, *Immunol. Rev.*, 101, 5, 1988.
61. **Taracha, E. L. N., Goddeeris, B. M., Scott, J. R., and Morrison, W. I.,** Standardization of a technique for analysing the frequency of parasite-specific cytotoxic T lymphocyte precursors in cattle immunized with *Theileria parva*, *Parasite Immunol.*, 14, 143, 1992.

62. **Goddeeris, B. M., Morrison, W. I., Toye, P. G., and Bishop, R.,** Strain specificity of bovine *Theileria parva*-specific cytotoxic T cells is determined by the phenotype of the restricting class I MHC, *Immunology*, 69, 38, 1990.
63. **Taracha, E. L. N., Morrison, W. I., and Goddeeris, B. M.,** The cytotoxic T cell: MHC restriction, strain specificity and role in immunity to *Theileria parva* infection, *Ann. Rech. Vet.*, 23, 311, 1992.
64. **Goddeeris, B.M.,** A bovine class I major histocompatibility complex molecule plus a *Theileria parva* antigen resembles mouse H-2K^d, *Anim. Genet.*, 21, 323, 1990.
65. **Garrett, T. P. J., Saper, M. A., Bjorkman, P. J., Strominger, J. L., and Wiley, D. C.,** Specificity pockets for the side chains of peptide antigens in HLA-Aw68, *Nature*, 342, 692, 1989.
66. **Guo, H.-C., Jardetzky, T. S., Garrett, T. P. J., Lane, W. S., Strominger, J. L., and Wiley, D. C.,** Different length peptides bind to HLA-Aw68 similarly at their ends but bulge out in the middle, *Nature*, 360, 364, 1992.
67. **Toye, P. G., MacHugh, N. D., Bensaid, A. M., Alberti, S., Teale, A. J., and Morrison, W. I.,** Transfection into mouse L cells of genes encoding two serologically and functionally distinct bovine class I MHC molecules from a MHC-homozygous animal: evidence for a second class I locus in cattle, *Immunology*, 70, 20, 1990.
68. **Chen, B. P. and Parham, P.,** Direct binding of influenza peptides to class I HLA molecules, *Nature*, 337, 743, 1989.
69. **Schwartz, R. H.,** A clonal deletion model for Ir gene control of the immune response, *Scand. J. Immunol.*, 7, 3, 1978.
70. **Schwartz, R. H.,** Review. Acquisition of immunologic self-tolerance, *Cell*, 57, 1073, 1989.
71. **Preston, P. M., Brown, C. G., and Richardson, W.,** Cytokines inhibit the development of trophozoite-infected cells of *Theileria annulata* and *Theileria parva* but enhance the proliferation of macroschizont-infected cell lines, *Parasite Immunol.*, 14, 125, 1992.
72. **DeMartini, J. C. and Baldwin, C. L.,** Effects of gamma interferon, tumor necrosis factor alpha, and interleukin-2 on infection and proliferation of *Theileria parva*-infected bovine lymphoblasts and production of interferon by parasitised cells, *Infect. Immun.*, 59, 4540, 1991.
73. **Dobbelaere, D. A. E., Coquerelle, T. M., Roditi, I. J., Eichhorn, M., and Williams, R. O.,** *Theileria parva* infection induces autocrine growth of bovine lymphocytes, *Proc. Natl. Acad. Sci. U.S.A.*, 85, 4730, 1988.
74. **Dobbelaere, D. A., Roditi, I. J., Coquerelle, T. M., Kelke, C., Eichhorn, M., and Williams, R. O.,** Lymphocytes infected with *Theileria parva* require both cell-cell contact and growth factor to proliferate, *Eur. J. Immunol.*, 21, 89, 1991.
75. **Heussler, V. T., Eichhorn, M., Reeves, R., Magnuson, N. S., Williams, R. O., and Dobbelaere, D. A.,** Constitutive IL-2 mRNA expression in lymphocytes, infected with the intracellular parasite *Theileria parva*, *J. Immunol.*, 149, 562, 1992.
76. **Dobbelaere, D. A., Prosperao, T. D., Roditi, I. J., Kelke, C., Baumann, I., Eichhorn, M., Williams, R. O., Ahmed, J. S., Baldwin, C. L., Clevers, H., and Morrison, W. I.,** Expression of Tac antigen component of bovine interleukin-2 receptor in different leukocyte populations infected with *Theileria parva* or *Theileria annulata*, *Infect. Immun.*, 58, 3847, 1990.
77. **Entrican, G., McInnes, C. J., Logan, M., Preston, P. M., Martinod, S., and Brown, C. G.,** Production of interferons by bovine and ovine cell lines infected with *Theileria annulata* or *Theileria parva*, *Parasite Immunol.*, 13, 339, 1991.
78. **Young, A. S., Maritim, A. C., Kariuki, D. P., Stagg, D. A., Wafula, J. M., Mutiga, J. J., Cummins, J. M., Richards, A. B., and Burns, C.,** Low-dose oral administration of human interferon alpha can control the development of *Theileria parva* infection in cattle, *Parasitology*, 101, 201, 1990.
79. **Mahoney, D. F., Kerr, J. D., Goodger, B. V., and Wright, I. G.,** The immune response of cattle to *Babesia bovis* (Syn. *B. argentina*). Studies on the nature and specificity of protection, *Int. J. Parasitol.*, 9, 297, 1979.

Chapter 10

Infectious Bovine Rhinotracheitis (Bovine Herpesvirus 1): Helper T Cells, Cytotoxic T Cells, and NK Cells

M. Denis, G. Splitter, E. Thiry, P.-P. Pastoret, and L. A. Babiuk

CONTENTS

I. Introduction 157
A. Characteristics of the BHV-1 Virus 157
B. Infectious Bovine Rhinotracheitis/Infectious Pustular Vulvovaginitis 158
C. Strategies for Control 159
II. BHV-1 Infection and Immunity 160
A. Nonspecific and Specific Immune Mechanisms 160
B. BHV-1-Induced Immunosuppression 160
C. Immune Response and Latency 162
III. Nonspecific Responses to BHV-1 Infection 163
A. Interferon 163
B. Macrophages 163
C. Polymorphonuclear Neutrophilic Granulocytes 163
D. Natural Killer Cells 163
IV. Specific Responses to BHV-1 Infection 164
A. Helper T Lymphocytes 165
B. Cytotoxic T Lymphocytes 165
V. Conclusion 166
References 167

I. INTRODUCTION

The goal of this review is to give a summary of the immune responses, especially cell-mediated immune responses, to bovine herpesvirus 1 (BHV-1) or infectious bovine rhinotracheitis (IBR) virus. IBR is a major worldwide threat to cattle. The biology of the infection is peculiar, due to the establishment of the virus in latent form and the further possibility of viral activation and reexcretion. The vaccines that are now commonly used are able to protect cattle against clinical disease. However, they still need to be improved and this could be achieved by taking different elements into account: prevention of latency, control of viral reactivation, discrimination between infected and vaccinated animals, and rapid development of protection after a single vaccination. Therefore, a more thorough study of the immune response to the virus appears necessary. Many aspects of the cellular immune response to BHV-1 infection are still poorly described, such as the response to the nonstructural antigens of the virus or the relative importance of the different cellular immune mechanisms during the course of a BHV-1 infection. Previous reviews dealt with the virus and its interactions with the host.[1-4] Our aim is to rely on both the well-known concepts of the infection and the more recent data available in this area.

A. CHARACTERISTICS OF THE BHV-1 VIRUS

BHV-1 is, like herpes simplex virus (HSV), a member of the Alphaherpesvirinae subfamily, closely related to varicella zoster virus, pseudorabies virus, and equine herpesvirus 1.[5] According to the proposal of the International Committee on the Taxonomy of Viruses,[6] alphaherpesviruses exhibit a variable host range, relatively short reproductive cycle, rapid spread in culture, efficient destruction of infected cells, and capacity to establish latent infections primarily but not exclusively in sensory ganglia.

The BHV-1 virion (Figure 10-1) has a diameter of 150 to 200 nm[2,4] and consists, like all other herpes virions, of four elements. The core is an electron-opaque structure representing the genome (a linear double-stranded DNA of 135 to 140 kb). It is surrounded by an icosahedral nucleocapsid, an electron-

0-8493-4952-4/94/$0.00+$.50

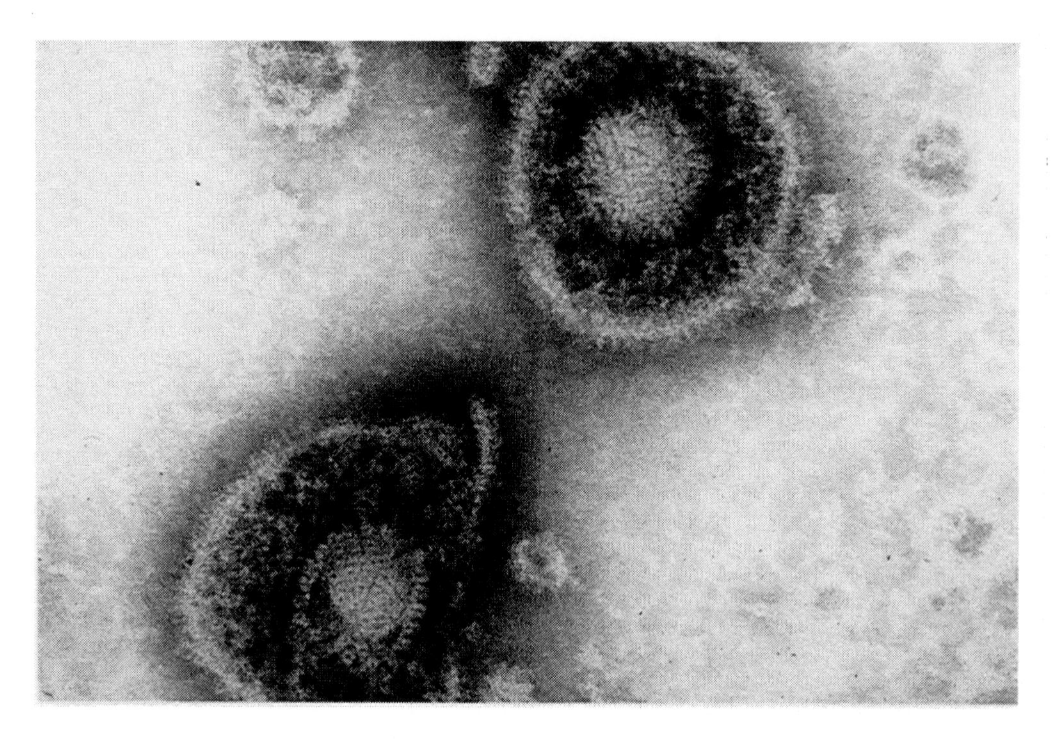

Figure 10-1 Electron micrograph of BHV-1 particles. (From G. Burtonboy. With permission.)

dense amorphous material called "tegument", and by the bilayer of the outer envelope in which viral glycoprotein spikes are embedded.

More than 50 different polypeptides are encoded by the virus in the infected cell.[2] Among them, 15 are nonstructural proteins (not represented in the virions).[7] Like other herpesviruses, BHV-1 exhibits temporal control of protein synthesis in the infected cells.[7,8] The cascade is divided into three steps.

Immediate early (IE) or α proteins are produced immediately after the release of viral DNA: they are nonglycosylated proteins that regulate viral transcription. Three major IE transcripts have been identified, encoding proteins of 180, 135, and 55 kDa homologous to HSV ICP4, ICP0, and ICP22, respectively.[9]

The second step produces early or β proteins, including most of the viral glycoproteins, such as gI, gIV (corresponding to HSV gB[10] and gD[11], respectively),[8] and other minor glycoproteins[12] and enzymes such as thymidine kinase.[13]

The third step, corresponding to the synthesis of the late or γ proteins, is delayed until after the synthesis of virion DNA. Glycoprotein gIII, which is a homologue of HSV gC,[14] is expressed during the late phase.[8]

The general mechanism of replication of HSV[15] can most probably apply to BHV-1 (Figure 10-2). As a first step in the infection process, the virus attaches to cell receptors and penetrates the susceptible cell following fusion of the viral envelope with the cell membrane. These two steps constitute the process of virus entry, which has been described in the HSV system in great detail.[16] Then, the de-enveloped capsid is transported to the nuclear pore, where viral DNA is released into the nucleus. Viral gene expression occurs in a cascade fashion. Viral DNA is replicated by a rolling circle mechanism, yielding concatemers that are cleaved and packaged into the preassembled capsids. Then, envelopment of the capsids takes place at patches in the inner lamellae of the nuclear membrane, representing aggregations of viral membrane proteins. The enveloped capsids accumulate in the endoplasmic reticulum and are transported into the extracellular space where mature infectious virus particles are released.

B. INFECTIOUS BOVINE RHINOTRACHEITIS/INFECTIOUS PUSTULAR VULVOVAGINITIS

BHV-1 causes IBR, a worldwide disease of cattle, characterized by fever, abortion, and respiratory symptoms.[1-4] The virus is easily transmitted directly from one animal to another by aerosol or contact. The lesions induced by a BHV-1 infection consist of microfocal epithelial necrosis located in the upper

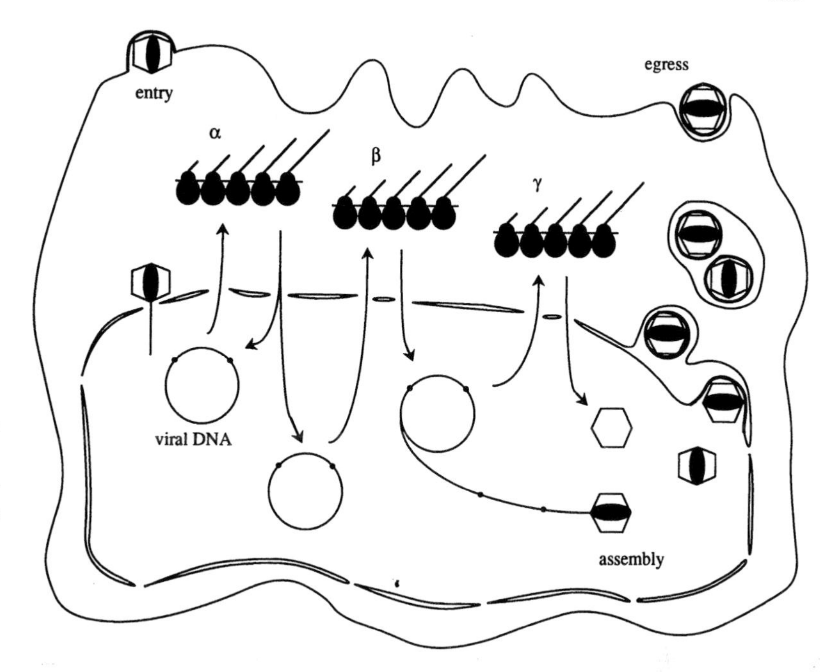

Figure 10-2 Replicative cycle of BHV-1 in infected cells. (Adapted from Reference 15.)

respiratory tract. These lesions normally resolve in a few days. However, secondary pneumonia is a frequent and often fatal complication of IBR. Infectious pustular vulvovaginitis (IPV, balanoposthitis in bulls) is a second syndrome caused by BHV-1. The characteristic lesions consist of small pustules disseminated over the genital mucosa which heal in a few days. However, secondary bacterial infections may occur. Encephalitis was also known to be induced by BHV-1, but the viral strains involved in this particular pathology have been recently classified as BHV-5.[5]

Although only one serotype of BHV-1 is detected in cross-neutralization assays,[17] restriction endonuclease digestion analysis,[18] use of subtype-specific monoclonal antibodies,[19] or SDS-PAGE profile of viral proteins[20] can identify two subtypes of virus, termed BHV-1.1 and 1.2. Differences between BHV-1.1 and 1.2 strains cannot be related to differences in pathology.[17,21,22]

One of the major characteristics of BHV-1, like other herpesviruses, is the establishment of the latent stage in the animal. Latency is defined as the silent persistence of the virus in the host, undetectable by conventional virological procedures.[23] It is established following acute infection, as well as after vaccination with an attenuated strain[24] or following superinfection. Subsequent intermittent episodes of reexcretion may occur.

C. STRATEGIES FOR CONTROL

The economic losses induced by the disease are serious, due to mortality, abortion, and drop in milk production in milking cows. Vaccination is widespread, using attenuated or inactivated viral strains. However, although these vaccines elicit good immune responses and succeed in inducing protection against the disease, they are usually unable to prevent infection with a wild strain, establishment of latency and reexcretion of virus particles upon reactivation of the latent virus. Presently, veterinary research is directed toward the design of new vaccines, aiming at both controlling the spread of the virus in herds and providing means to discriminate between vaccinated and infected cattle.

Two different types of approach are being considered. The first one consists of eliciting a local immune response that would act as a barrier to infection. This approach could appear as the ideal one, since it should prevent disease as well as latency and spread of virus. However, there is also a need for designing control strategies for already infected animals. In this case, the approach consists of improving the immune response so that reactivation does not lead to reexcretion of infectious virus.

BHV-1 subunit vaccines have been shown to induce a strong specific antibody response.[25,26] However, as long as these vaccines were injected intramuscularly, the systemic response that was induced did not always prevent the establishment of viral latency following challenge with virulent virus.[27] Only by delivering the subunit vaccine intranasally, together with cholera toxin B subunit as adjuvant, could an effective local barrier to infection be engendered.[28] Further, recent work has shown that recombinant truncated gI lacking the transmembrane domain can elicit BHV-1-specific IgA, IgG_1, and lymphocyte recognition. Nasally immunized calves were protected from clinical disease.[29] The other advantage of this approach is that the use of a single recombinant BHV-1 glycoprotein (namely gI) appears sufficient to elicit an effective mucosal immune response, therefore allowing a distinction between vaccinated and infected cattle.

Current efforts are also aimed at improving the attenuated vaccine strains. The main advantage of using attenuated viruses is that they replicate in the host and therefore can be administered intranasally without any adjuvant to engender local immunity. A glycoprotein (gIII)-deleted mutant virus has been proposed as a marker vaccine.[30] However, the single gIII gene deletion does not sufficiently attenuate the virus since the gIII-negative mutant is still able to cause clinical symptoms. The deletion in the glycoprotein gE gene seems more promising. Cattle exposed to this mutant do not develop fever.[31] The other possibility consists of combining a deletion of a gene (such as thymidine kinase[32]) for the purpose of attenuating the virus with a deletion in a gene coding for another nonessential protein that normally induces antibodies to obtain a marker vaccine.[31] Alternatively, the feasibility of using BHV-1 as a vaccine vector, expressing foreign DNA inserted into the BHV-1 thymidine kinase gene, has been demonstrated.[33] But the immune response elicited by the attenuated vaccines, even when administerd intranasally, does not completely prevent viral excretion when the animals are challenged. Moreover, attenuated vaccines are able to establish latency and be reactivated. Nevertheless, these disadvantages can be counterbalanced by the fact that such vaccines theoretically stimulate not only a humoral response, but also a $CD8^+$ cytotoxic T lymphocyte (CTL) response. Therefore, they might be used successfully in already infected cattle to increase cellular immune defenses and improve the immune control of reactivation. Indeed, until now, neither attenuated nor inactivated vaccines appear to prevent reexcretion (and thereby dissemination of the virus) in latently infected cattle. On the assumption that sufficiently immunized animals could completely control the reexcretion of the virus,[23] other vaccination strategies should be considered. For instance, cytokines, such as interleukin-1 (IL-1),[34] IL-2,[35] or interferon-γ (IFN-γ), are proposed as adjuvants. In the near future, vaccination against BHV-1 may be highly improved, preventing the spread of BHV-1 infection and stimulating the immune system to control current infection in cattle.

II. BHV-1 INFECTION AND IMMUNITY

A. NONSPECIFIC AND SPECIFIC IMMUNE MECHANISMS

The immune reponse to BHV-1 is both nonspecific and specific. Nonspecific defences include interferon-α (IFN-α) which is rapidly produced after infection, macrophages, natural killer (NK) cells, and polymorphonuclear neutrophilic granulocytes (PMN). Specific defenses are mediated by both T lymphocytes and antibodies. However, the immune mechanisms are often so interrelated that it is difficult to totally dissociate their effects. For example, IFN-γ, which is produced by activated T lymphocytes, is able to stimulate NK activity. Thus, even though specific and nonspecific mechanisms can be identified, the multiple interactions which occur between the different components of the host immune response need to be considered. Most likely, nonspecific mechanisms are primarily involved in protection against primary infection, while both nonspecific and specific defenses protect cattle from secondary infection or reactivation of latent virus.

Immune mechanisms can be further classified according to their time of appearance in BHV-1 infections (Table 10-1). In the case of respiratory infection with BHV-1, since infection occurs in the upper respiratory tract, the primary specific defense barrier the virus encounters is formed by local neutralizing antibodies in the nasal mucus to prevent spread in the respiratory system. However, upon primary infection or if mucosal antibodies are unable to neutralize the virus, replication occurs in epithelial cells resulting in the production of a large number of progeny viruses. Infection then becomes generalized through viremia, neural spread, and cell-to-cell transmission.[23] At this stage, an inflammatory response has developed, resulting in oedema, lymphocyte infiltration of the lamina propria, and accumulation of macrophages and PMN.

B. BHV-1-INDUCED IMMUNOSUPPRESSION

Immunosuppression has been reported in a large number of viral diseases and is rapidly induced following a BHV-1 infection. Several *in vitro* studies have demonstrated that the virus has immunosuppressive effects on leukocyte functions. *In vivo* observations related to changes in leukocyte trafficking

Table 10-1 **Immune Mechanisms Involved in Protection Against BHV-1 Infection**

Nonspecific mechanisms: primary and secondary infection		Specific mechanisms: secondary infection	
1°	IFN	1°	Mucosal IgA, IgG1 (neutralization)
2°	NK cells	2°	T lymphocytes (IFN-γ, CTL)
	Macrophages (IFN, ADCC, phagocytosis)		Serum antibodies (neutralization, ADCC, opsonization)
	PMN (ADCC, phagocytosis, IFN-like product)		

also give evidence for alteration of the integrity of the host immune system. The immunosuppressive mechanisms are considered relevant to the induction of secondary bacterial infections (namely, pneumonic pasteurellosis), although the viral-induced inflammatory processes per se may also create an environment conducive to bacterial growth.[36,37]

Various *in vitro* systems have been used to characterize the immunosuppressive effect of BHV-1 on mononuclear cells. Some have focused on the interaction between virus and proliferating immune cells. Live as well as UV-inactivated BHV-1 are able to inhibit the proliferative response of peripheral blood mononuclear cells (PBMC) to either IL-2 or antigen.[38-40] Live virus has also been shown to inhibit mitogen-induced proliferation.[38] In no instance could the inhibition be reversed by the addition of exogenous IL-2, but the presence of suppressive factors could be demonstrated in virus-free supernatants of adherent PBMC infected with BHV-1.[39] This observation is important regarding the mechanisms of immunosuppression, by pointing to an indirect effect mediated by the release of cellular mediators or cytokines rather than a direct viral interaction with the proliferating cells. However, this is not the only mechanism since BHV-1 infection of activated T lymphocytes results in lymphocytolysis.[40] Therefore, the possibility that some T cells, such as activated T cells, at the site of BHV-1 infection, may be susceptible to a direct effect of the virus cannot be ruled out.

In vitro studies have demonstrated that BHV-1 can also induce alteration of macrophage functions. Although replication of the virus in peripheral blood macrophages does not seem to alter their bactericidal capacity,[41] BHV-1 infection of alveolar macrophages results in impairment of immune receptor functions[42] and ADCC (antibody-dependent cell-mediated cytotoxicity) activity,[43] and reduction in tumor necrosis factor-α release after stimulation with endotoxin inducers.[44] Immune factors (most likely virus-specific antibodies and/or immune complexes) seem responsible for an indirect inhibitory effect of the virus on alveolar macrophage function. However, other hypotheses are not to be excluded; namely, the insertion of viral antigens in the membrane of infected alveolar macrophages may disrupt transmembrane signaling.[42]

In vivo studies have focused on modifications in leukocyte trafficking after BHV-1 infection, and thereby changes in the composition of peripheral blood and lung leukocytes populations. Marked changes in PBMC composition and function occurred as soon as 24 h after aerosol challenge of calves with BHV-1. Mitogen-induced PBMC proliferation was significantly depressed in the first days after BHV-1 infection.[45] Concomitantly, a depletion of circulating $CD8^+$ lymphocytes,[46] corresponding to an influx of $CD8^+$ cells in the lung parenchyma,[47] was observed. IFN-α, which is produced by lung macrophages and interstitial lymphocytes following BHV-1 infection, is involved in the migration of $CD8^+$ lymphocytes to the lung.[37] However, it has not been demonstrated that these tissue-localized lymphocytes mediate CTL activity. Rather, these lymphocytes may contribute indirectly to nonspecific cytotoxicity by producing IFN-γ, which activates MHC-nonrestricted cytotoxic cells.

Alveolar macrophages are the predominant phagocytes in the lung. Macrophages act as an important defense mechanism during the early stage of bacterial invasion as well as in clearing Gram-positive bacteria.[48] They may be protected *in vivo* from BHV-1 infection by IFN because less than 0.1% of alveolar macrophages retrieved by lung lavage from experimentally infected calves are infected with BHV-1.[49] In contrast to the results found after *in vitro* infection of alveolar macrophages, complement and Fc receptor functions are not impaired in cells retrieved from infected animals,[49] but selective inhibition of ADCC, generation of IL-1,[50] or production of neutrophil chemotactic factors have been reported.[36] Reduction of macrophage functions may result in impairment of some antimicrobial and immunoregulatory functions. However, stimulation of alveolar macrophages occurs after BHV-1 infection.[50] Regardless of the mechanism, the subsequent increased expression of MHC class II molecules on stimulated alveolar macrophages is a favorable event for the virus-specific immune response.

Neutrophils are the cell type responsible for bacterial clearance from the lung. *In vivo* studies demonstrated that BHV-1 infection can cause a delay in neutrophil infiltration of the lung in the presence of bacteria. Neutrophils retrieved from the experimentally infected calves appeared paralyzed *in vitro*, displaying little random migration or responsiveness to a chemotactic stimulus.[36] These observations suggest a mechanism where BHV-1 infection alters the host ability to control bacterial invasion of the lung.

Together, these data suggest that the mechanism of virus-induced immunosuppression is multifactorial. The dysfunction of the host immunological defense system involves various immune cell types and various interacting cellular mediators, which are important in the virus-specific immune response. Understanding how the virus induces the network of immunosuppression will be important in designing vaccines to prevent initial infection and recrudescence of existing infections.

C. IMMUNE RESPONSE AND LATENCY

The mechanism whereby BHV-1 establishes latency can be extrapolated from the model of HSV.[15] After multiplication at the local site of infection, which promotes contact between the virus and the axonal nerve endings, the virus enters the neurons and is rapidly transported within the axon to the trigeminal and sacral ganglia. Indeed, viral DNA can be detected in trigeminal ganglia of cattle following respiratory infection[51] and in sacral ganglia after intravaginal inoculation.[52] However, macrophages or epithelial cells are other putative latency sites.[2]

Although viral replication does not occur during latency, restricted transcription of the BHV-1 genome occurs in infected neurons in cattle and rabbits.[53] BHV-1 latency-related RNA (LR RNA) has been characterized;[54] like HSV latency-associated transcript 1 which is transcribed in the opposite direction to the immediate early gene ICP0 and overlaps its 3′ end, BHV-1 LR RNA is complementary to a large part of IE p135 transcript.[55] Nevertheless, the functional role of these transcripts in latency is still unknown. Data obtained in the HSV-1 system show that detectable levels of latency-associated transcripts are not required for the establishment and maintenance of latency, but may have a role in reactivation.[56]

Reactivation occurs either spontaneously or following particular stimuli such as superinfection with another virus, stress conditions (like transport and parturition), or treatment with corticosteroids.[2,23] A recent model for BHV-1 reactivation suggests downregulation of LR RNA expression to be a necessary but not sufficient event in the process of reactivation.[57] Viral reactivation often leads to the excretion of infectious virus without loss of the virus from the neuron at the primary site of infection. Reactivated virus undergoes a lytic cycle in the neuron and the progeny viruses go back to the primary site of infection. This induced reexcretion plays a major role in the epidemiology of the disease.

Although considerable information exists regarding the immune response to acute BHV-1 infection, relatively little is known about the involvement of immunological factors in the establishment and maintenance of latency. According to one hypothesis, some form of immune surveillance might account for the maintenance of the virus in the latent state.[58] Indeed, reactivation is induced by a variety of stimuli known to alter the immune response; namely, stress conditions or corticosteroid treatment. There is evidence from both *in vitro* and *in vivo* studies that corticosteroids impair both PMN and lymphocyte functions, leading to the concept that these cellular populations might be involved in the maintenance of latency. However, corticosteroids might also have a direct effect on latently infected cells.[59] This latter hypothesis, which does not involve the immune response, seems more likely. Indeed, in BHV-1 latently infected rabbits, very rapid changes are observed in trigeminal ganglia after a single dose of dexamethasone, with changes in viral transcription occurring within 15 to 18 h after treatment.[57] Similarly, in cattle, BHV-1 virions can be detected in nasal washings by electron microscopy 24 h after dexamethasone is given.[60] Evidence can also be obtained from *in vitro* studies. The fact that dexamethasone represses LR promoter activity in infected bovine cells[61] suggests a possible direct role for dexamethasone in BHV-1 reactivation. This observation also explains the transient decrease in the number of neurons expressing LR RNA following dexamethasone treatment. Thus, a nonimmunological mechanism of action for corticosteroids in BHV-1 reactivation appears highly probable.

Nevertheless, even if the immune response does not play a role in the maintenance of latency, it is of importance when reactivation occurs. The immune response must quickly eliminate both the infected cells and free virus to control viral pathogenesis. Indeed, the immune status of the latently infected animal is thought to influence the pattern of viral reexcretion.[23] Theoretically, to achieve optimal control of the reactivated virus, immune recognition of the infected cells should occur before the assembly of new virions. Therefore, various cytotoxic mechanisms including CTL or ADCC are expected to be effective if they are specific for proteins that are synthesized before replication of viral DNA (such as gI or gIV).

III. NONSPECIFIC RESPONSES TO BHV-1 INFECTION

A. INTERFERON

After a primary BHV-1 nasal infection, a rapid local protection is conferred by IFN-α,[62] which is produced by leukocytes present at the site of inflammation and appears in nasal secretions within a few hours of infection. Its activity peaks 3 to 4 days after infection and persists for 8 days. Since BHV-1 is not highly susceptible to IFN *in vitro*, some indirect mechanisms may explain the influence of IFN in gaining local protection early in infection. Indeed, several *in vivo* experiments[63-65] provide evidence that IFN can modulate leukocyte migration and phagocytosis, and increase NK cell activity.[66]

B. MACROPHAGES

Other nonspecific defenses that occur once the inflammatory response is established include: phagocytosis, production of IFN-α, ADCC, and cytolysis of virus-infected cells by macrophages. Macrophages are abundant in most body systems, including the alveoli of the lungs. They are known to express Fc receptors, allowing their participation in ADCC. Bovine alveolar macrophages from normal calves express little Fc receptor activity, but exposure to bovine IFN-α increases the ability of the cells to function as effectors of ADCC.[67] Interestingly, they are themselves protected from BHV-1-infection by IFN-α. Peripheral blood macrophages also lyse BHV-1 infected allogeneic target cells by an antibody-independent mechanism and this seems to be the most prominent cytotoxic function detected in cattle after BHV-1 challenge.[68,69] Since PBMC from BHV-1 nonimmune animals do not lyse BHV-1-infected target cells, the mechanism involved cannot be attributed to natural cytotoxicity. It was therefore suggested that immune T lymphocytes produce IFN-γ and possibly other factors responsible for monocyte activation.[69,70] This hypothesis is further supported by appearance of the macrophage response approximately 1 week after viral challenge, corresponding to the development of the specific proliferative T lymphocyte response. MHC-nonrestricted cytotoxic activity could also be detected in lung parenchyma leukocytes from BHV-1 inoculated cattle.[71] This activity, which appears as soon as 1 day after challenge, might also be due to activated macrophages. Indeed, stimulation of alveolar macrophages occurs after BHV-1 infection.[50] Macrophage activation can be explained either by an increased influx of monocytes into the lung or by stimulation of the resident alveolar macrophages (as a consequence of IFN release by virus-infected epithelial cells, alveolar macrophages themselves, or lymphocytes).

C. POLYMORPHONUCLEAR NEUTROPHILIC GRANULOCYTES

PMN are known to migrate to and accumulate in the bovine lung after respiratory tract infection with BHV-1. Various *in vitro* studies have demonstrated their role in antiviral immunity. It appears that in addition to acting as predominant effector cells in ADCC[72] and phagocytizing opsonized virus, PMN are able, upon exposure to herpesvirus antigens, to release a substance with interferon-like activity, rendering cells resistant to viral infection.[73,74]

D. NATURAL KILLER CELLS

NK cells are also known to mediate resistance to viral infections. In mice and man, they are defined as $CD3^-$, T cell receptor (TCR) $(\alpha, \beta, \gamma, \delta)^-$ large granular lymphocytes mediating cytolytic reactions that do not require expression of MHC class I or class II molecules on the target cells.[75] Although NK cells are a heterogenous population, they share several characteristics which allow their cytotoxic activity to be distinguished from either macrophage or lymphocyte cytotoxicity. Indeed, NK cells are nonadherent effector cells and, for optimal lysis, require a long incubation period with the target cells. The *in vitro* lysis of BHV-1-infected tumor target cells proved to be due to nonadherent, large, low-density, non-T/non-B PBMC which were thought to be NK, since no previous sensitization of the effector cells was needed.[76] Some reports described the need for a 3- to 7-day *in vitro* culture period to detect such activity.[77,78] Nevertheless, the *in vitro* assays used to detect their activity may influence the interpretation as to the role of NK cells in cytotoxicity. Indeed, fresh or cultured cells, cells cultured in the presence or absence of lymphokines, have been used as effector cells to define NK activity.

Regardless of species, minor fractions of $CD3^+$ T lymphocytes expressing αβ or γδ T cell receptor also display NK-like cytotoxicity.[75] However, although MHC-nonrestricted bovine CTL have been observed after *in vitro* culture in the presence of IL-2,[79] it remains to be determined whether bovine γδ T lymphocytes are also able to perform NK-like cytotoxicity. On this subject, a population of MHC-nonrestricted cytotoxic mononuclear cells showing no conventional B or T cell markers has been isolated in cattle and identified as $CD3^+$, $CD45^+$, and Fc $receptor^+$ lymphocytes.[81,82] This population is present in the thymus, in T cell areas of spleen and lymph nodes, and occasionally infiltrates mucosae, where it can reveal a large granular lymphocyte-like morphology. This killer cell population is not recognized by anti-WC1 (a differentiation antigen of ruminant γδ T cells) or anti-γδ TCR monoclonal antibodies. However, cytoplasmic RNA extracted from this cell population reacts positively with a bovine TCR δ chain probe in dot-blot assays, no reaction being detected with a TCR α chain probe. This latter result supports the hypothesis that this population represents a subset of $WC1^-$ γδ T cells.[82]

A major interest in the NK response to BHV-1-infected cells has been to define the target structure that is recognized. NK cells are indeed able to discriminate between virus-infected and normal cells, but the nature of the target antigens that are recognized remains controversial. A first assumption appears to be the need for active viral protein expression in the target cells. This was shown in the HSV system using target cells infected with UV-inactivated virus or by treating the target cells with inhibitors at the time of the infection to prevent viral protein expression.[83,84] Such targets were not recognized by NK cells. Bovine NK cells were also unable to lyse emetine- or actinomycin D-treated BHV-1-infected cells,[85] suggesting that the virion structural antigens left at the surface of the cell during the process of penetration were not sufficient to induce the target formation.

By blocking DNA replication in infected cells with phosphonoacetic acid to prevent synthesis of BHV-1 late proteins (like glycoprotein gIII), it was demonstrated that these late proteins were not necessary for recognition by NK cells.[86] Cells that were infected with BHV-1 and treated with cyclohex-imide and actinomycin D in sequence to allow only expression of immediate early proteins were not lysed by NK cells.[85] Palmer and Splitter confirmed these results by selectively transfecting tumor cells with the genes coding for the major glycoproteins of BHV-1.[87] They observed NK recognition of the gI and gIV expressing cells, but not gIII expressing cells. More interestingly, these results introduce the concept that single herpesvirus proteins can induce susceptibility to lysis by NK cells. If one considers the role of these glycoproteins in the interaction between viral and host cell membranes, an attractive hypothesis to explain the interaction between NK and BHV-1-infected target cells can be formulated. Indeed, gI and gIV are both involved in viral attachment and penetration,[88] a process involving fusion between the cell mem-brane and the virion envelope.[15] Thus, when these proteins are expressed at the cell surface, one may expect that they play a role in cell-cell interactions; indeed, cells transfected with gI have been shown to induce spontaneous cell fusion.[89] In contrast, gIII, which functions as a major viral attachment protein interacting with a heparin-like receptor at the cell surface, is not involved in the process of penetration.[90] Another interesting feature is that the same glycoproteins, gI and gIV, expressed at the cell membrane promote interaction with PMN, leading to the release of the antiviral product with IFN-like activity.[74] Therefore, it can be speculated that the same receptor, interacting with gI and/or gIV, is present on both NK cells and PMN and that a similar mechanism is responsible for the induction of NK and PMN activity.

The presence of viral products at the surface of the target cell has been considered important for recognition by NK cells in a variety of systems.[91,92] However, in these systems, the inability of antiviral antibodies to inhibit lysis by NK cells[83,91] introduces the need for other explanations. Moreover, the viral structural antigens left at the cell membrane during the process of penetration, which most likely contain gI and gIV glycoproteins, do not induce lysis by NK cells. Such observations would be in agreement with the "missing self" theory of NK recognition.[93] BHV-1 infection as well as glycoprotein gene transfection provides the target cells with viral peptides, which may cause self peptide displacement from MHC class I molecules. Displacement of self-peptides has been shown to occur in HSV-infected cells.[94] According to the hypothesis of Chadwick et al., incorporation of non-self-peptides into MHC class I molecules may render the cells sensitive to NK lysis.[95] Nevertheless, further investigations, including cloning of NK effector cells, need to be done to help us understand bovine NK cell recognition of BHV-1-infected cells.

IV. SPECIFIC RESPONSES TO BHV-1 INFECTION

Both cytotoxic and proliferative specific T lymphocyte responses are stimulated after BHV-1 infection and appear in peripheral blood about 8 days after infection. Since their presence is detected at the time of or immediately before the healing of lesions, they are thought to be effective in the process of recovery from both infection or recrudescence of the disease following reactivation.[59] Tissue-infiltrating T lympho-cytes are also considered to play an important role in the clearance of local HSV infections.[96] The mechanism whereby lymphocytes aid in recovery can involve both direct activity, namely CTL-mediated cytotoxicity, and indirect activity through the release of cytokines such as IL-2 or IFN-γ.

Upon secondary infection, antibodies present in the mucus (especially IgA and IgG_1) are particularly effective at neutralizing infective virus. Serum antibodies are involved in virus neutralization, but they do not act until local infection has been established and the first inflammatory lesions have occurred.[97] However, once the virus has infected epithelial cells, it can be disseminated to neighboring cells even in the presence of high titers of specific antibodies. Indeed, viral spread can occur through intercellular bridges, thereby avoiding antibodies present in the extracellular fluid.[23] This kind of spread may be important for viral propagation after reactivation. In this case, the protective effects of specific antibodies may rely on processes such as ADCC.[98]

A. HELPER T LYMPHOCYTES

The CD4+ T lymphocyte proliferative response to BHV-1 is known to be specific for gI, gIII, and gIV. Indeed, plasmids containing the gI, gIII, and gIV genes have been constructed, and recombinant proteins produced by this *in vitro* system were added directly to antigen presenting cells in T lymphocyte proliferation assays. T cell proliferation indicated that these major glycoproteins are recognized by BHV-1 immunized cattle.[99] Further, it has been possible to develop CD4+ lymphocyte cell lines specific, for example, for gIV. Another interesting method developed by Nataraj et al. allowed confirmation of these results.[101] This was achieved by fusing PBMC from a BHV-1 immune calf with the murine thymoma cell line BW 5147. Using this approach, stable T cell hybridomas were obtained. Most hybridomas were specific for gI, gIII, or gIV.

To identify amino acid sequences that comprise T cell epitopes, individual BHV-1 genes were inserted into a vector and digested using restriction enzymes specific for sites within the gene. The mRNA that was produced to the length of the restriction enzyme truncated gene was translated into the respective protein and cultured with lymphocytes. Therefore, depending on where the gene was digested, peptide fragments of varying length were produced. Using these peptides, epitopes within the gIII and gIV proteins recognized by bulk cultured T cells from one animal have been mapped. The gIII protein contains at least two epitopes (between amino acids 214-298 and 341-429)[99] and the gIV protein contains one epitope between 78-157.[101] Using additional restriction enzyme sites within the gIV gene, the gIV epitope comprises a 20-amino acid peptide between 78-98. The possibility now exists to characterize T cell recognition of these important viral proteins and their respective epitopes by a larger number of animals. These data will assess the degree of antigenic conservation within a population of infected animals and provide information related to why animals produce preferential immune responses that characterize individual host protection to BHV-1 disease.

The tegument protein, called VP8, which is the most abundant viral protein in infected cells, also stimulates a good T cell proliferation.[103] Interestingly, this protein only stimulates a weak antibody response. It is important to notice that some variations in the intensity of the response were observed from one animal to another.

Further studies must now identify additional proteins recognized by BHV-1-specific helper T lymphocytes. Not only minor surface glycoproteins of the virions, but also nonstructural antigens expressed in the infected cells might be antigens for the cell-mediated immune response. Furthermore, since the existence in cattle of TH1 and TH2 helper T cells, as shown in mice and human, is expected, further studies on BHV-1-specific T cells will require the identification of the cytokine pattern produced by helper T cell subpopulations. An IFN-γ assay has been recently developed to assess production of this cytokine by BHV-1-stimulated T lymphocytes.[104]

B. CYTOTOXIC T LYMPHOCYTES

The cytolytic activity of bovine mononuclear cells has previously been reviewed by Cook and Splitter.[104] Although various non-MHC-restricted mechanisms mediate cytotoxicity against BHV-1-infected cells, a limited number of studies were successful in demonstrating the presence of MHC and virally restricted bovine CTL in PBMC from BHV-1 immune cattle. Different methods have been used. These include *in vitro* cultures of PBMC in the presence of UV-inactivated virus and IL-2, limiting dilution in the presence of IL-2 with or without antigen, and long-term cloning.[79,105-107] These PBMC have been tested for cytolytic activity against both fibroblasts and lymphoblasts as targets. Unlike HSV-2,[108] BHV-1 infection of fibroblasts does not decrease MHC class I expression on the surface of the cells at the time of the assay, theoretically allowing the infected fibroblasts to be recognized by CD8+ CTL. Synthesis of MHC class I molecules in fibroblasts is dramatically reduced in the first hours after HSV infection, probably due to the effect of the HSV U_L41 gene product, which is thought to be responsible for the host shut-off function. A similar gene has not been identified in BHV-1, and cessation of host RNA synthesis and degradation of host RNA in BHV-1-infected cells do not seem to occur before the onset of viral DNA and RNA synthesis.[13]

Evidence of CTL activity was given by the absence of cytotoxicity against MHC-mismatched target cells and against target cells infected with a different virus. Effector cells were also shown to lose their cytotoxic activity after negative selection of CD8+ cells using monoclonal antibodies and complement.[107]

Investigations were conducted to define the target proteins recognized by BHV-1-specific CTL. Only glycoproteins gIII and gIV have been shown to be targets for bovine BHV-1-specific CD8+ CTL.[107] The method used to demonstrate this activity involved vaccinia recombinant viruses expressing the major glycoproteins of BHV-1. Using this method, gI did not appear as an important CTL target, while both gIII- and gIV-expressing target cells were recognized by CTL. Another observation that was made related to the differential *in vitro* secondary stimulation of CTL depending on the method of antigen presentation.[107] Indeed, CTL from immune animals which had been stimulated by UV-inactivated BHV-1 lysed predominantly gIII-expressing target cells, while BHV-1-infected autologous fibroblasts mainly stimulated gIV-specific CTL. This observation could be explained by a different processing pathway of BHV-1 antigens. Infected autologous fibroblasts present BHV-1 epitopes in association with MHC class I molecules, but UV-inactivated BHV-1 is processed and presented by cell types in the culture (most likely macrophages). Differential presentation by different cell types has been described in other systems. However, in the study described above, cloning of gIII- and gIV-specific CTL appears necessary to further define the differences in BHV-1 processing and presentation by fibroblasts and macrophages.

In mice, glycoproteins gI and gIII were shown to be CTL targets.[109] Nevertheless such results obtained in a different model system are not easily extended to the natural host. Indeed, from the HSV literature, it appeared that the MHC phenotype had a major influence on the protein identified as a dominant CTL target. Furthermore, the system used to express the viral protein may also play a role; therefore, identification of target antigens for the CTL response in BHV-1 immune cattle should not rely on a single method. Transfected cells expressing BHV-1 proteins, and recombinant and deleted viruses may provide evidence of the involvement of particular BHV-1 protein(s) as a CTL target.

V. CONCLUSION

The events which occur following a primary BHV-1 virus infection are summarized in Figure 10-3. Infection is initiated by the interaction between the virus and host cell receptors, followed by entry and activation of the viral α, β, and γ genes in sequence. This temporal control of gene expression provides for the effective programming and coordination of viral and host cell activities to ensure efficient production of infectious virus. Part of the host cell programming also includes the production of IFN by the infected cell. This IFN rapidly activates adjacent uninfected cells to become refractory to virus replication. One of the cells activated by IFN is the tissue macrophage which produces early cytokines (TNF-α, IFN, IL-5, IL-8, IL-1, and CSFs) responsible for initiating the inflammatory process and immune responses to the virus. As mentionned above, one of these early cytokines, IFN itself will act on adjacent uninfected cells to partially limit viral replication. Another initial target of the early cytokines are T helper cells which are stimulated to produce late cytokines (IL-2, IL-3, IL-4, IL-5, IFN-γ, and TNF-β), which activate and fine-tune the immune response cascade. Cytokine redundancy and balance ensures the appropriate and sequential activation of the cells involved in the immune response. Disturbance of this balance can result in the inability to clear the virus or immunopathology. These late cytokines can also specifically activate macrophages which are pivotal in killing BHV-1-infected cells as well as presenting antigen to ensure the proper development of an immune response. Late cytokines, produced by T helper cells, induce B cells to produce antibody which can neutralize the virus and inhibit extracellular spread of the virus. Second, antibody reduces the extent of intracellular spread by binding to viral antigens expressed on the virus-infected cells, leading to lysis of infected cells by activation of complement (complement-mediated cell lysis) or interaction with killer cells (ADCC). In BHV-1, the most effective killer cell has been shown to be the PMN, both in lysis of virus-infected cells and in being the first cell recruited to the site of virus infection, suggesting that the PMN is very important in limiting BHV-1 infection and spread *in vivo*. Late cytokines also activate NK cells and CTL to kill virus-infected cells. Whether any of these cytotoxic mechanisms will be important in limiting intracellular spread is totally

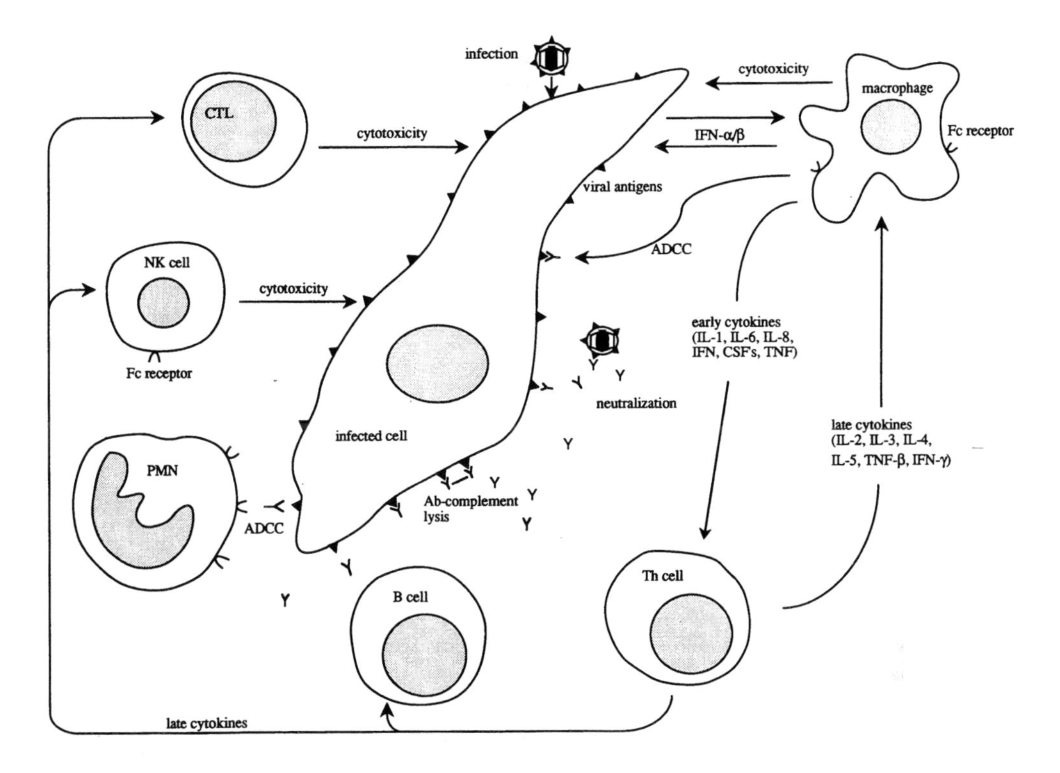

Figure 10-3 Defense mechanisms in BHV-1 infection.

dependent on how quickly the infected cells are recognized and killed following infection. Studies have demonstrated that the replication cycle of BHV-1 virus is complete by approximately 10 to 12 h post-infection with virus antigen expression detected as early as 4 to 6 h post-infection. Therefore, the defense mechanisms have an interval of approximately 4 to 5 h to recognize virus infected cells and kill them before peak viral replication is completed. Any interruption in the activation of specific cells to kill the infected cell will obviously favor the virus.

In this chapter, it is evident that there is a complex series of interactions between BHV-1, the various host cells that it infects, and the host immune system. The speed and effectiveness of each defense mechanism, acting in concert, will determine whether disease occurs or not. If the virus produces or induces factors that suppress either the magnitude or kinetics of any of these responses, it will favor the virus. Therefore, the understanding of the role of these factors in modulating the immune response will be crucial in developing strategies to either prevent primary infection or reduce the extent of virus replication associated with recrudescence.

REFERENCES

1. **Straub, O. C.,** Infectious bovine rhinotracheitis virus, in *Virus Infections of Ruminants*, Dinter, Z. and Morein, B., Eds., Elsevier Science, Amsterdam, 1990, chap. 11.
2. **Wyler, R., Engels, M., and Schwyzer, M.,** Infectious bovine rhinotracheitis/vulvovaginitis (BHV-1), in *Herpesvirus Diseases of Cattle, Horses and Pigs,* Wittmann, G., Ed., Kluwer Academic, Boston, 1989, chap. 1.
3. **Yates, W. D. G.,** A review of infectious bovine rhinotracheitis, shipping fever pneumonia and viral-bacterial synergism in respiratory disease of cattle, *Can. J. Comp. Med.*, 46, 225, 1982.
4. **Gibbs, E. P. J. and Rweyemamu, M. M.,** Bovine herpesviruses. Part I. Bovine herpesvirus 1, *Vet. Bull.*, 47, 317, 1977.
5. **Roizman, B., Desrosiers, R. C., Fleckenstein, B., Lopez, C., Minson, A. C., and Studdert, M. J.,** The family herpesviridae: an update, *Arch. Virol.*, 123, 425, 1992.

6. **Roizman, B., Carmichael, L. E., Deinhardt, F., de-The, G., Nahmias, A. J., Plowright, W., Rapp, F., Sheldrick, P., Takahashi, M., and Wolf, K.**, Herpesviridae. Definition, provisional nomenclature and taxonomy, *Intervirology*, 16, 201, 1981.
7. **Misra, V., Blumenthal, R. M., and Babiuk, L. A.**, Proteins specified by bovine herpesvirus 1 (Infectious Bovine Rhinotracheitis Virus), *J. Virol.*, 40, 367, 1981.
8. **Ludwig, G. V. and Letchworth, G. J.**, Temporal control of bovine herpesvirus 1 glycoprotein synthesis, *J. Virol.*, 61, 3292, 1987.
9. **Wirth, U. V., Vogt, B., and Schwyzer, M.**, The three major immediate-early transcripts of bovine herpesvirus 1 arise from two divergent and spliced transcription units, *J. Virol.*, 65, 195, 1991.
10. **Misra, V., Nelson, R., and Smith, M.**, Sequence of a bovine herpesvirus type-1 glycoprotein gene that is homologous to the herpes simplex gene for the glycoprotein gB, *Virology*, 166, 542, 1988.
11. **Tikoo, S. K., Fitzpatrick, D. R., Babiuk, L. A., and Zamb, T. J.**, Molecular cloning, sequencing, and expression of functional bovine herpesvirus 1 glycoprotein gIV in transfected bovine cells, *J. Virol.*, 64, 5132, 1990.
12. **Baranowski, E.**, personal communication, 1992.
13. **Seal, B. S., Whetstone, C. A., Zamb, T. J., Bello, L. J., and Lawrence, W. C.**, Relationship of bovine herpesvirus 1 immediate-early, early, and late gene expression to host cellular gene transcription, *Virology*, 188, 152, 1992.
14. **Fitzpatrick, D. R., Babiuk, L. A., and Zamb, T. J.**, Nucleotide sequence of bovine herpesvirus type 1 glycoprotein gIII, a structural model for gIII as a new member of the immunoglobulin superfamily, and implications for the homologous glycoproteins of other herpesviruses, *Virology*, 173, 46, 1989.
15. **Roizman, B. and Sears, A. E.**, Herpes simplex viruses and their replication, in *Virology*, Fields, Ed., Raven Press, New York, 1990, chap. 65.
16. **Fuller, A. O. and Lee, W.-C.**, Herpes simplex virus type 1 entry through a cascade of virus-cell interactions requires different roles of gD and gH in penetration, *J. Virol.*, 66, 5002, 1992.
17. **Gillespie, J. H., McEntee, K., Kendrick, J. W., and Wagner, W. C.**, Comparison of infectious pustular vulvovaginitis virus with infectious bovine rhinotracheitis virus, *Cornell Vet.*, 49, 288, 1959.
18. **Engels, M., Steck, F., and Wyler, R.**, Comparison of the genomes of infectious bovine rhinotracheitis and infectious pustular vulvovaginitis virus strains by restriction endonuclease mapping, *Arch. Virol.*, 67, 169, 1981.
19. **Metzler, A. E., Matile, H., Gassmann, U., Engels, M., and Wyler, R.**, European isolates of bovine herpesvirus 1: a comparison of restriction endonuclease sites, polypeptides, and reactivity with monoclonal antibodies, *Arch. Virol.*, 85, 57, 1985.
20. **Pastoret, P.-P., Burtonboy, G., Aguilar-Setien, A., Godart, M., Lamy, M. E., and Schoenaers, F.**, Comparison between strains of infectious bovine rhinotracheitis virus (bovid herpesvirus 1), from respiratory and genital origins, using polyacrilamide gel electrophoresis of structural proteins, *Vet. Microbiol.*, 5, 187, 1980.
21. **Edwards, S., White, H., and Nixon, P.**, A study of the predominant genotypes of bovid herpesvirus 1 found in the U.K., *Vet. Microbiol.*, 22, 213, 1990.
22. **Edwards, S., Newman, R. H., and White, H.**, The virulence of British isolates of bovid herpesvirus 1 in relationship to viral genotype, *Br. Vet. J.*, 147, 216, 1991.
23. **Pastoret, P.-P., Thiry, E., Brochier, B., and Derboven, G.**, Bovid herpesvirus 1 infection of cattle: pathogenesis, latency, consequences of latency, *Ann. Rech. Vet.*, 13, 221, 1982.
24. **Pastoret, P.-P., Babiuk, L. A., Misra, V., and Griebel, P.**, Reactivation of temperature-sensitive and non-temperature-sensitive infectious bovine rhinotracheitis vaccine virus with dexamethasone, *Infect. Immun.*, 29, 483, 1980.
25. **Lupton, H. W. and Reed D. E.**, Evaluation of experimental subunit vaccines for infectious bovine rhinotracheitis, *Am. J. Vet. Res.*, 41, 383, 1980.
26. **Babiuk, L. A., L'Italien, J., van Drunen Littel-van Den Hurk, S., Zamb T., Lawman, M. J. P., Hughes, G., and Gifford, G. A.**, Protection of cattle from bovine herpesvirus type 1 (BHV-1) infection by immunization with individual viral glycoproteins, *Virology*, 159, 57, 1987.
27. **Israel, B. A., Marshall, R. L., and Letchworth, G. J.**, Epitope specificity and protective efficacy of the bovine immune response to bovine herpesvirus-1 glycoprotein vaccines, *Vaccine*, 6, 349, 1988.
28. **Israel, B. A., Herber, R., Gao, Y., and Letchworth, G. J.**, Induction of a mucosal barrier to bovine herpesvirus 1 replication in cattle, *Virology*, 188, 256, 1992.
29. **Gao, Y. and Splitter, G.**, unpublished data, 1992.

30. **Liang, X., Babiuk, L. A., and Zamb, T. J.,** An in vivo study of a glycoprotein gIII-negative bovine herpesvirus 1 (BHV-1) mutant expressing β-galactosidase: evaluation of the role of gIII in virus infectivity and its use as a vector for mucosal immunization, *Virology*, 189, 629, 1992.
31. **Van Engelenburg, F. A. C., Kaashoek, M. J., Rijsewijk, F. A. M., Van Den Burg, L., Moerman, A., Gielkens, A. L. J., and Van Oirschot, J. T.,** Deletion of the glycoprotein E gene or thymidine kinase gene of bovine herpesvirus type 1 produces attenuated vaccine strains, Proc. 17th Int. Herpesvirus Workshop, Edinburgh, 1992.
32. **Kit, S., Qavi, H., Gaines, J. D., Billingsley, P. and McConnell, S.,** Thymidine kinase-negative bovine herpesvirus type 1 mutant is stable and highly attenuated in calves, *Arch. Virol.*, 86, 63, 1985.
33. **Bello, L. J., Whitbeck, J. C., and Lawrence, W. C.,** Bovine herpesvirus 1 as a live vector for expression of foreign genes, *Virology*, 190, 666, 1992.
34. **Reddy, D. N., Reddy, P. G., Minocha, H. C., Fenwick, B. W., Baker, P. E., Davis, W. C., and Blecha, F.,** Adjuvanticity of recombinant bovine interleukin-1β: influence on immunity, infection, and latency in a bovine herpesvirus-1 infection, *Lymphokine Res.*, 9, 295, 1990.
35. **Hughes, H. P. A., Campos, M., Godson, D. L., van Drunen Littel-van Den Hurk, S., McDougall, L., Rapin, N., Zamb, T., and Babiuk, L. A.,** Immunopotentiation of bovine herpes virus subunit vaccination by interleukin-2, *Immunology*, 74, 461, 1991.
36. **McGuire, R. L. and Babiuk, L. A.,** Evidence for defective neutrophil function in lungs of calves exposed to infectious bovine rhinotracheitis virus, *Vet. Immunol. Immunopathol.*, 5, 259, 1984.
37. **Bielefeldt Ohmann, H., Babiuk, L. A., and Harland, R.,** Cytokine synergy with viral cytopathic effects and bacterial products during the pathogenesis of respiratory tract infection, *Clin. Immunol. Immunopathol.*, 60, 153, 1991.
38. **Carter, J. J., Weinberg, A. D., Pollard, A., Reeves, R., Magnuson, J. A. , and Magnuson, N. S.,** Inhibition of T-lymphocyte mitogenic responses and effects on cell functions by bovine herpesvirus 1, *J. Virol.*, 63, 1525, 1989.
39. **Hutchings, D. L., Campos, M., Qualtiere, L., and Babiuk, L. A.,** Inhibition of antigen-induced and interleukin-2-induced proliferation of bovine peripheral blood leukocytes by inactivated bovine herpesvirus 1, *J. Virol.*, 64, 4146, 1990.
40. **Griebel, P. J., Bielefeldt Ohmann, H., Lawman, M. J. P., and Babiuk, L. A.,** The interaction between bovine herpesvirus type 1 and activated bovine T lymphocytes, *J. Gen Virol.*, 71, 369, 1990.
41. **Woldehiwet, Z. and Rowan, T. G.,** In vitro replication of bovid herpesvirus-1 in macrophages derived from peripheral blood leucocytes of calves, *J. Comp. Pathol.*, 103, 183, 1990.
42. **Brown,T. T. and Shin, K.,** Effect of bovine herpesvirus-1 or parainfluenza-3 virus on immune receptor-mediated functions of bovine alveolar macrophages in the presence or absence of virus-specific serum or pulmonary lavage fluids collected after virus infection, *Am. J. Vet. Res.*, 51, 1616, 1990.
43. **Forman, A. J. and Babiuk, L. A.,** Effect of infectious bovine rhinotracheitis virus infection on bovine alveolar macrophage function, *Infect. Immun.*, 35, 1041, 1982.
44. **Bienhoff, S.E., Allen, G. K., and Berg, J. N.,** Release of tumor necrosis factor-alpha from bovine alveolar macrophages stimulated with bovine respiratory viruses and bacterial endotoxins, *Vet. Immunol. Immunopathol.*, 30, 341, 1992.
45. **Bielefeldt Ohmann, H. and Babiuk, L. A.,** Viral-bacterial pneumonia in calves: effect of bovine herpesvirus-1 on immunologic functions, *J. Infect. Dis.*, 151, 937, 1985.
46. **Griebel, P. J., Qualtiere, L., Davis, W. C., Gee, A., Bielefeldt Ohmann, H., Lawman, M. J. P., and Babiuk, L. A.,** Peripheral blood T lymphocyte population dynamics and function following a primary bovine herpesvirus-1 infection, *Viral Immunol.*, 1, 287, 1988.
47. **Bielefeldt Ohmann, H., Campos, M., Griebel, P., Snider, M., Beskorwayne, T., Rapin, N., and Babiuk, L. A.,** Leukocyte trafficking and localization in bovine respiratory tract tissue during primary BHV-1 infection: implications for regulation by interferons, *EOS J. Immunol. Immunopharmacol.*, 9, 137, 1989.
48. **Green, G. M. and Kass, E. H.,** The role of the alveolar macrophage in the clearance of bacteria from the lung, *J. Exp. Med.*, 119, 167, 1964.
49. **Forman, A. J., Babiuk, L. A., Baldwin, F., and Friend, S. C. E.,** Effect of infectious bovine rhinotracheitis virus infection of calves on cell populations recovered by lung lavage, *Am. J. Vet. Res.*, 43, 1174, 1982.
50. **Bielefeldt Ohmann, H. and Babiuk, L. A.,** Alteration of alveolar macrophage functions after aerosol infection with bovine herpesvirus type 1, *Infect. Immun.*, 51, 344, 1986.

51. **Ackermann, M., Peterhans, E., and Wyler, R.**, DNA of bovine herpesvirus type 1 in the trigeminal ganglia of latently infected calves, *Am. J. Vet. Res.*, 43, 36, 1982.
52. **Ackermann, M. and Wyler, R.**, The DNA of an IPV strain of bovid herpesvirus 1 in sacral ganglia during latency after intravaginal infection, *Vet. Microiol.*, 9, 53, 1984.
53. **Rock, D. L., Hagemoser, W. A., Osorio, F. A., and Reeds, D. E.**, Detection of bovine herpesvirus type 1 RNA in trigeminal ganglia of latently infected rabbits by in situ hybridisation, *J. Gen. Virol.*, 67, 2515, 1986.
54. **Rock, D. L., Beam, S. L., and Mayfield, J. E.**, Mapping bovine herpesvirus type 1 latency-related RNA in trigeminal ganglia of latently infected rabbits, *J. Virol.*, 61, 3827, 1987.
55. **Wirth, U. V., Fraefel, C., Vogt, B., Vlcek, C., Paces, V., and Schwyzer, M.**, Immediate-early RNA 2.9 and early RNA 2.6 of bovine herpesvirus 1 are 3′ coterminal and encode a putative zinc finger transactivator protein, *J. Virol.*, 66, 2763, 1992.
56. **Steiner, I., Spivack, J. G., Lirette, R. P., Brown, S. M., MacLean, A. R., Subak-Sharpe, J. H., and Fraser, N. W.**, Herpes simplex virus type 1 latency-associated transcripts are evidently not essential for latent infection, *EMBO J.*, 8, 505, 1989.
57. **Rock, D., Lokensgard, J., Lewis, T., and Kutish, G.**, Characterization of dexamethasone-induced reactivation of latent bovine herpesvirus 1, *J. Virol.*, 66, 2484, 1992.
58. **Babiuk, L. A. and Rouse, B. T.**, Immune control of latency, *Can. J. Microbiol.*, 25, 267, 1979.
59. **Davies, D. H. and Carmichael, L. E.**, Role of cell-mediated immunity in the recovery of cattle from primary and recurrent infections with infectious bovine rhinotracheitis virus, *Infect. Immun.*, 8, 510, 1973.
60. **Pastoret, P.-P., Aguilar-Setien, A., Burtonboy, G., Mager, J., Jetteur, P., and Schoenaers, F.**, Effect of repeated treatment with dexamethasone on the re-excretion pattern of infectious bovine rhinotracheitis virus and humoral immune response, *Vet. Microbiol.*, 4, 149, 1979.
61. **Jones, C., Delhon, G., Bratanich, A., Kutish, G., and Rock, D.**, Analysis of the transcriptional promoter which regulates the latency-related transcript of bovine herpesvirus 1, *J. Virol.*, 64, 1164, 1990.
62. **Straub, O. C., and Ahl, R.**, Lokale Interferonbildung Beim Rind Nach Intranasaler Infektion Mit Avirulentem IBR-IPV — Virus und Deren Wirkung auf Eine Anschließende Infektion Mit Maul und Klauenseuche — Virus, *Zbl. Vet. Med. B.*, 23, 470, 1976.
63. **Babiuk, L. A., Bielefeldt Ohmann, H., Gifford, G., Czarniecki, Ch. W., Scialli, V. T., and Hamilton, E. B.**, Effect of bovine α 1 interferon on bovine herpesvirus type 1-induced respiratory disease, *J. Gen. Virol.*, 66, 2383, 1985.
64. **Babiuk, L. A., Lawman, M. J. P., and Gifford, G. A.**, Recombinant bovine $alpha_1$ interferon: use in reducing bovine herpesvirus-1 induced respiratory disease, *Antimicrob. Agents Chemother.*, 31, 752, 1987.
65. **Lawman, M. J. P., Gifford, G., Gyongyossy-Issa, M., Dragan, R., Heise, J., and Babiuk, L. A.**, Immunomodulation of polymorphonuclear (PMN) leukocyte functions by recombinant bovine interferon $alpha_1$ during bovine herpes virus-1 induced respiratory disease, *Antiviral Res.*, 8, 225, 1987.
66. **Jensen, J. and Schultz, R. D.**, Bovine natural cell mediated cytotoxicity (NCMC): activation by cytokines, *Vet. Immunol. Immunopathol.*, 24, 113, 1990.
67. **Bielefeldt Ohmann, H., Gilchrist, J. E., and Babiuk, L. A.**, Effect of recombinant $alpha_1$ interferon (BoIFN-α1) on the interaction between bovine alveolar macrophages and bovine herpesvirus-1, *J. Gen. Virol.*, 65, 1487, 1984.
68. **Rouse, B. T. and Babiuk, L. A.**, The direct antiviral cytotoxicity by bovine lymphocytes is not restricted by genetic incompatibility of lymphocytes and target cells, *J. Immunol.*, 118, 618, 1977.
69. **Campos, M. and Rossi, C. R.**, Cell-mediated cytotoxicity of bovine mononuclear cells to IBRV-infected cells: dependence on sephadex G-10 adherent cells, *Vet. Immunol. Immunopathol.*, 8, 363, 1985.
70. **Campos, M., Bielefeldt Ohmann, H., Hutchings, D., Rapin, N., Babiuk, L. A., and Lawman, M. J. P.**, Role of interferon-γ in inducing cytotoxicity of peripheral blood mononuclear leukocytes to bovine herpesvirus type 1 (BHV-1)-infected cells, *Cell. Immunol.*, 120, 259, 1989.
71. **Campos, M., Griebel, P., Bielefeldt Ohmann, H., and Babiuk, L. A.**, Cell-mediated cytotoxic responses in lungs following a primary bovine herpes virus type 1 infection, *Immunology*, 75, 47, 1992.
72. **Grewal, A. S., Rouse, B. T., and Babiuk, L. A.**, Mechanisms of resistance to herpesvirus: comparison of the effectiveness of different cell types in mediating antibody-dependent cell mediated cytotoxicity, *Infect. Immun.*, 15, 698, 1977.

73. **Rouse, B. T., Babiuk, L. A., and Henson, P. M.,** Neutrophils in antiviral immunity: inhibition of virus replication by a mediator produced by bovine neutrophils, *J. Infect. Dis.*, 141, 223, 1980.
74. **Bielefeldt Ohmann, H., Campos, M., Fitzpatrick, D. R., Rapin, N., and Babiuk, L. A.,** A neutrophil-derived antiviral protein: induction requirements and biologicals properties, *J. Virol.*, 63, 1916, 1989.
75. **Hercend, T. and Schmidt, R. E.,** Characteristics and uses of natural killer cells, *Immunol. Today*, 9, 291, 1988.
76. **Cook, C. G. and Splitter, G. A.,** Characterization of bovine mononuclear cell populations with natural cytotoxic activity against bovine herpesvirus-1 infected cells, *Cell. Immunol.*, 120, 240, 1989.
77. **Brigham, S. H. and Rossi, C. R.,** Cell-mediated cytotoxicity of peripheral blood mononuclear cells stimulated in vitro for infectious bovine rhinotracheitis virus-infected cells, *Vet. Immunol. Immunopathol.*, 13, 203, 1986.
78. **Campos, M. and Rossi, C. R.,** Cytotoxicity of bovine lymphocytes after treatment with lymphokines, *Am. J. Vet. Res.*, 47, 1524, 1986.
79. **Lawman, M. J. P., Griebel, P., Hutchings, D. L., Davis, W. C., Heise, J., Qualtiere, L., and Babiuk, L. A.,** Generation of IL-2 dependent bovine cytotoxic T lymphocyte clones reactive againt BHV-1 infected target cells: loss of genetic restriction and virus specificity, *Viral Immunol.*, 1, 163, 1987/1988.
80. **Amadori, M., Archetti, I. L., Verardi, R., and Berneri, C.,** Isolation of mononuclear cytotoxic cells from cattle vaccinated against foot-and-mouth disease, *Arch. Virol.*, 122, 293, 1992.
81. **Amadori, M., Archetti, I. L., Verardi, R., and Berneri, C.,** Target recognition by bovine mononuclear, MHC-unrestricted cytotoxic cells, *Vet. Microbiol.*, 33, 383, 1992.
82. **Amadori, M. and Archetti, I. L.,** unpublished data, 1992.
83. **Fitzgerald-Bocarsly, P., Howell, D. M., Pettera, L., Tehrani, S., and Lopez, C.,** Immediate-early gene expression is sufficient for induction of natural killer cell-mediated lysis of herpes simplex virus type 1-infected fibroblasts, *J. Virol.*, 65, 3151, 1991.
84. **Lopez-Guerrero, J. A., Alarcon, B., and Fresno, M.,** Mechanism of recognition of herpes simplex virus type 1-infected cells by natural killer cells, *J. Gen. Virol.*, 69, 2859, 1988.
85. **Denis, M,** unpublished data, 1992.
86. **Cook, C. G., Letchworth, G. J., and Splitter, G. A.,** Bovine naturally cytolytic cell activation against bovine herpes virus type 1-infected cells does not require late viral glycoproteins, *Immunology*, 66, 565, 1989.
87. **Palmer, L. D., Leary, T. P., Wilson, D. M., and Splitter, G. A.,** Bovine natural killer-like cell responses against cell lines expressing recombinant bovine herpesvirus type 1 glycoproteins, *J. Immunol.*, 145, 1009, 1990.
88. **Dubuisson, J., Israel, B. A., and Letchworth, G. J., III,** Mechanisms of bovine herpesvirus type 1 neutralization by monoclonal antibodies to glycoproteins gI, gIII and gIV, *J. Gen. Virol.*, 73, 2031, 1992.
89. **Fitzpatrick, D. R., Zamb, T., and Babiuk, L. A.,** Expression of bovine herpesvirus type 1 glycoprotein gI in transfected bovine cells induces spontaneous cell fusion, *J. Gen. Virol.*, 71, 1215, 1990.
90. **Okazaki, K., Matsuzaki, T., Sugahara, Y., Okada, J., Hasebe, M., Iwamura, Y., Ohnishi, M., Kanno, T., Shimizu, M., Honda, H., and Kono, Y.,** BHV-1 adsorption is mediated by the interaction of glycoprotein gIII with heparin-like moiety on the cell surface, *Virology*, 181, 666, 1991.
91. **Moller, J. R., Rager-Zisman, B., Quan, P.-C., Schattner, A., Panush, D., Rose, J. K., and Bloom, B. R.,** Natural killer cell recognition of target cells expressing different antigens of vesicular stomatitis virus, *Proc. Natl. Acad. Sci. U.S.A.*, 82, 2456, 1985.
92. **Bishop, G. A., Kümel, G., Schwartz, S. A., and Glorioso, J. C.,** Specificity of human natural killer cells in limiting dilution culture for determinants of herpes simplex virus type 1 glycoproteins, *J. Virol.*, 57, 294, 1986.
93. **Ljunggren, H.-G., Kärre, K.,** In search of the 'missing self': MHC molecules and NK cell recognition, *Immunol. Today*, 11, 237, 1990.
94. **Kuzushima, K., Isobe, K.-I., Morishima, T., Takatsuki, A., and Nakashima, I.,** Inhibitory effect of herpes simplex virus infection to target cells on recognition of minor histocompatibility antigens by cytotoxic T lymphocytes, *J. Immunol.*, 144, 4536, 1990.
95. **Chadwick, B. S., Sambhara, S. R., Sasakura, Y., and Miller, R. G.,** Effect of class I MHC binding peptides on recognition by natural killer cells, *J. Immunol.*, 149, 3150, 1992.

96. **Nash, A. A., Jayasuriya, A., Phelan, J., Cobbold, S. P., Waldmann, H., and Prospero, T.,** Different roles for L3T4+ and Lyt2+ T cell subsets in the control of an acute herpes simplex virus infection of the skin and nervous system, *J. Gen. Virol.*, 68, 825, 1987.
97. **Asso, J. and Le Jan, C.,** Viral infections of the respiratory tract of calves: local immunity, *Vet. Sci. Commun.*, 1, 297, 1978.
98. **Rouse, B. T., Wardley, R. C., and Babiuk, L. A.**, The role of antibody dependent cytotoxicity in recovery from herpesvirus infections, *Cell. Immunol.*, 22, 182, 1976.
99. **Leary, T. P. and Splitter, G. A.,** Recombinant herpesviral proteins produced by cell-free translation provide a novel approach for the mapping of T lymphocyte epitopes, *J. Immunol.*, 145, 718, 1990.
100. **Nataraj, C., Zamb, T., and Srikumaran, S.**, Bovine-murine T-cell hybridomas specific for bovine herpesvirus 1, Proc. 17th Int. Herpesvirus Workshop, Edinburgh, 1992.
101. **Leary, T. P. and Splitter, G. A.,** A method for the rapid identification of T lymphocyte epitopes, *Peptide Res.*, 3, 259, 1990.
102. **Hutchings, D. L., van Drunen Little-van Den Hurk, S., and Babiuk, L. A.,** Lymphocyte proliferative responses to separated bovine herpesvirus 1 proteins in immune cattle, *J. Virol.*, 64, 5114, 1990.
103. **Babiuk, L. A.,** unpublished data, 1992.
104. **Cook, C. G. and Splitter, G. A.,** Comparison of bovine mononuclear cells with other species for cytolytic activity against virally infected cells, *Vet. Immunol. Immunopathol.*, 20, 239, 1989.
105. **Campos, M. and Rossi, C. R.,** *In vitro* induction of cytotoxic lymphocytes from infectious bovine rhinotracheitis virus hyperimmune cattle, *Am. J. Vet. Res.*, 47, 2411, 1986.
106. **Splitter, G. A., Eskra, L., and Abruzzini, A. F.,** Cloned bovine cytolytic T cells recognize bovine herpes virus-1 in a genetically restricted, antigen-specific manner, *Immunology*, 63, 145, 1988.
107. **Denis, M., Slaoui, M., Keil, G., Babiuk, L. A., Ernst, E., Pastoret, P.-P., and Thiry, E.,** Identification of different target glycoproteins for bovine herpesvirus-1-specific cytotoxic T lymphocytes depending on the method of in vitro stimulation, *Immunology*, 78, 7, 1993.
108. **Jennings, S. R., Rice, P. L., Kloszewski, E. D., Anderson, R. W., Thompson, D. L., and Tevethia, S. S.,** Effect of herpes simplex virus type 1 and 2 on surface expression of class I major histocompatibility complex antigens on infected cells, *J. Virol.*, 56, 757, 1985.
109. **Fitzpatrick, D. R., Zamb, T., Parker, M. D., van Drunen Littel-van Den Hurk, S., Babiuk, L. A., and Lawman, M. J. P.,** Expression of bovine herpesvirus 1 glycoproteins gI and gIII in transfected murine cells, *J. Virol.*, 62, 4239, 1988.

Chapter 11

Foot-And-Mouth Disease (Aphthovirus): Viral T Cell Epitopes

T. Collen

CONTENTS

I. Introduction 173
II. Foot-and-Mouth Disease 174
A. Humoral Immunity and Protection 175
B. Cell-Mediated Immunity (CMI) 177
III. T Cell Epitope Prediction from Primary Protein Sequences 178
A. AMPHI 179
B. SOHHA 179
C. Rothbard and Taylor Motifs 180
D. The "Rarity Rule" 180
E. Allele-Specific Major Histocompatibility Complex (MHC) Binding Motifs 180
1. MHC Class I Motifs 180
2. MHC Class II Motifs 181
IV. T Cell Epitopes in Foot-and-Mouth Disease Virus 182
V. Implications for Foot-and-Mouth Disease Subunit Vaccine Development 186
VI. Summary 190
References 190

I. INTRODUCTION

Foot-and-mouth disease (FMD) is an acute and highly contagious vesicular disease which causes economically important epidemics in cattle, sheep, and pigs. The disease is present in most of Africa, the Middle East, Asia, and South America. Great Britain, New Zealand, Australia, North and Central America, Japan, Iceland, and Scandinavia are disease-free. Control measures vary from country to country but generally consist of some form of "stamping-out": that is, slaughter of infected and contact animals, disinfection of affected premises, implementation of livestock movement restrictions, and vaccination of all other susceptible animals within a certain distance of the outbreak focus. Routine vaccination is used in regions where the disease occurs with a low frequency, although routine vaccination can also restrict disease to young animals in areas where FMD is endemic if zoosanitary support is provided. Attempts to eradicate FMD have been generally unsuccessful for a variety of reasons, but mainly due to the disease status in the targeted region, the co-existence of wild and domestic hosts, economic, political, or logistic difficulties in implementing control measures, and the short duration of protection afforded by virus vaccines. However, eradication has been achieved in some geographically isolated and/or developed countries such as Chile and the member states of the European Community[1] where effective surveillance, diagnosis, and regulatory controls can maintain the disease-free status.[1,2]

The importance of humoral immunity in FMD is well documented,[3-6] and in the early 1980s it was widely expected that the next generation of vaccines would be based on a recombinant or a chemically synthesized analogue of a linear FMD virus (FMDV) determinant which had been shown to elicit neutralizing antibody in small animal models.[7-9] The failure to develop such a vaccine highlights the fact that we know more about the antigenic structure of FMDV than about the induction, regulation and, indeed, the precise mechanism(s) of immunity in the natural host. Moreover, there are two observations which question whether antibody alone is sufficient for protection. Firstly, during potency testing of virus vaccines, a variable number of vaccinated animals with low or no detectable neutralizing antibody will resist challenge with live virus, while others which have acceptable titers of neutralizing antibody will be susceptible. This lack of protection in the face of neutralizing antibody production is especially striking in peptide-vaccinated cattle. Secondly, inapparent but persistent infection is a common sequel to both

0-8493-4952-4/94/$0.00+$.50

clinical and sub-clinical disease. Paradoxically, these "carrier" animals have good antiviral immunity as assessed by antibody, and the establishment of persistent infection is equally likely to occur in vaccinated or nonvaccinated animals. Together, these observations suggest that the induction of virus-specific antibody is not, in itself, sufficient for protection and point toward a role for cell-mediated immunity and T cells in the successful elimination of virus.

The ability of T cells to distinguish "self" from "non-self" in the form of peptides derived from "self" and "non-self" proteins bound to major histocompatibility complex (MHC) proteins is central to the recognition and elimination of virally infected cells by cytolytic T cells (CTL). In addition, T cells provide "help" for effector cells such as CTL and B cells, and the importance of helper T cells in regulating humoral responses is now well established (for review, see Reference 10). T cell responses to small peptide antigens tend to be all-or-none.[11,12] Variation in the responsiveness to complex protein antigens also occurs, but in these cases, responses tend to be high or low rather than all-or-none.[11-14] The response of helper T cells has been shown to be focused on a limited number of discrete sites in a protein rather than all of the sites which can potentially be recognized, a phenomenon termed *immunodominance*.[14-18] Low responsiveness appears to be a qualitative rather than quantitative phenomenon resulting from the failure to respond to an immunodominant site even though other epitopes are recognized.[19]

The design of a subunit vaccine requires knowledge of the effector mechanisms which are protective and the immunogenic determinants which induce those mechanisms. In the case of FMD, it is not yet clear whether the variable protection elicited by FMD peptide vaccines is due to their failure to correctly regulate the antibody response, or failure to induce a requisite cell-mediated effector mechanism, or both. However, given that antibody is an important component of the protective response, the minimal design for a subunit vaccine should incorporate both B cell and immunodominant helper T cell epitopes. This chapter addresses the identification of MHC class II-restricted T cell epitopes in FMDV with a view to using the information in the design of a subunit vaccine.

II. FOOT-AND-MOUTH DISEASE

FMD has been known for at least 400 years[20] and its etiological agent, FMDV, was the first animal virus to be discovered.[3] FMDV is the sole member of the Aphthovirus genus of the family Picornaviridae. More than 60 strains of FMDV have been grouped into seven serologically distinct types (O, A, C, SAT1-3, and Asia)[21] on the basis of their ability to elicit cross-protective immunity in cattle. However, the extent of antigenic variation within a serotype can be almost as great as that observed between serotypes and cross-protection between subtypes is frequently poor or nonexistent. The infectious virion consists of 60 copies each of four structural proteins, VP1-4, which enclose a positive sense, single-stranded (ss)RNA genome. The ssRNA and VP4 are wholly internal to the 28-nm diameter icosahedral capsid formed by 12 subunits of VP1, VP2, and VP3. The three-dimensional structure of type O1 strain BFS 1860 has been recently resolved to 2.9 Å.[22]

FMDV naturally infects a wide variety of animals, including cattle, sheep, goats, pigs, buffalo, impala, red kangaroos, wombats, hedgehogs, coypu, water voles, moles, agouti and some species of deer. The clinical symptoms of FMD become apparent after 2 to 8 days incubation and are characterized by excessive salivation, fever, and the rapid formation of vesicular lesions in and around the mouth, on the coronary bands and the interdigitating spaces of the feet, and on the udder and the teats.[23,24] Affected animals may suffer from lameness, weight loss, and acute reduction in milk yield, any or all of which can persist for several years after the disease has resolved.

The severity of symptoms, susceptibility to infection, and excretion of virus vary with the host species, the age of the animal, and the level of herd immunity. Thus, whereas clinical symptoms in sheep and goats tend to be mild, pigs are severely affected, cattle are intermediate between sheep and pigs, and African buffalo (*Syncerus caffer*) rarely display any symptoms at all. FMD is not usually fatal in adult animals and mortality rarely exceeds 5% although morbidity may reach 100%. However, young animals are more severely affected and lesions in the myocardium can result in over 50% mortality due to heart failure. Pigs excrete large amounts of infectious virus and it has been calculated that the amount of virus excreted in a 24-h period by a single pig is equivalent to the amount excreted by 1500 to 3000 domestic cattle.[25] Cattle, on the other hand, are highly susceptible to airborne virus and the proximity of pigs and cattle is therefore an important factor in airborne spread of disease.

Infection occurs after inhalation of an aerosol containing virus into the upper respiratory tract[24,26] or through mechanical transmission. The primary site of virus entry and replication is thought to be the mucosae of the pharynx because of the large amounts of virus recovered from the dorsal surface of the soft palate, the lateral walls of the pharynx, and the lymphoid tissues draining this region (tonsils and retropharyngeal lymph nodes) prior to viremia.[24] However, infection does not produce characteristic lesions at these sites and entry through the tonsils and dissemination via the circulation remains a possible route of entry. Lesions begin with pycnosis of the nuclei of cells in the basal layer of the stratified epithelium and typically occur on the tongue and the coronary bands of the feet. Loss of the epithelium covering a lesion can occur within 24 h. Prior to the loss of the epithelium, pathological changes in the endothelium of the dermal blood vessels causes the leakage of lymph into the lesions. Typical mouth lesions are deep and the ulcer has a distinctive under-run edge. Less typical lesions cause necrosis and loss of superficial layers of epithelium, followed by rapid healing. Secondary bacterial infections delay healing and lead to scarring, severe damage to the mammary glands and, in extreme cases, loss of the horny covering of the feet. Lesions can be found in the rumen and other parts of the alimentary tract, pancreas, adrenal glands, thyroid, pineal body, pituitary gland, kidneys, lungs, and heart, and FMDV replication has been observed in lymph node cells, bone marrow, and testes. In muscle, and especially the heart, the acute degeneration of the fibers may give the tissue a "striped" appearance.

FMDV productively infects various cell types *in vivo,* including fibroblast, epithelial, and endothelial cells and it can be propagated *in vitro* in cultures of bovine tongue epithelium; bovine nasal turbinate; calf thyroid and calf; lamb; pig, and baby hamster kidney (BHK) cell lines. Typically, FMDV is released from BHK cells by lytic cell death. However, virus has also been observed budding from BHK cells in cytoplasmic "blebs" many hours prior to lysis,[27] and lytic release is less obvious in epithelial cell lines such as Vero cells and primary cultures of skin cells. Recently, persistently infected BHK cell lines were established.[28] The virus isolated from these cell lines has an increased virulence for BHK cells, including the parental BHK line, but is attenuated for mice and cattle.[29] This has a parallel with persistent infection *in vivo* in that persistently infected cattle can infect other susceptible animals even though they co-exist with the virus and show no clinical symptoms themselves.

The mechanism by which persistence is established *in vivo* is totally unclear. Persistence as a consequence of either ineffective or inappropriate immune pressure is unlikely since persistence can be established *in vitro* in the absence of such pressure.[28] There is some suggestion that the polyribocytidilic acid [poly(C)] tract of the 5′ noncoding region of the genome may be involved in attenuating virulence since virus recovered from "carrier" cattle infected with a cloned virus isolate of known poly(C) tract length was found to have an elongated poly(C) tract compared to the challenge virus and virus isolated from animals which cleared the infection.[30] Structural region mutations are also implicated since virus isolated from persistently infected BHK cell lines contains several fixed mutations which do not normally occur in natural isolates of FMDV[31] or at structurally analogous positions in any other picornavirus. Another possibility is that persistence is cell tropic. Immunologically "privileged sites" such as salivary glands favor persistence and it has been suggested that the serous cells of the mixed secretory glands in the lamina propria of the dorsal soft palate are persistently infected in the "carrier" state.[32]

A. HUMORAL IMMUNITY AND PROTECTION

Almost 100 years ago, cattle were shown to be protected by passive transfer of convalescent antibody and the major mechanism of protection against FMD[3] was identified. Subsequently, virus vaccines were developed which consist of chemically inactivated, purified virus either adsorbed to aluminium hydroxide in the presence of low concentrations of saponin or, in the case of pigs, emulsified in incomplete oil. In cattle, the success of these vaccines has been attributed to the induction of high titers of neutralizing antibody.[4-6] The duration of protection conferred by virus vaccines is variable, but typically lasts 2 to 6 months after a single dose of vaccine and up to 49 months after multiple doses.[33,34] The antibody response is dose related[6,35] and characterized by a rapid rise in IgM, IgG1, and IgG2. The IgM response reaches a peak at around 10 to 14 days and then rapidly declines. Typical IgG responses reach a peak between 14 and 28 days, with the titer of IgG1 being greater than that of IgG2. However, for some animals, IgG2 is greater. Generally, the neutralizing antibody titer and average affinity of vaccine-induced responses are lower than those induced by natural infection.

Infection of FMD-susceptible cattle causes a rapid rise in serum antibody which can be detected at around 4 days. This response is mainly IgM, is relatively heterotypic, and reaches a peak at around 10 to 14 days post-infection. Typically, the titer of IgM declines to preinfection levels within 30 to 40 days,[36] although some cattle still show high titers after more than 80 days.[37] Isotype switching occurs between 4 and 7 days post-infection and results in a predominantly IgG1 response with detectable levels of IgG2 and IgA. The development of IgG1 coincides with the resolution of lesions, termination of viraemia, and reduction of virus excretion. Serum levels of IgG peak at around 14 to 28 days and protection against challenge with the same virus isolate lasts for up to 1.5 years and possibly longer since, in one study, one animal out of a group of three cattle was found to be clinically protected 4.5 years after the original infection.[38]

The importance attributed to the humoral response has led to extensive characterization of the antigenic structure of FMDV recognized by antibodies, and various studies have established the involvement of VP1 in this respect. Trypsin treatment of intact FMDV was shown to result in the cleavage of VP1 with a concomitant reduction in immunogenicity as assessed by the induction of protective immunity in guinea pigs.[39,40] Of the isolated FMDV proteins, only VP1 was able to induce neutralizing antibody[41,42] and protection in pigs.[42] It was subsequently shown by analyzing the serological response of guinea pigs to overlapping CBrN and enzymatic cleavage products of isolated VP1 that the regions VP1 138-154 and VP1 200-213 represent linear antigenic determinants.[7,43] The antiviral antibody response elicited by chemically synthesized VP1 144-159[8] and VP1 141-160[9] conjugated to keyhole limpet hemocyanin confirmed that this region mimics a genuine and important viral determinant. More recently, neutralizing MAb (monoclonal antibodies) binding sites have been analyzed by selecting virus variants which can escape neutralization by MAb. Sequencing the structural protein coding region of the escape variants and parental virus genomes located differences in the derived amino acid sequences associated with the lack of neutralization. At least four neutralizing MAb clusters have been identified for FMDV type O1 strain Kaufbeuren[44,45] and type A10 strain Holland[46] which involve all three of the major structural proteins. These sites can be precisely located on the surface of the three-dimensional structure.[22]

The neutralizing MAb binding sites map either to a linear sequence located within the VP1 141-160 region or to noncontiguous sequences involving one or more of the structural proteins. Some escape variants selected using neutralizing MAb which recognize the linear site also generate variants with amino acid substitutions in the carboxy terminus of VP1. The region VP1 141-160 is part of a prominent surface-exposed loop spanning VP1 130-160 and sometimes referred to as the "FMDV loop". This loop is highly disordered in the three-dimensional structure of FMDV type O1 strain BFS 1860 and is not resolved by X-ray diffraction.[22] Nevertheless, the 17 carboxy-terminal residues of VP1 in one protomer can be seen to associate with the "FMDV loop" of VP1 on an adjacent protomer. Thus, antibody recognition of the linear site may involve either the "FMDV loop" on its own or in association with a second linear determinant,[7,22] VP1 197-213.

The ability of various "FMDV loop"-related peptides to induce high levels of serotype-specific neutralizing antibody suggested that this region constituted the minimum information which would be useful in a peptide vaccine. Neutralizing antibody induced in guinea pigs by truncated peptides indicated that the minimum length for an antigenic peptide was VP1 146-156, although VP1 137-160 was more effective.[47] DiMarchi et al.[48] synthesized a more complex antigen which had the general form CC-[VP1 200-213]-PPS-[VP1 141-158]-PCG. The FMDV sequences were linked by a diproline-serine spacer to introduce a secondary structural turn and terminated by a cysteine at or near each end to increase the likelihood that the two chains would associate with each other in a polymerized or cyclic form; this peptide was termed FMDV-15.

"FMDV loop" peptides linked to carrier molecules,[8,9,49,50] as free peptides[47,48,51-53] or as recombinant fusion proteins,[54-59] elicit serotype-specific neutralizing antibody in a variety of species including mice, rabbits, guinea pigs, cattle, sheep, and pigs. Nevertheless, protection of a natural host is variable. The protection conferred by FMDV-15[48] typically requires up to ten-fold higher concentrations of neutralizing antibody than would be expected for cattle immunized with a virus vaccine of the same serotype. Thus, a large proportion of the anti-peptide antibody which is capable of neutralizing virus *in vitro* is incapable of preventing viral replication and dissemination *in vivo*.

At present, there is insufficient evidence to define precisely why the anti-peptide response in cattle is less effective than antiviral responses. However, several observations point to the antibody response of nonprotected cattle being qualitatively different to that of protected cattle, irrespective of whether they are peptide or virus vaccinated. Firstly, FMDV-15 induces antibody with a higher average affinity for virus in protected cattle than non-protected cattle.[60] Secondly, data from cattle immunized with two different serotype viruses or peptides suggests that protected cattle develop a higher titer of IgG1 compared to nonprotected cattle where the tendency is for IgG2>IgG1.[61] Passive protection of mice using antisera from virus- and peptide-vaccinated cattle correlates well with the clinical status observed in the natural host.[62] Using this assay, the protective capacity of intact antibody and $F(ab)_2$ have been compared and show that the Fc of the antibody is required for passive protection, whereas it is redundant in the traditional *in vitro* neutralization assay. Similar results have been obtained using neutralizing MAb where depletion of monocytes by treatment of mice with colloidal silica as well as the removal of Fc both rendered the antibodies ineffective.[63,64] Thus, the clearance of virus in mice is mediated by antibody through Fc receptors on cells of the reticuloendothelial system. Third, the fine specificity of anti-peptide antibodies induced in cattle and guinea pigs is indistiguishable,[65] yet guinea pigs are more easily protected than cattle. Thus, although the specificity of antibody elicited by peptide vaccine is, by the nature of the antigen, more restricted than that induced by virus vaccines, specificity does not seem to account for the poorer efficacy of the peptide vaccine in cattle. On the contrary, and distinct from the serotype-specific immunity conferred by virus vaccines, FMDV-15 sometimes induces heterotypic protection[66] in cattle and guinea pigs and serotype cross-reactive antibodies which recognize the carboxy-terminal sequence of VP1.[65,66]

Not surprisingly, perhaps, protection in cattle appears to be determined by B cell differentiation and affinity maturation rather than by variation in the B cell repertoire. It is well documented that the qualitative aspects of an antibody response such as isotype switching, affinity maturation, and the generation of B cell memory are influenced by T cells and cytokines (for review see Reference 10), and there are indications that the TH1/TH2 phenotype of helper T cells is determined, in part, by the antigen presenting cell which elicits the response. Thus, the available evidence suggests that protective immunity to FMD is likely to depend on how effectively T cells are activated and by what cell type the antigen is presented. It follows, therefore, that the lower frequency of protection achieved with peptide vaccines may be due to an increased likelihood of low or nonresponsiveness due to MHC-encoded immune-response (Ir)-gene effects.

B. CELL-MEDIATED IMMUNITY (CMI)

T cells have important functions in protection against viral diseases.[67] Recognition of viral antigens by T cells requires presentation of antigenic determinants on MHC molecules.[68,69] Recognition of antigen by MHC class I-restricted $CD8^+$ T cells and MHC class II-restricted $CD4^+$ T cells stimulates cytokine release, proliferation, and effector function(s). In mouse and man, $CD4^+$ T cells are functionally heterogeneous and have been subdivided on the basis of their cytokine secretion:[70,71] TH1 cells are distinguished by secretion of interleukin-2 (IL-2) and gamma-interferon (IFN-γ), whereas TH2 cells secrete IL-4, IL-5, IL-6, and IL-10. Recent evidence from cloned human T cells suggests that $CD8^+$ T cells can be similarly subdivided.[72] Furthermore, while CTL function is usually associated with $CD8^+$ T cells, evidence from cloned mouse T cells suggests that TH1 $CD4^+$ T cells are also directly involved in CMI, including the lysis of virally infected cells.[70]

In mice, antiviral antibody responses are biased toward the IFN-γ regulated isotype IgG2a,[73] and it has recently been shown that high-level secretion of IFN-γ by NK cells is important although not sufficient to induce and regulate TH1-like responses.[74,75] The limited data available concerning FMDV-vaccinated cattle suggests that there is an IgG subclass bias associated with protection.[63] However, it is not yet possible to interpret this in terms of a TH1/TH2-like behavior since the regulatory influence of IFN-γ and IL-4 on the production of bovine IgG subclasses is unknown.

The presence of IgG2a in sera from FMDV-infected euthymic but not athymic Balb/c mice and the ability of T cell-depleted, FMDV-primed spleen cell cultures to produce antibody *in vitro* in the presence but not the absence of T cell-derived cytokines has defined FMDV as a T cell-dependent antigen.[76] Nevertheless, delayed-type hypersensitivity (DTH), an overt expression of a TH1-like CMI response, induced in guinea pigs by FMDV infection does not directly correlate with protection.[77] The regulatory functions of DTH, CTL, and natural killing (NK) have not been evaluated nor has their contribution to innate immunity by interferon production. However, a direct role for NK cells is suggested by the ability of a bovine NK-like cell line which produces high titers of IFN-γ to eliminate FMDV from infected tissue culture.[78]

Erythrocyte (E)-rosetting[79,80] and MAb-directed immunofluorescent labeling[81] of peripheral blood mononuclear cells have been used to examine changes in the lymphocyte subsets of Asian buffalo[79] and cattle[80,81] in response to vaccination[79-81] and/or infection.[81] Although the absolute number of T cells detected by E-rosetting gave an under-representation of the circulating T cell pool, it is safe to conclude that FMD vaccination[79,80] and infection[81] cause detectable fluctuations in both total and subsets of peripheral T cells. Vaccination of Asian buffalo was shown to induce a strong DTH response.[79]

Antigen-specific *in vitro* proliferation of cattle[82] and pig[83] peripheral blood mononuclear cells has been used to examined the specificity of the T cell response elicited by monovalent (single serotype) FMD vaccines. In both species, the response was found to be serotype cross-reactive and to recognize at least the three major structural proteins, of which VP3 stimulated the strongest cross-reactive response. In cattle,[84] these proliferative responses were inhibited by MAb specific for CD4 and MHC class II, indicating that the responding population was $CD4^+$ MHC class II restricted. In the pig,[83] proliferation was similarly shown to be $CD4^+$, MHC class II-restricted although a MHC class I-specific MAb also inhibited proliferation. The inhibitory effect of the MHC class I-specific antibody is difficult to reconcile with the accessory role of CD4, but since proliferation was not inhibited by CD8-specific antibodies, this phenomenon may relate to the CD4/CD8 double-positive phenotype often found on a high proportion of T cells in pigs[85] rather than to actual functional restriction.

III. T CELL EPITOPE PREDICTION FROM PRIMARY PROTEIN SEQUENCES

Reliable methods for predicting T cell epitopes from primary sequences would be invaluable where the sequence of the whole pathogen is known, as is the case for FMDV. Clearly, however, such an approach is of limited value for a complex, poorly defined organism. The expectation that T cell epitopes can be predicted from primary sequence data extends from the observation that T cells recognize short peptides in which only the primary and local secondary structure are important. Central to the issue of what defines a T cell epitope is the conceptual problem of how a single MHC combining site can accommodate a wide variety of ligands. One possibility is that the peptides that can bind share a number of structural features that allow them to bind in a preferred site with similar conformation. Until recently, the current view of T cell epitope structure was derived from comparison of known MHC class I- and class II-restricted stimulatory peptides for the presence of common structural or biophysical features.

Direct evidence concerning the structure of T cell epitopes has been obtained recently by sequencing peptides isolated from MHC molecules[86-92] and has identified consensus, allele-specific motifs for several different MHC class I molecules.[87,88] These motifs have a restricted length (8 to 10 residues) along which most positions can accommodate many different residues. However, one or a very few residues occur with a high frequency at certain positions which have been termed "anchor" residues. These anchor residues are characteristic of the different alleles studied. MHC class II binding motifs are less well defined due to the variable length of the isolated peptides (13 to 17 residues) and the lack of discrete anchor residues. Nevertheless, it is generally considered that class II binding motifs do exist,[89-92] but that binding is inherently more degenerate than for class I. Thus, although generalized algorithms are unlikely to predict more than a fraction of actual T cell determinants, the use of defined allele-specific motifs for use with ruminant species is not a practical alternative at the present time.

Structural analysis of several MHC class I molecules[93-97] has revealed that the α1- and α2-domains form eight anti-parallel ß-strands spanned by two long α-helical regions which form the floor and the walls of a peptide binding cleft, respectively. Sequence data and information derived from mutation experiments have been used to construct a model for the structure of MHC class II[98,99] in which segments of the α1- and ß1-domains, which map by homology to the α-helices of the class I α1- and α2-domains, respectively, also form the walls of a cleft with the remainder of the α1- and ß1-domains forming the strands on the floor of the cleft. The most striking feature of the class I structures are the deep conserved pockets at the ends and a polymorphic pocket in the middle of the groove into which complementary side chains of a peptide can bind. The restricted use of binding pockets appears to explain how MHC class I molecules and presumably class II molecules can bind such a broad range of peptides: assuming that each position in the peptide is independent, then the number of permutations for all positions within a peptide of 8 or 9 residues gives millions of unique peptide sequences, even allowing for fixing at the anchor positions. Jardetzky et al.[88] have calculated that there are about 13 million peptide permutations which fulfill the nonamer HLA-B27 binding motif.

The different lengths of the peptides eluted from MHC class I and class II molecules have important implications for antigen processing and the orientation of peptide within the binding groove. The short length of peptides which can be accommodated by the class I binding groove is necessary because the MHC class I binding site is "closed" at both ends: at one end by a tyrosine side chain and at the other end by a salt bridge between residues on the two α-helices. In contrast, the MHC class II binding site is open at one end and it is likely that the class II molecule acts as a template for processing,[100] and either the amino terminus of the peptide is generated independently and bound by the occluded end of the groove or that roughly cleaved peptides are bound and amino-terminally trimmed after binding. In both cases, the protruding carboxy terminus would be trimmed to variable but functionally convenient lengths.

In the main, algorithms for predicting T cell epitopes fall into two categories: algorithms which predict MHC contact residues and those which predict the side chains of importance for contact with MHC and the T cell receptor (TCR). In general, predictive algorithms assume that the structural constraints required for binding to MHC and TCR occur as a periodic distribution of residues in the protein sequence. As such, the favored structure is an amphipathic α-helix (3.6 residues per turn of a helix or 100° periodicity), although the authors of these algorithms recognize that T cell epitopes can also adopt 3_{10}-helix (three residues per turn of a helix or 120° periodicity) or ß-pleated strand conformations (180° periodicity). Some T cell epitopes do exist as helices in the native protein,[15,17] but many others do not and it is believed that the success of predictive algorithms such as AMPHI and SOHHA is because they predict the potential to form a helix after proteolytic excision from the protein. Analysis of the MHC and TCR contact residues of the I-A^d-restricted ovalbumin 323-339 epitope[101] (class II) and crystallographic analysis of peptide bound to HLA-B27[96] (class I) indicate that binding for these epitopes is in an extended ß-conformation rather than a helical structure. Despite their lack of accuracy, the ease with which predictive algorithms can be implemented makes them useful in initial screening procedures.

A. AMPHI

The finding that both of the immunodominant T cell sites of sperm whale myoglobin[15,17] are amphipathic α-helices suggested that this might be a more general property of antigens recognized by T cells. DeLisi and Berzofsky[102] described a method for searching primary sequence data to identify hydrophobic periodicity consistent with that of an α-helix. They proposed that an amphipathic α-helical conformation would allow one face of the structure to interact with an MHC molecule while leaving the other face free to interact with the TCR. A computer algorithm (AMPHI)[103] analyzes primary sequence data for helical periodicity in overlapping blocks of either 7 (two turns of a helix) or 11 (three turns of a helix) residues and assesses the stability of predicted amphipathic segments as a function of the length of the segment and the magnitude of the hydrophobic intensity around 100 or 120° for the individual blocks in the segment to give an amphipathic score. A segment which has an amphipathic score above a certain threshold is considered to be stable. AMPHI identifies immunodominant helper T cell epitopes with about 75% success. Amphipathicity and α-helical conformation contribute independently to T cell antigenicity and there is a strong propensity for a carboxy-terminal lysine[104]. Helix-breaking residues, glycosylation sites, turns, and coils are discouraged within the predicted sequences. Examples of the successful use of AMPHI to identify T cell epitopes include a helper T cell site in the circumsporozoite protein of *Plasmodium falciparum,*[105] a promiscuous fusion protein T cell epitope of measles virus[106] and a relatively conserved helper T cell site in the envelope protein of an HIV isolate.[107]

B. SOHHA

The strip-of-helix hydrophobicity algorithm (SOHHA)[108,109] is based on calculating the periodicity of hydrophobic residues in a protein sequence. Analysis is performed by averaging the Kyte-Doolittle hydropathic values of residues at position 1, 5, 8, 12, 15, 19, and 23 for each residue of the protein in turn. If the intermediate values are also calculated (i.e., the average of 1 and 5; 1, 5, and 8; etc.), the result is a table of indices for one to six turns of an α-helix which can be examined for the presence of hydrophobic strips spanning three or more turns. If a helix is represented as a four-quadrant cylinder, then in an ideal strip-of-helix, hydrophobic residues (such as isoleucine, leucine, phenylalanine, and valine) are non-randomly distributed to quadrant III, while charged residues (such as arginine, aspartic acid, glutamic acid, histidine, and lysine) are excluded from quadrant III. The termini of helices are taken as the first asparagine or the residue 1 or 2 places amino-terminal to a proline and the next helix-breaking residue such as a proline or glycine is taken as the end. The SOHHA was devised after observing that the invariant chain (Ii) of MHC class II has a strong hydrophobic strip near the carboxy terminus[108] which was proposed to be involved in the association of Ii with the peptide binding cleft during export of nascent MHC class II molecules into the late endosomal pathway for interaction with antigenic peptides. SOHHA and a related algorithm, SHA,[109] in which the SOHHA is applied to overlapping eight-residue segments and consecutive segments with strip-of-helix hydrophobicity greater than 2.5 are joined, have been shown to predict known α-helices and T cell epitopes with a greater accuracy than AMPHI and Rothbard and

Taylor motifs. Similar calculations can be performed for 3_{10}-helices and ß-pleated sheets by using 3- and 2-residue periodicity, respectively. The authors of these algorithms have shown that they correctly identify a number of previously defined T cell epitopes.

C. ROTHBARD AND TAYLOR MOTIFS

Rothbard[110] observed a pattern in the primary structure of 29 out of 30 known T cell epitopes which occurred significantly more often than would be expected in random peptides. The pattern consisted of either a four- or five-residue motif: (charge or glycine)-(hydrophobic)-(hydrophobic)-(polar or glycine) or (charge or glycine)-(hydrophobic)-(hydrophobic)-(hydrophobic or proline)-(polar or glycine). Subsequently, Rothbard and Taylor[111] extended the initial observation to show that 46 of 55 known epitopes contained one of these motifs and that the epitopes were not segregated by species or by the class of MHC restriction. Alignment of epitopes around the four- and five-residue patterns suggested the presence of putative allele-specific subpatterns. This method does not identify the termini of epitopes nor does it rank the selections as do the AMPHI and SOHHA algorithms. Nevertheless, it has been used successfully to predict various class I and class II-restricted epitopes, including two epitopes in the 65-kDa protein[112] and one in the 19-kDa protein[113] of *Mycobacterium tuberculosis*, a helper T cell epitope in the influenza matrix protein,[114] and four epitopes in the L1 protein of human papillomavirus type 16 and one in the E6 protein of the same virus.[115]

D. THE "RARITY RULE"

Kourilsky and Claverie[116] proposed that sequences from a pathogen which are most likely to be recognized by T cells will occur where the sequence of the pathogen differs markedly from host species proteins on the basis that T cells which would recognize similar sequences will have been deleted from the T cell repertoire during thymic development. Analysis is implemented by comparing tetrapeptides from the pathogen with tetrapeptides in a database of representative host species proteins; regions where the pathogen sequence is infrequent or missing from the database ("Rarity") are scored positive. This method has been used successfully to predict four cytolytic T cell epitopes in the human immunodeficiency virus *gag* protein, but the requirement for an appropriate database means that it is less readily implemented than the other methods described, particularly in farm animal species.

E. ALLELE-SPECIFIC MAJOR HISTOCOMPATIBILITY COMPLEX (MHC) BINDING MOTIFS

While algorithms which deal with the general principles of T cell epitope prediction help to assess the more immunodominant regions of a protein, they do not normally take into account the allele-specific nature of MHC restriction. Various methods have been used to look for more specific characteristics required for MHC binding, including sequence alignment according to the structural similarity of residues in epitopes which use the same restriction element, single amino acid substitution of stimulatory peptides, and more directly, sequencing of peptides eluted from a defined MHC molecule. The MHC class I motifs which have been defined generally conform to an eight- or nine-residue sequence which is anchored at two or three residues. By contrast, peptides eluted from MHC class II molecules display much more heterogeneity both in length, varying between 13 to 17 residues, and sequence such that there is no obvious reinforcement of anchor residues. At present, allele-specific class II binding motifs similar to those identified for class I have still to be clearly demonstrated.

1. MHC class I motifs

An eight- or nine-residue H-$2K^k$ binding motif consisting of glutamic acid or, less frequently, aspartic acid at position 2 and isoleucine in the carboxy-terminal position has been defined by alignment of known H-$2K^k$-restricted T cell epitopes from influenza and other viral antigens.[117] Similarly, a nine- or ten-residue H-$2K^d$ binding motif consisting of tyrosine in position 2 and a hydrophobic residue with a large aliphatic side chain in the carboxy-terminal position has been determined by alignment of H-$2K^d$-restricted sequences.[118] The accuracy of the putative binding motif has been tested by comparing the ability of amino acid-substituted and truncated peptides to compete with each other in a CTL assay. The same H-$2K^d$ binding motif was obtained directly by sequencing peptides isolated from H-$2K^d$ molecules.[87] H-$2K^b$, H-$2D^b$, HLA-A2.1,[87] and HLA-B27[88] motifs have also been derived by sequencing pooled peptides eluted from specific MHC molecules. Allele-specific motifs have been successfully used to predict a H-$2K^k$-restricted CTL epitope in the NS1 protein of influenza[119] and a H-$2K^b$-restricted

Table 11-1 **Consensus MHC Class I-Restricted Binding Motifs**

Restriction element	Position 1	2	3	4	5	6	7	8	9	0	Ref.
H-2K^{d}	*	**Y**	*	*	*	*	*	*	**a**		87,118
H-2K^{d}	*	**Y**	*	*	*	*	*	*	*	**a**	118
H-2K^{k}	*	**D**	*	*	*	*	*	**b**			117
H-2K^{k}	*	**D**	*	*	*	*	*	*	**b**		117
H-2K^{k}	*	**E**	*	*	*	*	*	**b**			117
H-2K^{k}	*	**E**	*	*	*	*	*	*	**b**		117
H-2K^{b}	*	*	*	*	**F**	*	*	**c**			87
H-2K^{b}	*	*	*	*	**Y**	*	*	**L**			120
H-2D^{b}	*	*	*	*	**N**	*	*	*	**d**		87
HLA-A2.1	*	**L**	*	*	*	*	*	*	**e**		87
HLA-B27	*	**R**	*	*	*	*	*	*	**f**		88

Note: Anchor positions are shown in bold: a = I,L,V,T; b = I; c = L,M,I,V; d = M,I,L; e = V,L; f = K,R,L,Y,A,N where the underscoring indicates a strong signal during sequencing. Abbreviations for the amino acid residues are: A, alanine; C, cysteine; D, aspartic acid; E, glutamic acid; F, phenylalanine; G, glycine; H, histidine; I, isoleucine; K, lysine; L, leucine; M, methionine; N, asparagine; P, proline; Q, glutamine; R, arginine; S, serine; T, threonine; V, valine; W, tryptophan; and Y, tyrosine

epitope in ovalbumin.[120] Thus, these motifs have potential for accurately predicting epitopes from primary protein sequences. Table 11-1 shows the MHC class I motifs that have been identified to date.

2. MHC Class II Motifs

Sette et al.[101,121] showed that proteins which bind to I-A^{d} share a structural motif based on the identification of residues in contact with the MHC molecules. These authors have subsequently defined an I-A^{d} binding motif[122] using the I-A^{d}-restricted ovalbumin 327-332 sequence valine-histidine-alanine-alanine-histidine-alanine as a template motif and assigning substitution values to experimentally tolerated (value = 3) or nontolerated (value = 1) residues according to how well the substituted peptide could bind to I-A^{d}. When the result of a particular substitution was not experimentally verified, the substitution value was extrapolated from the most structurally similar of the experimentally tested residues according to Dayhoff tables. Thus, a sequence in which substitutions are completely tolerated would generate a motif value of $3^6 = 729$, whereas a single substitution with moderately deleterious effect would reduce this to $(3^5 \times 2) = 486$. Motif values greater than 400 are operationally defined as I-A^{d} binding motifs.

Comparison of peptides that bind to I-E^{d} identified[123] a basic residue at either position 1, 2, or 3 and positions 4 and 6 with a noncharged and usually hydrophobic residue at position 5. Thus, the minimum motif for I-E^{d} binding[122] is (basic)-(basic)-(noncharged)-(basic), although (basic)-(*)-(basic)-(noncharged)-(basic) and (basic)-(*)-(*)-(basic)-(noncharged)-(basic) where (*) is any residue also fulfil this motif and can be identified by visual inspection of primary sequence data.

From a comparison of HLA-DR1-restricted sequences, Rothbard et al.[111,124] derived a motif where the residues at positions 1 and 8 are hydrophobic and charged, respectively, and residues at positions 4 and 5 are hydrophobic and fall at the centre of a Rothbard and Taylor motif.[111] HLA-DR1 binding of peptides substituted with alanine experimentally identified residues 1 and 8 as critical anchor positions[125] and independently validated the motif of Rothbard et al. A detailed analysis[126] of peptides which bind HLA-DR1, 2, 52a, and 5 in a fairly type-specific manner identified a motif consisting of aromatic or hydrophobic residues at positions 1 and 2; a positively charged residue at position 3; glutamine, aspartic acid, or asparagine at position 4 or 5; threonine, serine, alanine, proline or leucine at position 6 or 7; and a positively charged residue at position 8, 9, or 10. Subsequent analysis[127] of peptides which bind HLA-DR1, 2, 5, and 7 in a promiscuous manner identified a similar motif where position 1 is an aromatic or hydrophobic residue; position 6 is mainly alanine, cysteine, isoleucine, leucine, proline, serine, threonine, or valine; position 8 is a charged residue; and the residue at position 9 is another hydrophobic residue. Noninvasive fluorescence assays suggest that the minimum consensus motif for DR1 binding fits a two-residue-contact model with two hydrophobic residues of the peptide lying 14 to 16 Å apart in the bound state and that binding occurs via hydrophobic pockets involving tryptophan.[128] MHC class II-derived

Table 11-2 **Consensus MHC Class II-Restricted Binding Motifs**

Restriction element	Position												Ref.
	1	2	3	4	5	6	7	8	9	0	1	2	
I-A^d	*	**a**	*	*	*	*	**b**	**c**	*	*	*		91
I-A^s	*	**d**	**e**	*	*	*	*	**f**	*	*	*		90
I-A^b	*	**g**	*	*	*	*	*	**h**	*	*	*		90
I-E^b	**i**	**j**	**k**	*	*	*	*	**l**	**m**	*	*	**n**	90
HLA-DR1	**Y**	*	*	*	*	*	*	**K**	*	*	*		125
HLA-DR1,2,5,52a	**o**	*	**l**	**p**	*	**q**	*	**l**	*	*	*		92,111,126
HLA-DR1,2,5,52a	**o**	*	**l**	*	**p**	*	**q**	**l**	*	*	*		92,111,126
HLA-DR1,2,5,7	**o**	*	*	*	*	**q**	*	**l**	**r**	*			92,111,127

Note: Proposed anchor positions are shown in bold: a = A,I,L,V; b = A,F,M,S,T; c = E,I,N,Q,R,T; d = E,I,T,V; e = A,T,S; f = H,R; g = D,N,Q; h = I,P,S; i = Y; j = L,V; k = I,Y; l = K,R; m = H,R; n = F,Y; o = F,I,L,V,W,Y; p = D,Q,N; q = A,L,P,S,T; r = A,L,I,T,V,Y where the underscoring indicates consensus residues. Amino acid residue abbreviations are given in the legend to Table 11-1.

peptide sequences have been aligned in an attempt to define consensus motifs[90] for I-A^s, I-A^b, and I-E^b. Table 11-2 shows the proposed class II "motifs". Comparison of Tables 11-1 and 11-2 quickly reveals that the class II motifs are less clearly defined than class I motifs and, as yet, cannot be effectively used for epitope prediction as has been shown for the class I motifs.[119,120,129]

IV. T CELL EPITOPES IN FOOT-AND-MOUTH DISEASE VIRUS

Tables 11-3 through 11-7 list the results of predictions generated by AMPHI,[103] SOHHA,[108] the motifs of Rothbard et al.,[110,111,124] and the I-A^d/I-E^d [122] binding motifs for VP1-4 and the RNA-dependent RNA polymerase of FMDV type O1 strain Kaufbeuren. The different methods identify different but often overlapping regions and, to date, few of the predictions have been tested. From the available data, it is not yet possible to conclude which, if any, of the methods or combination of methods is most likely to be useful for ruminant studies.

For the purpose of defining viral T cell epitopes, T cell responses primed by virus should recognize a viral peptide and responses primed by the peptide should be restimulated by virus. The read-out for such analyses are usually T cell proliferation or cytokine production and in this respect T cell clones are useful because the read-out signal tends to be stronger by virtue of the monospecific nature of the responder. T cell clones are an important tool for analysis of the fine specificity of T cells, but the advantages of clones must be weighed against the disadvantage that established clones tend to be those which grow well in culture and may not objectively represent the important specificities *in vivo*. Antibody production has also been used where the peptide also contains a B cell epitope; a helper epitope within the carboxy terminus of the VP1 141-160 "FMDV loop" has been inferred from the antiviral antibody response of guinea pigs immunized with truncated peptides spanning VP1 134-160.[50] The VP1 141-160 region contains several α-helical predictions and an I-A^d motif which was compromised by including residue 161. Antibody production in mice has been reported to be MHC restricted by H-2^k for VP1 141-160CG,[51] and H-2^k and H-2^d for VP1 136-152,[53] such that antiviral antibody is not induced by these peptides in mice with other MHC haplotypes.

In general, "FMDV loop" peptides such as FMDV-15 are, at best, poorly recognized by proliferating polyclonal T cells[37,82,130,131] from virus- and peptide-vaccinated cattle. However, *in vitro* proliferation induced by synthetic peptides has identified various other structural and nonstructural protein determinants. These include VP1 21-40, VP1 96-116, VP1 161-213, VP2 54-72, VP3 97-110,[37,82] P3D 103-119, P3D 295-315, P3D 365-380, and P3D 443-470.[37] It should be emphasized that the analysis of VP1 is by no means complete and the peptides utilized in this analysis were synthesized for an entirely different reason. As such, the regions identified were colinearly synthesized with VP1 141-160 in all cases, but were identified as the T cell component because none of the cattle which recognized these peptides responded to VP1 141-160 on its own. Furthermore, the design of the peptides did not provide overlap between the regions and, as such, potentially important epitopes may have been missed.

Table 11-3 **T Cell Epitope Predictions for FMDV O1 Kaufbeuren VP1**

```
VP1          1         2         3         4         5         6
Seq TTSAGESADPVTTTVENYGGETQIQRRQHTDVSFIMDRFVKVTPQNQINILDLMQIPSHT
1   .....AAAAAAAAAAAAA..................AAAAAAAA.AAAAAAAAAA.AAAA
2   .....................................RRRRRRRR......RRRR...RR
3   ............................................................
4   ...SSSSSSS.................SSSSSSSSSSSSSSSSS................
5   ...................................SSSSSSSSS....SSSSSSSSS...
6   ............................................................
7   .........................DDDD...............................
Str                              aaaZaaa bbbBbbbb bbbCbb       a

             7         8         9        10        11        12
Seq LVGALLRASTYYFSDLEIAVKHEGDLTWVPNGAPEKALDNTTNPTAYHKAPLTRLALPYT
1   .AAAAAA..........................AAAAAAAAAAAAA..AAA.........
2   RRRRRRR.........RRRRR..............RRRR.....................
3   ............................................................
4   SSSSSSSSSSSSSSSSSSSSSSSS....................................
5   .SSSSSSSSSS.................................................
6   ...DDDDDD.............................................DDDDDD
7   ............................................................
Str  aaaaAaabbbbbDbbbbbbbb   bbEbbb            bbbFbb     bbG1bb

            13        14        15        16        17        18
Seq APHRVLATVYNGECRYNRNAVPNLRGDLQVLAQKVARTLPTSFNYGAIKATRVTELLYRM
1   AAAAAAAAAA......................AAAAAAAA........AAAAAAAAAA..
2   ...RRRRR.........................RRRRRRRR....RRRRRRRRRRRRRR.
3   ............................................................
4   ....SSSSSSSS..........SSSSSSSSSSSSSSSSSS...............SSSSS
5   .....SSSSSSS..................................SSSSSSSSSSSSSS
6   ...................................DDDDDD......DDDDDD.......
7   ............................................................
Str     bG2bb1111111111111111111111111           bbHbbb bbbbIbbb

            19        20        21
Seq KRAETYCPRPLLAIHPTEARHKQKIVAPVKQTL
1   ........AAAA.....................
2   .................................
3   .................................
4   SSSSSSSSSSSSSSSS........SSSSSSSSS
5   SSSSSSSS.........................
6   .................................
7   ....................DDDD.........
Str bbbbbbb
```

Note: Seq = protein sequence; 1 = AMPHI[63] mid point of blocks; 2 = Rothbard and Taylor motif;[67] 3 = Rothbard and Taylor HLA-DR1 motif;[71] 4 = α-helix SOHHA;[64] 5 = 3_{10}-helix SOHHA;[64] 6 = IA^d motif;[70] 7 = IE^d motif;[70] Str = structure.[11] Characters in the structure line refer to residues which take part in α-helix (aaa) or ß-sheet (bbb) structures where the underscored letters refer to location of the structure in relation to the eight-stranded ß-barrel structure adopted by picornavirus proteins. Amino acid residue abbreviations are given in the legend of Table 11-1.

Table 11-4 **T Cell Epitope Predictions for FMDV O1 Kaufbeuren VP2. Legend as for Table 11-3**

```
VP2          1         2         3         4         5         6
Seq DKKTEETTLLEDRILTTRNGHTTSTTQSSVGVTYGYATAEDFVSGPNTSGLETRVVQAER
1   ........................................AAAAAAA..........AAA
2   ............RRRR....RRRR......RRRRRRRR..RRRR.........RRRR..R
3   ................................RRRRRRRR.................RRR
4   ...............................SSSSSSSSSSSSSSS....SSSSSSSSSS
5   ......................................................SSSSSS
6   ................DDDDDD.......DDDDDD.................DDDDDD..
7   ...........................................................D
Str             bbA1bbb  bbbA2bbb                         aaaZaa

             7         8         9        10        11        12
Seq FFKTHLFDWVTSDSFGRCHLLELPTDHKGVYGSLTDSYAYMRNGWDVEVTAVGNQFNGGC
1   ........AAAAAAAAAAAA........AAAAAAAAAA..............AAAA....
2   RRR.RRRRRRR.......RRRR......RRRR............................
3   RRRRRRRRRRRRR.............RRRRRRRR..........................
4   SS..........................................SSSSSSSSSSSSSSSS
5   SSSSSSSSSS....SSSSSSSSSS...................................S
6   .............................DDDDDDDDDDD....................
7   DDDDD.......................................................
Str   bbbbBbbbb     bbCbbb     aaaaaAaaaaabbbbbbDbbbbbb     bbbb

            13        14        15        16        17        18
Seq LLVAMVPELYSIQKRELYQLTLFPHQFINPRTNMTAHITVPFVGVNRYDQYKVHKPWTLV
1   ...AAAAAAA.................................AAAAAAAAAAAAAA...
2   .......RRRR....RRRR.........................................
3   .............RRRRRRRR.......................................
4   SSSSSSS.........................SSSSSSSSS...................
5   SSSSSSS.........SSSSSSSS................................SSSS
6   ...................................DDDDDD..................D
7   ............................................................
Str bEbbbb         aaaBaaaabbbFbbb   bbbG1bbb       bG2     bbbb

            19        20        21
Seq VMVVAPLTVNTEGAPQIKVYANIAPTNVHVAGEFPSKE
1   ......................................
2   .................RRRRR......RRRR......
3   ......................................
4   .............SSSSSSSSSSSSSSSSSSSSSS...
5   SSSSSS..........SSSSSSSS...SSSSSSSS...
6   DDDDDDDD..............................
7   ......................................
Str bbHbbbbb        bbbbbbbIbbbbbbbb
```

In view of these criticisms, VP1 21-40 was synthesized, assayed for induction of T cell proliferation, and found to be recognized in a high proportion of vaccinated cattle.[37,82] This region of VP1 is not predicted by AMPHI nor Rothbard and Taylor motifs, but does contain an α-helical SOHHA prediction and an I-E^d motif. The SOHHA prediction would appear to be the more accurate in this case since CnBr cleavage of VP1 21-40 peptide identified the methionine at residue 36 as critical for T cell recognition.[37] VP1 161-213 was also strongly recognized and, although it contains several overlapping predictions, once again the SOHHA prediction appears to be the more accurate since prediction spans VP1 170-190;

Table 11-5 **T Cell Epitope Predictions for FMDV O1 Kaufbeuren VP3. Legend as for Table 11-3**

```
VP3          1         2         3         4         5         6
Seq GIFPVACSDGYGGLVTTDPKTADPVYGKVFNPPRNQLPGRFTNLLDVAEACPTFLRFEGG
1   ........AAAA...........AAAAAAAAAAA.AAAAAAAAAAAAAAAAAAAAA....
2   ............RRRR...RRRR...RRRR.........RRRR..RRRR...........
3   ..........RRRRRRRRRRRRRRR...................................
4   ..SSSSSSSSSSSS......................SSSSSSSSSSSSSSSS........
5   .....................................SSSSSSSSSSSSSSSSSSSSSS.
6   ............................................................
7   ............................................................
Str                                           aaaZaaaabbBbbb

            7         8         9        10        11        12
Seq VPYVTTKTDSDRVLAQFDMSLAAKQMSNTFLAGLAQYYTQYSGTINLHFMFTGPTDAKAR
1   .......................AAAAAAAAAAAAAAAAAA...................
2   ...........RRRRR...............RRRR.......RRRR.RRRRR........
3   ........................................RRRRRRRR............
4   ..........................................................SS
5   ............................................................
6   ................................DDDDDD..................DDDD
7   ............................................................
Str  bbBbbb       bbCbbb      aaaaaAaaaaaabbbbbbbDbbbbbb     bbb

           13        14        15        16        17        18
Seq YMVAYAPPGMEPPKTPEAAAHCIHAEWDTGLNSKFTFSIPYLSAADYAYTASGVAETTNV
1   ..........AAAAAAA........................AAA.........AAAAAA.
2   ................RRRRR............RRRRR.......RRRRR..RRRRRRR.
3   ............................................................
4   SSSSSSS.......SSSSSSSSSSS.....SSSSSSSSSS..................SS
5   ..................SSSSSSSSSSSS.....SSSSSSSSSSSSSSSSSSS......
6   DDDD...............DDDDDD...................DDDDDD..........
7   ............................................................
Str bEbbbb        aaaBaaaabbbFbbb   bbG1bbb        bG2

           19        20        21
Seq QGWVCLFQITHGKADGDALVVLASAGKDFELRLPVDARAE
1   ........................................
2   ........................................
3   ........................................
4   SSSSSSSSSSSSSSSSSSSSSSSSSSSSS...........
5   ...............SSSSSSSSSSS..............
6   .................DDDDDDDD...............
7   ........................................
Str  bbbbbbHbbbbb   bbbbbbbbIbbbbbbbbb
```

peptides VP1 141-180 and VP1 (141-160)-(181-200) were less well recognized than VP1 161-213; VP1 141-180 and VP1 (161-180)-(196-213) were equally well recognized; and VP1 141-160 and VP1 200-213 were not recognized. These results suggest that recognition involves the region around VP1 180. The regions VP1 21-40 and VP1 160-213 contain important T cell epitopes in that they can be recognized at concentrations which are equimolar with virus, thus suggesting that the peptides are processed to reveal naturally occurring epitopes.

Table 11-6 **T Cell Epitope Predictions for FMDV O1 Kaufbeuren VP4. Legend as for Table 11-3.**

```
VP4          1         2         3         4         5         6
Seq GAGQSSPATGSQNQSGNTGSIINNYYMQQYQNSMDTQLGDNAISGGSNEGSTDTTSTHTT
1   ...................AA....AAAAAAAAAAAAA......................
2   ....................................................RRRR.RRR
3   ............................................................
4   ............................................................
5   ............................................................
6   .....................................................DDDDDD.
7   ............................................................
Str                                                      aaaaaaa


             7         8
Seq NTQNNDWFSKLASSAFSGLFGALLA
1   ....AAAAAAAAAAAAAAAA.....
2   R....RRRRRRRR....RRRR....
3   .........................
4   SSSSSSSSSSSSSSSSSS.......
5   .........................
6   ........DDDDDDDDD........
7   .........................
Str aaaaaaaaaa
```

The RNA polymerase (P3D) peptide sequences all contained overlapping predictions and were recognized by several virus-vaccinated cattle and by several animals in a group of MHC-typed cattle infected with FMDV type O1 strain Kaufbeuren.[37] T cells isolated from these animals were found to be serotype cross-reactive against other virus preparations.

Several peptides were synthesized to specifically test the utility of the AMPHI and Rothbard and Taylor algorithms. The majority of the sequences selected for peptide synthesis were not recognized in proliferation assays by the available vaccinated cattle. However, two of the predicted sequences were stimulatory for several cattle. These were VP2 54-72 which was predicted by all seven of the algorithms but specifically contained three overlapping HLA-DR1 binding motifs within a sequence which gave a high amphipathic score, and VP3 97-110 which was selected because it contained Rothbard and Taylor predictions in a conserved region of three different serotype viruses: types O1, A24 and C1. Analysis using CD4$^+$ bovine T cell clones derived from virus-vaccinated cattle identified the VP1 21-40 and VP2 54-72 epitopes as FMDV type O1 serotype specific, and VP3 97-110 as a serotype cross-reactive epitope which was present in FMDV types O1, A24, and C1 but not a South African Territories SAT-2 isolate which is more distantly related to O1 than the other two viruses.[37]

V. IMPLICATIONS FOR FMD SUBUNIT VACCINE DEVELOPMENT

Until recently, it was not possible to address the impact of MHC-linked Ir gene effects on the response of cattle to a potential peptide vaccine such as FMDV-15. However, Glass et al.[132] have recently demonstrated Ir gene control of the response of cattle to ovalbumin using a one-dimensional isoelectric focusing technique to define polymorphism in bovine MHC (BoLA) class II antigens.[133] Applying this approach to the analysis of cattle immunized with FMDV-15, it was possible to demonstrate that the fine specificity and the magnitude of the T cell response was related to and restricted by the BoLA class II type.[131] The nonresponse by one animal and the low responsiveness of two other animals was directly associated with two BoLA class II types: EDF (Edinburgh DR-like Focusing) types 5 and 11. The other EDF types examined were associated with both high and low responsiveness. Whether these observations directly correlate with the quantity of specific antibody or the isotype profile generated is unknown, but both aspects are currently being examined.

Table 11-7 **T Cell Epitope Predictions for FMDV O1 Kaufbeuren RNA polymerase (P3D). Legend as for Table 11-3**

```
P3D          1         2         3         4         5         6
Seq GLIVDTRDVEERVHVMRKTKLAPTVAHGVFNPEFGPAALSNKDPRLNEGVVLDEVIFSKH
1   ............................................................
2   .............RRRR..RRRR...RRRR..................RRRRRRRRRR..
3   ...........RRRRRRRRRRRRRRRRRRRRRR..................RRRRRRRR.
4   ............................................................
5   ...........SSSSSSSSSSSSSSSSS................................
6   ............................................................
7   ................DDDD........................................

             7         8         9        10        11        12
Seq KGDTKMSEEDKALFRRCAADYASRLHSVLGTANAPLSIYEAIKGVDGLDAMEPDTAPGLP
1   ..............AAAAAAAAAAAAAAAA......AAAAAAAAAAAAAA..........
2   ..........RRRRR....RRRR......RRRR......RRRR.....RRRR.RRRRR..
3   ...................................................RRRRRRRR.
4   .................SSSSSSSSSSSSSSSSSS..SSSSSSSSSSSSSSSS.......
5   ..................SSSSSSSSSSSS..SSSSSSSSSSSS................
6   .....................DDDDDDDDD.DDDDDDDDDD...................
7   ............................................................

            13        14        15        16        17        18
Seq WALQGKRRGALIDFENGTVGPEVEAALKLMEKREYKFVCQTFLKDEIRPLEKVRAGKTRI
1   .........................AAA..............AAAAAAAAAAAA....AA
2   ........RRRRR...RRR....RRRRRRRR...........................RR
3   ......RRRRRRRR.......RRRRRRRRRRRR.......................RRRR
4   .......................................................SSSSS
5   ..........SSSSSSS.....SSSSSSSS..............................
6   ....................................................DDDDDD..
7   .....................................................DDDDDD.

            19        20        21        22        23        24
Seq VDVLPVEHILYTRMMIGRFCAQMHSNNGPQIGSAVGCNPDVDWQRFGTHFAQYRNVWDVD
1   AAAA............AAAAAAA......AAA.............AAA....AAA.....
2   RR.....RRRRRRRRRR...............................RRRR........
3   RRRR.RRRRRRRRRRRRR..........................................
4   SSSSS.......................................................
5   ............................................................
6   ............................................................
7   ............................................................
```

To determine whether the poor protection elicited by "FMDV loop" peptide vaccines was due to nonresponsiveness at the T cell level, Collen et al.[37,84] used primary *in vitro* T cell proliferation assays of naive cattle lymphocytes to identify animals which could or could not respond to VP1 21-40. Individual "responder" and "nonresponder" animals thus identified did not proliferatively respond to VP1 141-160 and were immunized with either commercial virus vaccine or a peptide antigen in which VP1 21-40 was colinearly synthesized with VP1 141-160 and delivered as an emulsion in Freund's incomplete adjuvant. This gave four possible priming scenarios: virus immunization of "responder" (1) or "nonresponder" (2), and peptide immunization of "responder" (3) or "nonresponder" (4), where the responder status was with reference to VP1 21-40. Antiviral and anti-peptide T cell and antibody responses were assessed and, after 21 days, the animals were challenged with live virus and examined for clinical symptoms of FMD.

Table 11-7 **(Continued)**

```
            25        26        27        28        29        30
Seq YSAFDANHCSDAMNIMTEEVFRTEFGFHPNAEWILKTLVNTEHAYENKRIYVGGGMPSGC
1   ........AAAAA..........................................AAAAA
2   ..........RRRR....RRRR.........RRRRRRRRR..RRRR..RRRRR.......
3   ................RRRRRRRR................RRRRRRRR............
4   ..........................SSSSSSSSSSSSSSSSSSS...........SSSS
5   ...SSSSSSSSSSS..............................................
6   ............................................................
7   ............................................................

            31        32        33        34        35        36
Seq SATSIINTILNNIYVLYALRRHYEGVELDTYTMISYGDDIVVASDYDLDFEALKPHFKSL
1   .AAA.AAAAAA........AAAA.....AAAAA......................AAAAA
2   ............................RRRR..................RRRR......
3   ..........................RRRRRRRR..........................
4   SSSSSSSSSSSSSSSS..SSSSSSSSSSSSSSSSSSS........SSSSSSSSSS.....
5   .........SSSSSSSSSSSSSSSS..SSSSSSSSSS.......................
6   ............DDDDDD.........................................D
7   .....................................................DDDDD..

            37        38        39        40        41        42
Seq GQTITPADKSDKGFVLGHSITDVTFLKRHFHMDYGTGFYKPVMASKTLEAILSFARRGTI
1   AAAA...............................AAAAA.....AAAAAAAA.......
2   ............RRRRR...................RRRR.....RRRRRRRR....RRR
3   ..........RRRRRRRRR...............RRRRRRRR..................
4   ...........................................SSSSSSSSSSSSSSS.S
5   ................SSSSSSSSSSSSSSSSSSS.....SSSSSSSSSSSSSSSSSS..
6   DDDDD..........................................DDDDDD.......
7   ...........................DDDD.............................

            43        44        45        46        47
Seq QEKLISVAGLAVHSGPDEYRRLFEPFQGLFEIPSYRSLYLRWVNAVCGDA
1   .....AAA.......AAAAAAAAAAAAAAAAAAAA....AAAAAA.....
2   R.RRRR..RRRRR.......RRRR...RRRR.........RRRR......
3   ..................................................
4   SSSSSSSSS.............................SSSSSSSSSS..
5   SSSSSSSSS.........................................
6   ..................................................
7   ..................................................
```

From 7 days post-immunization, the peptide- and virus-immunized "responders" recognized their respective immunizing antigen, VP1 21-40, and virus in an *in vitro* proliferation assay confirming that the epitope contained in VP1 21-40 represented a naturally processed viral determinant. Similarly, both animals developed a neutralizing antiviral antibody response. In contrast, both of the "nonresponder" cattle failed to develop an anti-peptide T cell response and the peptide-immunized "nonresponder" animal also failed to make an antibody response to either peptide or virus. However, antiviral T and B cell responses were induced in the virus-immunized "non responder", showing that the nonresponsiveness was not a general failure of immunization. The peptide-immunized "nonresponder" was the only animal not protected when the cattle were challenged with live virus. Similar responses could not be induced by the T and B cell peptides when they were used as separate peptides either *in vitro*[37] or *in vivo,* indicating that linked recognition may be an important requirement for priming a protective response (unpublished observations). These results demonstrate, albeit in a small number of animals, that nonresponsiveness in cattle to the VP1 141-160 "FMDV loop" peptide can be overcome without recourse to epitopes from heterologous antigens as was shown in mice by Francis et al.,[51] and that a strong T cell response is required for protective immunity. More importantly, protection was achieved with 0.5 mg of peptide which is ten-fold lower than the dose required for single dose protection with FMDV-15.[48,66] T cells isolated from the peptide-immunized "responder" were $CD4^+$ class II-restricted and produced large amounts of IFN-γ and IL-2 mRNA in response to *in vitro* stimulation with virus, suggesting that they might be capable of eliciting TH1 or at least TH0 effector functions *in vivo* (H. Takamatsu, personal communication).

The observation by Francis et al.[51] that the $H\text{-}2^k$ restriction of the murine antibody response to VP1 141-160 could be overcome by the addition of an $I\text{-}A^d$-restricted T cell epitope from another protein raises an important consideration for the design of a peptide vaccine, namely that the T cell help need not come from the same protein as the B cell epitope. A conserved T cell epitope with degenerate MHC binding would reduce the number of helper epitopes required for different serotype vaccines while at the same time increasing the probability that different individuals will be able to respond. Early studies using haptens and small proteins suggested that to stimulate an antibody response, the B and T cell epitopes must be on the same molecule (i.e., covalently linked)[134] and this was termed *intermolecular help.*[135] Subsequent studies identified that noncovalently linked proteins which are part of a more complex structure such as a virus or cell could also provide help[135] for remote B cell epitopes and this was termed *intrastructural help. Intermolecular intrastructural help* has been demonstrated for influenza virus[136,137] and hepatitis B virus.[138] Studies with influenza virus[137] have indicated that although T cells specific for surface or internal proteins can provide help for antibody to surface or internal proteins, help is most effectively provided by T cells specific for the same protein as the B cell. The ability of T cells which are specific for internal virus proteins (such as matrix to help B cells specific for surface protein such as hemagglutinin) suggests a role for B cells in the uptake and presentation of these proteins to T cells. In the case of FMDV, the most serotype-conserved protein is VP4 followed by the RNA-dependent RNA polymerase (P3D). VP3 and P3D have been shown to be cross-reactively recognized by T cells[37,82] although the response to VP4 has not yet been determined. Nevertheless, cognate intermolecular intrastructural help for structural protein-specific B cells might be effected by epitopes from P3D equally as well as from structural proteins VP3 or VP4.

Interestingly, the serotype-specific memory T cell epitopes VP1 21-40 and VP2 54-72 are contiguous or overlapping with B cell sites 3 and 2, respectively, in the three-dimensional structure of type O1 strain BFS, whereas the serotype cross-reactive epitope VP3 97-110 is not.[37,139] Similar observations have been made for memory helper T and B cell epitopes from influenza virus[140,141] in mice and poliovirus[142] in man. One mechanism which could account for this linkage would be cognate interaction, wherein the B cell serves a dual role as the antigen presenting cell (APC) and as the effector and the recipient of T cell help. The role of surface immunoglobulin in the function of B cells as APCs is well established,[143,144] it has been postulated that during secondary responses the most efficient APCs are memory B cells[145] with high-affinity surface-immunoglobulin and that this confers a selective advantage to the B cell in recruiting helper T cells when the antigen concentration becomes limiting. Assuming that no factors other than proteases influence the products of processing, then the class II isotype of the individual will be the major factor in determining which antigenic fragments are presented. However, there is some evidence to suggest that the specificity of the surface immunoglobulin influences the fragments generated by processing.[146] Together, these factors would favor the selection of high-affinity memory B cells and T cells which recognize the antigens presented by them. However, T cell epitopes recognized by poliovirus-specific mouse CTL are also located in the same regions as neutralizing B cell epitopes.[147] Thus, an alternative explanation for the juxtaposition of T cell and B cell epitopes could be that the very same exposed loops favored for recognition by B cells provide the most easily cleaved peptides during antigen processing. Either way, these data suggest that there should be a T cell epitope(s) located in the region of the "FMDV loop" which has not yet been detected.

Several studies have demonstrated that the immunogenicity of peptides can be improved by coupling T cell and B cell epitopes.[51,65,82,148-153] However, these studies also indicate that it is necessary to consider the orientation of the B cell and T cell epitopes within the construct. Some studies have found that antibody production to the B cell epitope requires the T cell determinant to be in the carboxy-terminal position relative to the B cell epitope,[51,151] whereas others have found the reverse.[65,82,152,153] In particular, Partidos *et al.*[153] found that when the T cell epitope was in the carboxy-terminal position, the antibody which was induced was specific for the T cell rather than the B cell epitope. Golvano et al.[152] have suggested that the orientation of the epitopes operates at two levels: one is related to the polarity of the peptide and its ability to adopt a particular conformation, and the other relates to the accessible cleavage sites created by that conformation. Specifically, the joining of two sequences could either mask, expose, or create cleavage sites, and the appropriate orientation would allow the T cell epitope to be released intact from the larger peptide after internalization by the APC.

VI. SUMMARY

The importance of antibody in the protective immune response of cattle against FMD was established almost 100 years ago.[3] However, the correlation established between neutralizing antibody and protection and the identification of the linear "FMDV loop" B cell site have led to a rather dogmatic approach to subunit vaccine development based on the ability of candidate antigens to induce *in vitro* neutralizing antibody. There is a growing body of circumstantial and direct evidence to indicate that cell-mediated immunity is required and may be crucial for protection, although the precise nature of this involvement still has to be defined.

One role which cell-mediated immunity might play is in the regulation of the humoral response. This chapter has attempted to illustrate how prediction and characterization of epitopes recognized by $CD4^+$ T cells can be used to study the functional significance of T cells in cattle, albeit in very few animals, and to overcome MHC restricted nonresponsiveness to a candidate peptide vaccine antigen. The number of epitopes required for vaccine development could be potentially reduced if serotype cross-reactive epitopes are used, although the major limitation will be their effective recognition in the context of different MHC molecules. A further consideration is that of memory. It is generally considered that good B cell memory requires an active cell-mediated response although persistence of antigen is also a major factor. However, if memory is improved by the introduction of T cell epitopes into a candidate peptide vaccine molecule, it may still be necessary to address whether the conserved arginine-glycine-aspartic acid (RGD) motif[154] present in the "FMDV loop" is acceptable or detrimental in this context. The RGD motif is found in many ligands of integrins involved in cell adhesion, and it has recently been shown that inoculation of the peptide glycine-arginine-glycine-aspartic acid-serine-proline containing the RGD motif into mice can abrogate T cell-mediated immune responses *in vivo.*[155] Finally, it is likely that non-antibody mediated effector mechanisms such as CTL and NK cells will prove to be important in the regulation of the humoral response and the elimination of virus *in vivo* due to cytokine production and killing of infected cells.

With the cessation of vaccination in Europe, the need for an effective strategic reserve subunit vaccine will increase. It is currently unclear what form such a vaccine will take, but it is essential that the development of subunit vaccines gives due consideration to the induction and maintenance of cell-mediated immunity as well as to the specificity of the antibody response.

REFERENCES

1. **Kitching, R. P.,** The application of biotechnology to the control of foot-and-mouth disease virus, *Br. Vet. J.*, 148, 375, 1992.
2. **Donaldson, A. I. and Doel, T. R.,** Foot-and-mouth disease: the risk for Great Britain after 1992, *Vet. Rec.*, 131, 114, 1992.
3. **Loeffler, F. and Frosch, P.,** Summarischer Bericht uber Deiergebnisse der Utersuchungen zur Erforschung der Maul-und-Klauenseuche, *Zentrablatt Bakteriol., Parasitenk., Infectionskrankh. Hyg., Abt. I, Original*, 22, 257, 1897.
4. **Mackowiak, C., Lang, C., Fontaine, J., Camand, R., and Petermann, H. G.,** Relationship between neutralising antibody titre and protection in animals immunised against foot-and-mouth disease, *Ann. Inst. Pasteur*, 103, 252, 1962.
5. **Van Bekkum, J. G.,** Correlation between serum antibody level and protection against challenge with FMD virus, presented at the Meeting of the Standing Tech. Committee of the Eur. Commission for the Control of Foot-and-Mouth Disease, Brescia, Italy, FAO Rome, 1969, 38.
6. **Pay, T. W. F. and Hingley, P.J.,** Correlation of 140S antigen dose with the serum neutralising antibody response and the level of protection induced in cattle by foot-and-mouth disease vaccines, *Vaccine*, 5, 60, 1987.

7. **Strohmaier, K., Franze, R., and Adam, K.-H.,** Location and characterisation of the antigenic portion of the FMDV immunising protein, *J. Gen. Virol.*, 59, 295, 1982.
8. **Pfaff, E., Mussgay, M., Böhm, H. O., Schulz, G. E., and Schaller, H.,** Antibodies against a preselected peptide recognise and neutralise foot-and-mouth disease virus, *EMBO J.*, 1, 869, 1982.
9. **Bittle, J. L., Houghton, R. A., Alexander, H., Shinnick, T. M., Sutcliffe, J. G., Lerner, R. A., Rowlands, D. J., and Brown, F.,** Protection against foot-and-mouth disease by immunisation with a chemically synthesised peptide predicted from the nucleotide sequence, *Nature*, 298, 30, 1982.
10. **Vitteta, E. S., Fernandez-Botran, R., Myers, C. D., and Sanders, V. M.,** Cellular interactions in the humoral immune response, *Adv. Immunol.*, 45, 1, 1990.
11. **Benacerraf, B. and McDevitt, H. O.,** Histocompatibility-linked immune response genes, *Science*, 175, 273, 1972.
12. **Berzofsky, J. A., Cease, K. B., Cornette, J. L., Spouge, J. L., Margalit, H., Berkower, I. R., Good, M. F., Miller, L. H., and DeLisi, C.,** Protein antigenic structures recognised by T cells: potential applications to vaccine design, *Immunol. Rev.*, 98, 9, 1987.
13. **Livingstone, A. M. and Fathman, C. G.,** The structure of T cell epitopes, *Annu. Rev. Immunol.*, 5, 477, 1987.
14. **Berzofsky, J. A.,** Ir genes: antigen-specific genetic regulation of the immune response, in *The Antigens*, Sela, M., Ed., Academic Press, New York, 1987, 1.
15. **Berkower, I., Buckenmeyer, G. K., Gurd, R. N., and Berzofsky, J. A.,** A possible immunodominant epitope recognised by murine T lymphocytes immune to different myoglobins, *Proc. Natl. Acad. Sci. U.S.A.*, 79, 4723, 1982.
16. **Manca, F., Clarke, J. A., Miller, A., Sercarz, E. E., and Shastri, N.,** A limited region within hen egg-white lysozyme serves as a focus for a diversity of T cell clones, *J. Immunol.*, 133, 2075, 1984.
17. **Berkower, I. J., Buckenmeyer, G. K., and Berzofsky, J. A.,** Molecular mapping of a histocompatibility-restricted immunodominant epitope with synthetic and natural peptides: implications for T cell antigenic structure, *J. Immunol.*, 136, 2498, 1986.
18. **Berzofsky, J. A.,** Structural features of protein antigenic sites recognised by helper T cells: what makes a site immunodominant? in *The Year in Immunology 1985-1986*, 2, Cruze, J. M. and Lewis, R. E., Jr., Eds., Karger, Basel, 1986, 28.
19. **Kojima, M., Cease, K. B., Buckenmeyer, G. K., and Berzofsky, J. A.,** Limiting dilution comparison of the repertoires of high and low responder MHC-restricted T cells, *J. Exp. Med.*, 167, 1100, 1988.
20. **Fracastorius, H.,** De sympathia et antipathia rerum liben unus, in *De Contagione et Contagiosis Morbis et Curatione Libri III.*, Venice, Heirs of L.A. Junta 1, 1549, 12. Translation in *Heironymus Fracastorius: Contagion, Contagious Diseases and Their Treatment*, Wright, W. C., Ed., Putnam's Sons, G. P., London, 1930, 52.
21. **Pereira, H. G.,** Subtyping of foot-and-mouth disease virus, in *Developments in Biological Standardisation 35*, Mackowiak, C. and Regemey, R. H., Eds., Karger, S., Basel, 1977, 167.
22. **Archarya, R., Fry, E., Stuart, D., Rowlands, D., and Brown, F.,** The three dimensional structure of foot-and-mouth disease at 2.9 Å resolution, *Nature*, 337, 709, 1989.
23. **Sard, D. M.,** Clinical aspects of FMD, *Vet. Rec.*, 102, 186, 1978.
24. **Burrows, R., Mann, J. A., Garland, A. J. M., Greig, A., and Goodridge, D.,** The pathogenesis of natural and simulated natural foot-and-mouth disease infection in cattle, *J. Comp. Pathol.*, 91, 599, 1981.
25. **Sellers, R. F.,** Quantitative aspects of the spread of foot-and-mouth disease, *Vet. Bull.*, 41, 431, 1971
26. **Gibson, C. F. and Donaldson, A. I.,** Exposure of sheep to natural aerosols of foot-and-mouth disease virus, *Res. Vet. Sci.*, 41, 45, 1986.
27. **Yilma, T., McVicar, J. W., and Breese, S. S.,** Pre-lytic release of foot-and-mouth disease virus in cytoplasmic blebs, *J. Gen. Virol.*, 41, 105, 1978.
28. **De la Torre, J. C., Martínez-Salas, E., Díez, J., Villaverde, A., Gebauer, F., Rocha, F., Dávila, M., and Domingo, E.,** Coevolution of cells and viruses in a persistent infection of foot-and-mouth disease virus in cell culture, *J. Virol.*, 62, 2050, 1988.
29. **Díez, J., Hofner, M., Domingo, E., and Donaldson, A. I.,** FMDV strains isolated from persistently infected cell cultures are attenuated for mice and cattle, *Virus Res.*, 18, 3, 1990.
30. **Costa Giomi, M. P., Gomes, I., Tiraboschi, B., Auge de Mello, P., Bergmann, I. E., Scodeller, E. A., and La Torre, J. L.,** Heterogeneity of the polyribocytidilic acid tract in aphthovirus: changes in the size of the poly(C) of viruses recovered from persistently infected cattle, *Virology*, 162, 58, 1988.

31. **Díez, J., Dávila, M., Escarmís, C., Mateu, M. G., Dominguez, J., Pérez, J. J., Giralt, E., Melero, J. A., and Domingo, E.,** Unique amino acid substitutions in the capsid proteins of foot-and-mouth disease virus from a persistent infection in cell culture, *J. Virol.*, 64, 5519, 1990.
32. **Donn, A.,** The Pathogenesis of Persistence of Foot-and-Mouth Disease Virus in Experimentally Infected Cattle and a Model Cell System, Ph.D. thesis, University of Hertfordshire, 1993.
33. **Van Bekkum, J. G., Fish, R. C., and Dale, C. N.,** Immunogenic studies in Dutch cattle vaccinated with foot-and-mouth disease vaccines under field conditions. I. Neutralising antibody responses to O and A types, *Am. J. Vet. Res.*, 24, 77, 1963.
34. **Van Bekkum, J. G., Fish, R. C., and Nathans, I.,** Immunogenic studies in Dutch cattle vaccinated with foot-and-mouth disease vaccines under field conditions: neutralising antibody responses to O, A and C types, *Am. J. Vet. Res.*, 30, 2125, 1969.
35. **Black, L., Nicholls, M. J., Rweyemamu, M. M., Ferrari, R., and Zunino, M. A.,** Foot-and-mouth vaccination: a multifactorial study of the influence of antigen dose and potentially competitive immunogens on the response of cattle of different ages, *Res. Vet. Sci.*, 40, 303, 1986.
36. **Brown, F., Cartwright, B., and Newman, J. F. E.,** Further studies of the early antibody in the sera of cattle and guinea pigs infected with foot-and-mouth disease virus, *J. Immunol.*, 93, 397, 1964.
37. **Collen, T.,** T-Cell Responses of Cattle to Foot-and-Mouth Disease Virus, Ph.D. thesis, Council for National Academic Awards, 1991.
38. **Cunliffe, H. R.,** Observations on the duration of immunity in cattle after experimental infection with foot-and-mouth disease virus, *Cornell Vet.* 54, 501, 1964.
39. **Brown, F., Cartwright, B, and Stewart, D. L.,** The effect of various inactivating agents on the viral ribonucleic acid infectivities of foot-and-mouth and on its attachment to susceptible cells, *J. Gen. Microbiol.*, 31, 179, 1963.
40. **Wild, T. F., Burroughs, J. N., and Brown, F.,** Surface structure of foot-and-mouth disease virus, *J. Gen. Virol.*, 4, 313, 1969.
41. **Laporte, J., Grosclaude, J., Wantygheim, J., Barnard, S., and Rouzé, P.,** Neutralisation en culture cellulaire infectieux du virus de la fièvre aphteuse par des sérums provenant de porcs immunisés à l'aide d'une protéine virale purifée, *Compt. Rend. Acad. Sci.*, 276D, 3399, 1973.
42. **Bachrach, H. L., Moore, D. M., McKercher, P. D., and Polatnick, J.,** Immune and antibody responses to an isolated capsid protein of foot-and-mouth disease virus, *J. Immunol.*, 115, 1636, 1975.
43. **Bachrach, H. L., Moore, D. M., McKercher, P. D., Moore, D. M., and Robertson, B. H.,** Foot-and-mouth disease virus: immunogenicity and structure of fragments derived from capsid protein VP3 and of virus containing cleaved VP3, *Vet. Microbiol.*, 7, 85, 1982.
44. **Stave, J. W., Card, J. L., Morgan, D. O., and Vakharia, V. N.,** Neutralisation sites of type O1 foot-and-mouth disease virus defined by monoclonal antibodies and neutralisation-escape virus variants, *Virology*, 162, 21, 1988.
45. **Kitson, J. D. A., MaCahon, D., and Belsham, G. J.,** Sequence analysis of monoclonal antibody resistant mutants of type O foot-and-mouth disease virus: evidence for the involvement of the three surface exposed capsid proteins in four antigenic sites, *Virology*, 179, 26, 1990.
46. **Thomas, A. A. M, Woortmeijer, R. J., Puijk, W., Bartling, S. J.,** Antigenic sites on foot-and-mouth disease type A10, *J. Virol.*, 62, 2782, 1988.
47. **Francis, M. J., Fry, C. M., Rowlands, D. J., Bittle, R. A., Houghten, R. A., Lerner, R. A., and Brown, F.,** Immune response to uncoupled peptides of foot-and-mouth disease virus, *Immunology*, 61, 1, 1987.
48. **DiMarchi, R., Brooke, G., Gale, C., Cracknell, V., Doel, T. R., and Mowat, N.,** Protection of cattle against foot-and-mouth disease by a synthetic peptide, *Science*, 232, 639, 1986.
49. **Murdin, A. D. and Doel, T. R.,** Synthetic peptide vaccines against foot-and-mouth disease. I. Duration of the immune response and priming in guinea-pigs, rabbits and mice, *J. Biol. Standard.*, 15, 39, 1987.
50. **Wiesmüller, K.-H., Jung, G., and Hess, G.,** Novel low-molecular weight synthetic vaccine against foot-and-mouth disease containing a potent B cell and macrophage activator, *Vaccine*, 7, 29, 1989.
51. **Francis, M. J., Hastings, G. Z., Syred, A. D., McGinn, B., Brown, F., and Rowlands, D. J.,** Non-responsiveness to a foot-and-mouth disease virus peptide overcome by addition of foreign helper T cell determinants, *Nature*, 330, 168, 1987.

52. **Yarov, A. V., Gel'fanov, V. M., Grechaninova, L. A., Yu Surovoi, A., Vol'pina, O. M., Ivanov, V. T., Chepurkin, A. V., and Ivanyushenkov, V. N.,** Antigenic structure of foot-and-mouth disease virus. V. Protection of susceptible animals with the use of synthetic peptide from the foot-and-mouth disease virus, *Biorganicheskaya Khimiya*, 15, 1313, 1989.
53. **Vol'pina, O. M., Yu Surovoi, A., and Ivanov, V. T.,** Protection against viral infections with the aid of synthetic peptides. Foot-and-mouth disease as an example, *Biomed. Sci.*, 1, 23, 1990.
54. **Winther, M. D., Allen, G., Bomford, R. H., and Brown, F.,** Bacterially expressed antigenic peptide from foot-and-mouth disease virus capsid elicits variable immunologic responses in animals, *J. Immunol.*,136, 1835, 1986.
55. **Broekuijsen, M. P., Blom, T., Kottenhagen, M., Pouwels, P. H., Meloen, R. H., Barteling, S. J., and Enger-Valk, B. E.,** Synthesis of fusion proteins containing antigenic determinants of foot-and-mouth disease virus, *Vaccine*, 4, 119, 1986.
56. **Broekuijsen, M. P., Van Rijn, J. M. M., Blom, M., Pouwels, P. H., Enger-Valk, B. E., Brown, F., and Francis, M. J.,** Fusion proteins with multiple copies of the major antigenic determinant of foot-and-mouth disease virus protect both the natural host and laboratory animals, *J. Gen. Virol.*, 68, 3137, 1987.
57. **Clarke, B. E., Newton, S. E., Carroll, A. R., Francis, M. J., Appleyard, G., Syred, A. D., Highfield, P. E., Rowlands, D. J., and Brown, F.,** Improved immunogenicity of a peptide epitope after fusion to hepatitis B core protein, *Nature*, 330, 381, 1987.
58. **Agterberg, M., Adriaanse, H., Lankhof, H., Meloen, R., and Tommassen, J.,** Outer membrane PhoE protein of *Escherichia coli* as a carrier for foreign antigenic determinants: immunogenicity of epitopes of foot-and-mouth disease virus, *Vaccine*, 8, 85, 1990.
59. **Morgan, D. O. and Moore, D. M.,** Protection of cattle and swine against foot-and-mouth disease, using biosynthetic peptide vaccines, *Am. J. Vet. Res.*, 51, 40, 1990.
60. **Steward, M. W., Stanley, C. M., DiMarchi, R., Mulcahy, G., and Doel, T. R.,** High-affinity antibody induced by immunisation with a synthetic peptide is associated with protection of cattle against foot-and-mouth disease, *Immunology*, 72, 99, 1991.
61. **Mulcahy, G., Gale, C., Robertson, P., Iyisan, S., DiMarchi, R. D., and Doel, T. R.,** Isotype responses of infected, virus-vaccinated and peptide-vaccinated cattle to foot-and-mouth disease virus, *Vaccine*, 8, 249, 1990.
62. **Mulcahy, G., Pullen, L., Gale, C., DiMarchi, R. D., and Doel, T. R.,** Mouse protection test as a predictor of the protective capacity of synthetic foot-and-mouth disease vaccines, *Vaccine*, 9, 19, 1991.
63. **McCullough, K. C., Crowther, J. R., Butcher, R. N., Carpenter, W. C., Brocchi, E., Capucci, L., and De Simone, F.,** Immune protection against foot-and-mouth disease virus studied using virus neutralising and non-neutralising concentrations of monoclonal antibodies, *Immunology*, 58, 421, 1986.
64. **McCullough, K. C., Parkinson, D., and Crowther, J. R.,** Opsonisation-enhanced phagocytosis of foot-and-mouth disease virus, *Immunology*, 65, 187, 1988.
65. **Doel, T. R., Doel, C. M. F. A., Staple, R. F., and DiMarchi, R.,** Cross-reactive and serotype-specific antibodies against foot-and-mouth disease virus generated by different regions of the same synthetic peptide, *J. Virol.*, 66, 2187, 1992.
66. **Doel, T. R., Gale, C., Do Ameral, C. M. C. F., Mulcahy, G., and DiMarchi, R.,** Heterotypic protection induced by synthetic peptides corresponding to three serotypes of foot-and-mouth disease virus, *J. Virol.*, 64, 2260, 1990.
67. **Mills, K. H. G.,** Recognition of foreign antigen by T cells and their role in immune protection, *Curr. Opin. Infect. Dis.*, 2, 804, 1989.
68. **Zinkernagel, R. M. and Doherty, P. C.,** Immunological surveillance against altered self components by sensitised T-lymphocytes in lymphocytic choriomeningitis, *Nature*, 251, 547, 1974.
69. **Brodsky, F. M. and Guagliardi, L. E.,** The cell biology of antigen processing and presentation, *Annu. Rev. Immunol.*, 9, 707, 1991.
70. **Mosmann, T. R. and Coffman, R. L.,** Heterogeneity of cytokine secretion patterns and functions of helper T cells, *Adv. Immunol.*, 46, 111, 1989.
71. **Romagnani, S.,** Human TH1 and TH2 subsets: doubt no more, *Immunol. Today*, 12, 256, 1991.

72. **Salgame, P., Abrams, J. S., Clayberger, C., Goldstein, H., Convit, J., Modlin, R. L., and Bloom, B. R.,** Differing lymphokine profiles of functional subsets of human CD4 and CD8 T cell clones, *Science*, 254, 279, 1991.
73. **Coutelier, J.-P., Van Der Logt, J. T. M., Heesen, F. W. A., Warner, G., and Van Snick, J.,** IgG_{2a} restriction of murine antibodies elicited by viral infections, *J. Exp. Med.*, 165, 58, 1987.
74. **Romagnani, S.,** Induction of TH1 and TH2 responses: a key role for the "natural" immune response?, *Immunol. Today*, 13, 379, 1992.
75. **Snapper, C. M. and Mond, J. J.,** Towards a comprehensive view of immunoglobulin class switching, *Immunol. Today*, 14, 15, 1993.
76. **Collen, T., Pullen, L., and Doel, T. R.,** T cell-dependent induction of antibody against foot-and-mouth disease virus in a mouse model, *J. Gen. Virol.*, 70, 395, 1989.
77. **Knudsen, R. C., Groocock, C. M., and Anderson, A. A.,** Immunity to foot-and-mouth disease virus in guinea pigs: clinical and immune responses, *Infect. Immun.*, 24, 787, 1979.
78. **Amadori, M., Archetti, I. L., Verardi, R., and Berneri, C.,** Isolation of mononuclear cytotoxic cells from cattle vaccinated against foot-and-mouth disease, *Arch. Virol.*, 122, 293, 1992.
79. **Sharma, R., Presad, S., Ahuja, K. L., Rahman, M. M., and Kumar, A.,** Cell mediated immune response following foot-and-mouth disease vaccination in buffalo calves, *Acta Virol.*, 29, 509, 1985.
80. **Soós, T. and Sándor, T.,** Három különbözö O1 típisú ragadós száj-és körömfájás vírustörzs immunogén tulajdonságának vizgálata. II. A celluláris immunitás vizsgálata, *Allatogyogyaszati Oltoanyagellenorzo Intezet Evkonye*, 167, 1986.
81. **Sigal, L. J., Gomez, G., and Braun, M.,** Changes in mononuclear peripheral blood in cattle with foot-and-mouth disease, *Vet. Immunol. Immunopathol.*, 30, 431, 1992.
82. **Collen, T. and Doel, T. R.,** Heterotypic recognition of foot-and-mouth disease virus by cattle lymphocytes, *J. Gen. Virol.*, 71, 309, 1990.
83. **Sáiz, J. C., Rodríguez, A., González, M., Alonso, F., and Sobrino, F.,** Heterotypic lymphoproliferative responses in pigs vaccinated with foot-and-mouth disease virus. Involvement of isolated capsid proteins, *J. Gen. Virol.*, 73, 2601, 1992.
84. **Collen, T., DiMarchi, R., and Doel, T. R.,** A T cell epitope in VP1 of foot-and-mouth disease is immunodominant for vaccinated cattle, *J. Immunol.*, 146, 749, 1991.
85. **Saalmüller, A., Reddehase, M. J., Bühring, H.-J., Jonjic, S., and Koszinowski, U. H.,** Simultaneous expression of CD4 and CD8 antigens by a substantial proportion of resting porcine T lymphocytes, *Eur. J. Immunol.*, 17, 1297, 1987.
86. **Van Bleek, G. M. and Nathenson, S. G.,** Isolation of an endogenously processed immunodominant viral peptide from the class I H-$2K^b$ molecule, *Nature*, 348, 213, 1991.
87. **Falk, K., Rötzschke, O., Stevanovic, S., Jung, G., and Rammensee, H.-G.,** Allele-specific motifs revealed by sequencing of self-peptides eluted from MHC molecules, *Nature*, 353, 290, 1991.
88. **Jardetzky, T. S., Lane, W. S., Robinson, R. A., Madden, D. R., and Wiley, D. C.,** Identification of self peptides bound to purified HLA-B27, *Nature*, 353, 326, 1992.
89. **Rudensky, A. Y., Preston-Hurlburt, P., Hong, S.-C., Barlow, A., and Janeway, C. A., Jr.,** Sequence analysis of peptides bound to MHC class II molecules, *Nature*, 353, 622, 1991.
90. **Rudensky, A. Y., Preston-Hurlburt, P., Al-Ramadi, B. K., Rothbard, J., and Janeway, C. A., Jr.,** Truncation variants of peptides isolated from MHC class II molecules suggest sequence motifs, *Nature*, 359, 429, 1992.
91. **Hunt, D. F., Michel, H., Dickinson, T. A., Shabanowitz, J., Cox, A. L., Sakaguchi, K., Appella, E., Grey, H. M., and Sette, A.,** Peptides presented to the immune system by murine class II major histocompatibility complex molecule I-A^d, *Science*, 256, 1817, 1992.
92. **Kropshofer, H., Max, H., Müller, C. A., Hesse, F., Stevanovic, S., Jung, G., and Kalbacher, H.,** Self-peptide released from class II HLA-DR1 exhibits a hydrophobic two-residue contact motif, *J. Exp. Med.*, 175, 1799, 1992.
93. **Bjorkman, P. J., Saper, M. A., Samraoui, B., Bennett, W. S., Strominger, J. L., and Wiley, D. C.,** Structure of the human class I histocompatibility antigen, HLA-A2, *Nature*, 329, 506, 1987.
94. **Bjorkman, P. J., Saper, M. A., Samraoui, B., Bennett, W. S., Strominger, J. L., and Wiley, D. C.,** The foreign antigen binding site and T cell recognition regions of class I histocompatibility antigens, *Nature*, 329, 512, 1987.

95. **Garrett, T. P. J., Saper, M. A., Bjorkman, P.J., Strominger, J. L., and Wiley, D. C.**, Specificity pockets for the side chains of peptide antigens in HLA-Aw68, *Nature*, 342, 692, 1989.
96. **Madden, D. R., Gorga, J. C., Strominger, J. L., and Wiley, D. C.**, The structure of HLA-B27 reveals nonamer self-peptides bound in an extended conformation, *Nature*, 353, 321, 1991.
97. **Fremont, D. H., Matsumura, M., Stura, E. A., Peterson, P. A., and Wilson, I. A.**, Crystal structures of two viral peptides in complex with murine MHC class I H-2K^b, *Science*, 257, 919, 1992.
98. **Brown, J. H., Jardetzky, T., Saper, M. A., Samraoui, B., Bjorkman, P. J., and Wiley, D. C.**, A hypothetical model of the foreign antigen binding site of class II histocompatibility molecules, *Nature*, 332, 845, 1988.
99. **Peccoud, J., Dellabona, P., Allen, P., Benoist, C., and Mathis, D.**, Delineation of antigen contact residues on an MHC class II molecule, *EMBO J.*, 9, 4215, 1990.
100. **Delovitch, T. L. and Lang, Y.**, MHC class II molecules may function as a template for the processing of a partially processed epitope into a T cell epitope, *J. Cell. Biochem.*, S17C, 67, 1993.
101. **Sette, A., Buus, S., Colon, S., Smith, J. A., Miles, C., and Grey, H. M.**, Structural characteristics of an antigen required for its interaction with Ia and recognition by T cells, *Nature*, 328, 395, 1987.
102. **DeLisi, C. and Berzofsky, J. A.**, T cell antigenic sites tend to be amphipathic structures, *Proc. Natl. Acad. Sci. U.S.A.*, 82, 7042, 1985.
103. **Margalit, H., Spouge, J. L., Cornette, J. L., Cease, K. B., DeLisi, C., and Berzofsky, J. A.**, Prediction of immunodominant helper T cell antigenic sites from the primary sequence, *J. Immunol.*, 138, 2213, 1987.
104. **Spouge, J. L., Guy, H. R., Cornette, J. L., Margalit, H., Cease, K., Berzofsky, J. A., and DeLisi, C.**, Strong conformational propensities enhance T cell antigenicity, *J. Immunol.*, 138, 204, 1987.
105. **Good, M. F., Maloy, W. L., Lunde, M. N., Margalit, H., Cornette, J. L., Smith, G. L., Moss, B., Miller, L. H., and Berzofsky, J. A.**, Construction of synthetic immunogen: use of new T-helper epitope on malaria circumsporozoite protein, *Science*, 235, 1059, 1987.
106. **Partidos, C. D. and Steward, M. W.**, Prediction and identification of a T cell epitope in the fusion proteins of measles virus immunodominant in mice and humans, *J. Gen. Virol.*, 71, 2099, 1990.
107. **Cease, K. B., Margalit, H., Cornette, J. L., Putney, S. D., Robey, W. G., Ouyang, C., Streicher, H. Z., Fischinger, P. J., Gallo, R. C., DeLisi, C., and Berzofsky, J. A.**, Helper T cell antigenic site identification in the acquired immunodeficiency syndrome virus gp120 envelope protein and induction of immunity in mice to the native protein using a 16-residue synthetic peptide, *Proc. Natl. Acad. Sci. U.S.A.*, 84, 4249, 1987.
108. **Elliot, W. L., Stille, C. J., Thomas, L. J., and Humphreys, R. E.**, An hypothesis on the binding of an amphipathic, α-helical sequence in Ii to the desetope of class II antigens, *J. Immunol.*, 138, 2949, 1987.
109. **Reyes, V. E., Phillips, L., Humphreys, R. E., and Lew, R. A.**, Prediction of protein helices with a derivative of the strip-of-helix hydrophobicity algorithms, *J. Biol. Chem.*, 264, 12854, 1989.
110. **Rothbard, J. B.**, Peptides and the cellular immune response, *Ann. Inst. Pasteur*, 137D, 518, 1986.
111. **Rothbard, J. B. and Taylor, W. R.**, A sequence pattern common to T cell epitopes, *EMBO J.*, 7, 93, 1988.
112. **Lamb, J. R., Ivanyi, J., Rees, A. D. M., Rothbard, J. B., Howland, K., Young, R. A., and Young, D. B.**, Mapping of T cell epitopes using recombinant antigens and synthetic peptides, *EMBO J.*, 6, 1245, 1987.
113. **Lamb, J. R., Rees, A. D. M., Bal, V., Ikeda, H., Wilkinson, D., De Vries, R. R. P., and Rothbard, J. B.**, Prediction and identification of an HLA-DR-restricted T cell determinant in the 19-kDa protein of Mycobacterium tuberculosis, *Eur. J. Immunol.*, 18, 973, 1988.
114. **Rothbard, J. B., Pemberton, R. M., Bodmer, H. C., Askonas, B. A., and Taylor, W. R.**, Identification of residues necessary for clonally specific recognition of a cytotoxic T cell determinant, *EMBO J.*, 8, 2321, 1989.
115. **Strang, G., Hickling, J. K., McIndoe, G. A., Howland, K., Wilkinson, D., Ikeda, H., and Rothbard, J. B.**, Human T cell responses to human papillomavirus type 16 L1 and E6 synthetic peptides: identification of T cell determinants, HLA-DR restriction and virus type specificity, *J. Gen. Virol.*, 71, 423, 1990.
116. **Claverie, J.-M., Kourilsky, P., Langlande-Demoyen, P., Chalufour-Prochnicka, A., Dadaglio, G., Tekaia, F., Plata, F., and Bougueleret, L.**, T-immunogenic peptides are constituted of rare sequence patterns. Use in the identification of T epitopes in the human immunodeficiency virus *gag* protein, *Eur. J. Immunol.*, 18, 1547, 1988.

117. **Gould, K. G., Scotney, H., and Brownlee, G. G.,** Characterisation of two distinct major histocompatibility complex class I K^k-restricted T cell epitopes within the influenza A/PR/8/34 virus hemagglutinin, *J. Virol.*, 65, 5401, 1991.
118. **Romero, P., Coradin, G., Luescher, I. F., and Maryanski, J. L.,** H-2K^d-restricted antigenic peptides share a simple binding motif, *J. Exp. Med.*, 174, 603, 1991.
119. **Cossins, J., Gould, K. G., Smith, M., Driscoll, P., and Brownlee, G. G.,** Precise prediction of a K^k-restricted cytotoxic T cell epitope in the NS1 protein of influenza virus using an MHC allele-specific motif, *Virology*, 193, 289, 1993.
120. **Rötzschke, O., Falk, K., Stevanovic, S., Jung, G., Walden, P., and Rammensee, H.-G.,** Exact prediction of a natural T cell epitope, *Eur. J. Immunol.*, 21, 2891, 1991.
121. **Sette, A., Buus, S., Colon, S., Miles, C., and Grey, H. M.,** I-A^d-binding peptides derived from unrelated protein antigens share a common structural motif, *J. Immunol.*, 141, 45, 1988.
122. **Sette, A., Buus, O., Appella, E., Smith, J. A., Chesnut, R., Miles, C., Colon, S. M., and Grey, H. M.,** Prediction of major histocompatibility complex binding regions of a protein antigens by sequence pattern analysis, *Proc. Natl. Acad. Sci. U.S.A.*, 86, 3296, 1989.
123. **Sette, A., Adorini, L., Buus, S., Appella, E., Colon, S. M., Miles, C., Doria, G., Nagy, Z. A., Tanaka, S., and Grey, H. M.,** Structural requirements for the interaction between peptide antigens and I-E^d molecules, *J. Immunol.*, 143, 3289, 1989.
124. **Rothbard, J. B., Lechler, R. I., Howland, K., Bal, V., Eckels, D. D., Sekaly, R. P., Long, E. O., Taylor, W. R., and Lamb, J. R.,** Structural model of HLA-DR1 restricted T cell antigen recognition, *Cell*, 52, 515, 1988.
125. **Jardetzky, T. S., Gorga, J. C., Busch, R., Rothbard, J., Strominger, J. L., and Wiley, D. C.,** Peptide binding to HLA-DR1: a peptide with most residues substituted to alanine retains MHC binding, *EMBO J.*, 9, 1797, 1990.
126. **O'Sullivan, D., Sidney, J., Del Guercio, M.-F., Colón, S. M., and Sette, A.,** Truncation analysis of several DR binding epitopes, *J. Immunol.*, 146, 1240, 1991.
127. **O'Sullivan, D., Arrhenius, T., Sidney, J., Del Guercio, M.-F., Albertson, M., Wall, M., Oseroff, C., Southwood, S., Colón, S. M., Gaeta, F. C. A., and Sette, A.,** On the interaction of promiscuous antigenic peptides with different DR alleles, *J. Immunol.*, 147, 2663, 1991.
128. **Kropshofer, H., Bohlinger, I., Max, H., and Kalbacher, H.,** Self and foreign peptide interact with intact and disassembled MHC class II antigen HLA-DR via tryptophan pockets, *Biochemistry*, 30, 9177, 1991.
129. **Falk, K., Rötzschke, O., Deres, K., Metzger, J., Jung, G., and Rammensee, H. -G.,** Identification of naturally processed viral nonapeptides allows their quantification in infected cells and suggests an allele-specific T cell epitope forecast, *J. Exp. Med.*, 174, 425, 1991.
130. **Van Lierop, M.-J. C., Van Maanen, K., Meloen, R. H., Rutten, V. P. M. G., Jong, M. A. C., and Hensen, E. J.,** Proliferative lymphocyte responses to foot-and-mouth disease virus and three FMDV peptides after vaccination or immunisation with these peptides in cattle, *Immunology*, 75, 406, 1992.
131. **Glass, E. J., Oliver, R. A., Collen, T., Doel, T. R., DiMarchi, R., and Spooner, R. L.,** MHC class II restricted recognition of FMDV peptides by bovine T cells, *Immunology*, 74, 594, 1991.
132. **Glass, E. J., Oliver, R. A., and Spooner, R. L.,** Variation in T cell responses to ovalbumin in cattle: evidence for Ir gene control, *Anim. Genet.*, 21, 15, 1990.
133. **Glass, E. J., Oliver, R. A., and Spooner, R. L.,** Bovine T cells recognise antigen in association with MHC class II haplotypes defined by one-dimensional isoelectric focussing, *Immunology*, 72, 380, 1991.
134. **Mitchison, N. A.,** The carrier effect in secondary responses to hapten protein conjugates. II. Cellular cooperation, *Eur. J. Immunol.*, 1, 18, 1971.
135. **Lake, P. and Mitchison, N. A.,** Regulatory mechanisms in the immune response to cell-surface antigens, in *Cold Spring Harbor Symp. Quant. Biol.*, 41, 589, 1976.
136. **Russell, S. M. and Liew, F. Y.,** T cells primed by influenza virion internal components can cooperate in the antibody response to haemagglutinin, *Nature*, 280, 147, 1979.
137. **Scherle, P. A. and Gerhard, W.,** Differential ability of B cells specific for external vs. internal influenza virus proteins to respond to help from influenza virus-specific T cell clones *in vivo*, *Proc. Natl. Acad. Sci. U.S.A.*, 85, 4446, 1988.

138. **Milich, D. R., McLachlan, A., Thornton, G. B., and Hughes, J. L.,** Antibody production to the nucleocapsid and envelope of the hepatitis B virus primed by a single synthetic T cell site, *Nature*, 329, 547, 1987.
139. **Collen, T. and Doel, T. R.,** Analysis of specificity of T cells reactive with foot-and-mouth disease virus suggests that B cell presentation influences the memory repertoire in cattle, presented at Europic '91, VIIth Meeting of the European Study Group on Picornaviruses, University of Kent at Canterbury, England, August 24 to 30, 1991.
140. **Barnett, B. C., Graham, C. M., Skehel, J. J., and Thomas, D. B.,** The immune response of Balb/c mice to influenza haemagglutinin: commonality of the B cell and T cell repertoires and their relevance to antigenic drift, *Eur. J. Immunol.*, 19, 515, 1989.
141. **Graham, C. M., Barnett, B. C., Hartlmayr, I., Burt, D. S., Faulkes, R., Skehel, J. J., and Thomas, D. B.,** The structural requirements for class II (I-A^d)-restricted T cell recognition of influenza haemagglutinin: B cell epitopes define T cell epitopes, *Eur. J. Immunol.*, 19, 523, 1989.
142. **Graham, S., Wang, E. C. Y., Jenkins, O., and Borysiewicz, L. K.,** Analysis of the human T cell response to picornaviruses: identification of T cell epitopes close to B cell epitopes in poliovirus, *J. Virol.*, 67, 1627, 1993.
143. **Lanzavecchia, A.,** Antigen-specific interaction between T and B cells, *Nature*, 314, 537, 1985.
144. **Abbas, K. A.,** A reassessment of the mechanisms of antigen-specific T cell-dependent B cell activation, *Immunol. Today*, 9, 89, 1988.
145. **Malynn, B. A., Romeo, D. T., and Wortis, H. H.,** Antigen-specific B cells efficiently present low doses of antigen for induction of T cell proliferation, *J. Immunol.*, 135, 980, 1985.
146. **Davison, H. W. and Watts, C.,** Epitope-directed processing of specific antigen by B lymphocytes, *J. Cell Biol.*, 109, 85, 1989.
147. **Kutubuddin, M., Simons, J., and Chow, M.,** Poliovirus-specific major histocompatibility complex class I-restricted cytolytic T cell epitopes in mice localise to neutralising antigenic regions, *J. Virol.*, 66, 5967, 1992.
148. **Leclerc, C., Przewlocki, G., Schutze, M.-P., and Chadid, L.,** A synthetic vaccine constructed by copolymerisation of B and T cell determinants, *Eur. J. Immunol.*, 17, 269, 1987.
149. **Borras-Cuesta, F., Petit-Camurdan, A., and Fedon, Y.,** Engineering of immunogenic peptides by co-linear synthesis of determinants recognised by B and T cells, *Eur. J. Immunol.*, 17, 1213, 1987.
150. **Borras-Cuesta, F., Fedon, Y., and Petit-Camurdan, A.,** Enhancement of peptide immunogenicity by linear polymerisation, *Eur. J. Immunol.*, 18, 199, 1988.
151. **Cox, J. H., Ivanyi, J., Young, D. B., Lamb, J. R., Syred, A. D., and Francis, M. J.,** Orientation of epitopes influences the immunogenicity of synthetic peptide dimers, *Eur. J. Immunol.*, 18, 2015, 1988.
152. **Golvano, G., Lasarte, J. J., Sarobe, P., Gullón, A., Prieto, J., and Borras-Cuesta, F.,** Polarity of immunogens: implications for vaccine design, *Eur. J. Immunol.*, 20, 2363, 1990.
153. **Partidos, C. D., Stanley, C. M., and Steward, M. W.,** Immune responses in mice following immunisation with chimeric synthetic peptides representing B and T cell epitopes of measles virus proteins, *J. Gen. Virol.*, 72, 1293, 1991.
154. **Ruoslahti, E. and Pierschbacher, M. D.,** Arg-Gly-Asp: a versatile cell recognition signal, *Cell*, 44, 517, 1986.
155. **Ferguson, T. A., Mizutani, H., and Kupper, T. S.,** Two integrin-binding peptides abrogate T cell mediated immune responses *in vivo*, *Proc. Natl. Acad. Sci. U.S.A.*, 88, 8072, 1991.

Chapter 12

Pathogenesis and Immunity in Lentivirus Infections of Small Ruminants

B. A. Blacklaws, P. Bird, and I. McConnell

CONTENTS

I. Introduction 199
II. Virus Life Cycle 200
III. Infected Cell 200
IV. Immune Responses to Maedi-Visna Virus 201
A. Natural Resistance 201
B. Early Specific Immune Events 201
C. Immune Responses in Persistently Infected Sheep 202
1. Specific Antiviral Immune Responses 202
2. Immune Changes in Persistently Infected Sheep 202
V. Pathology 202
A. Central Nervous System 203
B. Lung 203
C. Mammary Gland 203
D. Joints 203
E. Pathogenesis 203
VI. Viral Variants 204
A. Immune Selection 204
B. Tropism 205
VII. Viral Antigens 205
VIII. Immune Potentiation of Disease 205
IX. Synergy with Other Infectious Agents 206
X. Autoimmunity 206
XI. Summary 206
References 206

I. INTRODUCTION

Maedi-visna virus (MVV) is a retrovirus of the subfamily lentivirinae which includes human, simian, feline, and bovine immunodeficiency viruses (HIV, SIV, FIV, and BIV, respectively), caprine arthritis encephalitis virus (CAEV), and equine infectious anemia virus (EIAV).[1] MVV was originally described in Iceland as the etiological agent of chronic interstitial pneumonia and demyelinating leukoencephalomyelitis in sheep.[2-4] The virus was first isolated in 1957 and its study led to the concept of slow viral diseases[5,6] characterized by long incubation periods, and slow progressive pathology which is eventually fatal to the host. MVV is transferred from one infected animal to another[7] and major routes of transmission are thought to be in nasal discharges from closely housed animals or ingestion of colostrum from infected ewes. Spread of infection within a flock is more rapid if sheep are kept in close contact for prolonged periods of time and where there is additional infection of the respiratory tract by other agents, most notably co-infection with the sheep pulmonary adenomatosis virus.[8-10] There is no evidence of the infection being transferred transplacentally from infected ewes to fetuses.[11,12] Infectious secretions probably contain the virus in the form of infected cells rather than cell-free virions.[13-16] MVV appears to be specific for infection of sheep, but can also infect rabbits and goats experimentally. In the latter species, MVV will cause arthritis in a similar manner to CAEV.[15]

The major target cell in MVV infection is the monocyte/macrophage cell lineage[17,18] which disseminates infection in the host that develops lesions in multiple organs: lung (maedi: interstitial pneumonitis),

0-8493-4952-4/94/$0.00+$.50

central nervous system (visna: ataxia and wasting), mammary glands (mastitis),[19] joints (arthritis),[20,15] lymph nodes (lymphadenopathy), and gut. Although the immunodeficiency viruses cause severe immune dysfunction due in part to their infection of T lymphocytes, these viruses also cause infection of the monocyte/macrophage lineage. Lesions similar to those seen in MVV (e.g., meningoencephalomyelitis) are present in HIV- and SIV-infected hosts. In HIV, central nervous system disease is well documented and there is evidence for pneumonitis and mastitis.[21]

The ruminant lentiviruses provide a unique insight into lentiviral pathogenesis. The restriction in ruminant lentivirus tropism to cells of the macrophage lineage allows the consequences of macrophage/monocyte infection in lentiviral pathogenesis[22,23] to be studied in the absence of the T lymphocyte dysfunction seen in the immunodeficiency-inducing lentiviruses such as HIV.

II. VIRUS LIFE CYCLE

MVV is an enveloped virus which has a diploid RNA genome of approximately 9000 b bound by repeat sequences.[24-26] It has a typical retroviral genome containing three structural genes for *gag*, *pol*, and *env* in this order 5′ - 3′. *Gag* (p55) contains the major matrix (p17), capsid (p25), and nucleoprotein (p14) proteins; *pol,* the reverse transcriptase (protease and integrase); and *env,* the viral envelope glycoprotein (gp135) which is thought to be cleaved to external gp110 and transmembrane gp41.[27,28] As with all lentiviruses, MVV contains small open reading frames for regulatory proteins between the genes for *pol* and *env* and 3′ to *env*: *tat*, MVV *vif* protein (open reading frame Q), and *rev.*[26,29,30]

Virus binds to the cell surface through its envelope proteins and it is thought that gp41 then mediates membrane fusion between the virus envelope and plasma membrane of the cell.[30-36] Upon entering the cell, the virion is uncoated and the RNA genome reverse transcribed to a DNA provirus which contains duplications of the 5′ and 3′ terminals of the genome to give the provirus identical terminal regions: the long terminal repeats (LTRs).[37-41] These contain the viral promoter and enhancer sequences. The provirus moves to the nucleus where integration may occur.[42,43] It is transcribed to produce mRNAs which may undergo two or more splicing events to produce regulatory proteins (*tat, rev, vif*).[44,45] During productive infection, the mRNAs are unspliced or singly spliced to produce structural proteins (*gag, pol, env*).[46,47] Progeny virions are thought to bud through cellular membranes which have an accumulation of *env;*[48] then cleavage of precursor *gag* p55 allows the condensation of the virion core to produce the typical bar shape of lentiviruses.[49,50]

III. INFECTED CELL

The viremia seen in MVV infection is all cell associated[51] and so cell-to-cell spread of the virus will be the major mode of transfer within the host. The major infected cell type *in vivo* is of the monocyte/macrophage lineage.[17,18,52] with no clear evidence of lymphocyte infection.[53,54] Recent work has suggested that peripheral blood dendritic cells are also infected, but until good dendritic cell markers are available in the sheep, this must await confirmation.[55] *In vitro* MVV will infect skin fibroblasts and choroid plexus cells as well as cells derived from rabbit, rat, and human sources.[56,57]

Three replication states have been suggested for MVV: latent, restricted, and productive. *In vitro*, infection of fibroblasts and choroid plexus cells leads to productive infection with approximately 5000 copies of viral RNA per cell and expression of viral protein.[58] *In vivo,* there is a much more complex situation with most infected cells expressing viral RNA but no protein.[59-60] In blood, it has been estimated that 1 in 10^5 to 10^6 leukocytes are infected.[18,51,52] Expression of viral antigens in these infected monocytes is blocked at the level of transcription as they contain 50 to 100 copies of MVV RNA but do not express antigen (restricted replication state).[18,59,60,62] When the monocytes leave the blood and move into tissue, they differentiate into macrophages and it is in mature (activated) macrophages that productive MVV infection is found.[52,62,63] Tissue macrophages have been shown to contain abundant copies of MVV RNA and to express viral antigens (productive replication),[18,52,59,60,] although RNA levels are two orders of magnitude down compared to macrophages *in vitro.*[62,64] The switch from restricted to productive replication seems to be heavily dependent on the maturation state of the monocyte/macrophages and may reflect the binding of cellular transcription factors to MVV promoter sequences sharing homology to cell promoters.[65] In restricted replication, regulatory proteins may be the only proteins expressed while these and structural proteins are expressed during productive infection. There is little evidence of true latency in MVV (the presence of viral DNA but no RNA or antigen) with the number of cells positive for viral

DNA and RNA being similar by *in situ*.[59,60] However, with recent advances in PCR-based *in situ* hybridization technology, bronchial lining epithelial cells have been shown to contain viral DNA, a cell type not previously suspected of being infected *in vivo*.[66]

IV. IMMUNE RESPONSES TO MVV

A. NATURAL RESISTANCE

Natural resistance mechanisms of the host require no prior exposure to foreign antigen for induction and thus are immediately available against infection. These defense mechanisms therefore help to create genetic resistance to infection. Cellular effector mechanisms include macrophages, natural killer cells, and interferon (IFN) and are relatively nonspecific in their actions. The complement system is also important in terms of nonspecific immunity since the classical pathway of complement activation can be directly triggered by retroviruses through the interaction of C1q with viral polypeptides.[67]

Different sheep breeds show different levels of resistance to MVV. For example, early work involved Icelandic strains of MVV which caused rapid brain disease in Icelandic sheep but, at 10 times the infectious dose, only a self-limiting encephalitis in American sheep.[51,63] There is now thought to be a linkage between goat MHC class II alleles and disease susceptibility to CAEV infection.

Although macrophages have antiviral effects (no reports in MVV), studies on these are complicated in MVV since these cells are the main target of infection by MVV. The effect of MVV infection on macrophage function is uncertain and may not be involved in early resistance to MVV but more in disease pathogenesis. There has been one report that CAEV infection of macrophages reduces their antigen presenting cell function[68] and MVV-infected macrophages have a decreased response to chemotactic factors.[69] In both these reports, *in vitro* systems have been used where MVV infection is more productive and thus may not reflect the macrophage function in the *in vivo* state. Preliminary studies on the phagocytic activity of *in vitro* MVV-infected cultured blood monocytes and *in vivo* infected alveolar macrophages showed contrasting effects. MVV infection *in vitro* reduced the phagocytic activity of cultured monocytes after 5 days of infection, while alveolar macrophages from infected sheep without lung lesions had increased phagocytic activity for *Pasteurella haemolytica* but not for sheep red blood cells when compared with uninfected controls (Wei Cheng Lee, unpublished observations). The activation state of macrophages has been studied in alveolar lung macrophages from naturally infected sheep. These are washed from the lung and then tested *in vitro*. The macrophages show increased activation states in that they have increased expression of class II molecules and secrete chemotactic factors and fibronectin.[70] Increased levels of MHC class II on macrophages from infected joints have also been seen.[71] Whether this is a direct result of viral infection or arises from chronic inflammation in affected tissues is not known.

The major study on cytokine production in infected sheep has looked at the production of IFN by lymphocytes cultured with MVV-infected macrophages.[72] This IFN has the properties of both IFN-α and IFN-γ,[73] and its production is blocked by anti-MHC class II antibodies.[74] It has been shown to inhibit the production of MVV from macrophages *in vitro*[75-77] and thus may help to limit viral spread, perhaps by decreasing the activation of monocytes to macrophages.[77] This lentivirus-induced IFN has also been shown to induce the expression of class II *in vitro*, which may be relevant to the phenotype of macrophages seen from persistently infected animals.

B. EARLY SPECIFIC IMMUNE EVENTS

Early specific anti-MVV immune events can be recorded in the sera of experimentally infected sheep. Within the first 1 to 2 weeks of infection, peripheral blood lymphocytes will make a proliferative response to MVV antigen (lymphocyte subset unknown) which declines within 6 weeks and then becomes intermittently detectable.[78-80] A serum antibody response can also be found within the first 1 to 2 months after MVV infection[51] which contains neutralizing antibody by the first 3 to 6 months post-infection.[51,81,82]

Using lymphatic cannulation techniques, the very early immune response to MVV infection within lymphoid tissue has been studied in draining efferent lymph.[83] A wave of $CD8^+$ lymphoblasts leaves the node between 4 and 15 days after local subcutaneous infection with MVV. The peak and magnitude of the $CD8^+$ lymphoblast response varies between individual sheep, and direct MVV-specific cytotoxic activity was only seen in one of five animals. MVV-specific lymphocyte proliferative activity is not detected in efferent lymph until day 15 post-infection, while preliminary evidence suggests that specific proliferation to MVV by cells teased from the acutely infected node can be detected much earlier than this. MVV-specific antibody is detected as early as day 4 post-infection, while neutralizing antibody is not found until day 12 post-infection and correlates with the appearance of IgG_1 anti-*env* antibody.

In this study,[83] the timing of the specific MVV immune response was correlated with viral replication and dissemination. A wave of infected cells, albeit at a very low frequency (maximum 11 in 10^6), leaves the node around day 9 to 18 post-infection, disseminating virus systemically. At this period, a specific MVV immune response is already detectable but it also coincides with the time when a decrease in response of efferent lymphocytes to both IL-2 and concanavalin A is seen. At this very early time point, there is therefore already evidence for evasion of the immune response by the virus.

C. IMMUNE RESPONSES IN PERSISTENTLY INFECTED SHEEP

1. Specific Antiviral Immune Responses

Serum antibody responses are detected in persistently infected sheep and are of IgM or IgG_1 isotype with no IgG_2 response.[84,85] Neutralizing antibody titers remain high in persistently infected sheep[81] and are presumably to the *env* proteins of the virus. Antibody has also been detected in cerebrospinal fluid,[51] but whether this is production of antibody from B cells present in the central nervous system or passage of serum antibody into the cerebrospinal fluid is unknown. It is known that anti-MVV neutralizing antibodies are of lower affinity than the virus for its cellular receptor[86] and so neutralization may be relatively ineffective *in vivo*. Complement-fixing antibodies are present in sera,[51,82,87] but antibody involved in antibody-dependent cell-mediated cytotoxicity (ADCC) may be absent.[85] The major mechanism for virus spread *in vivo* is probably cell to cell and very rarely involves free virus. Therefore, antibody-dependent complement fixation and ADCC may be more important defense mechanisms than antibody neutralization of free virus.

The cell-mediated immune response is important in clearing viral infections and in regulating the humoral immune response.[88,89] Both $CD4^+$ and $CD8^+$ T lymphocytes have been implicated in protection in a murine retroviral system.[90] MVV-specific cellular immune responses are detected in peripheral blood lymphocytes. Using intracerebral or intranasal inoculation of virus, several groups have shown intermittent proliferative responses to viral antigen.[79,91a,91b] However, in our experiments with both long-term naturally and experimentally (subcutaneous) infected sheep, proliferative responses to purified viral antigens were always present in peripheral blood lymphocytes.[92] This response was due mainly to $CD4^+$ lymphocytes, although in some sheep there was a minor contribution by $CD8^+$ lymphocytes and was due in part (the contribution varied within individual sheep as would be expected in an outbred population) to a response specific for the MVV p25 *gag* protein.[92]

In assays in which exogenous protein is added, it is not surprising that $CD4^+$ lymphocyte responses predominate. We have also shown that MVV-specific $CD8^+$ lymphocytes are present in the blood and efferent lymph of persistently infected animals. After culture *in vitro* on live infected cells, these lymphocytes can be shown to act as cytotoxic T lymphocytes (CTL), lysing MVV-infected autologous fibroblasts, but not infected heterologous fibroblasts.[93] Depletion of $CD8^+$ cells from these cultures results in the loss of this MHC-restricted CTL activity.

2. Immune Changes in Persistently Infected Sheep

Unlike HIV infection of humans, there is no gross immunodeficiency in MVV persistently infected sheep. MVV-infected sheep have no grossly abnormal levels of serum immunoglobulin and apparently normal T lymphocyte responses to protein antigens and mitogens *in vitro*. However, *in vivo* in naturally and experimentally infected animals, there is evidence of milder forms of immune dysfunction. Delayed-type hypersensitivity reactions in the skin to challenge antigens are reduced in intensity and there is an increased incidence of *Pasteurella multocida* infection, which is not a common pathogen in sheep.[94] Lower antibody responses to specific challenge antigens have also been noticed. These examples of immune dysfunction *in vivo* may be an indication that accessory cell function in immune responses is impaired due to the infection of accessory cells by MVV. There has also been one report of decreased ratios of CD4/CD8 in peripheral blood and synovial fluid in late-stage disease, but this has not been substantiated by other groups.[95]

V. PATHOLOGY

Maedi-visna is not an oncogenic retrovirus. The central pathological lesion in MVV infection is a chronic active inflammatory process occurring at many tissue sites. This process is characterized by extensive lymphoproliferation, with accumulation of lymphocytes and macrophages and development of classical germinal center-like structures typical of reactive lymphoid tissue.[19,96,97] The organ-specific changes include smooth muscle hyperplasia (lungs), fibrosis (mammary glands), and gliosis (central nervous system). The chronic inflammatory infiltrate eventually destroys normal tissue architecture and severely compromises function so that the major disease presentations after MVV infection are interstitial pneumonitis, central nervous system disease, and mastitis.[14,19] Other organs are also involved (e.g., gut, heart, skin, and joints) but the latter is more typical of the disease caused by CAEV in goats.[98]

A. CENTRAL NERVOUS SYSTEM

In the central nervous system (CNS), MVV induces a demyelinating leukoencephalomyelitis in the brain and spinal cord. There is perivascular cuffing of lymphocytes in white and grey matter, which eventually leads to diffuse immunopathology[99] of the white matter in the cerebrum, cerebellum, pons, medulla oblongata, and spinal cord. The lesions are often near the ventricles and aqueduct. Primary demyelination occurs[100] and within the lesion the oligodendrocytes and astrocytes are infected.[61] Glial nodules due to glial cell proliferation are seen in less severe lesions. The precise mechanism of cell destruction in the CNS is not known and is generally assumed to be secondary to the development of inflammatory lesions. However, it has recently been shown that the visna *tat* gene encodes a protein which contains a peptide motif which is directly neurotoxic both *in vivo*[101] and *in vitro* (G. Harkiss, personal communication).

B. LUNG

The lungs in maedi are increased in weight and show a chronic interstitial inflammatory response with dense cellular infiltration, hyperplasia of the smooth muscle in the alveolar septae, and slight fibrosis. Peribronchial and perivascular lymphoid hyperplasia and epithelial proliferation in small bronchi and bronchioli are also seen.[96] Phenotypic analysis of lymphocytic changes in the lungs and associated lymphoid tissue in uncomplicated maedi show that there is an increased number of $CD4^+$ and $CD8^+$ lymphocytes, both in the bronchus associated lymphoid tissue and interalveolar septae. The regional mediastinal lymphoid tissue shows marked germinal center development and invasion of germinal centers by $CD8^+$ T cells similar to that seen in HIV infections in man.[102,103]

C. MAMMARY GLAND

The mammary gland shows a lympho-plasmacytic mastitis where the lymphocytes and plasmacytes infiltrate glandular interstitium and lymphocytes infiltrate ductal and acinar lumens. There is also the appearance of lymphatic nodules within lobular interstitium and around ductules with some degeneration of acinar and ductal epithelium.[19] An indurative mastitis with extensive fibrosis is a common feature in MVV mastitis.

D. JOINTS

Clinical features of MVV-induced arthritis in sheep are similar to those seen in goats with CAEV.[104] The carpal joints are most commonly affected, followed by the tarsal, then other joints. The arthritis begins as a synovitis with considerable soft tissue swelling around affected joints followed by diffuse mineralization. There is hyperplasia of the synovium and it is infiltrated by large numbers of lymphocytes and macrophages and occasional plasma cells. The proliferating synovium is strongly MHC class II positive and both $CD4^+$ and $CD8^+$ lymphocytes are markedly increased in the hyperplastic synovium.[105] In severe cases, there are lymphocytic nodules. As inflammation persists and progresses, there is eventual destruction of the articular surfaces of the joints.[106]

E. PATHOGENESIS

The pathogenesis of MVV is very complex. There are two inputs into lesion formation: first, the immune response to the virus and infected cells and, second, the effect of the virus on the infected organ. There is little doubt that the immune system has a major part to play in the formation of lesions in MVV infection. Early experiments by Nathanson and co-workers[99] showed that immunosuppressed sheep (by anti-sheep thymocyte serum and cyclophosphamide) infected intracerebrally with MVV had fewer lesions in the CNS than normal sheep, but apparently the extent of infection was similar as the number of viral isolations remained the same. However, when infected sheep were vaccinated intradermally and intramuscularly with purified virion antigen, there was only a slight increase in the severity of CNS lesions and number of viral isolations compared to control infected sheep after 8 weeks.[107] Therefore, the immune response is unlikely to account for all MVV pathology.

The above authors also showed that there was a correlation between virus infectious dose given and CNS lesion severity and frequency of virus isolations,[107] suggesting that the extent of target antigen expression (i.e., spread of infection and virus expression) is directly involved in disease progression. This is further supported using naturally infected goats and CAEV antigen given intraarticularly. In this system, it has been possible to show that injecting CAEV antigen at the site of lesion formation increases the severity of disease.[108]

A central problem in lesions caused by infection with MVV is the apparent lack of viral antigen or viral-infected cells in the lesions, especially as the amount of antigen has been implicated in the severity of lesions seen (see above). Only one to two viral antigen positive cells are found per section in infected lung and CNS. Syncytia formation is a property of MVV infection and giant cells are occasionally seen in MVV-infected tissues. It may be that the specific immune response seen in persistently infected sheep clears many of the productively infected cells from the tissues. However, constant recruitment of more infected monocyte/macrophages to the tissue allows a chronic active inflammatory lesion to persist. It has been postulated that MVV is disseminated by the spread of monocytes/macrophages harboring virus in a restricted or latent replication state (The Trojan Horse Hypothesis).[16,59,109] Once in the organ, the virus-infected cell becomes productively infected and lesions start to form due to expression of viral antigen. Such recruitment would require a reservoir of latently infected cells elsewhere in the sheep. There is already evidence of MVV infection of the monocyte/ macrophage lineage in the bone marrow,[18] and *in situ* PCR has revealed a large number of latently infected cell types *in vivo*[66] in the bronchial epithelium. Uninfected macrophages may also be recruited to the affected tissues and become infected within the tissue via free virus or cell-to-cell spread.

Natural infection with MVV is thought to occur through the lungs, which may account for involvement of this organ in the naturally infected sheep. However, even in experimentally infected sheep (intradermally and intraarticularly), lung pathology is a prominent finding (unpublished observations). One explanation as to why lungs, CNS, and other organs are usually most involved in MVV disease is that pathology may predominate in tissues where monocytes mature to true macrophages, allowing the switch from restricted MVV replication to productive replication and chronically activating an MVV-specific immune response.

It is likely that both inflammatory and growth promoting cytokines are involved in the pathogenesis. The extensive proliferation of fibroblasts which occurs in mammary gland mastitis or the smooth muscle proliferation seen in the lung could well be induced through increased synthesis of macrophage cytokines (e.g., IL-1, TGF-β, and TNF-α) resulting in the tissue-reactive changes. The LTR of lentiviruses contains a number of AP1 elements which are responsive to cytokines generated within inflammatory lesions, and it may well be that the interplay between cytokines and infected cells not only maintains the infection, but provides the correct tissue microenvironment for a chronic state of infection.

VI. VIRAL VARIANTS

Variation in the retrovirus family is high and mainly ascribed to the high error rate of both the reverse transcriptase of these viruses and cellular RNA polymerase (genome to provirus and then provirus to genome), both of which have no proofreading capability. Within the restraints of maintaining functional proteins for viable progeny, the mutated virions are fixed in the population if there is a growth advantage. This may take the form of escape from the immune system (immune selection) or improve the tropism of the virus for a particular tissue. Retroviruses therefore exist within their hosts as "quasispecies", a polymorphic population of the same virus. Sequencing studies between different strains of MVV show that *env* and *rev* are the most variable proteins.[26,29,30] Most point mutations accumulate in *env*[110-116] in particular regions of the gene.[30]

A. IMMUNE SELECTION

If viral mutants arise which can escape existing immune surveillance, then they may express antigen which can induce another immune response and thus more activation and pathology at the site of infection. Neutralizing antibody-resistant variants of MVV are known to arise, both *in vitro*[117] and *in vivo*,[118] although the original infecting strain of virus is never replaced completely. The presence of neutralizing antibody does not correlate with disease progression,[1] throwing doubts on whether neutralization-resistant viral mutants are important in disease pathogenesis.

In MVV, no work has been published on the selection of T lymphocyte epitope mutants, but there is evidence from a model system, LCMV infection of transgenic mice, that escape mutants for CTL determinants may arise.[119] Indeed, with HIV there is some data suggesting loss of recognition of certain CTL epitopes during the course of infection with the concomitant isolation of viruses mutated in these epitopes.[120] However, lentiviruses have multiple T cell epitopes[120] and so fixing of a mutant of one epitope in the viral population would be difficult as there would still be recognition of other nonmutated epitopes.

B. TROPISM

It is well documented that there are strains of virus which show different tropisms in tissue culture cells. Not only do virus strains grow better or worse on different tissue culture cells, but they may show varying cytopathic effects: syncytial formation or lytic infection. Virus isolates have been divided into different groups on fibroblasts[121] or synovial cells.[122] Some may cause a lytic infection of macrophages *in vitro* but this does not correlate with pathogenicity *in vivo*.[74] Selection of variants has also been performed *in vivo* with the selection of CNS-tropic variants by passage in brains.[123] Strains of MVV tropic for the CNS would be expected to cause visna and CNS lesions rather than maedi in sheep.

VII. VIRAL ANTIGENS

It is now established in HIV infection that viral proteins may affect cells in the absence of the viral genome. The extracellular protein of HIV, gp120, interacts with CD4, thus inhibiting signaling to $CD4^+$ T lymphocytes. This may increase the extent of immunosuppression seen in this disease. Gp120 also acts directly on B cells to increase IL-6 production, helping in the induction of B cell hyperplasia and hypergammaglobulinemia.[124] In addition, there is evidence in HIV that *tat* may act directly as a growth factor.[125] Other retroviral proteins have been implicated in disease using transgenic mice. Mice with the complete HTLV-1 genome have been judged to be tolerant to viral proteins but develop a chronic inflammatory arthritis similar to that seen in HTLV-1 infections in humans.[126] The authors suggest that the pathogenic mechanism here could involve transactivation of cellular genes by $p40^{tax}$. MVV and CAEV *tat* could also mediate this type of effect if the activation occurs through AP1 sites.

VIII. IMMUNE POTENTIATION OF DISEASE

We have postulated above that the immune response to MVV contributes to lesion formation during its action in clearing viral antigen and infected cells. However, the immune response may also help in the spread of infection within the animal, both at the time of and after infection. In sheep given immunizations of MVV proteins before infection, there was an increased incidence of arthritis. This may be due to the presence of anti-MVV antibody which coats the virus and allows uptake by Fc receptor-bearing cells. As the major target cell (monocyte/macrophage) is Fc receptor-positive, the presence of preexisting anti-MVV antibody may allow an increased infection rate of macrophages and therefore increase the effective infectious dose of virus.[127-129] As arthritis was the disease studied in this case, the actual known causes of joint lesions must be considered. It is well known that antibody/antigen immune complexes are part of the pathogenic process in arthritis, and thus the presence of anti-MVV antibody from the beginning of infection may allow the formation of these complexes, thereby predisposing the infected sheep to arthritis.

Cytotoxic T lymphocytes are known to act by "programmed cell death" of the target and one of the first events in this process is apoptosis,[130] DNA fragmentation of the cellular DNA. The stage in the viral replication cycle at which the host cell is killed may determine how the viral DNA reacts: polyomavirus DNA undergoes apoptosis with the host cell DNA after interaction with specific CTL,[131] while packaged reovirus DNA is resistant to apoptosis[132] and is therefore not affected by CTL lysis of its host. Only the provirus of MVV is DNA and, since it is present in the nucleus, it is likely that it will undergo fragmentation with the host cell genome; however, the rest of the viral life cycle involves RNA genome intermediates. Few virions bud from the cytoplasmic membrane of MVV-infected macrophages but they tend to accumulate in intracellular vacuoles. A similar phenomenon is seen in HIV-infected macrophages. Therefore, if an infected macrophage is killed by CTL, ADCC, complement, or NK cells, all that may result is the release of virus into the surrounding tissue and further spread of infection.

The activation state of the macrophage may affect the expression of MVV within the cell. In mice transgenic for the LTR of MVV, expression from the MVV promoter was only seen in mature activated macrophages.[133] This is presumably through the action of cellular transcription factors on the viral promoter. Therefore, when the immune system becomes activated to clear viral antigen, the monocyte/macrophage cell lineage is also activated, thus increasing the production of more viral antigen[52]. It is this very cycle which probably causes the chronic inflammatory lesions which are seen.

IX. SYNERGY WITH OTHER INFECTIOUS AGENTS

Co-infection of sheep with MVV and another retrovirus, sheep pulmonary adenomatosis (SPA, Jaagsiekte) virus, causes increased spread of the virus and more severe lesions in the lung.[8-10] SPA causes smooth muscle hyperplasia of the lung and the production of excess amounts of bronchial and nasal discharges.

X. AUTOIMMUNITY

Many aspects of MVV lesions are reminiscent of lesions induced by autoimmune phenomena. The demyelination of neurons in the CNS is similar to that seen in multiple sclerosis and the arthitis is typical of lesions seen in rheumatoid arthritis. There may well be an autoimmune component involved in lesion formation once the virus has triggered tissue damage and immune activation. There is evidence of reaction to self brain antigens[134] and there may be reactions to heat shock proteins in MVV-infected animals (G. Harkiss, personal communication).

XI. SUMMARY

MVV induces chronic inflammatory lesions in which lymphocytes accumulate and may form nodular structures reminiscent of germinal centers of lymph nodes. The pathogenesis of disease is very complex, with the interaction of several components: the immune response to viral antigens, the virus itself, and autoimmune reactivity. The immune response and viral gene expression are interdependent, leading to the accumulated pathology seen. In persistently infected animals, the immune response is not sufficient to clear the virus infection; whether this is due to a dysfunction in some component of the immunity or the nature of the virus infection itself (through a reservoir of latently infected cells or cells with virus in a restricted replication state) is unknown. Further studies to define these problems may well help to illuminate the pathogenetic mechanisms and the role of the immune response in preventing spread of infection which will be of comparative value to studies on the human and simian lentiviruses.

REFERENCES

1. **Narayan, O. and Clements, J. E.,** Lentiviruses, in *Virology*, Fields, B. N., Knipe, D. M., et al., Eds., Raven Press, New York, 1990, 1571.
2. **Sigurdsson, B.,** Maedi, a slow progressive pneumonia of sheep: an epizoological and pathological study, *Br. Vet. J.*, 110, 255, 1954.
3. **Sigurdsson, B. and Palsson, P. A.,** Visna of sheep — a slow demyelinating infection, *Br. J. Exp. Pathol.*, 39, 519, 1958.
4. **Sigurdsson, B., Grimsson, H., and Palsson, P. A.,** Maedi, a chronic progressive infection of sheep's lungs, *J. Infect. Dis.*, 90, 233, 1952.
5. **Sigurdsson, B.,** Maedi, a slow progressive pneumonia of sheep: an epizoological and pathological study, *Br. Vet. J.*, 110, 255, 1954.
6. **Sigurdsson, B.,** Rida, a chronic encephalitis of sheep, *Br. Vet. J.*, 110, 341, 1954.
7. **Cross, R. F., Smith, C. K., and Moorhead, P. D.,** Vertical transmission of progressive pneumonia of sheep, *Am. J. Vet. Res.*, 36, 465, 1975.
8. **Palsson, P. A.,** Maedi and visna in sheep, in *Frontiers of Biology: Slow Virus Diseases of Animals and Man,* Kimberlin, R. H., Ed., North Holland Publishing, Co., Amsterdam, 1976, 44, 17.
9. **Houwers, D. J. and Terpstra, C.,** Sheep pulmonary adenomatosis, *Vet. Rec.*, 114, 23, 1984.
10. **DeMartini, J. C., Rosadio, R. H., Sharp, J. M., et al.,** Experimental co-induction of type D retrovirus-associated pulmonary carcinoma and lentivirus-associated lymphoid interstitial pneumonia in lambs, *J. Natl. Cancer Inst.*, 79, 167, 1987.

11. **Cutlip, R. C., Lehmkuhl, H. D., and Jackson, T. A.**, Intrauterine transmission of ovine progressive pneumonia virus, *Am. J. Vet. Res.*, 42, 1795, 1981.
12. **Houwers, D. J., Konig, C. D., DeBoer, G. F., and Schaake, J., Jr.**, Maedi-visna control in sheep. I. Artificial rearing of colostrum deprived lambs, *Vet. Microbiol.*, 8, 179, 1983.
13. **De Boer, G. F., Terpstra, C., Houwers, D. J., and Hendriks, J.**, Studies in epidemiology of maedi-visna in sheep, *Res. Vet. Sci.*, 26, 202, 1979.
14. **Gudnadottir, M.**, Visna-maedi in sheep, *Prog. Med. Virol.*, 18, 336, 1974.
15. **Narayan, O. and Cork, L. C.**, Lentiviral diseases of sheep and goats: chronic pneumonia, leukoencephalomyelitis and arthritis, *Rev. Infect. Dis.*, 7, 89, 1985.
16. **Haase, A. T.**, Pathogenesis of lentivirus infections, *Nature*, 322, 130, 1986.
17. **Narayan, O., Wolinsky, J. S., Clements, J. E., Strandberg, J. D., Griffin, D. E., and Cork, L. C.**, Slow virus replication: the role of macrophages in the persistence and expression of visna viruses of sheep and goats, *J. Gen. Virol.*, 59, 345, 1982.
18. **Gendelman, H. E., Narayan, O., Molineaux, S., Clements, J. E., and Ghotbi, Z.**, Slow, persistent replication of lentiviruses: role of tissue macrophages and macrophage precursors in bone marrow, *Proc. Natl. Acad. Sci. U.S.A.*, 82, 7086, 1985.
19. **Deng, P., Cutlip, R. C., Lehmkuhl, H. D., and Brogden, K. A.**, Ultrastructure and frequency of mastitis caused by ovine progressive pneumonia virus infection in sheep, *Vet. Pathol.*, 23, 184, 1986.
20. **Oliver, R. E., Gorham, J. R., Parish, S. F., et al.**, Ovine progressive pneumonia: pathologic and virologic studies on the naturally occurring disease, *Am. J. Vet. Res.*, 42, 1554, 1981.
21. **Georgsson, G., Palsson, P. A., and Petursson, G.**, Some comparative aspects of Visna and AIDS, in *Modern Pathology of AIDS and Other Retroviral Infections*, Racz, P., Haase, A. T., and Gluckman, J. C., Eds., Karger, Basel, 1990, 82.
22. **Meltzer, M. S., Skillman, D. R., Gomatos, P. J., Kalter, D. C., and Gendelman, H. E.**, Role of mononuclear phagocytes in the pathogenesis of human immunodeficiency virus infection, *Annu. Rev. Immunol.*, 8, 169, 1990.
23. **Embretson, J., Zupancic, M., Ribas, J. L., Burke, A., Racz, P., Tenner-Racz, K., and Haase, A. T.**, Massive covert infection of helper T lymphocytes and macrophages by HIV during the incubation period of AIDS, *Nature*, 362, 359, 1993.
24. **Beemon, K. L., Faras, A. J., Haase, A. T., et al.**, Genome complexities of murine leukaemia and sarcoma, reticuloendotheliosis, and visna viruses, *J. Virol.*, 17, 525, 1976.
25. **Vigne, R., Brahic, M., Filippi, P., and Tamalet, J.**, Complexity and polyadenylic acid content of visna virus 60S-70S RNA, *J. Virol.*, 21, 386, 1977.
26. **Sonigo, P., Alizon, M., Staskus, K., et al.**, Nucleotide sequence of the visna lentivirus: relationship to the AIDS virus, *Cell*, 42, 369, 1985.
27. **Haase, A. T. and Baringer, J. R.**, The structural polypeptides of RNA slow viruses, *Virology*, 57, 238, 1974.
28. **Vigne, R., Filippi, P., Querat, G., Sauze, N., Vitu, C., Russo, P., and Delori, P.**, Precursor polypeptides to structural proteins of visna virus, *J. Virol.*, 42, 1046, 1982.
29. **Querat, G., Audoly, G., Sonigo, P., and Vigne, R.**, Nucleotide sequence analysis of SA-OMVV, a visna related ovine lentivirus: phylogenetic history of lentiviruses, *Virology*, 175, 434, 1990.
30. **Sargan, D. R., Bennet, I. D., Cousens, C., Roy, D. J., Blacklaws, B. A., Dalziel, R. G., Watt, N. J., and McConnell, I.**, Nucleotide sequence of EV1, a British isolate of maedi-visna virus, *J. Gen. Virol.*, 72, 1893, 1991.
31. **Crane, S. E., Kanda, P., and Clements, J. E.**, Identification of the fusion domain in the visna virus transmembrane protein, *Virology*, 185, 488, 1991.
32. **Gallaher, W. R.**, Detection of a fusion peptide sequence in the transmembrane protein of human immunodeficiency virus, *Cell*, 50, 327, 1987.
33. **Freed, E. O., Myers, D. J., and Risser, R.**, Characterisation of the fusion domain of the human immunodeficiency virus type 1 envelope glycoprotein, gp41, *Proc. Natl. Acad. Sci. U.S.A.*, 87, 4650, 1990.
34. **White, J., Kielian, M., and Helenius, A.**, Membrane fusion proteins of enveloped animal viruses, *Quart. Rev. Biophys.*, 16, 151, 1983.
35. **Bosch, M. L., Earl, P. L., Fargnolik, K., et al.**, Identification of the fusion peptide of primate immunodeficiency viruses, *Science*, 244, 694, 1989.

36. **Kowalski, M., Potz, J., Basiripour, L., et al.,** Functional regions of the human immunodeficiency virus envelope glycoprotein, *Science,* 237, 1351, 1987.
37. **Marsh, M. and Helenius, A.,** Virus entry into animal cells, *Adv. Virus Res.,* 36, 107, 1989.
38. **Maddon, P. J., McDougal, J. S., Clapham, P. R., et al.,** HIV infection does not require endocytosis of its receptor, CD4, *Cell,* 54, 865, 1988.
39. **Stein, B. S., Gowda, S. D., Lifson, J. D., et al.,** pH-independent HIV entry into CD4-positive T cells via virus envelope fusion to the plasma membrane, *Cell,* 49, 659, 1987.
40. **McClure, M. O., Marsh, M., and Weiss, R. A.,** Human immunodeficiency virus infection of CD4-bearing cells occurs by a pH-independent mechanism, *EMBO J.,* 7, 513, 1988.
41. **Coffin, J. M.,** Retroviridae and their replication, in *Virology*, Fields, B. N., Knipe, D. M., et al., Eds., Raven Press, New York, 1990, 1437.
42. **Harris, J. D., Blum, H., Scott, J. V., et al.,** The slow virus visna: reproduction *in vitro* of virus from extrachromosomal DNA, *Proc. Natl. Acad. Sci. U.S.A.,* 81, 7212, 1984.
43. **Hirsch, V. M., Zack, P. M., Vogel, A. P., and Johnson, P. R.,** Simian immunodeficiency virus infection of macaques: end-stage disease is characterised by widespread distribution of proviral DNA in tissues, *J. Infect. Dis.,* 163, 976, 1991.
44. **Davis, J. L. and Clements, J. E.,** Characterisation of a cDNA clone encoding the visna virus transactivating protein, *Proc. Natl. Acad. Sci. U.S.A.,* 86, 414, 1989.
45. **Gourdou, I., Mazarin, V., Querat, G., Sauze, N., and Vigne, R.,** The open reading frame S of visna virus is a trans-activating gene, *Virology,* 171, 170, 1989.
46. **Davis, J. L., Molineaux, S., and Clements, J. E.,** Visna virus exhibits a complex transcriptional pattern: one aspect of gene expression shared with the acquired immunodeficiency syndrome retrovirus, *J. Virol.,* 61, 1325, 1987.
47. **Vigne, R., Barban, V., Querat, G., et al.,** Transcription of visna virus during its lytic cycle: evidence for a sequential early and late gene expression, *Virology,* 161, 218, 1987.
48. **Dubois-Dalcq, M., Reese, T. S., and Narayan, O.,** Membrane changes associated with assembly of visna virus, *Virology,* 74, 520, 1976.
49. **Thormar, H.,** An electron microscope study of tissue cultures infected with visna virus, *Virology,* 14, 463, 1961.
50. **Gonda, M. A., Wong-Staal, F., Gallo, R. C., et al.,** Sequence homology and morphologic similarity of HTLV-III and visna virus, a pathogenic lentivirus, *Science,* 227, 173, 1985.
51. **Petursson, G., Nathanson, N., Georgsson, G., et al.,** Pathogenesis of visna. I. Sequential virologic, serologic and pathologic studies, *Lab. Invest.,* 35, 402, 1976.
52. **Gendelman, H. E., Narayan, O., Kennedy-Stoskopf, S., Kennedy, P. G. E., Ghotbi, Z., Clements, J. E., Stanley, J., and Pezeshkpour, G.,** Tropism of sheep lentiviruses for monocytes: susceptibility to infection and virus gene expression increase during maturation of monocytes to macrophages, *J. Virol.,* 58, 67, 1986.
53. **Georgsson, G., Houwers, D. J., Palsson, P. A., and Petursson, G.,** Expression of viral antigens in the central nervous system of visna infected sheep: an immunohistochemical study of experimental visna induced by virus strains of increased neurovirulence, *Acta Neuropathol.,* 77, 299, 1989.
54. **Narayan, O., Zink, M. C., Gorrell, M., et al.,** Lentivirus induced arthritis in animals, *J. Rheumatol.,* 19 (32), 25, 1992.
55. **Gorrell, M. D., Brandon, M. R., Sheffer, D., Adams, R. J., and Narajan, O.,** Ovine lentivirus is macrophagetropic and does not replicate productively in T lymphocytes, *J. Virol.,* 66, 2679, 1992.
56. **Jolly, P. E. and Narayan, O.,** Evidence for interference, coinfections, and intertypic virus enhancement of infection by ovine-caprine lentiviruses, *J. Virol.,* 63, 4682, 1989.
57. **Gilden, D. H., Devlin, M., and Wroblewska, Z.,** The use of vesicular stomatitis virus (visna virus) pseudotypes to demonstrate visna virus receptors in cells from different species, *Arch. Virol.,* 67, 181, 1981.
58. **Haase, A. T., Stowring, L., Harris, J. D., et al.,** Visna DNA synthesis and the tempo of infection *in vitro, Virology,* 119, 399, 1982.
59. **Haase, A. T., Stowring, L., Narayan, O., et al.,** Slow persistent infection caused by visna virus: role of host restriction, *Science,* 195, 175, 1977.
60. **Brahic, M., Stowring, L., Ventura, P., and Haase, A. T.,** Gene expression in visna virus infection in sheep, *Nature*, 292, 240, 1981.

61. **Stowring, L., Haase, A. T., Petursson, G., et al.,** Detection of visna virus antigens and RNA in glial cells in foci of demyelination, *Virology,* 141, 311, 1985.
62. **Geballe, A. P., Ventura, P., Stowring, L., and Haase, A. T.,** Quantitative analysis of visna virus replication in vivo, *Virology,* 141, 148, 1985.
63. **Narayan, O., Kennedy-Stoskopf, S., Sheffer, D., et al.,** Activation of caprine arthritis-encephalitis virus expression during maturation of monocytes to macrophages, *Infect. Immun.,* 41, 67, 1983.
64. **Haase, A. T., Retzel, E. F., and Staskus, K. A.,** Amplification and detection of lentiviral DNA inside cells, *Proc. Natl. Acad. Sci. U.S.A.,* 87, 4971, 1990.
65. **Hess, J. L., Small, J. A., and Clements, J. E.,** Sequences in the visna virus LTR that control transcriptional activity and respond to viral *trans*-activation: involvement of AP-1 sites in basal activity and *trans*-activation, *J. Virol.,* 63, 3001, 1989.
66. **Staskus, K. A., Couch, L., Bitterman, P., Retzel, E. F., Zupancic, M., List, J., and Haase, A. T.,** *In situ* amplification of visna virus DNA in tissue sections reveals a reservoir of latently infected cells, *Microbial Pathogenesis,* 11, 67, 1991.
67. **Cooper, N. R., Jensen, F. C., Welsh, R. M., and Oldstone, M. B. A.,** Lysis of RNA tumor viruses by human serum: direct antibody-independent triggering of the classical complement pathway, *J. Exp. Med.,* 144, 970, 1976.
68. **Banks, K. L., Adams, D. S., McGuire, T. C., and Carlson, J.,** Experimental infection of sheep by caprine-arthritis encephalitis virus and goats by progressive pneumonia virus, *Am. J. Vet. Res.,* 44, 2307, 1983.
69. **Myer, M. S., Huchzermeyer, H. F. A. K., York, D. F., et al.,** The possible involvement of immunosuppression caused by a lentivirus in the aetiology of Jaagsiekte and pasteurellosis in sheep, *Onderstepoort J. Vet. Res.,* 55, 127, 1988.
70. **Cordier, G., Cozon, G., Greenland, T., et al.,** *In vivo* activation of alveolar macrophages in ovine lentivirus infection, *Clin. Immunol. Immunopathol.,* 55, 355, 1990.
71. **Harkiss, G. D., Watt, N. J., King, T. J., et al.,** Retroviral arthritis: phenotypic analysis of cells in the synovial fluid of sheep with inflammatory synovitis associated with visna virus infection, *Clin. Immunol. Immunopathol.,* 60, 106, 1991.
72. **Narayan, O., Sheffer, D., Clements, J. E., and Tennekoon, G.,** Restricted replication of lentiviruses. Visna viruses induce a unique interferon during interaction between lymphocytes and infected macrophages, *J. Exp. Med.,* 162, 1954, 1985.
73. **Zink, M. C., Narayan, O., Kennedy, P. G. E., and Clements, J. E.,** Pathogenesis of visna/maedi and caprine arthritis-encephalitis: new leads on the mechanism of restricted virus replication and persistent inflammation, *Vet. Immunol. Immunopathol.,* 15, 167, 1987.
74. **Lairmore, M. D., Butera, S. T., Callahan, G. N., and DeMartini, J. C.,** Spontaneous interferon production by pulmonary leukocytes is associated with lentivirus-induced lymphoid interstitial pneumonia, *J. Immunol.,* 140, 779, 1988.
75. **Carroll, D., Ventura, P., Haase, A., et al.,** Resistance of visna virus to interferon, *J. Infect. Dis.,* 138, 614, 1978.
76. **Kennedy, P. G. E., Narayan, O., Ghotbi, Z., et al.,** Persistent expression of Ia antigen and viral genome in visna-maedi virus-induced inflammatory cells. Possible role of lentivirus-induced interferon, *J. Exp. Med.,* 162, 1970, 1985.
77. **Zink, M. C. and Narayan, O.,** Lentivirus-induced interferon inhibits maturation and proliferation of monocytes and restricts the replication of caprine arthritis-encephalitis virus, *J. Virol.,* 63, 2578, 1989.
78. **Griffin, D. E., Narayan, O., and Adams, R. J.,** Early immune responses in visna, a slow viral disease of sheep, *J. Infect. Dis.,* 138, 340, 1978.
79. **Larsen, H. J., Hyllseth, B., and Krogsrud, J.,** Experimental maedi-virus infection in sheep: early cellular and humoral immune response following parenteral inoculation, *Am. J. Vet. Res.,* 43, 379, 1982.
80. **Sihvonen, L.,** Early immune responses in experimental maedi, *Res. Vet. Sci.,* 30, 217, 1981.
81. **Gudnadottir, M. and Palsson, P. A.,** Host-virus interaction in visna infected sheep, *J. Immunol.,* 95, 1116, 1965.
82. **DeBoer, G. F.,** Antibody formation in zwoegerziekte, a slow infection in sheep, *J. Immunol.,* 104, 414, 1970.

83. **Bird, P., Blacklaws, B., Reyburn, H. T., Allen, D., Hopkins, J., Sargan, D., and McConnell, I.,** Early events in immune evasion by the lentivirus maedi-visna occurring within infected lymphoid tissue, *J. Virol.,* 67, 5187, 1993.
84. **Petursson, G., Douglas, B. M., and Lutley, R.,** Immunoglobulin subclass distribution and restriction of antibody response in visna, in *Slow Viruses in Sheep, Goats and Cattle,* Sharp, J. M. and Hoff-Jorgensen, R., Eds., ECSC-EEC-EAEC, Brussels, Luxembourg, 1983.
85. **Reyburn, H. T., Bird, P., Blacklaws, B., Watt, N., and McConnell, I.,** Restriction in antibody subclass and specificity antibody in the responses to maedi-visna virus (MVV) in PI sheep, *Eur. Wkshop Small Rum. Lent.,* 1992.
86. **Kennedy-Stoskopf, S. and Narayan, O.,** Neutralising antibodies to visna lentivirus: mechanism of action and possible role in virus persistence, *J. Virol.,* 59, 37, 1986.
87. **Gudnadottir, M. and Kristinsdottir, K.,** Complement-fixing antibodies in sera of sheep affected with visna and maedi, *J. Immunol.,* 98, 663, 1967.
88. **Mims, C. A. and White, D. O.,** *Viral Pathogenesis and Immunology,* Blackwell Scientific, Oxford, 1984.
89. **Vitetta, E. S., Berton, M. T., Burger, C., et al.,** Memory B and T cells, *Annu. Rev. Immunol.,* 9, 193, 1991.
90. **Hom, R. C., Finberg, R. W., Mullaney, S., and Ruprecht, R. M.,** Protective cellular retroviral immunity requires both $CD4^+$ and $CD8^+$ immune T cells, *J. Virol.,* 65, 220, 1991.

91a. **Larsen, H. J., Hyllseth, B., and Krogsrud, J.,** Experimental maedi virus infection in sheep: cellular and humoral immune response during three years following intranasal inoculation, *Am. J. Vet. Res.,* 43, 384, 1982.

91b. **Sihvoven, L.,** Late immune responses in experimental maedi, *Vet. Microbiol.,* 9, 205, 1984.

92. **Reyburn, H. T., Roy, D. J., Blacklaws, B. A., Sargan, D. R., Watt, N. J., and McConnell, I.,** Characteristics of the T cell-mediated immune response to maedi-visna virus, *Virology,* 191, 1009, 1992.
93. **Blacklaws, B. A., Bird, P., Allen, D., and McConnell, I.,** Circulating cytotoxic T lymphocyte precursors in Maedi-visna virus infected sheep, *J. Gen. Virol.,* submitted.
94. **Watt, N. J., King, T. J., Collie, D., McIntyre, N., Sargan, D., and McConnell, I.,** Clinicopathological investigation of primary, uncomplicated maedi-visna virus infection, *Vet. Rec.,* 131, 455, 1992.
95. **Kennedy-Stoskopf, S., Zink, M. C., and Narayan, O.,** Pathogenesis of ovine lentivirus-induced arthritis: phenotypic evaluation of T lymphocytes in synovial fluid, synovium and peripheral circulation, *Clin. Immunol. Immunopathol.,* 52, 323, 1989.
96. **Georgsson, G. and Palsson, P. A.,** The histopathology of maedi, *Vet. Pathol.,* 8, 63, 1971.
97. **Sigurdsson, B.,** Atypically slow infectious diseases, in *Liore Jubilaire due Dr Ludo van Bogoert, Acta Med.,* Belgica, Bruxelles, 1962, 738.
98. **Michaels, F. H., Banks, K. L., and Reitz, M. S.,** Lessons from caprine and ovine retrovirus infections, *Rheum. Dis. Clin. N. Am.,* 17, 5, 1991.
99. **Nathanson, N., Panitch, H., Palsson, P. A., et al.,** Pathogenesis of visna. II. Effect of immunosuppression upon early central nervous system lesions, *Lab. Invest.,* 35, 444, 1976.
100. **Georgsson, G., Martin, J. R., Klein, J., Palsson, P. A., Nathanson, N., and Petursson, G.,** Primary demyelination in visna: an ultrastructural study of Icelandic sheep with clinical signs following experimental infection, *Acta Neuropathol. (Berl.),* 57, 171, 1982.
101. **Hayman, M., Arbuthnott, G., Harkiss, G., Brace, H., Filippi, P., Philippon, V., Thomson, D., Vigne, R., and Wright, A.,** Neurotoxicity of peptide analogues of the transactivating protein *tat* from maedi-visna virus and human immunodeficiency virus, *Neuroscience,* 53, 1, 1993.
102. **Watt, N. J., MacIntyre, N., Collie, D., Sargan, D., and McConnell, I.,** Phenotypic analysis of lymphocyte populations in the lungs and regional lymphoid tissue of sheep naturally infected with maedi visna-virus, *Clin. Exp. Immunol.,* 90, 204, 1992.
103. **Racz, P., Tenner-Racz, K., Kahl, C., et al.,** Spectrum of morphologic changes of lymph nodes from patients with AIDS or AIDS-related complexes, *Prog. Allergy,* 37, 81, 1986.
104. **Crawford, T. B., Adams, D. S., Cheevers, W. P., and Cork, L. C.,** Chronic arthritis in goats caused by a retrovirus, *Science,* 207, 997, 1980.
105. **Anderson, A., Harkiss, G., and Watt, N.,** Quantitative analysis of immunohistological changes in the synovial membrane of sheep infected with maedi-visna virus, *Clin. Immunol. Immunopathol.,* submitted.

106. **Kennedy-Stoskopf, S.,** Pathogenesis of lentivirus-induced arthritis, *Rheumatol. Int.,* 9, 129, 1989.
107. **Nathanson, N. and Martin, J. R.,** The effect of post-infection immunisation on the severity of experimental visna, *J. Comp. Pathol.,* 91, 185, 1981.
108. **McGuire, T. C., Adams, D. S., Johnson, G. C., Klevjer-Anderson, P., Barbee, D. D., and Gorham, J. R.,** Acute arthritis in caprine arthritis-encephalitis virus challenge exposure of vaccinated or persistently infected goats, *Am. J. Vet. Res.,* 47, 537, 1986.
109. **Peluso, R., Haase, A. T., Stowring, L., et al.,** A trojan horse mechanism for the spread of visna virus in monocytes, *Virology,* 147, 231, 1985.
110a. **Staskus, K. A., Retzel, E. F., Lewis, E. D., et al.,** Isolation of replication competent molecular clones of visna virus, *Virology,* 181, 228, 1991.
110b. **Scott, J. V., Stowring, L., Haase, A. T., Narayan, O., and Vigne, R.,** Antigenic variation in visna virus, *Cell,* 18, 321, 1979.
111. **Narayan, O., Griffin, D. E., and Clements, J. E.,** Virus mutation during 'slow infection'. Temporal development and characterisation of mutants of visna virus recovered from sheep, *J. Gen. Virol.,* 41, 343, 1978.
112. **Lutley, R., Petursson, G., Palsson, P. A., et al.,** Antigenic drift in visna: virus variation during long term infection of Icelandic sheep, *J. Gen. Virol.,* 64, 1433, 1983.
113. **Thormar, H., Barshatzky, M. R., Arnesen, K., and Kozlowski, P. B.,** The emergence of antigenic variants is a rare event in long-term visna virus infection *in vivo, J. Gen. Virol.,* 64, 1427, 1983.
114. **Stanley, J., Bhaduri, L. M., Narayan, O., and Clements, J. E.,** Topographical rearrangements of visna virus envelope glycoprotein during antigenic drift, *J. Virol.,* 61, 1019, 1987.
115. **Clements, J. E., Pedersen, F. S., Narayan, O., and Haseltine, W. A.,** Genomic changes associated with antigenic variation of visna virus during persistent infection, *Proc. Natl. Acad. Sci. U.S.A.,* 77, 4454, 1980.
116. **Clements, J. E., D'Antonio, N., and Narayan, O.,** Genomic changes associated with antigenic variation of visna virus II: common nucleotide sequence changes detected in variants from independent isolations, *J. Mol. Biol.,* 158, 415, 1982.
117. **Narayan, O., Griffin, D. E., and Chase, J.,** Antigenic shift of visna virus in persistently infected sheep, *Science,* 197, 376, 1977.
118. **Narayan, O., Clements, J. E., Griffin, D. E., and Wolinsky, J. S.,** Neutralising antibody spectrum determines the antigenic profiles of emerging mutants of visna virus, *Infect. Immun.,* 32, 1045, 1981.
119. **Pircher, H., Moskophidis, D., Rohrer, U., et al.,** Viral escape by selection of cytotoxic T cell resistant virus variants *in vivo, Nature,* 346, 629, 1990.
120. **Nixon, D. F.,** The cytotoxic T cell response to HIV, in *Immunology and Medicine: Immunology of HIV Infection,* Bird, A. G., Ed., Kluwer, Boston, 1992, 17, 59.
121. **Querat, G., Barban, V., Sauze, N., et al.,** Highly lytic and persistent lentiviruses naturally present in sheep with progressive pneumonia are genetically distinct, *J. Virol.,* 52, 672, 1984.
122. **Lairmore, M. D., Akita, G. Y., Russell, H. I., and DeMartini, J. C.,** Replication and cytopathic effects of ovine lentivirus strains in alveolar macrophages correlate with *in vivo* pathogenicity, *J. Virol.,* 61, 4038, 1987.
123. **Lutley, R., Petursson, G., Georgsson, G., Palsson, P. A., and Nathanson, N.,** Strains of visna virus with increased neurovirulence, in *Slow Viruses in Sheep, Goats and Cattle,* Sharp, J. M., and Jorgensen, R. H., Eds., Commission of the European Communities, Luxembourg, 1985, 45.
124. **Oyaizu, N., Chirmule, N., Ohnishi, Y., Kalyanaraman, V. S., and Pahwa, S.,** Human Immunodeficiency Virus Type 1 envelope glycoproteins gp120 and gp160 induce Interleukin-6 production in CD4+ T-cell clones, *J. Virol.,* 65, 6277, 1991.
125. **Ensoli, B., Barillari, G., and Gallo, R. C.,** Pathogenesis of AIDS-associated Kaposi's sarcoma, *Hematol. Oncol. Clin. N. Am.,* 5, 281, 1991.
126. **Iwakura, Y., Tosu, M., Yoshida, E., et al.,** Induction of inflammatory arthropathy resembling rheumatoid arthritis in mice transgenic for HTLV-1, *Science,* 253, 1026, 1991.
127. **Jolly, P. E., Huso, D. L., Sheffer, D., and Narayan, O.,** Modulation of lentivirus replication by antibodies: Fc portion of immunoglobulin molecule is essential for enhancement of binding, internalisation and neutralisation of visna virus in macrophages, *J. Virol.,* 63, 1811, 1989.
128. **Takeda, A., Tuazon, C. U., and Ennis, F. A.,** Antibody-enhanced infection by HIV-1 via Fc receptor-mediated entry, *Science,* 242, 580, 1988.

129. **Robinson, W. E., Montefiori, D. C., Gillespie, D. H., and Mitchell, W. M.,** Complement-mediated, antibody-dependent enhancement of HIV-1 infection *in vitro* is characterised by increased protein and RNA synthesis and infectious virus release, *J. Acquired Immune Defic.*, 2, 33, 1989.
130. **Duke, R. C., Chervenak, R., and Cohen, J. J.,** Endogenous endonuclease-induced DNA fragmentation: an early event in cell-mediated cytolysis, *Proc. Nat. Acad. Sci. U.S.A.*, 80, 6361, 1983.
131. **Sellins, K. S. and Cohen, J. J.,** Polyomavirus DNA is damaged in target cells during cytotoxic T-lymphocyte-mediated killing, *J. Virol.*, 63, 572, 1989.
132. **Howell, D. M. and Martz, E.,** Intracellular reovirus survives cytotoxic T lymphocyte-mediated lysis of its host cell, *J. Gen. Virol.*, 68, 2899, 1987.
133. **Small, J. A., Bieberich, C., Ghotbi, Z., et al.,** The visna virus long terminal repeat directs expression of a reporter gene in activated macrophages, lymphocytes, and the central nervous systems of transgenic mice, *J. Virol.*, 63, 1891, 1989.
134. **Panitch, H., Petursson, G., Georgsson, G., Palsson, P. A., and Nathanson, N.,** Pathogenesis of visna. III. Immune response to central nervous system antigens in experimental allergic encephalomyelitis and visna, *Lab. Invest.*, 35, 452, 1976.

Chapter 13

Cell-Mediated Responses Against Gastrointestional Nematode Parasites of Ruminants

S. J. McClure and D. L. Emery

CONTENTS

I. Introduction 213
II. Gastroenteric Mucosal Immunity 214
III. Cellular Immune Responses to Non-Nematode Antigens in the Gut 215
IV. Features of Parasite Immunity 215
V. Evidence for CMI in Immunity to Parasite Infections 216
VI. Induction of CMI During Gastrointestinal Nematode Infections 217
A. Systemic Responses 217
B. Local CMI Responses 218
1. Lymph Responses 218
2. Mucosal Responses 218
VII. Enteric CMI Effector Responses Against Gastrointestinal Nematodes 219
A. First Phase of Parasite Rejection 220
B. Second Phase of Parasite Rejection 221
VIII. Concluding Remarks 222
References 222

I. INTRODUCTION

Nematode parasites occur along the gastrointestinal (GI) tract of ruminants from the abomasum (*Haemonchus* spp., *Ostertagia* spp., *Trichostrongylus axei*) and small intestine (*T. colubriformis*, *Nematodirus*) to the large colon (*Oesophagostomum* spp., *Chabertia* spp.). During evolution, an equilibrium for survival of the host-parasite relationship in *open-range* grazing has been determined for each species in different climatic zones. However, with the advent of intensive livestock production, the impact of nematode parasitism has been more profoundly felt by the host. This impact is affected by the species and quantity of parasites ingested and the age and species of the host.

An acquired immunity is one mechanism for regulating parasite burdens within the GI of ruminants. Until recently, the study of specific immunological responses used antigenically complex parasite extracts to assay responses, and host lymphocytes (blood, lymph) to monitor the sequential development of immunity. Three recent advances have added precision and relevance to subsequent work: (1) the development of monoclonal antibodies (MAbs) and enzyme immunoassays (EIAs) to identify the role of subpopulations of host lymphoid and nonlymphoid cells and to detect their products, (2) the availability of recombinant parasite antigens for assays, and (3) the methodology to cultivate, differentiate, and clone *in vitro* lymphoid and nonlymphoid cells of ruminants.

In this chapter, we briefly overview features of mucosal immunity and parasite immunity in ruminants. Then we examine the cell-mediated immune (CMI) reactions involved in both the induction of parasite-specific responses and parasite expulsion. Where possible, mucosal immune responses receive most attention. To this end, CMI is defined broadly as reactions involving antigen recognition by thymus-derived (T) lymphocytes, ultimately producing overt cellular immunity (delayed-type hypersensitivity, DTH; cytotoxic T cells, CTL) or antibody responses of immunoglobulin (Ig) G, A, or E isotypes. This review will principally concentrate on three main helminth parasites of economic importance in ruminants: *Haemonchus contortus*, *Trichostrongylus colubriformis,* and *Ostertagia ostertagi* and *O. circumcincta. H. contortus* (Barber's pole worm) attach to the mucosa of the abomasum by their mouth parts and imbibe blood and serum, while *T. colubriformis* (black scour worm) and *O. ostertagi* (small

0-8493-4952-4/94/$0.00+$.50

brown stomach worm) reside in tunnels eroded into the epithelial surfaces of the small intestine and abomasum, respectively.[1]

II. GASTROENTERIC MUCOSAL IMMUNITY

The structure and cellular constituents of the ruminant mucosal immune system have been reviewed recently[2,3] and will not be discussed here. This system has several functions. In ontogeny it is responsible for the expansion of Ig diversity and the production of B cells for the entire immune system[4,5] and may also have a role in extrathymic T cell development. Throughout life it is a major site of peripheral expansion of lymphocytes[6] and its initial response to antigens can influence later encounters with the systemic system. Its function most relevant to gastrointestinal tract disease is that of local protection.

The principles of gut-associated immune functions with particular reference to ruminants, and the evidence for the interaction between mucosal sites involving mucosally restricted cell circuitry known as the "common mucosal immune theory", have been reviewed.[7-9] However, most functional aspects of cellular immunity in the gut have received little attention. In the absence of detailed studies in ruminants, the hypotheses for local sensitization by and response to gut lumenal antigen are extrapolated from other species. Briefly, it appears as follows: lymphocytes migrate through the normal gut, entering via blood vessels in the interfollicular area of the Peyer's patch and at the base of the villus, and leaving by the efferent lymph and possibly also by blood vessels. Antigen in the intestinal lumen is taken up by the specialized epithelial cells (M cells) overlying Peyer's patches, where it is processed and presented by subepithelial macrophages or dendritic cells to T cells. Soluble antigen may also be taken up by, or penetrate between, epithelial cells and presented to lamina propria or intraepithelial T cells by epithelial cells or subepithelial macrophages and dendritic cells. The resulting T cell activation induces B cell differentiation, and stimulated B cells and T cells migrate via afferent lymphatics to the draining node. Here, they proliferate and differentiate, to recirculate via efferent lymphatics and blood to lamina propria throughout the GI tract. A second exposure to antigen induces a local, antigen-specific IgA response, cytokine production, inflammation, and lymphocyte responses.

Not all of these details have been supported by studies in ruminants. Antigen uptake by M cells of rectal and jejunal Peyer's patches of calves and lambs has been demonstrated, and soluble ovalbumin administered intraduodenally has been detected in jejunal lamina propria and beneath the follicle-associated epithelium of the Peyer's patch[10,11] (authors' observations). However, in ruminants, the classical presentation of antigen by epithelial cells may not occur since these cells do not express MHC-II molecules.[12-14] The concentration on locally produced IgA may also need to be widened, as much of the local B cell differentiation in ruminants is toward IgG_1 production.[8]

Some data exist on gut-associated lymphocyte migration in ruminants, and recent reviews have dealt with the common mucosal system and the origin and fate of lymphocytes emigrating from sheep gut.[15,16] The concept of a common mucosal system has limits, as lymphocytes do not appear to migrate efficiently between gut and mammary gland of ruminants.[17] The common mucosal concept has been extended to involve the systemic immune response since mycobacterial infections in the lung or oral Bacillus-Calmette-Guerin (BCG) vaccination give rise to DTH reactions in the skin of ruminants. Early studies in sheep showed that lymphocytes leaving the gut returned to it preferentially, but the molecular basis for this non-random migration has yet to be determined. In other species, it is at least partly explained by the selective expression on site-specific vascular endothelium and circulating lymphocytes of addressins and their receptors. In addition, inflammation may increase immigration of T cells by adhesion to extracellular matrix proteins or by cytokine modulation of the expression or avidity of addressins and receptors. Such mechanisms are likely to be relevant to the GI tract, which can be regarded as chronically inflamed. Activation by antigen may also modulate expression of surface molecules, and in sheep it is those lymphocytes leaving the gut with memory rather than naive phenotype which preferentially return.[18] Lymphocyte migration through the gut may also be influenced directly or indirectly by local neuropeptides.

With the advent of MAbs to cell adhesion molecules, mucosal lymphocytes are now being characterized with respect to their expression of these molecules. The $CD4^+$ cells in the mucosal lamina propria of normal cattle do not express L-selectin (Mel 14) (C. Howard, personal communication). The majority of lymphocytes observed in jejunal lamina propria of immune sheep during the response to *Trichostrongylus colubriformis* are also Mel 14^-, but most of the intraepithelial lymphocytes are Mel 14^+.[14]

It should be noted that local protection is also partly mediated by the systemic immune system, chiefly via circulating Ig and complement proteins. In suckling neonates, milk provides passively transmitted maternal immunity. This is particularly important because of its relevance to disease and its effects on emergent acquired immunity and response to vaccination in young ruminants. Immunoglobulin G, especially G_1, IgA, and IgM isotypes are present in milk, with IgG the predominant isotope in colostrum[8]. However, with the advent of reagents for detecting bovine IgE, significant colostral transfer of IgE has now been demonstrated in calves.[19] Colostral protection of the neonate may also be aided by reactions induced by histoincompatible maternal lymphoid cells which are absorbed intercellularly from the digestive tract of neonatal lambs and transported to the mesenteric lymph nodes.[20] Calves fed complete colostrum showed enhanced bactericidal capacity against *Escherichia coli* and higher specific antibody titers than did calves fed cell-depleted colostrum.[21] In calves, maternal immunity can promote, probably via antigen/IgG_1 complexes, the systemic and local immune response to dietary antigens (reviewed in Reference 22). In a separate study, the ability to vaccinate lambs with irradiated *T. colubriformis* L_3 was not affected by colostrum from immune ewes.[23]

III. CELLULAR IMMUNE RESPONSES TO NON-NEMATODE ANTIGENS IN THE GUT

Our current knowledge of the induction of local effector mechanisms (cellular, humoral, and neural) in the gut is insufficient to permit prediction of the optimal methods of enteric vaccination. Therefore, we have compared routes of delivery and adjuvants for the induction of enteric cell-mediated responses in sheep to the model protein, ovalbumin.[24] Primary vaccination by the intraperitoneal, intra-Peyer's patch, or transepithelial routes was followed 2 weeks later by intraduodenal challenge. Of the adjuvants used, incomplete Freund's adjuvant resulted in the highest serum antibody titers, while Quil A gave the most consistent local cellular and humoral responses. While i.p. vaccination induced higher serum antibody titers, enteric vaccination induced greater numbers of specific IgA-containing cells in the enteric lamina propria. Transepithelial delivery induced the greatest increase in enteric T cells and $CD1^+$ dendritic cells. The only group to show increased numbers of mucosal mast cells and $CD56^+$ nerve fibers had been vaccinated transepithelially with Quil A, suggesting that mucosal inflammation may be a prerequisite for the induction of immediate hypersensitivity responses to soluble proteins. This may explain the correlation between helminth infections and high IgE responses. The chronic mucosal inflammation induced by sustained or repeated helminth infections may be responsible for the induction of mast cell-binding immunoglobulins.

We have also studied DTH reactions elicited by PPD in the intestinal mucosa of sheep vaccinated orally or intraperitoneally with BCG. Biopsies of the reactive area in the small intestine of animals necropsied 72 h after subserosal inoculation of 50 μg PPD showed typical edema and infiltration by small lymphocytes. MAbs to IFN-γ did not reduce the response substantially (D. Emery, unpublished data), indicating that cytokines additional to IFN-γ could effectively induce DTH responses in sheep.

IV. FEATURES OF PARASITE IMMUNITY

Gastrointestinal nematode infections are chronic, with "complete" immunity against all parasitic stages taking, in general, several months to develop. There are four broad overlapping effects of acquired immunity on nematode parasites: (1) rejection of incoming larvae (L_3), (2) retardation of larval development ("arrested L_3 or L_4" or "hypobiosis"), (3) depression of fecundity (decreased egg output by adult female worms and possibly also affecting males), and (4) expulsion of adult worms. Each manifestation develops sequentially with the timing and degree being dependent on the parasite, numbers of L_3 ingested, and age, sex, and breed/species of the host. In previously naive sheep, initially around 70% of *T. colubriformis* L_3 develop to adult worms,[25] but establishment rates for *O. circumcincta* and *H. contortus* are lower. For *T. colubriformis*, this initial rate of establishment can be enhanced by treating sheep with corticosteroids, suggesting an inflammatory/immunological screen reminiscent of innate immunity. After exposure of sheep to trickle infections of *T. colubriformis*, *H. contortus*, or *O. circumcincta* (1000–2000 L_3 per day) for 5 to 6 weeks, the percentage of incoming L_3 establishing is progressively reduced to less than 5% over the next 3 weeks.[25] During this time and subsequently, development of L_3 is retarded (particularly *Ostertagia*). Declining fecundity of *T. colubriformis* adults occurs after 10 to 12 weeks and expulsion of adult burdens occurs from 16 to 20 weeks.[25] With *H. contortus*, egg output and worm numbers are closely linked, with both declining after 20 to 24 weeks of continual larval intake.[26] Specific anti-worm immunity can be induced by "vaccination" with irradiated L_3, parasite extracts,[27] or large doses of L_3 followed by drenching.[28] In these cases, immunity against adult parasites is much reduced compared with exposure to "complete" infections, and the reciprocal situation was found in sheep immunized only with adult worms.[29]

Given that solid immunity takes 4 to 6 months to develop, it is not surprising that young stock (weaners) are highly susceptible to helminthiasis, and are more difficult to immunize with irradiated L_3 vaccines than older sheep.[30] This has a nutritional component, as infusion of supplementary protein into the abomasum during immunization of lambs against *H. contortus* enhanced their levels of acquired immunity[31] and lambs fed protein but not energy supplements gained immunity to *H. contortus* more rapidly.[32] In addition, substantial individual and breed differences occur in the expression of helminth immunity (reviewed in Reference 33) and the generation of these animals for studies on mechanisms of protective immunity is extremely valuable.

On pasture, mixed nematode infections occur. It has been demonstrated recently in naive Merino sheep that concurrent infections with *T. colubriformis* and *O. circumcincta* do not compromise the development of immunity to either parasite, but when *T. colubriformis* or *O. circumcincta* are given together with *H. contortus*, the establishment of the latter is reduced by up to 90%[34,35] (R. Dobson, personal communication). This effect may be reproduced by raising the abomasal pH.[34]

V. EVIDENCE FOR CMI IN IMMUNITY TO PARASITE INFECTIONS

The economically important GI nematodes of ruminants do not have a tissue migratory phase in their parasitic cycle as occurs with several rodent host/parasite models. Thus, great care must be exercised in mechanistic extrapolations, since naturally acquired immunity in ruminants is predominantly a mucosal, enteric immunity.

The evidence for CMI in parasite immunity has been largely deduced from the histopathology of parasite rejection and attempts to block the final effector response. As an example, administration of dexamethasone to immune sheep abrogates rejection of *T. colubriformis* L_3, *H. contortus* L_3[36-38] and enhances the survival of surgically transferred adult worms.[29] The greater susceptibility of lactating ewes to helminth infection, and the increased fecundity of established adult worms during the periparturient period,[39] was not associated with changes (decreases) in lymphoblast output and IgA titres in lymph from the abomasum of ewes infected with *O. circumcincta*[40] and thus may have a direct hormonal basis.

Attempts to define the inductive requirements for parasite immunity have come from the use of MAbs *in vivo* in rodent models. For example, the eosinophilia induced by helminthiasis can be specifically abrogated with MAbs to IL-5[41-43] without affecting the development of protective immunity to *Nippostrongylus braziliensis*, *Schistosoma mansoni*, *Heligmosomoides polygyrus,* or *Strongyloides venezuelensis* in mice.[41,42,42a] Administration of MAbs to IL-2 or IL-5[44] had no effect on helminth-induced mastocytosis in the murine intestine, whereas MAbs against IL-3 and IL-4 caused reductions.[44] Despite the reduced mastocytosis, *Nippostrongylus* was rejected normally. In contrast, immunity against a challenge infection with *H. polygyrus*, a completely enteric nematode similar to *Ostertagia*, *Trichostrongylus,* and many other ruminant helminths, was reduced by 30 or 90% by inoculation of MAbs against IL-4 or both IL-4 plus IL-4R, respectively.[42a] MAbs against IL-4 (BSF-1) specifically abolished the IgE response but failed to affect protective immunity to *Schistosoma* and *Nippostrongylus,*[42,44,45] while the administration of MAbs against IgE had no effect on immunity to *Strongyloides ratti* in mice.[46] However, MAbs against IFN-γ reduced protection against migrating *Schistosoma,*[41] and administration of anti-eosinophil serum to guinea pigs reduced protection against *T. colubriformis.*[47] These studies have emphasized that the effector mechanisms of parasite rejection are either more complex or more difficult to perturb with specific reagents than originally thought, and that the host-parasite relationship must be appreciated for extrapolation of relevant results from rodent to ruminant studies. Administration of MAbs against c-kit ligand (stem cell factor) interferes with normal rejection of *Trichinella* in mice,[48] suggesting that blocking effector cell development very early at the hemopoietic stem cell stage can disrupt immunity. In a different approach, immunity against the cestode, *Mesocestoides corti,* has been adoptively transferred in mice with a T cell clone recognizing a tetrathyridial antigen and secreting IL-1, IL-2, IFN-γ, GM-CSF, IL-5, and mast cell growth factor.[49] The importance of T lymphocytes in parasite immunity is reinforced by the prolonged parasitosis in mice treated with anti-CD4 MAb.[50] Similar studies are being attempted in ruminants as reagents become available. In sheep, administration of MAbs against CD4, CD8, T19, or IFN-γ did not affect rejection of *T. colubriformis* L_3 by immune sheep (S. McClure, unpublished data) and MAbs against T19 (WC1) T cells did not affect the development of immunity to *O. circumcincta* in partially immune sheep (D. Haig, unpublished data). By comparison, an anti-CD4 MAb (but not CD8) abolished the accelerated rejection of *Haemonchus contortus* normally exhibited by selected resistant lambs following an initial sensitising infection with the parasite.[51] In these studies, depletion of the targeted cells from blood was not complete, and depletion from gut mucosa was not assessed. Current studies indicate that 95% depletion of WCI cells from blood is not accompanied by detectable depletion from jejunal mucosa (S. McClure, unpublished data). Studies involving prolonged administration of MAbs are possible in sheep, but are limited in cattle by the development of anaphylactic reactions to murine Ig within 1 week.[52] However, the most important confirmation of the role of CMI in immunity to *Haemonchus* and *Ostertagia* has come from adoptive transfer studies. Protection, together with induction of IgA and abomasal MMC responses in recipients, was transferred with efferent lymphatic lymphoblasts and lymphocytes from the gastric lymph of immunized/challenged twin lambs to their susceptible co-twins.[53,54] The nature of the rejection mechanism transferred/induced by the donor lymphocytes was not known. In contrast, resistance against *Cooperia oncophera* or *T. colubriformis* could not be passively transferred with bovine colostrum or ovine intestinal lymph plasma obtained from normally infected sheep.[55,56] However, immunity against *H. contortus* has been transferred passively to lambs using serum from sheep vaccinated with "concealed/covert/novel" gut antigens.[57]

VI. INDUCTION OF CMI DURING GASTROINTESTINAL NEMATODE INFECTIONS

The habitat of GI nematodes adjacent or attached to the gut mucosa has restricted the study of local immunity to changes in blood, abomasal or intestinal efferent and pseudo-afferent lymph, repeated tissue biopsy, or functional and immunohistological analysis of gut and local lymph nodes obtained by sequential necropsies of parasitised animals.

A. SYSTEMIC RESPONSES

Antigen-reactive lymphocytes and serum IgG titers have been quantified for several parasitic infections in assays with soluble parasite extracts, usually from homogenized L_3. In animals more than 6 months old, antigen-reactive peripheral blood mononuclear cells were detected between 7 and 35 days after a single-dose or trickle infection with *T. colubriformis*, *H. contortus*, or *O. circumcincta*[58-60] and may persist up to 72 days.[55] The responses are variable in degree, duration, and consistency during a primary infection,[27] but occur more regularly 7 to 14 days after challenge infections with *Cooperia*, *H. contortus*, *O. ostertagi*, and *T. colubriformis*. For *T. colubriformis*, blood lymphocytes reacted with larval and adult extracts and adult excretory-secretory (ES) antigens from 7 to 28 days after challenge. In immune animals, responses before challenge were usually significant but low, and increased 2 days after challenge to peak at 7 to 8 days before decreasing.[14] Frequently, a second increase occurred around 14 days after challenge. However, if the immunized sheep had been recently infected with *T. colubriformis* and treated with anthelmintic, their response was usually very high before challenge and fell sharply by 2 days after challenge before increasing (S. McClure, unpublished data). The evidence of a rapid loss from blood of memory and antigen-reactive T cells suggests that these cells (in the gut or mesenteric nodes) may initiate the local DTH reaction described below. Following challenge of immune ruminants with *T. colubriformis*,[60] *Cooperia*,[58] or *Haemonchus*,[59] spontaneous proliferation of blood lymphocytes increases regularly 7 to 21 days later, with the effect of lowering the stimulation index (SI) of test responses to mitogens and parasite antigens. Blood leukocytes from immune sheep have been expanded *in vitro* with parasitic antigen and recombinant human IL-2 to produce cell lines and cell clones.[60,61] These have detected by proliferation, immunogenic parasite antigens which were converted to microparticles after SDS-PAGE and electroblotting.

Other CMI responses are also variably induced by a primary infection. At 7 to 30 days post-infection with *Cooperia punctata*, 48- to 72-h-delayed skin reactions were elicited intradermally with larval antigen in 2/6 calves.[58] Over the same period, Snider et al.[63] induced DTH reactions in the skin of *Ostertagia*, but not *Cooperia*-infected calves. Lymphocytes from calves infected with *Ostertagia* produced supernates which were chemotactic for eosinophils *in vitro*,[64] while Adams and Colditz[65] obtained an eosinophil-rich exudate within 24 h of an infusion of *Haemonchus* antigen into mammary glands of immune ewes or after multiple infusions into nonimmune ewes. The effect with *O. ostertagi* may be due, in part, to intrinsic chemotactic activity of components in organelles of the L_3 parasite.[66] Other supernates derived from incubation of sensitized ovine blood or mesenteric lymph node cells with worm antigen have contained eosinophil differentiation factors (IL-5) and colony-stimulatory activity (including IL-3),[61] IL-2, but little IFN-γ (D. Emery unpublished data). *In vivo*, serum levels of worm-specific IgG[60,61] and IgE[67] have been detected within 8 weeks or 9 to 12 weeks after single and trickle infections, respectively.

The situation is more complex in lambs less than 6 months old. Antigen-reactive blood lymphocytes in lambs infected with *Haemonchus contortus* have not been detectable in animals less than 4 months of age (see Reference 33), but appeared in lambs which were greater than 4 months old when infected. In nearly all instances in lambs and older animals, the presence and magnitude of lymphocyte proliferative responses did not presage protective immunity[33,68,70] (but see Reference 71), although the complexity of responses to L_3 extracts requires reevaluation with purified antigens.

The lack of consistent data from blood reflects the likelihood that parasite-specific, antigen-reactive cells in blood are recirculating from the site of mucosal parasitosis. The direction of current research is to monitor events during infection and challenge in the gut and lymph nodes by sequential necropsy and biopsy, and in the gut lumen and lymph by indwelling cannulae.

B. LOCAL CMI RESPONSES

1. Lymph Responses

Proliferative responses indicative of CMI have been detected in lymph draining the abomasum and small intestine of parasitised sheep. Lymphatic lymphocytes harvested as early as 12 days after infection and consistently at 35 days proliferated when incubated with *T. colubriformis* L_3 antigen.[73] Although antigen-reactive cells were not sought, increases in cytoplasmic IgA-positive cells or IgA titers in lymph were not detected until 35 days after infection with *Ostertagia* in sheep.[72] Recently, four- to five-fold increases in the levels of $CD4^+$ and $CD8^+$ efferent lymphatic lymphoblasts have been detected 5 to 9 days after infection of susceptible Suffolk sheep with 50,000 *O. circumcincta*. The increases were accompanied by rising quantities of mRNA for ovine IL-3 and GMCSF in lymph-borne cells and two-fold increases in the colony-stimulating activity of lymph supernates 5 to 7 and around 15 days after infection (D. Haig, unpublished data). Antigen-reactive cells have been consistently detected 3 to 7 days after challenge of immune sheep, and even at the time of challenge if immune sheep were recently challenged and then drenched with anthelmintics.[73] Boosts in lymphoblast output and IgA titers were also detected 3 to 6 days after challenge with *Ostertagia*[69] and the magnitude was dependent on the degree of immunity and the level of challenge. Immune sheep gave secondary responses in lymph after challenge with 50,000 *Ostertagia* L_3, but not 1000 L_3,[69] although the latter sheep expelled 50% more parasites than did control animals. Lymphadenectomy of the mesenteric lymph nodes either before or after immunization did not affect expression of immunity against *T. colubriformis* in sheep (D. Emery unpublished data) and only transient increases in lymphoblast output and dendritic cells were recorded 4 to 6 days after challenge with 30,000 *T. colubriformis* L_3. The differences in lymphoblast responses between intestinal and abomasal lymph observed after challenge of immune sheep with parasites probably reflect the vastly reduced microbial content of the abomasum where resting efferent lymph contains 1 to 2% lymphoblasts and intestinal lymph around 10 to 12%. Studies by McClure et al.[14] have indicated that if "highly immune" sheep expel more than 75% of a *T. colubriformis* L_3 challenge infection within 24 h, reactions typical of anamnestic responses at 5 to 8 days are reduced substantially. Heightened secondary responses appear to expel an increasing proportion of the challenge infection 5 to 14 days after challenge as the efficacy of the rapid expulsion wanes with time after immunization.[14,27] These secondary responses are recorded in lymph as antigen-reactive cells, and increases in lymphoblasts and parasite-specific antibody.

2. Mucosal Responses

As stated above, immune sheep in which the immediate hypersensitivity state has lapsed show a marked DTH response in the jejunal lamina propria after challenge. In animals challenged intraduodenally, this increased lymphocyte population was present from 3 to 7 days after challenge and was maximal at D5.[14] It involved increased numbers of $CD4^+$, $CD8^+$, and Tcrγδ $^+$ (including $T19^+$) cells in the lamina propria and increased numbers of $CD8^+$ and $\gamma\delta^+$ cells within the epithelium. The increase in $CD5^+$ cells lagged by 2 to 3 days, suggesting that the expanding T cell population did not express CD5 and was therefore activated. Particularly noticeable was the increase in $T19^-$ $\gamma\delta^+$ cells which appeared to move from the base of the villus into the epithelium and thence to be lost to the lumen. Increased lymphocyte numbers were accompanied by increased numbers of $CD1^+$ dendritic cells (whether increased cell number or increased expression) in lamina propria and epithelium, which had not returned to normal by 14 days after challenge. Local lymphoid accumulation has also been reported after challenge of immune cattle with *Ostertagia*.[74] Lymphocyte output, total and subset-specific, in efferent intestinal lymph of immune sheep increased within 2 days of challenge with *T. colubriformis*, while T cells other than those expressing CD8 also increased in blood (Reference 56; author's observations).

Continuing studies in our laboratory are assessing the proliferative response *in vitro* to *Trichostrongylus* antigens, of lymphocytes from gut-associated tissues of immune sheep. In susceptible sheep given a trickle infection with L_3, mesenteric lymphnode cells proliferated during incubation with L_3 extracts *in vitro* at 14 days after infection. The SI ranged from 0.6 to 17.1 in 20 sheep and there was a significant negative correlation with worm counts at 14 days (authors' unpublished data). Thus, the early detection of local responses in lymph and mesenteric node within the first 2 weeks of infection occur more reliably before similar antigen-reactive cells can be detected in blood. Before challenge of immune sheep with *T. colubriformis* L_3, significant antigen-specific proliferation occurs in cells from mesenteric node, jejunal lamina propria and jejunal epithelium. Specific responses by lymph node lymphocytes increased by 3 days after challenge and were significant at days 3, 10, and 25, but not at D7 or days 40 to 50 when spontaneous proliferation was extremely high. Specific responses were obtained from lamina propria lymphocytes collected 1, 2, 3, and 11 days after challenge, while intraepithelial lymphocytes responded to ConA (Concanavalin A) but not parasite antigens. With the systemic *Trichostrongylus* experiments referred to above, these results indicate that intestinal challenge of the locally primed immune system induces a rapid increase in T cell availability to the gut and rapid sequestration in the lamina propria and activation of memory T cells. This is followed by increased activation in the draining node and return of memory cells, initially activated and later (>7 days) resting, to the blood. This pattern is disturbed by giving oil adjuvants i.p. less than 4 days before challenge, as oil granulomas appear to attract T cells (especially $\gamma\delta^+$) to the early reaction.[75]

Specific T cell responses can also be inferred from the specific antibody responses in mucus detected from 3 to 4 days after challenge of immune sheep.

The role of local nerves in the immune response to GI nematodes is not known. Animals undergoing a secondary local immune response to *Trichostrongylus* had greatly increased numbers of nerve fibers expressing CD56 in the lamina propria (M. Stewart, personal communication) within 4 days after challenge. A similar increase was observed within 7 days of boosting immunized sheep transepithelially with ovalbumin, and both groups of sensitized sheep also had increased numbers of mast cells and T cells in the lamina propria. Although neuropeptides, endocrine hormones, and lymphokines interact in gastrointestinal physiology and immunity, IL-6 is the only lymphokine thus far accredited with nerve growth activity.[75a] A speculative involvement of nerve fibers in parasite immunity might occur through local axon reflex arcs or CNS-derived impulses to increase peristalsis, mast cell and eosinophil degranulation, and to regulate lymphocyte recirculation and local lymphocyte responses. We are currently exploring these possibilities.

These areas of interaction between the nervous system and the mucosal immune response are critical to understanding the way in which parasite immunity is induced and operates, and why helminth immunity takes so long to develop in young animals by comparison with conventional antibody responses to bacterial and protein antigens. With the development of reagents and probes for ruminant lymphocytes, cytokines, and neuropeptides, valuable insights into mucosal immunobiology should be forthcoming in the next few years.

VII. ENTERIC CMI EFFECTOR RESPONSES AGAINST GASTROINTESTINAL NEMATODES

Reviews of acquired immunity to GI nematode parasites in ruminants reveal that the protective effector mechanisms are still not well understood.[33,76,77] Although Smith et al.[69] found significant negative correlations between IgA titres in abomasal lymph and worm length, and incubation *in vitro* of *T. colubriformis* in host serum IgG1 depressed feeding[78] and oviposition,[79] two attempts to transfer immunity against GI tract nematodes with Ig have not been successful.[55,56] In contrast, protection against both *Haemonchus* and *Ostertagia* have been transferred adoptively with lymphatic lymphocytes and lymphoblasts.[53,54] Despite these results, still no single effector mechanism thus far described has been found sufficient or necessary to mediate protective immunity on its own.

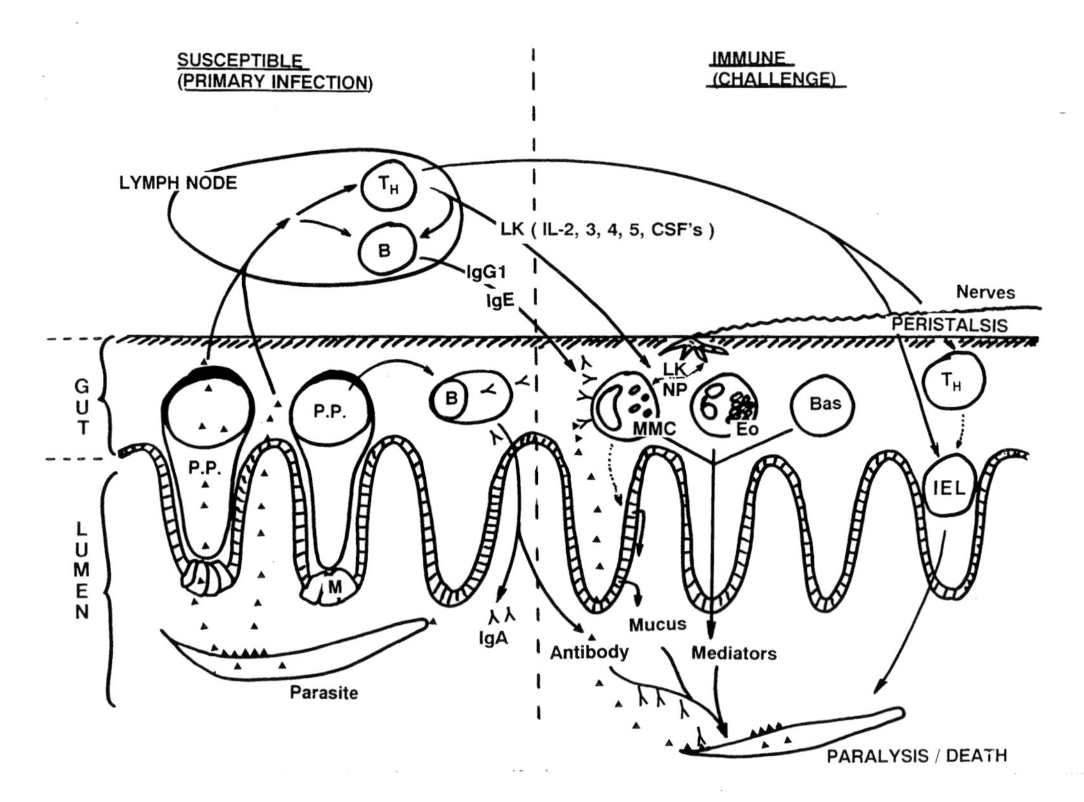

Figure 13-1 Schematic diagram of interactions during parasite rejection. Parasite antigen (▲) taken up by M cells (M) overlying Peyer's patches (P.P.) or diffusing through the gut epithelium activates antigen-specific B cells (B) and T cells (TH) to produce parasite-specific isotypes of antibody (IgG_1, IgA, and IgE) and lymphokines (LK). LKs activate mesenteric nerves and attract and differentiate non-lymphoid effector cells (mast cells, MMC; eosinophils, Eo; and basophils, Bas) in the lamina propria of the gut. MMC are sensitized with parasite-specific cytophilic antibody. Second and later contact with parasite antigen leads to an initial hypersensitivity response involving nerves, mucus, preformed antibody, and inflammatory mediators released from non-lymphoid cells. Larvae are paralyzed, disoriented, and expelled by peristalsis. Later reactions which also reject parasites from 5 days onward have features of secondary immune responses. The role of intraepithelial lymphocytes (IEL) is not known.

A. FIRST PHASE OF PARASITE REJECTION

The rejection of nematodes by immune sheep is a complex phenomenon mediated at two or more time points after challenge (Figure 13-1). Immunization with *Trichostrongylus* L_3 antigen generates responses which reject L_3 1 to 2 days after challenge,[28] but is less effective against adults. This type of response, termed "rapid rejection", can be precipitous and prevents establishment of *H. contortus* L_3 within 30 min of inoculation into the abomasum[37] or removes exsheathed *T. colubriformis* L_3 from the entire 15 m of the ovine small intestine within 4 h after inoculation through the pylorus (B. Wagland, unpublished results). In contrast, sheep immunized with adult *T. colubriformis* reject the majority of a larval challenge during the 2nd week of infection.[29] The rapid rejection of established adult *H. contortus* parasites after ingestion of *Haemonchus* L_3 has been termed "self-cure".[80,81] One mechanism for rapid rejection is suggested by its specificity and features. When combined infections (*T. colubriformis, O. circumcincta, H. contortus,* and *N. spathiger* L_3) were given to *H. contortus* immune sheep, all parasites were rejected; whereas in *T. colubriformis*-immune sheep, the abomasal nematodes were unaffected[81,82] (D. Emery, unpublished data). The effect was promulgated distally but not proximally and consisted of a specific recognition of parasite antigen followed by a nonspecific effector phase.[82] Taken together with the recent detection of a large increase in sheep mast cell protease in duodenal contents of immune sheep 1 day after challenge with *Trichostrongylus* L_3 (W. Jones, unpublished data) and in gastric lymph 1 to 3 days after challenge of immune sheep with *Haemonchus* L_3,[83] inflammatory mediators released from mucosal mast cells are considered largely responsible.[77] Evidence for the involvement of a mast cell reaction, which has features of anaphylaxis and bronchial asthma, is supported by the triggering of protease release *in vitro* by incubation of mucosal mast cells isolated from immune sheep with *T. colubriformis, H. contortus, T. vitrinus,* or *O. circumcincta* extracts and recombinant *T. colubriformis* and *H. contortus* antigens[84] (W. Jones unpublished data).

Histological data also supports mucosal mast cell involvement following induction of CMI leading to production of homocytotropic Ig. Increased numbers of mast cell/globule leukocytes (globule leukocytes derived from mast cells)[85] do not occur in intestinal sections until 5 or 6 weeks after a primary infection with *H. contortus* or *T. colubriformis,* respectively,[1,86] and mast cell numbers and larval migratory inhibitory activity exhibit an inverse relationship with fecal egg output from 6 to 8 weeks after infection with *T. colubriformis* (M. Stewart and W. Jones, unpublished data). These results suggest that mucosal mast cells may contribute to the changes in quality of the intestinal environment thought responsible for the ultimate rejection of established adult worms.[82] Increased numbers of mast cells/globule leukocytes have been described locally in animals immune to *Haemonchus, Ostertagia,* and *Trichostrongylus* species,[14,69,74,86-91] with the time of sampling post-challenge and the degree of host immunity being the most important determinants of the increases. Not only are mast cells increased locally, but mast cell/ globule leukocyte numbers in *Trichostrongylus*-immune sheep are increased in the abomasum, jejunum, and terminal ileum (M. Stewart, unpublished data), and mucosal mast cells from all these locations were triggered by larval antigen to release proteases in relatively comparable amounts (W. Jones and T. Bendixsen, unpublished data). In contrast, accumulation of eosinophils is more variable[14,74,87,88] and may reflect more closely the presence of parasites.[92]

Although ovine IgE remains to be isolated and assayed, the release, by specific parasite antigens, of proteases from mucosal mast cells of immune but not naive sheep suggests that cytotropic antibody is responsible. Release of protease from mast cells grown *in vitro* from ovine bone marrow has only been achieved in preliminary studies after sequential incubation of bone marrow mast cells with mesenteric lymph node cell supernates and parasite antigen (L. Hawke, unpublished data). However, BMMC were coated with cytophilic Ig (which was detected by two MAbs against ovine IgG1) after incubation in serum from *H. contortus*-immune, but not naive, sheep.[93]

Ultimately, the role of "IgE"-mediated chemical expulsion of nematodes must be confirmed *in vivo.* Assays have detected increases in platelet-activating factor (PAF), thromboxane, leukotrienes C_4 and B_4, histamine, and 5-hydroxytryptamine (5HT)[94-96] in gut contents within 1 week after challenge of immune sheep. However, increased levels of leukotrienes have not been detected in efferent lymph from the gut (J. Huntley, personal communication; authors' observations). It has also been shown that *T. colubriformis* L_3 and *H. contortus* L_3 can be paralyzed by similar gut contents in motility assays.[94,95,97] Serotonin (SRS-A) activity has been suggested as responsible,[94] but PAF and LTC_4 have been the only mediators from the above range with activity at physiological concentrations *in vitro* in larval migration inhibition assays (W. Jones, unpublished data). The amines, histamine, and 5HT did not affect the synthesis of acetylcholinesterase by *T. colubriformis,* but 5HT reduced the production of propionate by 50%.[98] *In vivo,* kinetic studies of the changes in mediator concentrations in the gut contents have been difficult to reconcile with worm rejection and immunity, such that more than a single mediator may be ultimately responsible. Attempts to define important mediators using drug antagonists *in vivo* have not been conducted in ruminants. Steroid antagonists (e.g., dexamethasone) extensively abrogate the early rejection of *H. contortus* in challenged immune sheep[36-38] and the intensity of the mast cell infiltration.[91] However, dexamethasone abolishes acquired immunity of genetically resistant sheep,[99] as well as stabilizing mast cell granules[100] and decreasing eosinophilia. Although more specific antagonists need to be examined, it is likely that the events of rapid rejection (*viz.* release of mediators, plasma leakage, protein loss, mast cell degranulation, mucus secretion, and goblet cell hyperplasia) are elicited by interactive elements.

B. SECOND PHASE OF PARASITE REJECTION

The second phase of parasite rejection by immune hosts occurs between 5 and 10 days for *T. colubriformis*[14] and comprises reactions typical of a secondary immune response. These are discussed above and include changes in local T and B lymphocyte populations, nerve fibers, mast cells/globule leukocytes, mucus antibody concentrations, and epithelial and lamina proprial integrity. There is no direct evidence for the role of any one of these in the second phase of rejection, although in partially immune Merino sheep, anti-CD4 MAbs could block accelerated rejection of *H. contortus* and the attendant accumulation of mast cells.[51] This result indicates that whereas solidly immune sheep reject a challenge infection within 1 day after challenge irrespective of administration of a range of MAbs (CD4, CD8, IFN-γ), partially immune animals (see Reference 51) require additional T helper cell activity to effect parasite rejection, which probably occurs during the second phase 5 to 10 days after challenge.

Despite the complexity of anti-nematode responses, it is likely that most progress in analysis of immunity to parasites will utilize drugs, MAbs to cytokines and key cell populations (e.g., T cells and mast cells), and exogenous cytokines to define important mechanisms of protection. Most "high responder" (resistant) animals actually acquire immunity faster to field or experimental infections; innate immunity to parasites has not been demonstrated to date.

VIII. CONCLUDING REMARKS

From the omissions in the above discussion it is evident that a predictive understanding of the local cell-mediated immunity to gastroenteric nematodes still requires investigation of a number of questions. These include the contribution of different antibody isotypes[101,102] in effector mechanisms and how they are induced; the role, if any, of "unrestricted" cytotoxic cells (e.g., "natural killer" cells, eosinophils) in nematode rejection; and the production of cytokines and the role in the induction and regulation of local GI tract responses of cytokines, T cell subpopulations, and local neural responses. The development of new reagents (drugs and MAbs) to block specific responses and the availability of recombinant parasite antigens to add precision and specificity to immunological assays will certainly promote knowledge of parasitic immunity in coming years.

REFERENCES

1. **Barker, I. K.,** A study of the pathogenesis of *Trichostrongylus colubriformis* infection in lambs with observations on the contribution of gastrointestinal plasma loss, *Int. J. Parasitol.,* 3, 743, 1973.
2. **Landsverk, T., Halleraker, M., Aleksandersen, M., McClure, S., Hein, W., and Nicander, L.,** The intestinal habitat for organised lymphoid tissues in ruminants. Comparative aspects of structure, function and development, *Vet. Immunol. Immunopathol.,* 28, 1, 1991.
3. **McClure, S. J. and Emery, D. L.,** Recent advances in veterinary immunology, particularly mucosal immunity, in Veterinary Vaccines, Peters, A. R., Ed., Butterworth, Oxford, 1993.
4. **Reynolds, J. D.,** Peyer's patches and the early development of B lymphocytes, *Curr. Top. Microbiol. Immunol.,* 135, 43, 1987.
5. **Reynaud, C.-A., Mackay, C. R., Müller, R. G., and Weill, J.-C.,** Somatic generation of diversity in a mammalian primary lymphoid organ; the sheep ileal Peyer's patch, *Cell,* 64, 995, 1991.
6. **McClure, S. J., Dudler, L., Thorpe, D., and Hein, W. R.,** Analysis of cell division among subpopulations of lymphoid cells in sheep. II. Peripheral lymphocytes, *Immunology,* 65, 401, 1988.
7. **Husband, A. J.,** Perspectives in mucosal immunity: a ruminant model, *Vet. Immunol. Immunopathol.,* 17, 357, 1987.
8. **Lascelles, A. K., Beh, K. J., Mukkur, T. K., and Watson, D. L.,** The mucosal immune system with particular reference to ruminant animals, in *The Ruminant Immune System in Health and Disease,* Morrison, W. I., Ed., Cambridge Press, Cambridge, 1986, 429.
9. **Sheldrake, R. F.,** Mucosal immunology: concepts and applications, in *Post Graduate Committee of Veterinary Science Proceedings 118,* (clinical immunology). University of Sydney, Sydney, 1989, 259.
10. **Liebler, E. M., Paar, M., and Pohlenz, J. F.,** M cells in the rectum of calves, *Res. Vet. Sci.,* 51, 107, 1991.
11. **Landsverk, T.,** The follicle-associated epithelium of the Ileal Peyer's patch in ruminants is distinguished by its shedding of 50 nm particles, *Immunol. Cell. Biol.,* 65, 251, 1987.
12. **Press, C., McClure, S. J., and Landsverk, T.,** Computer assisted morphometric analysis of absorptive and follicle associated epithelia of Peyer's patches in sheep fetuses and lambs indicates the presence of distinct T and B cell compartments, *Immunology,* 72, 368, 1991.
13. **Gorrell, M. D., Willis, G., Brandon, M. R., and Lascelles, A. K.,** Lymphocyte phenotypes in the intestinal mucosa of sheep infected with *Trichostrongylus colubriformis*, *Clin. Exp. Immunol.,* 72, 274, 1988.
14. **McClure, S. J., Emery, D. L., Wagland, B. M., and Jones, W. O.,** A serial study of rejection of *Trichostrongylus colubriformis* by immune sheep, *Int. J. Parasitol.,* 22, 227, 1992.

15. **Husband, A. J.,** Mucosal immune interactions in intestine, respiratory tract and mammary gland, *Prog. Vet. Microbiol. Immunol.,* 1, 25, 1985.
16. **Reynolds, J. D.,** Lymphocyte traffic associated with the gut: a review of *in vivo* studies in sheep, in *Migration and Homing of Lymphoid Cells* Husband, A. J., Ed., 2, 113, CRC Press, Boca Raton, 1988.
17. **Harp, J. A., Runnels, P. L., and Pesch, B. A.,** Lymphocyte recirculation in cattle: patterns of localisation by mammary and mesenteric lymph node lymphocytes, *Vet. Immun. Immunopathol.,* 20, 31, 1988.
18. **Mackay, C. R., Marston, W. L., Dudler, L., Spertini, O., Tedder, T., and Hein, W. R.,** Tissue-specific migration pathways by phenotypically distinct subpopulations of memory T cells, *Eur. J. Immunol.,* 22, 887, 1992.
19. **Thatcher, E. F. and Gershwin, L. J.,** Colostral transfer of bovine immunoglobulin E and dynamics of serum IgE in calves, *Vet. Immunol. Immunopathol.,* 20, 325, 1989.
20. **Sheldrake, R. F. and Husband, A. J.,** Intestinal uptake of intact maternal lymphocytes by neonatal rats and lambs, *Res. Vet. Sci.,* 39, 10, 1985.
21. **Riedel-Caspari, G. and Schmidt, F. W.,** Additional protection of neonatal calves against infection by colostral cells?, *Immunobiology,* 4 (Suppl.), 83, 1987.
22. **Porter, P., Powell, J. R., and Barratt, M. E. J.,** Inter-relationships between mucosal and systemic immunity determining the balance between damage and defence in the bovine gut in response to environmental antigens, in *Recent Advances in Mucosal Immunology,* McGhee, J. R., Mestecky, J., Ogra, P. L. and Bienenstock, J., Eds., Plenum Press, New York, Part B, 1987, 901.
23. **Dineen, J. K., Gregg, P., and Lascelles, A. K.,** The response of lambs to vaccination at weaning with irradiated *Trichostrongylus colubriformis* larvae; segregation into 'responders' and 'non-responders', *Int. J. Parasitol.,* 8, 59, 1978.
24. **McClure, S. J., Emery, D. L., and Husband, A. J.,** Effects of adjuvant and route of administration on the intestinal immune response of sheep to ovalbumin, *Immunology,* in press.
25. **Dobson, R. J., Waller, P. J., and Donald, A. D.,** Population dynamics of *Trichostrongylus colubriformis* in sheep: the effect of infection rate on the establishment of infective larvae and parasite fecundity, *Int. J. Parasitol.,* 20, 347, 1990.
26. **Barger, I. A., LeJambre, L. F., Gerogi, J. R., and Davies, H. I.,** Regulation of *Haemonchus contortus* populations in sheep exposed to continuous infection, *Int. J. Parasitol.,* 15, 529, 1985.
27. **Adams, D. B.,** The induction of selective immunological unresponsiveness in cells of blood and lymphoid tissue during primary infection of sheep with the abomasal nematode *Haemonchus contortus, Aust. J. Exp. Biol. Med. Sci.,* 56, 107, 1978.
28. **Emery, D. L., McClure, S. J., Wagland, B. M., and Jones, W. O.,** Studies of stage-specific immunity against *Trichostrongylus colubriformis* in sheep: immunization with normal and truncated infections, *Int. J. Parasitol.,* 22, 215, 1992.
29. **Emery, D. L., McClure, S. J., Wagland, B. M., and Jones, W. O.,** Studies of stage-specific immunity against *Trichostrongylus colubriformis* in sheep: immunization with adult parasites, *Int. J. Parasitol.,* 22, 221, 1992.
30. **Smith, W. D. and Angus, K. W.,** *Haemonchus contortus*: attempts to immunize lambs with irradiated larvae, *Res. Vet. Sci.,* 29, 45, 1980.
31. **Smith, W. D.,** Moredun Annual Report, 1991.
32. **Brown, M. D., Poppi, D. P., and Sykes, A. K.,** The effect of post-rumenal infusion of protein or energy on the pathophysiology of *Trichostrongylus colubriformis* infection and body composition in lambs, *Aust. J. Agric Res.,* 42, 253, 1991.
33. **Lloyd, S. and Soulsby, E. J. L.,** Immunobiology of gastro-intestinal nematodes of ruminants, in *Immune Responses in Parasitic Infections: Immunology, Immunopathology and Immunoprophylaxis,* Vol I, CRC Press, Boca Raton, FL, 1987, 1–41.
34. **Honda, C. and Bueno, L.,** Haemonchus contortus: egg laying influenced by abomasal pH in lambs, *Exp. Parasitol.,* 54, 371, 1982.
35. **Blanchard, J. L. and Westcott, R. B.,** Enhancement of resistance of lambs to *Haemonchus contortus* by previous infection with ostertagia circumcincta, *Am. J. Vet. Res.,* 46, 2136, 1985.
36. **Adams, D. B.,** The effect of dexamethasone on a single and a superimposed infection with *Haemonchus contortus* in sheep, *Int. J. Parasitol.,* 18, 575, 1988.
37. **Jackson, F., Miller, H. R. P., Newlands, G. F. J., Wright, S. E., and Hay, L. A.,** Immune exclusion of *Haemonchus contortus* larvae in sheep: dose dependency, steroid sensitivity and persistence of the response, *Res. Vet. Sci.,* 44, 320, 1988.

38. **Newlands, G. F. J., Miller, H. R. P., and Jackson, F.,** Immune exclusion of *Haemonchus contortus* larvae in the sheep: effects on gastric mucin of immunization, larval challenge and treatment with dexamethasone, *J. Comp. Pathol.,* 102, 433, 1990.
39. **O'Sullivan, B. M. and Donald, A. D.,** Responses to infection with *Haemonchus contortus* and *Trichostrongylus colubriformis* in ewes of different reproductive status, *Int. J. Parasitol.,* 3, 521, 1973.
40. **Smith, W. D., Jackson, F., and Williams, J.,** Studies on the local immune response of the lactating ewe infected with *Ostertagia circumcincta, J. Comp. Pathol.,* 93, 295, 1983.
41. **Korenaga, M., Hitoshi, Y., Yamaguchi, N., Sato, Y., Takatsu, K., and Tada, I.,** The role of interleukin-5 in protective immunity to *Strongyloides venezuelensis* infection in mice, *Immunology,* 72, 502, 1991.
42. **Sher, A., Coffman, R. L., Hieny, S., and Cheever, A. W.,** Ablation of eosinophil and IgE responses with anti-IL-5 or anti-IL-4 antibodies fails to affect immunity against *Schistosoma mansoni* in the mouse, *J. Immunol.,* 145, 3911, 1990.
42a. **Urban, J. R. Katona, I. M., Paul, W. E., and Finkelman, F. D.,** Interleukin 4 is important in protective immunity to a gastroinestinal nematode infection in mice, *Proc. Natl. Acad. Sci. U.S.A.,* 88, 5513, 1991.
43. **Coffman, R. L., Seymour, B. W. P., Hudak, S., Jackson, J., and Rennick, D.,** Antibody to interleukin-5 inhibits helminth-induced eosinophilia in mice, *Science,* 245, 308, 1989.
44. **Madden, K. B., Urban, J. F., Ziltener, H. J., Schroder, J. W., Finkelman, F. D., and Katona, I. M.,** Antibodies to IL-3 and IL-4 suppress helminth-involved mastocytosis, *J. Immunol.,* 147, 1387, 1991.
45. **Finkelman, F. D., Katona, I. M., Urban, J. F., Snapper, C.M., Ottara, J., and Paul, W. F.,** Suppression of *in vivo* polyclonal IgE responses by monoclonal antibody to the lymphokine B cell stimulatory factor 1 (IL-4), *Proc. Natl. Acad. Sci. U.S.A.,* 83, 9675, 1986.
46. **Korenga, M., Watanabe, N., and Toda, I.,** Effects of anti-IgE monoclonal antibody on a primary infection of *Strongyloids ratti* in mice, *Parasitol. Res.,* 77, 362, 1991.
47. **Gleich, G. J., Olson, G. M., and Herlich, H.,** The effect of antiserum to eosinophils on susceptibility and acquired immunity of the guinea pig to *Trichostrongylus colubriformis*, *Immunology,* 37, 873, 1979.
48. **Grencis, R. K., Else, K. J., Huntley, J. F., and Nishikawa, S. I.,** The *in vivo* role of stem cell factor (c-kit ligand) on mastocytosis and host-protective immunity to the intestinal nematode *Trichinella spiralis* in mice, *Parasite Immunol.,* 15, 55, 1993.
49. **Lammas, D. A., Mitchell, L. A., and Wakelin, D.,** Adoptive transfer of enhanced eosinophilia and resistance to infection in mice by an *in vitro* generated T cell line specific for *Mesocestoides corti* larval antigen, *Parasite Immunol.,* 9, 591, 1987.
50. **Katona, I. M., Urban, J. F., and Frakelman, F. D.,** The role of $L_3T_4^+$ and Lyt-2^+ T cells in the IgE response and immunity to *Nippostrongylus braziliensis*, *J. Immunol.,* 140, 3206, 1988.
51. **Gill, H. S., Watson, D. L., and Brandon, M. R.,** Monoclonal antibody to $CD4^+$ T cells abrogates genetic resistance to *Haemonchus contortus* in sheep, *Immunology,* 78, 43, 1992.
52. **Howard, C. J., Clarke, M. C., Sopp, P., and Brownlie, J.,** Immunity to bovine virus diarrhoea virus in calves: the role of different T cell subpopulations anlaysed by specific depletion *in vivo* with monoclonal antibodies, *Vet. Immunol. Immunopathol.,* 32, 303, 1992.
53. **Smith, W. D., Jackson, F., Jackson, E., Williams, J., Willadsen, S. M., and Fehilly, C. B.,** Resistance to *Haemonchus contortus* transferred between genetically histocompatible sheep by immune lymphocytes, *Res. Vet. Sci.,* 37, 199, 1984.
54. **Smith, W. D., Jackson, F., Jackson, E., Graham, R., Williams, J., Willadsen, S. M., and Fehilly, C. B.,** Transfer of immunity to Ostertagia circumcincta and IgA memory between identical sheep by lymphocytes collected from gastric lymph, *Res. Vet. Sci.,* 41, 300, 1986.
55. **Kloosterman, A., Benedictus, J., and Aghina, H.,** Colostral transfer of anti-nematode antibodies in cattle and its significance for protection, *Vet. Parasitol.,* 7, 133, 1980.
56. **Adams, D. B., Merritt, G. C., and Cripps, A. W.,** Intestinal lymph and the local antibody and immunoglobulin response to infection by *Trichostrongylus colubriformis* in sheep, *Aust. J. Exp. Biol. Med. Sci.,* 58, 167, 1980.
57. **Smith, W. D.** Protection in lambs immunised with *Haemonchus contortus* gut membrane proteins, *Res. Vet. Sci.,* 54, 94, 1993.

58. **Hanrahan, L. A., Benz, G. W., and Schultz, R. D.,** Experimentally induced *Cooperia oncophora* infection in calves: lymphocyte blastogenic and delayed hypersensitivity responses, *Am. J. Vet. Res.,* 45, 855, 1984.
59. **Zajac, A. M., Krakowka, S., Herd, R. P., and McClure, K. E.,** Experimental *Haemonchus contortus* infection in three breeds of sheep, *Vet. Parasitol.,* 36, 221, 1990.
60. **Emery, D. L., Bendixsen, T., and McClure, S. J.,** The use of electroblotted antigen of *Trichostrongylus colubriformis* to induce proliferative responses in sensitized lymphocytes from sheep, *Int. J. Parasitol.,* 21, 179, 1991.
61. **Haig, D. M., Windon, R. G., Blackie, W., Brown, D., and Smith, W. D.,** Parasite-specific T cell responses of sheep following live infection with the gastric nematode *Haemonchus contortus, Parasite Immunol.,* 11, 463, 1989.
62. **Snider, T. G., Williams, J. C., Karns, P. A., Romaire, T. L., Trammal, H. E., and Kearney, M. T.,** Immunosuppression of lymphocyte blastogenesis in cattle infected with *Ostertagia ostertagi* and/or *Trichostrongylus axei, Vet. Immunol. Immunopathol.,* 11, 251, 1986.
63. **Snider, T. G., Klesius, P. H., and Haynes, T. B.,** Dermal responses to *Ostertagia ostertagi* in *O. ostertagi* and *Coopera punctata*-inoculated calves, *Am. J. Vet. Res.,* 46, 887, 1985.
64. **Washburn, S. M. and Klesius, P. H.,** Leukokinesis in bovine ostertagiasis: stimulation of leukocyte migration by *Ostertagia, Am. J. Vet. Res.,* 45, 1095, 1984.
65. **Adams, D. B., and Colditz, I. G.,** Immunity to *Haemonchus contortus* and the cellular response to helminth antigens in the mammary gland of non-lactating sheep, *Int. J. Parasitol* 21, 631, 1991.
66. **Klesius, P. H., Snider, T. G., Horton, L. W., and Crowder, C. H.,** Visualisation of eosinophil chemotactic factor in abomasal tissue of cattle by immuno-peroxidase staining during *Ostertagia ostertagi* infection, *Vet. Parasitol.,* 31, 49, 1989.
67. **Thatcher, E. F., Gershwin, L. J., and Baker, N. F.,** Levels of serum IgE in response to gastrointestinal nematodes in cattle, *Vet. Parasitol.,* 32, 153, 1989.
68. **Shubber, A. H., Lloyd, S., and Soulsby, E. J. L.,** Immunological unresponsiveness of lambs to infection with *Haemonchus contortus.* Effect of infection in the ewe on the subsequent responsiveness of lambs, *Z. ParasitenKde,* 70, 219, 1984.
69. **Smith, W. D., Jackson, F., Jackson, E., Williams, J., and Miller, H. R. P.,** Manifestations of resistance to ovine ostertagiasis associated with immunological responses in the gastric lymph, *J. Comp. Pathol.,* 94, 591, 1984.
70. **Berezhko, V. K., Akulin, N. A., Buzmakova, R. A., and Kurochkina, K. G.,** Immunoallergic reactions and a phenomenon of self-cure in experimental haemonchosis (*Haemonchus contortus* of sheep depending on host's age), *Helminthologia,* 24, 119, 1987.
71. **Riffkin, G. G. and Yong, W. K.,** Recognition of sheep which have innate resistance to Trichostrongylid nematode parasites, in *Immunogenetic Approaches to the Control of Endoparasites, with Particular Reference to Parasites of Sheep,* Dineen, J. K. and Outteridge, P. M., Eds., CSIRO and AWC, Melbourne 1984, 30.
72. **Smith, W. D., Jackson, F., Jackson, E., and Williams, J.,** Local immunity and *Ostertagia circumcincta*: changes in the gastric lymph of immune sheep after a challenge infection, *J. Comp. Pathol.,* 93, 479, 1983.
73. **Adams, D. B. and Cripps, A. W.,** Cellular changes in the intestinal lymph of sheep infected with the enteric nematode, *Trichostrongylus colubriformis, Aust. J. Exp. Biol. Med. Sci.,* 55, 509, 1977.
74. **Snider, T. G., Williams, J. C., Knox, J. W., Marbury, K. S., Crowder, C. H., and Willis, E. R.,** Sequential histopathologic changes of Type 1, pre-Type II and Type III Ostertagiasis in cattle, *Vet. Parasitol.,* 27, 169, 1988.
75. **McClure, S. J., Wagland, B. M., and Emery, D. L.,** Effects of Freund's adjuvants on local, draining and circulating lymphocyte populations in sheep, *Immunol. Cell Biol.,* 69, 361, 1991.

75a. **Hoffman, F. M. and Hinton, D. R.,** Cytokine interactions in the central nervous system, *Regional Immunol.,* 3, 268, 1991.

76. **Rothwell, T. L. W.,** Immune expulsion of parasitic nematodes from the alimentary tract, *Int. J. Parasitol.,* 19, 139, 1989.
77. **Miller, H. R. P.,** The protective mucosal response against gastrointestinal nematodes in ruminants and laboratory animals, *Vet. Immunol. Immunopathol.,* 6, 167, 1984.
78. **Bottjer, K. P., Klesius, P. H., and Bone, L. W.,** Effects of host serum on feeding by *Trichostrongylus colubriformis* (nematoda), *Parasite Immunol.,* 7, 1, 1985.

79. **Bone, L. W. and Klesius, P. H.,** Effect of host serum on *in vitro* oviposition by *Trichostrongylus colubriformis* (Nematoda), *Int. J. Invert. Reprod. Dev.,* 10, 27, 1986.
80. **Stewart, D. F.,** Studies on the resistance of sheep to infestation with *Haemonchus contortus* and *Trichostrongylus* spp. and on the immunological reactions of sheep exposed to infection. V. The nature of the self-cure phenomenon, *Aust. J. Agric. Res.,* 4, 100, 1953.
81. **Gordon, H. McL.,** The self-cure reaction, *Vet. Med. Rev.,* 174, 1968.
82. **Dineen, J. K., Gregg, P., Windon, R. G., Donald, A., and Kelly, J. D.,** The role of immunologically specific and non-specific components of resistance in cross protection to intestinal nematodes, *Int. J. Parasitol.,* 7, 211, 1977.
83. **Huntley, J. F., Gibson, S., Brown, D., Smith, W. D., Jackson, F., and Miller, H. R. P.,** Systemic release of a mast cell proteinase following nematode infections in sheep, *Parasite Immunol.,* 9, 603, 1987.
84. **Jones, W. O., Huntley, J. F., and Emery, D. L.,** Isolation and degranulation of mucosal mast cells from the small intestine of parasitised sheep, *Int. J. Parasitol.,* 22, 519, 1992.
85. **Huntley, J. F., Newlands, G., and Miller, H. R. P.,** The isolation and characterisation of globule leucocytes: their derivation from mucosal mast cells in parasitised sheep, *Parasite Immunol.,* 6, 371, 1984.
86. **Sommerville, R. I.,** The histology of the ovine abomasum and the relation of the globule leucocyte to nematode infestations, *Aust. Vet. J.,* 32, 237, 1956.
87. **Douch, P. G. C., Harrison, G. B. L., Elliot, D. C., Buchanan, L. L., and Greer, K. S.,** Relationship of gastrointestinal histology and mucus antiparasitic activity with the development of resistance to trichostrongyle infections in sheep, *Vet. Parasitol.,* 20, 315, 1986.
88. **Ritchie, J. D. S., Anderson, N. Armour, J., Jarrett, W. F. H., Jennings, F. W., and Urquhart, G. M.,** Experimental ostertagia ostertagi infections in calves: parasitology and pathogensis of a single infection, *Am. J. Vet. Res.,* 27, 659, 1966.
89. **Klesius, P. H.,** Immunity to *Ostertagia ostertagi, Vet. Parasitol.,* 27, 159, 1988.
90. **Miller, H. R. P.,** Mucosal mast cells, basophils, immediate hypersensitivity reactions and protection against gastrointestinal nematodesin *The Ruminant Immune System in Health and Disease,* Morrison, W. I., Ed., Cambridge University Press, Cambridge, 1986.
91. **Huntley, J. F., Newlands, G. F. J., Jackson, R., and Miller, H. R. P.,** The influence of challenge dose, duration of immunity or steroid treatment on mucosal mast cells and on the distribution of sheep mast cell proteinase in Haemonchus infected sheep, *Parasite Immunol.,* 14, 1992.
92. **Dineen, J. K. and Windon R. J.** The effect of sire selection on the response of lambs to vaccination with irradiated *Trichostrongylus colubriformis* larvae, *Int. J. Parasitol.,* 10, 189, 1980.
93. **Huntley, J. H., Haig, D. M., Irvine, J., Inglis, L., MacDonald, A., Rance, A., and Moqbel, R.,** Characterisation of ovine mast cells derived from *in vitro* culture of haemopoietic tissue, *Vet. Immunol. Immunopathol.,* 32, 47, 1992.
94. **Douch, P. G. C., Harrison, G. B. L., Buchanan, L. L., and Greer, K. S.,** *In vitro* bioassay of sheep gastrointestinal mucus for nematode paralysing activity by substances with some properties characteristic of SRS-A, *Int. J. Parasitol.,* 13, 207, 1983.
95. **Kimambo, A. E. and MacRae, J. C.,** Measurement *in vitro* of a larval migration inhibitory factor in gastrointestinal mucus of sheep made resistant to the roundworm, *Trichostrongylus colubriformis, Vet. Parasitol.,* 28, 213, 1988.
96. **Jones, W. O. and Emery, D. L.,** Demonstration of a range of inflammatory mediators released in trichostrongylosis of sheep, *Int. J. Parasitol.,* 21, 361, 1991.
97. **Wagland, B. M., Jones, W. O., Hribar, L., Bendixsen, T., and Emery, D. L.,** A new simplified assay for larval migration inhibition, *Int. J. Parasitol.,* in press.
98. **Rothwell, T. L. W., Prichard, R. K., and Love, R. J.,** Studies on the role of histamine and 5-hydroxytryptamine in immunity against the nematode *Trichostrongylus colubriformis.* I. *In vivo* and *in vitro* effects of the amines, *Int. Arch. Allergy,* 46, 1, 1974.
99. **Presson, B. L., Gray, G. D., and Burgess, S. K.,** The effect of immunosuppression with dexamethasone on *Haemonchus contortus* infections in genetically resistant Merino sheep, *Parasite. Immunol.,* 10, 675, 1988.
100. **Perdue, M., Masson, S., Williams, K., Kosecka-Janiszewsk, U., and Crowe, S.,** The mucosal mast cell response and resultant functional abnormalities in the intestine, in *Mucosal Immunology*, Cripps, A. W., Ed., Newey & Beath, Newcastle, 1990, 14.

101. **Charley-Poulain, J., Luffau, G., and Pery, P.** Serum and abomasal antibody response of sheep to infections with *Haemonchus contortus, Vet. Parasitol.*, 14, 129, 1984.
102. **Gill, H. S., Husband, A. J., and Watson, D. L.**, Localisation of immunoglobulin-containing cells in the abomasum of sheep following infection with *Haemonchus contortus*, *Vet. Immunol. Immunopathol.*, 31, 179, 1992.

Chapter 14

Cytokine Applications in Infectious Diseases

M. Campos, D. L. Godson, H. P. A. Hughes and L. A. Babiuk

CONTENTS

I. Introduction 229
II. Prevention of Infectious Diseases 230
A. Nonspecific Stimulation During Stages of Increased Disease Susceptibility 230
B. Adjuvant/Co-Adjuvant Activities of Cytokines 232
1. Adjuvant Activity Exerted by Early Cytokines 232
2. Adjuvant Activity Exerted by Late Cytokines 232
3. Combining Early and Late Cytokines for Optimal Effect 233
III. Treatment of Infectious Diseases 234
IV. Disease Diagnosis and Prognosis 235
V. Epilogue 236
References 237

I. INTRODUCTION

Cytokines (CK) play a central role in the regulation of immune responses mounted against infectious agents and thus have considerable potential as immunotherapeutic agents. With the production of recombinant CK, information regarding the biology of CK has become available. This information is of great value in designing immunomodulatory strategies using these mediators. In general, the application of CK in veterinary medicine has closely paralleled advances achieved in human medicine. However, the major emphasis in immunoregulation in ruminants is in the area of infectious diseases, in contrast to tumor therapy in human medicine.

Cytokine research is evolving from the study of a few functions induced by poorly characterized factors to investigations of a complicated network of well-defined molecules that are essential for the maintenance of homeostasis. Unfortunately, most of the research in CK is confined to model systems where particular functions and interactions of a single CK are examined in isolation. As a result of this narrow approach, attempts to place CK in the context of infectious diseases are frequently illustrated as a network of secretory products which can induce the production of other CK and interact with multiple target cells and organs. These models can be misleading since they represent a pictorial table of possibilities rather than the actual host response to pathogens.

The *in vivo* use of these proteins faces a series of problems. For instance, the potency and multiple biological actions of CK can lead to unpredictable, undesirable, and even toxic effects following systemic administration of these products. As well, the interactions characteristic of the CK network makes it necessary to establish experimental disease models that mimic field situations to test adequately CK applications. However, significant advances have been made toward understanding the biological significance of CK production during infectious diseases. We believe that this knowledge regarding the role of CK in the pathogenesis and recovery from disease should form the basis for any possible clinical application of these potent mediators.

To understand how CK function within the global picture of pathogenesis and immunity, where their production assures that certain host responses necessary to maintain homeostasis are executed in a timely and coordinated fashion, we present the following model. A typical initial response to pathogens is characterized by early inflammatory signs, followed by infiltration and activation of inflammatory cells (i.e., neutrophils and macrophages), and initiation of specific immune responses by B and T lymphocytes. These responses are orchestrated by the production of CK in a sequential manner in response to the invading pathogens (Figure 14-1). Thus, according to their time of production, CK can be grouped as early and late CK. In general, early CK are produced by cells present at the site of infection and they are responsible for initiation of the inflammatory response, recruitment and activation of inflammatory cells, and in conjunction with antigen, effective T and B cell stimulation. In contrast, late CK are produced

0-8493-4952-4/94/$0.00+$.50

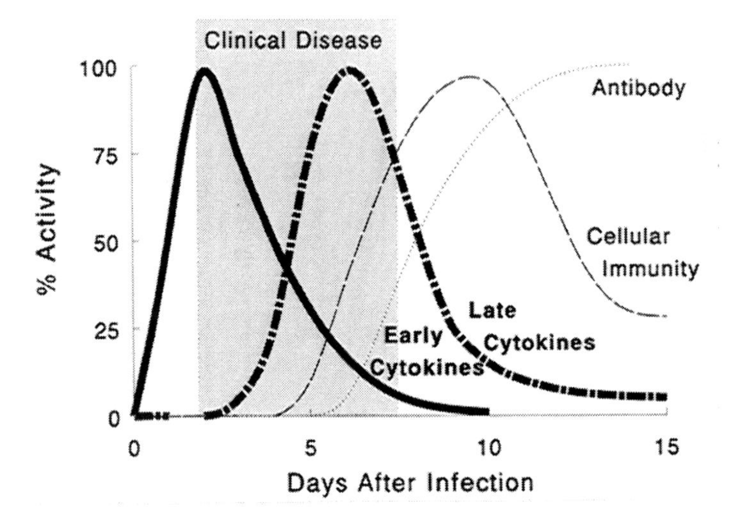

Figure 14-1 Coordination of the host response to infection. During the course of a typical infection, the host responds with the production of a number of cytokines. There is a temporal relationship between the production of each cytokine such that they can be grouped into two broad categories: early and late. The kinetics of cytokine production can be related to the development of clinical manisfestations of disease and to the elicitation of humoral and cell-mediated immunity.

mainly by T lymphocytes after activation by antigen and early CK, and they are responsible for the differentiation, amplification, and fine tuning of the immune response directed toward a given pathogen. Thus, the challenge which now confronts researchers is to understand not only the mechanisms of action of CK, but also how the host ensures the elaboration of these mediators in the proper sequence to preserve homeostasis. With this knowledge, we should be able to find clinical relevance for these potent biological response modifiers. In the interim, we can use the above-mentioned model of CK actions during infectious disease to propose uses for these mediators in the manipulation of the immune response in a variety of situations. These situations are illustrated in Figure 14-2 and include the detection of CK in serum or other samples for disease diagnosis and prognosis, the administration of early and late CK to prevent or treat diseases, and the regulation of early CK effects by blockers or antagonists. In this chapter, we will describe some of the experiences obtained in the veterinary field and outline the possible directions for future applications of these mediators.

II. PREVENTION OF INFECTIOUS DISEASES

The concept of using CK alone or in combination with current vaccination protocols to prevent disease has received considerable attention in the veterinary field. The possible applications for CK in this area can be divided in two general groups: one group refers to the use of CK as nonspecific stimulators of the immune system during stages of increased disease susceptibility, and the other refers to the use of CK as adjuvants or co-adjuvants to enhance specific immune functions during vaccination.

A. NONSPECIFIC STIMULATION DURING STAGES OF INCREASED DISEASE SUSCEPTIBILITY

Initial attempts to apply CK in the veterinary field were mainly directed toward exploiting the antiviral properties of IFNs.[1,2] However, broader immunoregulatory properties of these CK were soon demonstrated in experimental trials evaluating the effect of recombinant bovine (rBo) IFN-α and γ on bovine respiratory disease (BRD).[3] This BRD model consists of an aerosol challenge with bovine herpesvirus-1 (BHV-1) followed by an aerosol challenge with *Pasteurella haemolytica* 4 days later. The resultant disease mimics the clinical picture and pathological changes associated with the naturally occurring fibrinous pneumonia (shipping fever) observed in feedlots. Consequently, this model has been used extensively to understand the pathogenesis of this disease complex as well as modulation of the disease process by CK.[4] Under these experimental conditions, prophylactic treatment with either IFN-α or IFN-γ prior to exposure to BHV-1 reduced the severity of clinical signs and mortality associated with the subsequent bacterial infection.[3,5] As well, it was observed that the amount of virus recovered from the upper respiratory tract of calves infected with BHV-1 was not reduced by treatment with rBoIFN-α.[5] It was therefore concluded that the reduction in clinical disease and mortality was more likely due to the beneficial immunomodulatory properties of rBoIFN-α rather than to its direct antiviral property. IFN-α was found to be effective at doses ranging from 1 to 10 mg per animal, whereas IFN-γ was effective at approximately 10- to 100-fold lower doses.[4] Whether the beneficial results obtained with IFN-α and γ

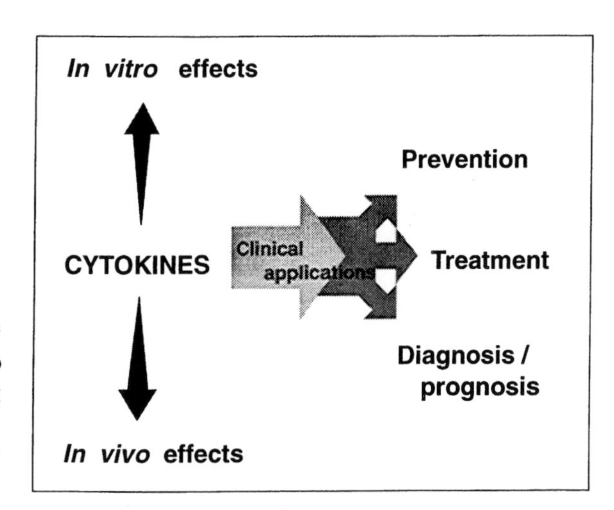

Figure 14-2 Clinical applications of cytokines. Based on *in vitro* and *in vivo* experiments, we can anticipate that cytokines will be used in the prevention, treatment, and diagnosis/prognosis of infectious diseases.

were related to the same biological activity or to different mechanisms has not been established. In either case, these experiments provided a clear indication that IFNs were capable of ameliorating clinical disease associated with bacterial infections in instances where the immune system had been previously compromised.

The effect of rBoIFN-α in alleviating BRD has also been tested in the field. Treatment of steers at feedlot entry with IFN-α (5 mg) reduced mortality caused by respiratory disease (unpublished results). Nevertheless, its beneficial effects under field conditions were not as striking as those observed under experimental situations. This disparity could be at least partially explained by the fact that not all animals in the field became infected at the same time relative to IFN treatment.[4] Another form of BRD in which rBoIFN-α has been shown to elicit a beneficial effect is "crowding disease".[6] This type of bovine respiratory disease occurs in young dairy calves housed in a closed environment. Similar to shipping fever, crowding disease has been associated with infectious and environmental conditions that facilitate bacterial colonization of the respiratory tract. Intramuscular administration of rBoIFN-α at doses similar to those used in the shipping fever model reduced the incidence of crowding disease.[6] As with shipping fever, the clinical benefits obtained from rBoIFN-α treatment in crowding disease also are thought to be due to the immunomodulatory capacity of this cytokine.

The immunoprophylactic effects of rBoIFN-γ have also been shown in an experimental model of dexamethasone-induced immunosuppression.[7] Dexamethasone-treated calves challenged with *Haemophilus somnus* had increased severity of pneumonia when compared with control calves. Administration of rBoIFN-γ prior to challenge resulted in reduction of pneumonic lung volume and severity of pneumonia in immunosuppressed calves ($p < 0.05$), but it did not influence the severity of disease in calves not treated with dexamethasone.[7] These findings support the hypothesis that IFN-γ can ameliorate the severity of disease in instances where the immune system of the host has been compromised. Nevertheless, there is evidence to suggest that IFN may induce clinical benefits in severe acute bacterial disease models in domestic animals in the absence of immunosuppression. For instance, administration of rBoIFN-α to calves prior to infection with *Salmonella typhimurium* significantly reduced the degree of septicemia.[8] Similarly, prophylactic treatment of pigs with recombinant porcine (rPo) IFN-α or rPoIFN-γ has been proven to alleviate clinical manifestations and mortality associated with *Actinobacillus pleuropneumoniae* infection.[9]

The application of CK in the prophylaxis of mastitis in dairy cattle is also under investigation. Mastitis is a clinical condition often associated with a reduction in the competence of normal defense mechanisms to cope with infection.[10] Intramammary infusion of rBoIFN-γ in periparturient dairy cows 24 h prior to challenge with *Escherchia coli* reduced the number of infected quarters, prevented death, diminished clinical signs, and shortened the duration of infection.[11] It is postulated that IFN-γ may provide nonspecific protection against a wide range of organisms to which the mammary gland is continuously exposed by modulating a variety of immune cells locally. Mastitis is an attractive field for CK applications given the economic significance of this disease and the fact that the mammary gland constitutes a closed organ system where the systemic toxicity associated with certain CK can be minimized. Recently, intramammary application of rBoIL-1β, rBoIL-2, and rBoGM-CSF have all been shown to protect the mammary gland from subsequent intramammary challenge with *Staphylococcus aureus*.[12]

B. ADJUVANT/CO-ADJUVANT ACTIVITIES OF CYTOKINES

A successful adjuvant could potentiate the immune response to vaccine antigens at three levels.[13] It must have the ability to (1) localize or target the antigen, (2) induce a state of enhanced and/or appropriate antigen expression, and (3) activate lymphocytes. While most conventional adjuvants (e.g., bacterial products and mineral oils) are able to carry out these functions, many of them are being restricted or banned due to their toxic effect and/or problems associated with residues in food producing animals.[14] Cytokines should be an effective replacement if formulated in a biologically inert release system.

1. Adjuvant Activity Exerted by Early Cytokines

The induction of IL-1 production in macrophages by bacteria, or bacterial extracts such as muramyldipeptide and lipopolysaccharide, has long been considered a contributing factor to the success of these compounds as adjuvants. In fact, the adjuvant effects of IL-1 have been well documented. It was suggested that the immuno-enhancing effects of IL-1 were due to its indirect effect, i.e. that it was enhancing the release of other CK such as IL-2 from antigen-specific T lymphocytes.[15] However, the ability of IL-1 to enhance the B cell responses to both carrier and hapten was not abolished by the stimultaneous administration of anti-IL-2 neutralizing antibody.[16]

In cattle, the adjuvant effects of IL-1 have been assessed using a modified live vaccine against BHV-1. Following both primary and secondary immunization with BHV-1 and rBoIL-1β treatment, there was a significant rise in cytotoxic responses against BHV-1-infected cells.[17] This augmented response was apparent at all doses tested (33, 100, and 330 ng/kg), whereas serum virus neutralization was slightly ($p < 0.09$) enhanced in only those calves treated with 100 ng/kg of rBoIL-1β. The lower and higher doses had little or no effect. These results support the findings of previous studies in which the adjuvant activity of IL-1 was extremely dose dependant.[15,16] However, they also indicate that IL-1 is able to augment cell-mediated responses more effectively than humoral responses in this system. This conclusion is supported by other studies in which it was shown that IL-1 enhanced T helper activity rather than antibody production.[16] Low doses of tumor necrosis factor-α (TNF-α) are also able to enhance immune responses to T-dependent antigens to the same extent as IL-1β; but interestingly, TNF-α was not able to enhance the response to T-independent antigens such as pneumococcal polysaccharide.[18]

When evaluating the adjuvant potential of CK, the beneficial effect of CK administration has to be weighed against the possible adverse effects. This is especially the case with IL-1 and TNF-α, which are primary mediators of inflammation. The toxicity of IL-1 has been avoided by using a synthetic peptide fragment of human IL-1β (VQGEESNDK, fragment 163-171). This peptide has been shown to have immunostimulatory, but not pyrogenic activity.[19] This activity can be blocked by a monoclonal antibody against both the peptide and the native molecule. However, this antibody cannot block the pyrogenic activity of IL-1β.[20] Further studies by this group have indicated that this peptide does not bind to either form of the IL-1 receptor, and that its activity is mediated by internalization and cytoplasmic activation of intracellular pathways.[21] Thus, by using a nonpyrogenic IL-1 peptide, adjuvancy can be retained without the harmful side effects of the native molecule. This region of the IL-1 molecule is extremely well conserved between species, and the bovine sequence at this region is almost 100% homologous to the human sequence.[22] This peptide may therefore act as an adjuvant in a number of different domestic animal species.

Site- or cell-specific delivery of the CK may also be important with regard to their ability both to act at low concentrations and with minimal toxicity. Live vector delivery may be one manner in which this is carried out; but again, careful evaluation of the response against the vector is essential. Cases in which the vector may induce a mild clinical response may become more severe if the vector is actively secreting "endogenous" pyrogens.

2. Adjuvant Activity Exerted by Late Cytokines

The pleiotropic effects that IL-2 has on T and B cells make it an excellent candidate for an adjuvant. Further, IL-2, along with IFN-γ was among the first CK to be characterized, cloned, and expressed in high-level expression systems. IL-2 and IFN-γ were also among the first CK to undergo clinical testing, and so there was a significant amount of data available regarding their *in vivo* use prior to their selection as potential adjuvants.

It appears that IL-2 can affect the response to immunizing agents in at least two ways. First, it is able to overcome genetic low- or nonresponsiveness; and second, it is able to enhance an existing response to immunization. Although experiments evaluating the adjuvant potential of IL-2 were carried out using

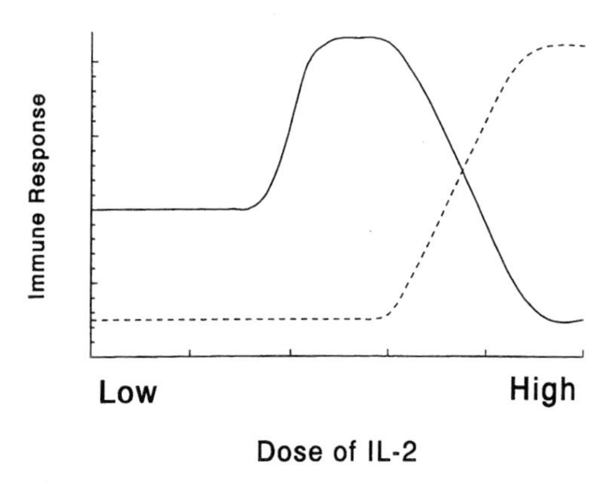

Figure 14-3 The effect of IL-2 dose on adjuvancy and overcoming non- or low responsiveness. At very low doses, IL-2 has no effect on the normal immune response to an antigen (solid line). There is then a dose range at which adjuvancy occurs; but if the range is exceeded, suppression is induced. It is proposed that concurrent with the suppression noticed by Nunberg et al.[28] it is at these high doses that *Ir* gene nonresponsiveness can be overcome (broken line).

a variety of animal models, a number of trends are evident. Figure 14-3 depicts the hypothetical relationship between IL-2 effects and dose based on the reported information. However, it would be presumptuous to put even approximate numbers on the dose of IL-2 required for each effect, as this may depend on the number of IL-2 treatments given, total dose, and animal model used.

Most studies have used large doses of IL-2, and very few have carried out a systematic titration of the effects of IL-2 on the immune response to vaccination. In those instances where IL-2 has been titrated, the adjuvant effects of IL-2 follow a dose response. Extremely low doses (i.e., <0.5 μg/kg) have little or no effect on bovine humoral or cellular responses to vaccines against BHV-1 or *Pasteurella haemolytica*.[23-26] Studies both in ruminants and in other animal models have indicated that there is a dose optimum at which IL-2 may improve the efficacy of a vaccine.[27,28] In contrast, higher doses can abrogate the adjuvant effect to the extent of being suppressive.[28] However, high doses of IL-2, by providing an excess of CK in draining lymph nodes have been reported as being effective in overcoming *Ir* gene-associated nonresponsiveness to malaria peptides.[29] These observations are highly significant because they indicate that different IL-2 doses act in vastly different ways (Figure 14-3).

In contrast to IL-2, where the mechanism of adjuvancy is believed to be through the ability of this CK to promote clonal and polyclonal expansion, the ability of IFN-γ to enhance vaccine efficacy is believed to be through activation of macrophages. The expression of MHC class II and presumably antigen on antigen presenting cells can be enhanced by the co-administration of IFN-γ with antigen.[30] In cattle, IFN-γ was an effective adjuvant when administered with immunizing doses of vesicular stomatitis virus "G" protein in the absence of any other adjuvant.[31] The degree of protection afforded by this immunization was comparable to antigen emulsified in complete Freund's adjuvant. When given in the absence of any co-adjuvant, the effective dose of IFN-γ was 500 units/ml; lower doses had no detectable effect. Unfortunately, higher doses were not assessed using this model and thus it is difficult to evaluate the optimal dose of IFN-γ which may be used without inducing toxic side effects. The adjuvant potential of IFN-γ has also been investigated in a murine model of malaria. IFN-γ did not simply augment the antibody and DTH responses to antigens, but it also decreased the parasite load in low-responding mice and mice that were depleted of $CD4^+$ T cells.[32,33] These studies have clearly indicated that small concentrations of IFN-γ may be able to act as a vaccine adjuvant and overcome suppression, thus having the potential to decrease vaccine failure rates.

3. Combining Early and Late Cytokines for Optimal Effect

Above, we have attempted to indicate that either early or late CK may be able to exert an adjuvant effect when used in the proper dose and formulation. However, many experiments using CK as adjuvants have also included some other proprietary adjuvant which may by itself modulate an immune response. Therefore, it is essential that CK adjuvant activity be examined in light of the other adjuvants used in the vaccine. For example, when IL-2 was given at low doses with simple mineral oils (e.g., emulsigen), it had little or no effect on primary immune responses in cattle. This is in contrast to when a co-adjuvant such as Avridine was used. Avridine is a lipid amine which induces IFN-γ and IL-1 production and it can be proposed that for IL-2 to be an effective adjuvant, co-stimulation with other CK may be necessary.[25,27] Similar conclusions can be drawn from experiments in which CK have been assessed for their ability to modulate the immune response against foot-and-mouth disease virus antigen (FMDV). IL-1 could enhance the antibody response against FMDV antigens, particularly when administered 24 h prior to immunization.[34] However, this response was not as strong as when muramyldipeptide was used as the adjuvant. In contrast, IL-2 appeared to be as efficient as muramyldipeptide when applied with the antigen. Significantly, when both CK were used together, a marked enhancement of antibody to FMDV antigen was observed compared to the use of either CK alone.[34] These data further support the concept of using mixtures of early and late CK to mimic the immune response that occurs during an infection, or when complex adjuvants containing bacterial products (i.e., Freund's complete adjuvant) are administered.

III. TREATMENT OF INFECTIOUS DISEASES

In contrast to the advances made in the area of disease prevention, the use of CK for treatment of infectious diseases in ruminants has been less fruitful. Nevertheless, *in vitro* and *in vivo* observations suggest that, under the appropriate conditions, CK could be of benefit in the treatment of certain infections in the veterinary field. However, when discussing the therapeutic potential of CK, we should understand that CK do not appear to be the ultimate panacea they once were thought to be. Early predictions of the potential therapeutic wonders that could be achieved through CK treatments must now be replaced by a more realistic expectation based on combining the beneficial effects of CK with those of conventional treatments. This principle has been well demonstrated in human patients where IFN-γ has been licensed for use in conjunction with antibiotics to treat chronic granulomatous disease.[35]

In ruminants, a number of CK [i.e., IL-1, IFNs, and colony stimulating factors (CSF)] are known to enhance the phagocytic and killing capacity of macrophages and neutrophils, thus enhancing normal clearance mechanisms.[12,36,37] This activity led to the suggestion that treatment of certain bacterial infections could potentially be improved with CK administration. This hypothesis has been tested in dairy cows in experimental trials using a *Staphylococcus aureus* mastitis model. In this model, chronically infected mammary glands were infused with either rBoIL-2 or rBoIL-1β. While treatment of infected quarters with rBoIL-2 resulted in temporary clearance of infection in 54% of the mammary glands, only 38% of treated quarters remained free of *S. aureus* at 14 days post-treatment. Similarly, treatment of infected glands with rBoIL-1β induced a transient clearance of bacteria in 83% of the treated quarters, yet only 42% of treated quarters remained free of infection after 14 days following treatment. Bacterial clearance was positively correlated with the ability of both CK treatments to enhance PMN function.[12] It was suggested that enhancement in PMN activation observed following treatment with rBoIL-2 was probably due to the ability of this CK to initiate a secondary CK response that was ultimately responsible for neutrophil activation.[12] These experiments demonstrated only a partial benefit of CK therapy in a localized bacterial infection model. Nevertheless, these studies clearly demonstrated the ability of CK treatments to enhance clearance mechanisms, thus facilitating recovery. This partial therapeutic effect may be improved by combining CK with antibiotics.

Cytokines such as the CSF and IL-3 are capable of promoting the differentiation and maturation of leukocytes and they are presently undergoing extensive clinical testing in human beings. Several CSF and combinations thereof are being used in cancer patients to restore bone marrow function after chemotherapy and irradiation therapy,[38] and to reduce the risk of infection in human immunodeficiency virus (HIV) infected individuals.[39] In the veterinary field, the aforementioned CK have not been investigated for their capacity in aiding the process of recovery from infectious illnesses. However, given the importance of hematopoietic integrity in the recovery from infections, it can be anticipated that this group of CK could be useful, in combination with antibiotics, to improve recovery and reduce convalescent time.

In addition to the above-mentioned benefits some CK such as TNF and IFN-γ have been shown to raise the intracellular concentration of certain antibiotics.[40] Therefore, the possible benefits of CK-antimicrobial combinations could very well exceed our expectations.

One of the consequences of the determination of the role of CK in the pathogenesis of disease is the realization that the production of early CK is responsible for the manifestation of numerous clinical features in the disease process. For instance, studies on the pathophysiology of septic shock induced by bacterial endotoxins have revealed that CK such as TNF, IL-1, and IFN-γ play a crucial role in the development of several destructive host responses distinctive to many bacterial infections.[41-44] Administration of monoclonal antibodies directed against some of these CK[45] or their receptor, as well as specific antagonists,[46,47] can successfully inhibit some of these pathophysiological events. Clinical trials are being conducted in human beings using some of these anti-CK approaches to reduce the mortality associated with Gram-negative sepsis.[48,49] In addition, extensive resources and efforts are being applied in the study of synthetic products that could block the production or action of certain CK, in particular IL-1 and TNF.[50-53] Although the feasibility of using similar strategies in the treatment of infectious disease of ruminants is still speculative, it is reasonable to anticipate that such approaches will find valuable applications in the veterinary setting.

IV. DISEASE DIAGNOSIS AND PROGNOSIS

The detection of disease and assessment of its severity is currently dependent on clinical observation and the measurement of body temperature and alterations in the blood leukogram, while the diagnosis of specific immune status is, for the most part, limited to serum antibody levels. The former methods are frequently insufficient to facilitate detection of disease early in its genesis, and to allow accurate estimation of disease outcome. Since the clinical signs being measured are consequent to the production of early CK such as IL-1, TNF, and IL-6, direct determination of the levels of these CK may provide an earlier and more valuable measurement of the severity of disease. While useful in determining exposure to particular pathogens, serum antibodies can only be detected in the later stages of disease and are thus not useful in formulating an early diagnosis and prognosis. As well, antibody levels may not accurately reflect the level of cellular immunity present, which can be of importance in a number of diseases. Cellular immunity can be more accurately assessed if the CK produced to regulate this response could be measured.

In intensive livestock management systems such as feedlots, labor costs for visual inspection and temperature monitoring are high. As well, these procedures are not particularly successful predictors of disease onset or outcome, so costly antibiotic therapy is often applied *en masse* to prevent outbreaks of disease. Thus, development of simple rapid tests to quantitate early CK would provide criteria to determine the severity of disease and consequently the intensiveness of therapy required. Early and accurate detection of disease should result in reduced overall treatment costs. As well, CK monitoring could also be used to assess the response to therapy. For instance, plasma TNF levels in patients with rheumatoid arthritis declined following steroid therapy, in parallel with clinical improvement.[54]

Accurate detection of disease could be beneficial in areas other than the clinical setting. Food safety is becoming an increasingly important factor in our society, which, together with intensive animal management systems and high volume food processing plants, places considerable importance on the meat inspection system. The development of simple rapid tests to evaluate early CK levels during antemortem inspection could eliminate diseased animals from entering the processing line. The tests may also be adapted for post-mortem determination of systemic involvement of disease and disposition of the carcass. These tests would significantly improve meat quality while reducing inspection costs.[55] Cytokine testing may be extended to evaluate the intensity of adverse responses to biologicals or even the induction of stress to certain management practices.

Assays to detect the early CK IL-1, TNF, and IL-6 in serum are currently available for human CK and are being used not only to detect disease, but to determine disease severity and predict outcome. In humans, elevated levels of TNF in the plasma have been correlated with severity of disease and mortality in septic shock,[56,57] meningococcal septicemia,[58] and falciparum malaria.[59] Increased levels of IL-6 were also found to correlate with the incidence of shock and risk of death in adults with sepsis.[60] In contrast, the relationship between plasma IL-1 elevation and clinical outcome is equivocal. Studies have found positive,[58] negative,[56] or no[61] correlation with death. While IL-1 and TNF are both early CK important for initiating the inflammatory response, their production appears to be independently regulated.[56] Both the time course of production and the elaboration of CK binding proteins may influence detection of CK in serum. In a study of children with sepsis, increases in TNF and IL-1 were transient, while elevation of IL-6 was both delayed and prolonged.[62] In this study, the magnitude of CK production, particularly IL-6 was better able to predict outcome than a standard severity of illness scoring test. In critically ill adult patients, TNF levels increased acutely during episodes of septic shock, but remained elevated only for a few hours. In contrast, IL-6 was always increased in the serum of acutely ill patients. Moreover, IL-6 correlated well with clinical score and mortality rate. Thus, IL-6 appears to be a good marker of severity during bacterial infection[63] and may be the most useful predictor of outcome.[62]

In veterinary medicine, there are also examples of the relationship between the levels of CK and disease. Increases in TNF concentration were detected in both serum and milk following intramammary *Escherichia coli* challenge.[64] Cattle that succumbed to acute endotoxic shock had higher levels of serum TNF than survivors.[64] Serum TNF levels were also correlated with clinical signs and mortality in foals with sepsis.[65] In horses with colic, serum TNF levels were higher in nonsurvivors and were correlated with inflammatory or strangulating intestinal obstruction rather than nonstrangulating intestinal obstruction.[66]

Further information regarding disease prognosis may be obtained by measurement of CK binding proteins or inhibitors, in addition to CK levels. For instance, in patients with severe meningococcemia, an increase in the ratio of serum TNF relative to two fragments derived from the TNF receptor was correlated with fatal outcome.[67]

As well as monitoring the severity of overwhelming systemic infections, CK detection has also been used to aid diagnosis of particular diseases. In some cases, the production of CK at the site of infection is sufficient for the CK to be detected in the serum, but higher levels are usually detected in local fluids, which offers an additional aid to the diagnosis of the infection. For instance, IL-1, TNF, and IL-6 were detected in high levels in the cerebrospinal fluid of patients with meningitis,[68] TNF levels in fecal samples could be used to monitor disease activity in patients with inflammatory bowel disease,[69] IL-6 was present in synovial fluids of arthritis patients,[70] and levels of TNF and IL-1 in bronchoalveolar fluid increased during clinical development of respiratory distress syndrome.[71]

The diagnosis of disease could be aided by elucidation of the pattern of CK that are produced by the host in response to the pathogen. For instance, the relative levels of IFN-α and TNF may allow discrimination between bacterial and viral infections. Even this simple type of diagnosis could be beneficial in the feedlot situation, where bacterial pneumonia secondary to viral infection is a major cause of disease. Traditionally, all sick animals are removed from their pen and moved to a "hospital" pen. However, this practice allows transmission of bacteria from animals with bacterial pneumonia to those that are only infected with virus. If these two groups of calves were kept separate, disease progression may be limited. As we learn more about the types of CK that are produced in specific disease situations, the CK profile may be used to provide more detailed diagnosis. For instance, using both TNF and IL-6 measurements, Kawasaki disease, an acute febrile illness of children, could be differentiated from anaphylactoid purpura and measles, diseases which have similar pathological and clinical signs.[72] Similarly, elevated levels of IFN-γ in ascitic fluid could be used to distinguish tuberculous peritonitis from other causes of ascites.[73]

The elaboration of late CK such as IL-2, IL-4, and IFN-γ is associated with the development of the immune response to the specific pathogen. Currently, the assessment of specific cellular immunity can be related to CK production upon stimulation of leukocytes with the pathogen *in vitro*. This method for the detection of cellular immunity has been used successfully to develop a test for the production of IFN-γ as a method for diagnosis of bovine tuberculosis.[74] We have also found the production of IFN-γ to be an indicator of immunity to BHV-1 since the production of IFN-γ by PBMC from calves following immunization was correlated with the development of virus specific cytotoxicity.[75] Using cytokine production as a means to monitor cell-mediated immunity may become an important tool to evaluate vaccine efficacy in addition to disease diagnosis. In some cases, late CK can be detected in serum samples. Elevated levels of IL-2 and IFN-γ have been reported in the acute sera of patients with dengue virus infection,[76] while increased IL-2 and IL-4 levels were detected in patients with scleroderma.[77] As we become cognizant of the range of CK elaborated *in vivo* during various diseases, this profile of CK is sure to become essential for diagnostic purposes.

V. EPILOGUE

The approach to disease management is being revolutionized by the availability of large quantities of purified CK and our ability to assay their biological functions. However, this revolution can only be sustained in the presence of continued vigilence which ensures that the approaches used are founded on a thorough understanding of complex interactions between the pathogens and the host. Armed with this new knowledge, industry should be in the position to rapidly exploit the beneficial properties of these molecules. However, whether industry embraces or discards this technology will depend on factors ranging from public and regulatory agency acceptance to perceived economic benefits. Therefore, it is imperative that all of these factors are taken into consideration when designing protocols for CK use. Due to the complexity of CK interactions, care must be taken that results from a single experiment do not form the basis for empirical testing in the field. Currently, it is known that CK can act either in a beneficial or a detrimental manner depending on the kinetics and magnitude of their production and their multiple interactions. For this reason, understanding the complex interactions between the pathogen, the host, and its environment (stressors) must be taken into consideration before treatment or preventive measures using CK can be optimized. For example, if the desired effect is to enhance a specific immune response to a vaccine, a different constellation of CK may be required as compared to those needed to stimulate nonspecific immune responses early in the infection process. In other situations, anti-cytokine therapy may be used to reduce the deleterious effect of some CK. However, before choosing a specific CK or anti-cytokine therapy, one requires a thorough understanding of which specific response is desired for a certain situation.

To date, applications of CK have focused on the prevention of infectious disease. However, as we gain more knowledge about the host immune responses and reactions to specific insults, it is becoming evident that CK will also become useful in disease diagnosis and treatment strategies. Based on these new developments, it is envisaged that a novel approach to accurate disease diagnosis will become common place in the near future. However, we should not assume that this will happen overnight without unforeseen impediments. The complexity of the CK network indicates that the road to applying the knowledge gained in CK biology into clinical practice will be full of surprises and difficulties. It is our strong belief that clinical medicine no longer can ignore the importance of these mediators. We are confident that the application of CK will become an essential component of disease management in the future.

REFERENCES

1. **Cummins, J. M. and Rosenquist, B. D.,** Protection of calves against rhinovirus infection by nasal secretion interferon induced by infectious bovine rhinotracheitis virus, *Am. J. Vet. Res.,* 41, 161, 1980.
2. **Roney, C. S., Rossi, C. R., Smith, P. C., Laureman, L. C., Spano, J. S., Hanrahan, L. A., and William, J. C.,** Effect of human leukocyte A interferon on prevention of infections bovine rhinotracheitis virus infection in cattle, *Am. J. Vet. Res.,* 46, 1251, 1985.
3. **Babiuk, L. A., Lawman, M. J. P., and Gifford, G. A.,** Use of recombinant bovine alpha interferon in reducing respiratory disease induced by bovine herpesvirus type 1, *Antimicrob. Agents Chemotherapy,* 31, 752, 1987.
4. **Babiuk, L. A., Sordillo, L. M., Campos, M., Hughes, H. P. A., Rossi-Campos, A., and Harland, R.,** Application of interferons in the control of infectious diseases of cattle, *J. Dairy Sci.,* 74, 4385, 1991.
5. **Babiuk, L. A., Bielefeldt Ohmann, H., Gifford, G., Czarniecki, C. W., Scialli, V. T., and Hamilton, E. G.,** Effect of bovine alpha interferon on herpesvirus type 1 induced respiratory disease, *J. Gen. Virol.,* 66, 2383, 1985.
6. **Bielefeldt Ohmann, H. and Martinod, S. R.,** Interferon immunomodulation in domestic food animals, in *Immunomodulation in Domestic Food Animals,* Blecha, F. and Charley, B., Eds. Acadamic Press, New York, 1990, 215.
7. **Chiang, Y., Roth, J. A., and Andiens, J. J.,** Influence of recombinant bovine interferon gamma and dexamethasone on pneumonia attributable to *Haemophilus somnus* in calves, *Am. J. Vet. Res.,* 51, 759, 1990.
8. **Peel, J. E., Kolly, C., Siegenthaler, B., and Martinod, S. R.,** Prophylactic effects of recombinant bovine interferon-gamma$_I$ on acute *Salmonella typhimurium* infection in calves, *Am. J. Vet. Res.,* 51, 1095, 1990.
9. **Campos, M., Rossi-Campos, A., Potter, A., Harland, R., Bielefeldt Ohmann, H., and Babiuk, L. A.,** Parenteral administration of interferons protects pigs from acute death caused by the gram-negative bacterium, *Actinobacillus (Haemophilus) pleuropneumoniae, Immunobiology,* Suppl. 4, 24, 1989.
10. **Paape, M. J. and Guidry, A. J.,** Effects of fat and casein on intracellular killing of *Staphylococcus aureus* by milk leukocytes, *Proc. Soc. Exp. Biol. Med.,* 135, 588, 1977.
11. **Sordillo, L. M. and Babiuk, L. A.,** Controlling acute *Escherichia coli* mastitis during the periparturient period with recombinant bovine interferon gamma, *Vet. Microbiol.,* 28, 189, 1991.
12. **Daley, M. J., Coyle, P. A., Williams, T. J., Forda, G., Doughery, R., and Hayes, P.,** *Staphylococcus aureus* mastitis: pathogenesis and treatment with bovine interleukin-1 beta and interleukin-2, *J. Dairy Sci.,* 74, 4413, 1991.
13. **Lise, L. D. and Audibert, F.,** Immunoadjuvants and analogs of immunodulatory bacterial structures, *Curr. Opin. Immunol.,* 2, 269, 1989.

14. **Stewart-Tull, D. E. S.,** The assessment and use of adjuvants in vaccines, in *Vaccines,* Gregoriadis, G., Ed., Pleunum Press, London, 1991, 85.
15. **Staruch, M. J. and Wood, D. D.,** The adjuvanticity of IL-1, *J. Immunol.,* 130, 2191, 1983.
16. **Reed, S. G., Pihl, D. L., Conlon, P. J., and Grabstein, K. H.,** IL-1 as an adjuvant. Role of T cells in the augmentation of specific antibody production by recombinant human IL-1 alpha, *J. Immunol.,* 142, 3129, 1989.
17. **Reddy, D. N., Reddy, P. G., Minocha, H. C., Fenwick, B. W., Baker, P. E., Davis, W. C., and Blecha, F.,** Adjuvanticity of recombinant bovine interleukin-1: influence on immunity, infection, and latency in a bovine herpesvirus-1 infection, *Lymphokine Res.,* 9, 295, 1990.
18. **Ghiara, P., Boraschi, D., Nenciono, L., Ghezzi, P., and Tagliabue, A.,** Enhancement of *in vivo* immune response by tumour necrosis factor, *J. Immunol.,* 139, 3676, 1987.
19. **Antoni, G., Presentini, R., Perin, F., Tagliabue, A., Ghiara, P., Censini, S., Volpini, G., Villa, L., and Boraschi, D.,** A short synthetic peptide fragment of human interleukin-1 with immunostimulatory but not inflammatory activity, *J. Immunol.,* 137, 3201, 1986.
20. **Boraschi, D., Volpini, G., Villa, L., Nenciono, L., Scapigliati, G., Nucci, D., Antoni, G., Matteucci, G., Cioli, F., and Tagliabue, A.,** A monoclonal antibody to the IL-1 beta peptide 163-171 blocks adjuvanticity but not pyrogenicity of IL-1 beta *in vivo, J. Immunol.,* 143, 131, 1989.
21. **Boraschi, D., Ghiara, P., Scapigliti, G., Villa, L., Sette, A., and Tagliabue, A.,** Binding and internalization of the 163-171 fragment of human IL-1 beta, *Cytokine,* 4, 201, 1992.
22. **Maliszewski, C. R., Baker, P. E., Schoenborn, M. A., Davis, B. S., Cosman, D., Gilles, S., and Cerretti, D. P.,** Cloning, sequence and expression of bovine interleukin 1 alpha and interleukin 1 beta complimentary DNAs, *Mol. Immunol.,* 25, 429, 1988.
23. **Reddy, P. G., Blecha, F., Minocha, H. C., Anderson, G. A., Morrill, J. L., Fedorka-Cray, P. J., and Baker, P. E.,** Bovine recombinant interleukin-2 augments immunity and resistance to bovine herpesvirus infection, *Vet. Immunol. Immunopathol.,* 23, 61, 1989.
24. **Campos, M., Hughes, H. P. A., Godson, D. L., Sordillo, L. M., Rossi-Campos, A., and Babiuk, L. A.,** Clinical and immunological effects of single bolus administration of recombinant interleukin-2 in cattle, *Can. J. Vet. Res.,* 56, 10, 1992.
25. **Hughes, H. P. A., Campos, M., Godson, D. L., van Drunen Littel-van den Hurk, S., McDougall, L., Rapin, N., Zamb, T., and Babiuk, L. A.,** Immunopotentiation of bovine herpesvirus subunit vaccination by interleukin-2, *Immunology,* 74, 461, 1991.
26. **Hughes, H. P. A., Campos, M., Potter, A. A., and Babiuk, L. A.,** Molecular chimerization of Pasteurella haemolytica leukotoxin to interleukin-2: effects on cytokine and antigen function, *Infect. Immun.,* 60, 565, 1992.
27. **Hughes, H. P. A., Campos, M., van Drunen Littel-van den Hurk, S., Zamb, T., Sordillo, L. M., Godson, D., and Babiuk, L. A.,** Multiple administration with cytokines potentiates antigen-specific responses to subunit vaccination with bovine herpesvirus-1 glycoprotein IV, *Vaccine,* 10, 226, 1992.
28. **Nunberg, J. H., Doyle, M. V., York, S. M., and York, C. J.,** Interleukin-2 acts as an adjuvant to increase the potency of inactivated rabies virus vaccine, *Proc. Natl. Acad. Sci., U.S.A.,* 86, 4240, 1989.
29. **Good, M. F., Pombo, D., Lunde, M. N., Maloy, W. L., Halenbeck, R., Koths, K., Miller, L. H., and Berzofsky, J. A.,** Recombinant human IL-2 overcomes genetic nonresponsiveness to malaria sporozoite peptides: correlation of effect with biologic activity of IL-2, *J. Immunol.,* 141, 972, 1988.
30. **Nakamura, M., Manser, T., Pearson, G. D. W., Daley, M. J., and Gefter, M. L.,** Effect of IFN on the immune response *in vivo* and on gene expression *in vitro, Nature,* 307, 381, 1984.
31. **Anderson, K. P., Fennie, E. H., and Yilma, T.,** Enhancement of a secondary antibody response to vesicular stomatitis virus "G" protein by IFN-gamma treatment at primary immunization., *J. Immunol.,* 140, 3599, 1988.
32. **Playfair, J. H. L. and De Souza, J. B.,** Recombinant gamma interferon is a potent adjuvant for a malaria vaccine in mice, *Clin. Exp. Immunol.,* 67, 5, 1987.
33. **Heath, A. W. and Playfair, J. H. L.,** Conjugation of interferon-gamma to antigen enhances its adjuvanticity, *Immunology,* 71, 454, 1990.
34. **McCullough, K. C., Pullen, L., and Parkenson, D.,** The immune response against foot-and-mouth disease virus: influence of the T lymphocyte growth factors IL-1 and IL-2 on the murine humoral response *in vivo, Immunology,* 31, 41, 1991.
35. **Baron, S., Tyring, S. K., Fleischmann, W. R. J., Coppenhaver, D. H., Niesel, D. W., Klimpel, G. R., Stanton, G. J., and Hughes, T. K.,** The interferons. Mechanisms of action and clinical applications, *J. Am. Med. Assoc.,* 266, 1375, 1991.

36. **Bielefeldt Ohmann, H., Lawman, M. J. P., and Babiuk, L. A.,** Bovine interferon: its biology and application in veterinary medicine, *Antiviral Res.,* 7, 187, 1989.
37. **Lawman, M. J. P., Campos, M., Bielefeldt Ohmann, H., Griebel, P., and Babiuk, L. A.,** Recombinant cytokines and their therapeutic value in veterinary medicine, in *Animal Biotechnology,* Babiuk, L. A., Phillips, J. P., and Moo-Young, M., Eds., Pergamon Press, New York, 1989, 63.
38. **Kanz, L., Brugger, W., Bross, K., and Mertelsmann, R.,** Combination of cytokines: current status and future prospects, *Br. J. Haematol.,* 79, 96, 1991.
39. **Mitsuyasu, R. T.,** Use of recombinant interferons and hematopoietic growth factors in patients infected with human immunodeficiency virus, *Rev. Infect. Dis.,* 13, 979, 1991.
40. **Bermudez, L. E., Inderlied, C., and Young, L. S.,** Stimulation with cytokines enhances penetration of azithromycin into human macrophages, *Antimicrob. Agents Chemother.,* 35, 2625, 1991.
41. **Beutler, B. and Cerami, A.,** Tumor necrosis, cachexia, shock, and inflammation: a common mediator, *Annu. Rev. Biochem.,* 57, 505, 1988.
42. **Billiau, A.,** Interferons and inflammation, *J. Interferon Res.,* **7**, 559, 1987.
43. **Heinzel, F.,** The role of IFN-gamma in the pathology of experimental endotoxemia, *J. Immunol.,* 145, 2920, 1990.
44. **Okusawa, S., Gelfand, J. A., Ikejima, T., Connolly, R. J., and Dinarello, C. A.,** Interleukin-1 induces a shock-like state in rabbits, *J. Clin. Invest.,* 81, 1162, 1988.
45. **Beutler, B., Milsark, I. W., and Cerami, A.,** Passive immunization against cachectin/tumor necrosis factor (TNF) protects mice from the lethal effects of endotoxin, *Science,* 229, 869, 1985.
46. **Alexander, H. R., Doherty, G. M., Buresh, C. M., Venzon, D. J., and Norton, J. A.,** A recombinant human receptor antagonist to interleukin 1 improves survival after lethal endotoxemia in mice, *J. Exp. Med.,* 173, 1029, 1991.
47. **McIntyre, K. W., Stepan, G. J., Kolinsky, K. D., Benjamin, W. R., Polcinski, J. M., Kaffka, K. L., Campen, C. A., Shizzonite, R. A., and Kilian, P. L.,** Inhibition of interleukin 1 (IL-1) binding and bioactivity *in vitro* and modulation of acute inflammation *in vivo* by IL-1 receptor anatgonist and anti-IL-1 receptor monoclonal antibody, *J. Exp. Med.,* 173, 931, 1991.
48. **Spooner, C. E., Markowtiz, N. P., and Saravolatz, L. D.,** The role of tumor necrosis factor in sepsis, *Clin. Immunol. Immunopathol.,* 2, S11, 1992.
49. **Dinarello, C.A.,** Anti-cytokine strategies, *Eur. Cytokine Net.,* 3, 7, 1992.
50. **Rosenthal, G. J., Craig, W. A., Corsini, E., Taylor, M., and Luster, M. I.,** Pentamidine blocks the pathophysiologic effects of endotoxemia through inhibition of cytokine release, *Toxicol. Appl. Pharmacol.,* 112, 222, 1992.
51. **Bedrosian, I., Sofia, R. D., Wolff, S. M., and Dinarello, C. A.,** Taurolidine, an analogue of the amino acid taurine, suppresses interleukin-1 and tumor necrosis factor synthesis in human peripheral blood mononuclear cells, *Cytokine,* 3, 568, 1991.
52. **Seow, W. K., Ferrante, A., Summors, A., and Thong, Y. H.,** Comparative effects of tetrandrine and berbamine on production of the inflammatory cytokines interleukin-1 and tumor necrosis factor, *Life Sci.,* 50, PL53, 1992.
53. **Zabel, P., Schonharting, M. M., Schade, U. F., and Schlaak, M.,** Effects of pentoxifylline in endotoxinemia in human volunteers, *Prog. Clin. Biol. Res.,* 367, 207, 1991.
54. **Espersen, G. T., Vestergaard, M., Ernst, E., and Grunnet, N.,** Tumour necrosis factor alpha and interleukin-2 in plasma from rhematoid arthritis patients in relation to disease activity, *Clin. Rheumatol.,* 10, 374, 1991.
55. **Saini, P. K. and Webert, D. W.,** Application of acute phase reactants during antemortem and postmortem meat inspection, *J. Am. Vet. Med. Assoc.,* 198, 1898, 1991.
56. **Cannon, J. G., Gelfand, J. A., Stanford, G. J., Endres, S., Lonnemann, G., Corsetti, J., Chernow, B., Wilmore, D. W., Wolff, S. M., Burkey, J. F., and Dinarello, C. A.,** Circulating interleukin-1 and tumor necrosis factor in septic shock and experimental endotoxin fever, *J. Infect. Dis.,* 161, 79, 1990.
57. **Suputtamongkol, Y., Kwiatkowski, D., Dance, D. A. B., Chaowagul, W., and White, N. J.,** Tumor necrosis factor in septicemic melioidosis, *J. Infect. Dis.,* 165, 561, 1992.
58. **Waage, A., Brandtzaeg, P., Halstensen, A., Kierulf, P., and Espevik, T.,** The complex pattern of cytokines in serum from patients with meningococcal septic shock, *J. Exp. Med.,* 169, 333, 1989.
59. **Kwiatkowski, D., Hill, A. V., Sambou, I., Twumasi, P., Castracane, J., Manogue, K. R., Cerami, A., Brewster, D. R., and Greenwood, B. M.,** TNF concentration in fatal cerebral, non-fatal cerebral and uncomplicated *Plasmodium falciparum* malaria, *Lancet,* 336, 1201, 1990.

60. **Hack, C. E., De Groot, E. R., Felt-Bersma, R. J., Nuijens, J. H., Strack Van Schijndel, R. J., Eerenberg-Belmer, A. J., Thijs, L. G., and Aaarden, L. A.,** Increased plasma levels of interleukin-6 in sepsis, *Blood,* 74, 1704, 1989.
61. **Damas, P., Reuter, A., Gysen, P., Demonty, J., Lamy, M., and Franchimont, P.,** Tumor necrosis factor and interleukin-1 serum levels during severe sepsis in humans, *Crit. Care Med.*, 17, 975, 1989.
62. **Sullivan, J. S., Kilpatrick, L., Costarino, T., Lee, S. C., and Harris, M. C.,** Correlation of plasma cytokine elevations with mortality rate in children with sepsis, *J. Pediatr.,* 120, 510, 1992.
63. **Damas, P., Ledoux, D., Nys, M., Vrindts, Y., De Groote, D., Franchimont, P., and Lamy, M.,** Cytokine serum level during severe sepsis in human; IL-6 as a marker of severity, *Ann. Surg.*, 215, 356, 1992.
64. **Sordillo, L. M. and Peel, J.,** Effects of interferon gamma on the production of tumor necrosis factor during acute Escherichia coli mastitis, *J. Dairy Sci.,* 75, 2119, 1992.
65. **Morris, D. D. and Moore, J. N.,** Tumor necrosis factor activity in serum from neonatal foals with presumed septicemia, *J. Am. Vet. Med. Assoc.,* 199, 1584, 1991.
66. **Morris, D. D., Moore, J. N., and Crowe, N.,** Serum tumor necrosis factor activity in horses with colic attributable to gastrointestinal tract disease, *Am. J. Vet. Res.,* 52, 1565, 1991.
67. **Girardin, E., Lombard-Roux, P., Grau, G. E., Suter, P., and Gallati, H.,** Imbalance between tumour necrosis factor alpha and soluble TNF receptor concentrations in severe meningococcaemia, *Immunology,* 76, 20, 1992.
68. **Waage, A., Halstensen, A., Shalaby, R., Brandtzaeg, P., Kierulf, P., and Espevik, T.,** Local production of tumor necrosis factor alpha, interleukin 1 and interleukin 6 in meningococcal meningitis. Relation to the inflammatory response, *J. Exp. Med.,* 170, 1859, 1989.
69. **Braegger, C. P., Nicholls, S., Murch, S. H., Stephens, S., and MacDonald, T. T.,** Tumour necrosis factor alpha in stool as a marker of intestinal inflammation, *Lancet,* 339, 89, 1992.
70. **Sawada, T., Hirohata, S., Inoue, T., and Ito, K.,** Correlation between rheumatoid factor and IL-6 activity in synovial fluids from patients with rheumatoid arthritis, *Clin. Exp. Rheumatol.,* 9, 363, 1991.
71. **Suter, P. M., Suter, S., Girardin, E., Roux-Lombard, P., Grau, G. E., and Dayer, J. M.,** High bronchoalveolar levels of tumor necrosis factor and its inhibitors, interleukin-1, interferon, and elastase, in patients with adult respiratory distress syndrome after trauma, shock or sepsis, *Am. Rev. Respir. Dis.,* 145, 1016, 1992.
72. **Furukawa, S., Matsubara, T., Yone, K., Hirano, Y., Okumura, K., and Yabuta, K.,** Kawasaki disease differs from anaphylactoid purpura and measles with regard to tumour necrosis factor-alpha and interleukin-6 in serum, *Eur. J. Pediatr.,* 151, 44, 1992.
73. **Ribera, E., Martinez Vasquez, J. M., Ocana, I., Ruis, I., Jiminez, J. G., Encabo, G., Segura, R. M., and Pascual, C.,** Diagnostic value of ascites gamma interferon levels in tuberculous peritonitis. Comparison with adenosine deaminase activity, *Tubercle,* 72, 193, 1991.
74. **Wood, P. R., Corner, L. A., Rothel, J. S., Baldock, C., Jones, S. L., Cousins, D. B., McCormick, B. S., Francis, B. R., Creeper, J., and Tweddle, N. E.,** A field comparison of the interferon-gamma assay and the intradermal tuberculin test for the diagnosis of bovine tuberculosis, *Aust. Vet. J.,* 168, 286, 1991.
75. **Campos, M., Bielefeldt Ohmann, H., Hutchings, D., Rapin, N., Babiuk, L. A., and Lawman, M. J. P.,** Role of interferon gamma in inducing cytotoxicity of peripheral blood mononuclear leukocytes to bovine herpesvirus type 1 (BHV-1) -infected cells, *Cell. Immunol.,* 120, 259, 1989.
76. **Kurane, I., Innis, B. L., Nimmannitya, S., Nisalak, A., Meager, A., Janus, J., and Ennis, F. A.,** Activation of T lymphocytes in dengue virus infections. High levels of soluble interleukin-2 receptor, soluble CD4, interleukin 2 and interferon-gamma in sera of children with dengue, *J. Clin. Invest.,* 88, 1473, 1991.
77. **Needleman, B. W., Wigley, F. M., and Stair, R. W.,** Interleukin-1, interleukin-2, interleukin-4, interleukin-6, tumor necrosis factor alpha and interferon gamma levels in sera from patients with scleroderma, *Arthritis Rheum.,* 35, 67, 1992.

INDEX

A

Accessory cells
  induction of T cell-mediated immunity, 96–103
  Langerhans cells, 96, 97, 99
Acquired immunity, GI parasites, 213
Activated cells, migration to gut, 119–120
Activation, see also Induction of T cell-mediated immune responses
  cytokine production, 77
  MVV infection, 201
Activation antigens, 6–7
ADCC, see Antibody-dependent cell-mediated cytotoxicity
Addressin, 111, 119
Adhesion cells, 97
Adhesion molecules, 5–6, see also L-selectin
  gut endothelium, 119
  lymphocyte homing
    to lymph nodes, 114–116
    to skin, 117
  naive T cells, 118
  TCR ligand recognition and signalling, 69
  VCAM, 6, 111, 117
Adjuvants, cytokines as, 232–234
Adoptive transfer
  nematode immunity, 217
  Theileria infection, 146, 148
Afferent lymph veiled cells (ALVC), 5
  induction of T cell-mediated immunity, 97–103
  CD1 expression in cattle, 4
  in *Theileria* infection, 144
Age-related changes in neutrophil function, 131–132
Ala-Pro start sequence, 79
Alkaline phosphatase, neutrophils, 129
Allele-specific motifs, FMD, 178
Alloantisera, class II typing, 45
αβ-Heterodimers, ligand recognition, 68
αβ T cells, 60
  in BHV-1 infections, 163
  development, 30–31
    fetal, 24–25
    intrathymic maturation, 22–26
  migration and homing
    to peripheral lymph node, 115–116
    to skin, 117
  TCR ligand recognition and signalling, 69
  in *Theileria* infection, 144
  in thymus, 9
Alpha chains
  amino acid residues, 63
  heterodimers, 63–64, 68
Alpha interferon, see Interferon-α
Alveolar macrophages
  BHV-1 infections, 161
  cytokine sources, 77
AMPHI, 179, 182–186
Anaphylactoid purpura, 236
Antibodies
  BHV-1 infections, 161, 165
  in nematode infections, 220
  in *Theileria* infection, 145–146
Antibody-dependent cell-mediated cytotoxicity (ADCC), 130–131, 134
  BHV-1 infections, 161, 162, 166
  cytokines and, 76
  enhancement, 134
  K cells, 135
  LAK and, 136
  MVV cells, 205
  neutrophil function, age-related changes, 132
Antibody specificity, FMD vaccine, 177
Antigen presenting cells (APC), 189
  interferon-γ and, 233
  Theileria infection, 147
Antigen processing and presentation, see also T cell receptor
  by ALVCs, 101–102
  FMDV, 178
  in gut mucosa, 214
  induction of T cell-mediated immunity, 94–96
AP-1, 83
Aphthovirus, see Foot-and-mouth disease
Arachidonic acid, 130
Arginine-glycine-aspartic acid (RGD) motif, 190
Ascorbic acid, and neutrophil function, 134
Aspartic protease, 94
Assays, MHC polymorphism, 47
Attenuated vaccine strains, BHV-1 infections, 160
A23187 calcium ionophore, 77
Autoimmunity
  CD5+ cells in, 8
  MHC polymorphisms and, 48
  MVV lesions, 206
Autologous MLR27, 10
Avridine, 133, 233
Azurocidin, 129

B

Babesia bovis, 152
Bactenecins, 129–130
Bactericidal permeability inducing protein, 129
Barber's pole worm, 213
Basophils, in nematode infections, 220
B cell epitopes, intrastructural help, 189–190
B cell oxidase, 130
B cells
  antigens expressed on, 8
  CD1 expression in cattle, 4
  CD45 on, 7
  cytokines and, 76
  L-selectin expression, 6
  in nematode infections, 220
  postnatal peripheral pools, 30
Beta chains, see also αβ T cells; αβ T cell receptor
  amino acid residues, 63
  heterodimers, 64–65
Beta interferon, 77, 78
Bf gene, 43
BHV-1, see Bovine herpesvirus 1
Biological response modifiers, 128
Black scour worm, 213
BoLA, see Major histocompatibility complex
Bovine herpesvirus 1 (BHV-1), 236
  characteristics of virus, 157–158
  cytokine antiviral properties, 230–231
  infection and immunity, 160–162
    BHV-1-induced immunosuppression, 160–162
    immune response and latency, 162
    nonspecific and specific immune mechanisms, 160
  infectious bovine rhinotracheitis/infectious pustular vulvovaginitis, 158–159
  MHC polymorphism and, 48
  and neutrophil function, 133
  NK activity, 135
  nonspecific responses to infection, 163–164
  specific responses to infection, 164–166
  strategies for control, 159–160
Bovine immunodeficiency-like virus, 133
Bovine leukemia virus (BLV), 49
  CD5+ cells in, 8

MHC and, 49
NK activity, 135
Bovine Leukocyte Adhesion Deficiency (BLAD), 6
Bovine Lymphocyte Antigen (BoLA), see Major histocompatibility complex
Bovine lymphocyte subpopulations
Theileria infection, 148
Bovine viral diarrhea (BVD)
MHC and, 49
and neutrophil function, 132
Brucella abortus, 96, 133

C

Calcium ionophore, 131
CAP 37, 129
Caprine arthritis encephalitis virus (CAEV), 49, 199, 201, 204
Caprine disease, MHC and, 49
Caprine lymphocyte antigen (CLA) complex, 47–48
Carrier animals, FMD, 174
Catalase, neutrophil, 129
Catecholamines, and neutrophil function, 133
Cathepsin B, 94
Cathepsin D, 94
Cathepsin G, 129
CC76- cells, 8
CD1 antigen, 4
fetal development, 21
thymus-lymphocytes, 8, 9
CD1+ cells, enteric, 215
CD1 specificity, ALVCs, 97, 99, 102
CD2 antigen, 4
immature T cells, 8, 9
lymphocyte recirculation and homing, 111, 117
CD2+ cells
BoWC6 antigen, 8
LAKs, 136
CD2 specificity, ALVCs, 102
CD3 antigen
immature T cells, 8, 9
as "pan T" marker, 2
TCR structure, 61–62
thymocyte development, 23
CD3+ cells
BHV-1 infections, 163
functional heterogeneity, 11–12
immature T cells, 8, 9
CD3δ, EMBL database accession numbers, 71
CD3 epsilon, 71
CD3γ, EMBL database accession numbers, 71
CD4 antigen, 3
thymocyte development, 23
fetal, 23–24
immature T cells in thymus, 8, 9
CD4- cells, LAK cells, 136
CD4+ cells
ALVCs and, 96–103
CD11a, 5
CD45 isoform, 10–11
CD45 on, 7
cytokine release, 95
development
differential *in vivo* homing to lymph nodes, 114–115
peripheral pool expansion, 28–30
T cell subset development, 27
functional heterogeneity, 11–12
gut, 214
interferon-γ effects, 233
L-selectin expression, 11
MHC class I molecules and, 39
MHC class I polymorphism, 45
in MVV infection, 202, 203
in nematode infections, 218
TCR ligand recognition and signalling, 69
TH1 and TH2, 75–76
in *Theileria* infection, 144, 147–148, 151
tissue distribution, 9
CD4+/CD8+ cells, 24
intrathymic selection, 26
L-selectin expression, 6
in *Theileria* infection, 144
CD4+/CD8- cells, *Theileria* infection, 146
CD4+/CD8- helper T cell, intrathymic selection, 26
CD4+/CD8+ killer cells, intrathymic selection, 26
CD5 antigen, 3, 5
cysteine-rich domains, 8
immature T cells, 8, 9
T cell subset development, 27
CD6, 3, 8, 9
CD8 antigen
development
fetal, 23–24
immature T cells, 8, 9
heterodimeric, 3
TCR ligand recognition and signalling, 69
CD8- cells, LAK cells, 136
CD8+ cells
BHV-1 infections, 160
CD45 isoform, 10–11
CD45 on, 7
class I molecules and, 39
development, 8, 9, 23–24
differential *in vivo* homing to lymph nodes, 114–115
peripheral pool expansion, 28–30
T cell subsets, 27
tissue distribution, 9
heterodimeric CD8, 3
homing to gut in fetus, 118
L-selectin expression, 11
in MVV infection, 201–203
in nematode infections, 218
TCR ligand recognition and signalling, 69
in *Theileria* infection, 144, 147–148
CD11a, 5, 99, 102
CD11a/CD18, lymphocyte recirculation and homing, 111
CD11b, 5, 8, 99, 102
CD11c, 5, 99, 102
CD18, 5
β2 integrins, 97, 111
bovine LAD, 132
lymphocyte recirculation and homing, 111
CD21, 8
CD25, 6
CD44, 6
lymphocyte recirculation and homing, 111
T cell homing to skin, 117
CD45, 7, 70
ALVCs, 99, 102
fetal development, 21
CD45 isoforms, 10–11
CD45R, 10, 11
CD45R+ cells
lymphocyte recirculation and homing, 111, 117
naive T cells, 118
CD45R- cells, T cell homing to skin, 117
CD45R0+ cells, 8
CD45RA, 7, 116
CD45RB, 7, 8
CD49d/CD29, lymphocyte recirculation and homing, 111
CD54, 5
CD56, 134
CD58, lymphocyte recirculation and homing, 111
cDNA, TCR loci, 70
Cell adhesion molecules, see Adhesion molecules
Cell assays, MHC polymorphism, 47
class I gene, 43–44
class II gene, 44–45
Cell culture, persistent FMD infection, 175

Cell-mediated cytotoxicity, *Theileria* infection, 147
Cell proliferation, *Theileria parva*-infected cells, 146
Cellular mediators, BHV-1 infections, 162
Central nervous system, MVV infection, 203, 205, 206
Cestode, 216
c-fms, 85
Chabertia spp., 213
Chemotaxis
  BHV-1 infections, 161
  neutrophil, 131, 132
  TNF-mediated, 86
Chimeric twin animals, *Theileria* infection, 146
Chloramphenicol, 133
Chronic granulomatous disease, therapeutic applications, 234
Chronic pulmonary infections, TNFα and, 85
CK1, 83
CK2, 83
c-kit, 85
Class I and II, see Major histocompatibility complex
Cloning, cytokine genes, 76, 77
CM-CSF, see Granulocyte-macrophage colony-stimulating factor (GM-CSF)
Cobalt, and neutrophil function, 133
Coccidia, and neutrophil function, 133
Co-infection
  MVV, 199, 206
  nematodes, 216
Colonization of fetal thymus, 19–21
Colony-stimulating factors (CSF), 80, 82, see also specific factors
  BHV-1 infections, 166
  in vitro biology, 83
  and neutrophil function, 134
  therapeutic applications, 234
Colony-stimulation activity, in nematode infections, 218
Colostrum
  neutrophil function, age-related changes, 132
  protection from nematode infection, 215
Complementarity determining region (CDR3), 68
Complement components, 6, 43, 131, 215
Complement-mediated lysis, *Theileria* infection, 146
Concanavalin-A, 77, 131
Concealed antigen, gut, 217
Concurrent infections, see Co-infection
Consensus motives, FMDV, 178
Constant (C) region, TCR, 31, 63, 71
Cooperia oncophera, 217
Cooperia punctata, 217
Copper, and neutrophil function, 133
Corticosteroids
  and GI parasite establishment, 215
  and neutrophil function, 133, 134
Covert antigens, gut, 217
Cross-neutralization assays, BHV-1 infections, 159
Cross-protection, *Theileria* strain specificity of CTL, 148–149
Cross reactivity
  CD3, 2
  IL-2, 80
CYP21, 43
Cysteine protease, 94
Cytochalasin B, 131
Cytokine applications in infectious diseases
  diagnosis and prognosis, 235–236
  early versus late function, 229–230
  prevention, 230
  treatment, 234–235
Cytokine receptors, 6–7, 85
Cytokines
  activation antigens, 6–7
  and LAK cells, 128
  assays, 79, 83
  BHV-1 infections, 160, 164–166
  cloning strategies, 76, 77
  control elements, 82–83
  hemopoietin (colony-stimulating factors) genes, 80, 82
  interferon genes, 76, 78
  interleukins and tumor necrosis factor genes, 78–81
  in vitro biology, 83–84
  in vivo biology, 84–85
  and neutrophil function, 134
  and neutrophil-mediated toxicity, 130–131
  parasite rejection, 222
  receptors, 85–86
  and *Theileria*-infected cells, 151–152
  tumor necrosis factor (TNF-α), 79, 81
Cytoplasmic calcium, neutrophil, 131, 133
Cytotoxicity, see also Antibody-dependent cell-mediated cytotoxicity
  neutrophil-mediated, 130–131
Cytotoxic T lymphocytes
  BHV-1 infections, 160, 162, 165–166
  class I molecules and, 39
  FMD, 174
  MHC class I gene polymorphism, 43–44
  MVV cells, 205
  Theileria infection, 147–151
Cytotoxin, and neutrophil function, 133

D

Defensins, 129
Degranulation
  and neutrophil function, 133
  neutrophils, 131
Delta chains, 61–62, see also γδ TCR
  amino acid residues, 63
  heterodimers, 67–68
Dendritic cells, 96, 97, 99, 117
  CD1 expression in cattle, 4
  enteric, 215
  gut mucosa, 214
  T cell interactions in absence of antigen, 103
Depletion experiments, 9
Dermal dendritic cells, see Dendritic cells
Development, see Differentiation
Dexamethasone, 221
  immunosuppression, 231
  and neutrophil function, 133, 134
DIB, 42
Dietary antigens, 215
Differentiation, see also Fetus; Induction of T cell-mediated immune responses; T cell ontogeny
  eosinophil, in nematode infections, 218
  T cell, 19–33
  T cell homing, 118–120
Differentiation antigens, lymphocyte subpopulations, 2–8, see also specific CD antigens
Dihydroheptaprenol, 134
Disease states
  cytokines and, 85
  MHC polymorphisms and, 48–49
Diversity (D) segment, 63
DMA, 42
DMB, 42
DNA, 42
DNA gene, 42
DNA replication
  BHV-1 infections, 164
  MVV, 200
DOB, 42
DP, 42
DQ, 42
DQB gene, 42
DQ-DR region, 42
DR, 42
DRB genes, 42
Drugs, and neutrophil function, 133

DYA, 42
DYB, 42

E

Early cytokines, 229–230
  adjuvant activity, 232
  and disease diagnosis and prognosis, 235, 236
East Coast fever, see *Theileria parva*
Ectoderm, T cell differentiation sites, 33
Effector mechanisms, in nematode infections, 219–220
Efferent lymph, *Theileria* infection, 148
Eimeria spp., 133
Elastase, neutrophil, 129, 131
EMBL accession numbers, 71, 77
Embryo splitting, 150
Encephalitis, MVV, 201
Endocrine hormones, in nematode infections, 219
Endoderm, T cell differentiation sites, 33
Endogenous antigens, *Theileria* infection, 147
Endogenous opiates, and neutrophil function, 133
Endoplasmic reticulum retention signal (ERRS), 61
Endothelium, see Vascular endothelium
Endotoxic shock, 85, 234–236
Enhancer sequences, MVV, 200
Enteric vaccination, 215
Enzyme immunoassays, GI parasites, 213
Enzymes, neutrophil
  nonoxidative, 128–130
  oxidative, 130, 131
Eosinophils, 128, 218, 220
Epidermal dendritic cells, see Dendritic cells
Epidermal growth factor, 3
Epidermal Langerhans cells, 96, 97, 99
Epitopes, 150
Epsilon chains, 61–62
Estrous cycle, and neutrophil function, 133
Eta chain, 61, 62–63
Excretion of virus, FMD, 174
Excretory-secretory (ES) antigens, 217
Exogenous antigens, *Theileria* infection, 147
Extrathymic T cell development, 32

F

Falciparum malaria, 235
Fc receptors
  ALVCs, 102
  neutrophil, 131
Fertility, MHC and, 49
Fetus
  lymphocyte recirculation and homing, 110–111, 118–120
  MHC and, 50
  morphogenesis and colonization of thymus, 19–21
  neutrophil function, age-related changes, 131
FMD, see Foot-and-mouth disease
FMDV-15, 189
Foot-and-mouth disease, 174
  cell-mediated immunity (CMI), 177–178
  humoral immunity and protection, 175–177
  infection, 173–175
  MHC polymorphism and, 48
  subunit vaccine development, 186–190
  T cell epitope prediction from primary protein sequences, 178–182
    AMPHI, 179
    rarity rule, 180
    Rothbard and Taylor motifs, 180
    SOHHA, 179–180
  T cell epitopes in virus, 182–188
Foot-and-mouth disease virus (FMDV) antigens
  IL-1 adjuvant activity, 233–234
  loop peptides, 176
  loop peptide vaccines, 187–190
Formylmethionylleucylphenylalanine (FMLP), 131
Formyl peptides, 131

G

β-Galactosidase, neutrophils, 129
Gamma chains, 61–62
  amino acid residues, 63
  heterodimers, 65–67
γδ T19⁺ cells, 114
γδ T cells
  BHV-1 infections, 163
  fetal development
    peripheral pool expansion, 28–30
    T cell receptor gene rearrangement and expression, 24–25
  homing
    in fetal and post-natal animals, 120
    to peripheral lymph node, 114–116
    to skin, 117
  in nematode infections, 218
  ontogeny of, 30–31
    development and tissue localization of, 27
    intrathymic maturation of, 26, 27
    thymus, 9
  TCR studies, 71
  in *Theileria* infection, 144
  tissue distribution, 9–10
γδ T cell receptor, 2, 60
Gamma interferon (IFN-γ), see Interferon-γ
Gastrointestinal nematode parasites
  cellular immune responses to non-nematode antigens in the gut, 215
  enteric effector responses, 219–222
  evidence for CMI in immunity, 216–217
  features of parasite immunity, 215–216
  gastroenteric mucosal immunity, 214–215
  induction of CMI during infections, 217
Genetic variation in neutrophil function, 132
Genome, 145
  in bovine LAD, 132
  MHC RFLP, 41, 145
    class I gene, 44
    class II gene, 45–47
    ovine and caprine, 47
  TCR loci, 70
Glucocorticoids, see Corticosteroids; Dexamethasone
β-Glucuronidase, neutrophils, 129
Glutathione peroxidase, 129
Glutathione reductase, 129
Glycoproteins, BHV-1 infections, 164
Granulocyte-colony-stimulating factor (G-CSF), 82, 134
Granulocyte-macrophage colony-stimulating factor (GM-CSF), 82
  Ala-Pro start sequence, 79
  cloning strategies, 77
  in vitro biology, 83–84
  and neutrophil function, 134
  receptors, 85
Granulocytes, see Neutrophils
Growth of animals, MHC and, 49
gt11 expression library of genomic DNA, 145
GTP/GDP binding, 62
Gut mucosa, see also Peyer's patches; Gastrointestinal nematode parasites
  homing of lymphocytes, *in vivo*, 118–119
  WC1⁺ populations, 10

H

Haemonchus, 213, 220
Haemonchus contortus, 214, 215, 217, 218, 220, 221

Haemophilus somnus, 133
Hassal's corpuscle, 27
Heat shock proteins, 43, 206
Heligmosomoides polygyrus, 216
Helminthiasis, see Gastrointestinal nematode parasite
Helper epitopes, FMDV, 189, 190
Helper T cells
BHV-1 infections, 165, 166
intrathymic selection, 26
MHC class I molecules and, 39
parasite rejection, 222
Theileria infection, 147
Hemopoietin family of receptors, 85
Hemopoietins, 80, 82
Heterodimers, 63–68, see also αβ T cells; γδ T cells
Heterotypic protection, FMDV-15, 177
High endothelial venules (HEV), 29, 113, 114, 119
Histamine, 221
Histocompatibility complex, FMD, 174
HLA-A2, 95
HLA-B27, 179
HMG-1, 83
Homing, see Recirculation and homing of lymphocytes
Homing receptors, 5–6
Homozygous typing cell (HTC), 45
Hormones, and neutrophil function, 133
Human CD3 cross reactivity, 2
Human immunodeficiency virus, 234
Humoral immunity, 202
BHV-1 infections, 160
FMD, 175–177
Hydroxyl radical, 130
5-Hydroxytryptamine, 221
Hypersensitivity response, in nematode infections, 220

I

ICAM-1, 5
ICAM-2, 5
Icelandic sheep, MVV, 201
IL, see specific interleukins
ILA53, 97
Immature T cells, see T cell ontogeny
development, 30–31
fetal development, 24–25
intrathymic maturation, 22–26
tissue distribution, 8–9
Immune potentiation of disease, MVV infection, 205
Immune response, MHC polymorphism and, 48
Immune selection, MVV variants, 204–205
Immunization
BHV-1 infections, 159–160
cytokine adjuvancy, 232–234
in *Theileria* infection, 144–145
Immunodominance, 151, 174
Immunoglobulin A
gut mucosa, 214, 215
in nematode infections, 220
Immunoglobulin E
enteric, 215
in nematode infections, 218, 220, 221
Immunoglobulin G
ALVCs, 101
enteric, 215
in nematode infections, 218
Immunoglobulin G_1, gut mucosa, 214
Immunoglobulin M, ALVCs, 101
Immunoglobulin M receptors, neutrophil, 131
Immunoglobulins
gut immune system, 215
isotypes
cytokines and, 76
gut immune system, 215
in nematode infections, 220
Immunoglobulin superfamily, lymphocyte recirculation and homing, 111
Immunosuppression, dexamethasone-induced, 231
Indolicidin, 129, 130
Induction of T cell-mediated immune responses
accessory cells for MHC class II-restricted immune responses *in vivo*, 96–103
afferent lymph veiled cells (ALVCs), 97–103
ALVC function *in vivo*, 102
relative efficiency of ALVCs and other antigen-presenting cell populations, 102–103
uptake of antigen by ALVCs, 101
antigen processing and presentation, 94–96
T cell recognition, 94
INF, see specific interferons
Infectious bovine rhinotracheitis, see Bovine herpesvirus 1
Infectious diseases
cytokines in treatment of, 234–235
MHC polymorphisms and, 48
and neutrophil function, 132
Infectious pustular vulvovaginitis, 158–159
Inflammatory mediators, in nematode infections, 220
In situ expression of GM-CSF, 84, 85
Integrins, 5, 6, 97, 116, 132
genetic defects in neutrophil function, 132
gut homing, 119
lymphocyte recirculation and homing, 111
naive versus memory T cells, 116
TCR ligand recognition and signalling, 69
Interdigitating cells, 96, 97, 99
Interferon-α
antiviral properties, 230–231
BHV-1 infections, 160
cDNAs and genes, 77
in vitro biology, 84
and LAK activity, 135, 136
MVV infection, 201
and neutrophil function, 134
Theileria-specific, 152
Interferon-β, cloning strategies, 77, 78
Interferon-γ
adjuvant activity, 232–233
and TNFa receptors, 86
assays, 79
BHV-1 infections, 160, 161, 164–166
cDNA and genes, 76—78
and disease diagnosis and prognosis, 236
in vitro biology, 83, 84
and LAK activity, 135, 136
MVV infection, 201
in nematode infections, 218
and neutrophil-mediated toxicity, 130–131
TH1 cell production, 75
Theileria-infected cell, 151
therapeutic applications, 234
Interferon-γ gene promoter region, 83
Interferon-ω, 76, 77
Interferons, 76–78
antiviral properties, 230–231
assays, 79
BHV-1 infections, 161, 163, 166
MVV infection, 201
neutrophil function, age-related changes, 132
therapeutic applications, 234
Interleukin-1, 78
adjuvant activity, 232
BHV-1 infections, 160, 166
and disease diagnosis and prognosis, 235, 236
gene control region, 83
induction of, 96
and neutrophil function, 134
Theileria-specific, 152
therapeutic applications, 234
Interleukin-1α

amino acid sequences, 80
assays, 79
cloning strategies, 77
in vitro biology, 84
Interleukin-1β
adjuvant activity, 232
Ala-pro start sequence, 79
amino acid sequences, 80
assays, 79
cloning strategies, 77
in vitro biology, 84
and neutrophil function, 134
therapeutic applications, 234
Interleukin-1 peptide, adjuvancy, 232
Interleukin-10, TH2 cell production, 76
Interleukin-2, 6, 79–81, 134
adjuvant activity, 232–234
and TNFα receptors, 86
assays, 79
BHV-1 infections, 160, 164, 166
cloning strategies, 77
and disease diagnosis and prognosis, 236
in vitro biology, 84
and LAK activity, 128, 135, 136
in nematode infections, 218
Theileria infection, 146–147, 151
therapeutic applications, 234
Interleukin-2 gene, 83
Interleukin-2-like activity, *Theileria parva*-infected cells, 146
Interleukin-2 receptor
activation antigens, 6–7
Theileria-specific, 151
Interleukin-3, 82
Ala-pro start sequence, 79
assays, 79
BHV-1 infections, 166
in vitro biology, 83–84
in nematode infections, 218
receptors, 85
therapeutic applications, 234
Interleukin-3 gene control region, 83
Interleukin-4, 80, 81
BHV-1 infections, 166
cloning strategies, 77
and disease diagnosis and prognosis, 236
and LAK activity, 135, 136
TH2 cell production, 76
Interleukin-5
BHV-1 infections, 166
in nematode infections, 218
receptors, 85
TH2 cell production, 76
Interleukin-6, 79
cloning strategies, 77
and disease diagnosis and prognosis, 235, 236
in nematode infections, 219
receptors, 85
Interleukin-7, 82
cloning strategies, 77
receptors, 85
Interleukin-8
BHV-1 infections, 166
neutrophil recruitment, 85
Interleukins, 78–81
Intermolecular help, 189
Intermolecular intrastructural help, 189
Intracellular signalling, TCR and, 68–70
Intrastructural help, 189
Isoelectric focusing
MHC class I gene polymorphism, 44
MHC class II polymorphism, 45, 47
Isotype selection
cytokines and, 76
in nematode infections, 215, 220

J

Jaagsiekte virus, 206
Joining (J) segment, 63
Joints, MVV-induced arthritis in, 203

K

Kawasaki disease, 236
Killer cells (K), 26, 127, 134–136
Killing capacity, therapeutic applications, 234

L

Lactoferrin, 128, 129
LAK, see Lymphokine-activated killer cells
LAM-1, see L-selectin
Lamina propria, 214, 220
Langerhans cells, 96, 97, 99
Large granular lymphocyte, 128, 134
Large Multifunctional Protease (LMP), 39
Late cytokines, 229–230
adjuvant activity, 232–233
and disease diagnosis and prognosis, 236
Latency
BHV-1 infections, 159, 162–163
MVV, 200, 204
LCFA-1, see CD11a
Leader (L) segment, 63
LECAM-1, see L-selectin
Lentivirus
and neutrophil function, 133
pathogenesis, see Maedi-visna virus
Leukocyte adhesion deficiency (LAD), 132
Leukocyte adhesion molecules, 97
Leukocyte common antigen, see CD45
Leukocyte integrins, see Integrins
Leukocyte markers, tissue distribution, and functional characteristics
differentiation antigens, lymphocyte subpopulations, 2–8
adhesion molecules and homing receptors, 5–6
antigens expressed on B cells, 8
definitive markers of T cells, 2
leukocyte common antigen, 7–8
lymphocyte activation antigens, 6
molecules defining, 2–4
other markers, 4–5
nomenclature, 2
tissue distribution and function of T cell subpopulations, 8–11
defined by CD45 isoforms, 10–11
defined by L-selectin, 11
immature, 8–9
mature, 9–11
$WC1^+$ γδ T cells, 9–10
Leukotoxin, 131, 133
Leukotrienes, 221
LFA-1, 5, see also CD18
LFA-3
ALVCs, 102
lymphocyte recirculation and homing, 111
T cell homing to skin, 117
Ligand recognition, TCR, 68–70
Limiting dilution assay, *Theileria* infection, 147–149
Lipid amine, 233
Lipopolysaccharide
cytokine stimulation, 77
and neutrophil function, 133
Lipoxygenase, and neutrophil function, 133
Liver, fetal, 21
LMP genes, 42–43
Local protection
gut immune system, 215
systemic immune system, 215

Long terminal repeats, MVV, 200
Low responsiveness, FMD, 174
L-selectin, 6, 8, 116
γδ T cells in fetal and post-natal animals, 120
lymphocyte recirculation and homing, 111
naive T cells, 118
as peripheral lymph node homing receptor, 114
T cell homing to skin, 117
L-selectin⁺ cells
Peyer's patches of young lambs, 119
T cell subset distribution, 11
Lung, MVV infection, 203
Lymphatics
cannulation of, 93
cytokine studies, 84
lymphocyte migration, 112–116
Lymph node cells, cytokine sources, 77
Lymph nodes
fetal development, peripheral pool expansion, 28
homing of lymphocytes, *in vivo*, 113–116
interdigitating cells, 96
post-thymic maturation, 28
Lymphoblasts, in nematode infections, 218
Lymphocytes, see also Recirculation and homing of lymphocytes; specific lymphocytes
BHV-1 infections, 163
cytokine sources, 77
gut mucosa, 214
interferons and recruitment of, 84–85
Theileria infection, 146
Lymphoid tumors, TCR probes, 71
Lymphokine-activated killer (LAK), 127, 135–136
Lymphokines, see also Cytokines; specific lymphokines
in nematode infections, 219, 220
Lymphotoxin, TH1 cell production, 75
Lymph responses, 218
Lysozyme, neutrophils, 129

M

Macrophages, 96, 128
antigen processing, 94, 96
BHV-1 infections, 160, 161, 163
cytokine sources, 77
GM-CSF, see Granulocyte-macrophage colony-stimulating factor
gut mucosa, 214
MVV, 201
MVV infection, 201
therapeutic applications, 234
Macrophage scavenger receptor, 3
Maedi-visna virus
autoimmunity, 206
immune potentiation of disease, 205–206
immune responses, 201–202
infected cell, 200–201
pathology, 202–204
synergy with other infectious agents, 206
viral antigens, 205
viral variants, 204–205
virus life cycle, 200
Major histocompatibility complex (MHC), 26
ALVCs, 102
antigen processing and recognition, 94–96
class I gene polymorphism, 43–44
class II gene polymorphism, 44–47
FMD, 186–190
and neutrophil function, 132
organization of the bovine MHC, 41–43
ovine and caprine, 47–48
polymorphism, significance of, 48–50
associations with physiological traits, 49–50
disease associations, 48–49
immune response associations, 48
MHC restriction, 48
reproduction and fetal allograft, 50
structure of MHC molecules, 38–40
T cell epitopes, 179
TCR ligand recognition and signalling, 68–70
Theileria infection, 148–150
Major histocompatibility complex class I molecules
in BHV-1 infections, 165
CD8 and, 3
polymorphism, 43–44
fetal development, 21
FMD studies, 178, 180–181
γδ T cell development, 26
and neutrophil function, 132
thymic emigrants, 28
Major histocompatibility complex class II molecules
BHV-1 infections, 161
FMD response, 178
interferon-γ and, 233
MVV, 201
polymorphism, 44–47
Theileria infection, 147
Major histocompatibility complex class II-restricted responses
accessory cells, 96–103
FMD response, 178
Mast cells, enteric, 220, 221
Mastitis, 234
cytokines in prophylaxis of, 231
MHC and, 49
MVV infection, 203
Maternal immunity, 215
Maturation, see Recirculation and homing of lymphocytes
M cells
gut mucosa, 214
in nematode infections, 220
Measles, 236
Mel 14, 214
Memory, 10
Memory helper T and B cell epitopes, 189
Memory T cells
adhesion molecules expression, 116
CD45 isoform, 10
homing to lymph nodes, 113–114
migration to gut, 119–120
Meningococcal septicemia, 235
Mesenteric lymph nodes, 215
Mesocestoides corti, 216
MHC, see Major histocompatibility complex
β-Microglobulin, 21
$β_2$-Microglobulin structure, 38
Migration, see also Recirculation and homing of lymphocytes
to gut, 119–120, 214
homing to lymph nodes, 113–114
in *Theileria* infection, 143
Milk production, MHC and, 49
Missing self, BHV-1 infections, 164
Mixed leukocyte reaction (MLR), *Theileria parva*-infected cells, 146, 147
Molybdenum, and neutrophil function, 133
Monoclonal antibodies
BHV-1 infections, 159
GI parasites, 213
leukocyte differentiation antigens, 1, 2
Monocyte colony-stimulating factor, 85
Monocyte/macrophage lineage, MVV infection, 200, 204
Monocytes, 97
CD1 expression in cattle, 4
L-selectin expression, 6
Moraxella bovis, 133
Morphogenesis, thymus, 19–21
Mucosa, CD4⁺ T cells, 9
Mucosal addressin, 119
Mucosal immunity

BHV-1 infections, 161
gastroenteric, 214–215
GI parasites, 213
in nematode infections, 218–219
Mucosal system, 214
Multipotential colony-stimulating factor (interleukin-3), 82
Mutation, MVV, 204–205
MVV, see Maedi-visna virus
Myeloid markers, 21
Myeloperoxidase, neutrophil, 128–130, 132

N

Naive T cells
adhesion molecules expression, 116
CD11a antigen, 5
homing to gut versus lymph nodes, 119
surface antigens, 118
Natural killer (NK) cells, 3, 127, 134–135
BHV-1 infections, 160, 161, 163–164
MVV infections, 205
Natural resistance, MVV, 201
Nematodes, see Gastrointestinal nematode parasite
Nematodirus, 213
Neonates, see also T cell ontogeny
colostral protection, 215
neutrophil function, age-related changes, 131
Nerves, in nematode infections, 219, 220
Neuropeptides
and lymphocyte migration, 214
in nematode infections, 219
Neutralization, BHV-1 infections, 161
Neutrophilia, in LAD, 132
Neutrophils, 127
BHV-1 infections, 160–163, 166
deficits in function, 131–133
enhancement of function, 133–134
interferons and recruitment of, 84–85
L-selectin expression, 6
morphology, physiology, and biochemistry, 128–131
therapeutic applications, 234
Nippostrongylus braziliensis, 216
Nippostrongylus spathiger, 220
NK cells, see Natural killer cells
Nomenclature, 2
Nonresponsiveness, FMD vaccines, 173, 187–188
Non T, non B cells, see K cells
Novel gut antigen, 217
Nutritional deficiencies, and neutrophil function, 133

O

OCT-1, 83
Ocular squamous cell carcinoma, 49
Oesophagostomum, 213
Oesophagostomum radiatum, 48
Omega interferon (IFN-ω), 76
One-dimensional isoelectric focusing (1D-IEF), 44
Ontogeny of T cells, see T cell ontogeny
Opsonization, BHV-1 infections, 161
Orf virus, 84, 85
Ostertagia, 213, 219
Ostertagia circumcincta, 214, 215, 217, 220
Ostertagia ostertagi, 213, 217
Ovine disease association, MHC and, 49
Ovine lymphocyte antigen (OLA) complex, 47–48
Oxidase activating factor, 130
Oxidase enzyme complex, 130
Oxidative metabolism, neutrophil, 130, 131, 134

P

Parasite immunity, 213, 215–216
Parasite infections, see also Gastrointestinal nematode parasite; *Theileria parva*
Passive protection
FMD, 176
MVV, 215
Pasteurella haemolytica, 131, 133
Pasteurella multocida, 202
Pathogenesis, MVV, 203–204
Pathogens, and neutrophil function, 133
Pathology
cytokines and, 85
MVV infection, 202–204
Peripheral blood cells
mononuclear (PBMC), BHV-1 infections, 161
T cell development, 27–33
Peripheral lymph nodes
addressing, 114
homing of lymphocytes, *in vivo*, 113–116
Peripheral tissues, homing of lymphocytes, *in vivo*, 116–118
Persistent infection
FMD, 173–174
MVV, 202
Peyer's patches, 9, 119
antigen uptake, 214
L-selectin$^+$ cells, 11
lymphocyte migration, 111, 112, 114
in nematode infections, 220
of young lambs, 119
p56, 70
Phagocytosis
BHV-1 infections, 161
MHC class I molecules and, 39
Phagolysosomes, 39
Pharmaceuticals, and neutrophil function, 133
Phorbol myristate acetate (PMA), 77, 130, 131
Phosphoinositide (PI) breakdown, 70
Phospholipase C (PLC), 70
Platelet-activating factor, 131, 221
Pneumonia, 230
Polymerase chain reaction, 46–47, 77
Polymorphic immunodominant molecule (PIM), 144
Polymorphonuclear neutrophilic granulocytes, see Neutrophils
p110/75, ALVCs, 87, 99, 102
Post-capillary venules, lymphocyte migration, 112
Post-natal animals, homing of T cells, 118–120
Predictor of disease outcome, IL-6 as, 235
Probactenecins, 129
Progesterone, and neutrophil function, 133
Promoter, MVV, 200
Proteasome, 39, 42
Protective immunity
FMDV, 175–177
in *Theileria* infection, 145–146
Protein kinase C, 132
Proto-oncogene products, 85

R

Random migration, neutrophil, 132
Rarity rule, 180
Recirculating TCR αβ repertoire, 31
Recirculation and homing of lymphocytes, 29–30
γδ T cells in fetal and post-natal animals, 120
to gut, 118–120
major pathways of lymphocyte migration, 112–113
migration streams, 110–113
to peripheral lymph nodes, 113–116
to skin and peripheral tissues, 116–118
the sheep model, 110
Recombinant IFN-γ, enhancement, 134
Recombinant IFN-α, *Theileria*-specific, 151–152
Regression of thymus, 21–22
Regulatory proteins

CD4 and CD8 molecules and, 69–70
MVV, 200
Reproduction, MHC and, 50
Respiratory viruses
BHV 1, see Bovine herpesvirus 1
and neutrophil function, 132
Restriction endonuclease analysis, BHV-1 infections, 159
Restriction fragment length polymorphism (RFLP), 145
MHC polymorphism
class I gene, 44
class II gene, 45–47
ovine and caprine, 47
RNA, 182
Rothbard and Taylor motifs, 180

S

Schistosoma mansoni, 216
Schizont antigens, *Theileria*, 145
Scrapie, 49
SDS-PAGE, BHV-1 infections, 159
Selectin, see L-selectin
Selenium, 133
Septic shock, 234–236
Serine proteases, and bactenecins, 129–130
Serology, MHC polymorphism
BHV-1 infections, 159
class I gene, 43
class II gene, 45, 47
memory T cell epitopes, 189
ovine and caprine, 47
Serotonin, 221
Sheep
disease association, MHC and, 49
gut mucosa, 214
lymphocyte recirculation and homing in fetus, 110–111
ovine lymphocyte antigen (OLA) complex, 47–48
Sheep pulmonary adenomatosis (SPA) virus, 206
Shipping fever, 230
Shutdown, 113
Signaling pathways, TCR, 68–70
Singlet oxygen, 130
Skin
accessory cells, 96, 97, 99
homing of lymphocytes, *in vivo*, 116–118
T cell preferred routes of entry, 113
WC1$^+$ populations, 10
SOHHA, 179–180, 182–186
Southern blot analysis, class I genes, 41
Species specificity, IL-2, 80
Specific immunity to MVV, 201–202
Spleen, T cell ontogeny, 29
Splicing events, MVV, 200
Sporozites, *Theileria*, 143
Sporozoite antigens, *Theileria*, 145
Sporozoite-neutralizing antibodies, 145–146
Staphylococcus aureus mastitis, 234
Strain specificities, *Theileria* infections, 144–145, 148–149
Stress, and neutrophil function, 133
Strip-of-helix hydrophobicity algorithm, see SOHHA
Strongyloides ratti, 216
Strongyloides venezuelensis, 216
Structural analysis, FMD studies, 178
Structural proteins, MVV, 200
Subunit vaccine development, FMD, 186–190
Sulfur, and neutrophil function, 133
Superoxide anion, neutrophil function, 130–132
Susceptibility to infection, FMD, 174
Symptoms, FMD, 174

T

T19, 3, 5, 6, 9–10, 120
intrathymic maturation of T cells, 26
lymphocyte recirculation and homing, 111, 120
T19/WC1 molecules, tissue-specific homing of γδT19$^+$ T cells, 120
Tac antigen, 151
TAP1, 42
TAP2, 42, 43
T cell clones, *Theileria* infection, 146
T cell development, see also Induction of T cell-mediated immune responses; T cell ontogeny
extrathymic, 32
fetal, see Fetus
homing and recirculation, see Recirculation and homing of lymphocytes
T cell epitope prediction, FMDV, 178–182
T cell epitopes
BHV-1 infections, 165
in foot-and-mouth disease, 178–188
Theileria strain specificity of CTL, 149
T cell growth factor, 151
T cell-mediated responses, see Induction of T cell-mediated immune responses
T cell ontogeny, 19–33
development of peripheral T cells, 27–33
emigration from thymus, 27–28
expansion of the peripheral T cell pool, 28–30
extrathymic T cell development, 32–33
ontogeny of the peripheral T cell repertoire, 30–32
development of thymus, 19–22
growth and regression of the thymus, 21–22
morphogenesis and colonization of the fetal thymus, 19–21
intrathymic maturation of γδ T cells, 26–27
intrathymic maturation of αβ T cells, 22–26
intrathymic selection, 25–26
pathways of thymocyte development, 23–24
T cell receptor gene rearrangement and expression, 24–25
T cell receptors, see also CD3
γδT cells, 71
gene rearrangement and expression in fetus, 24–25
genomic structure of loci, 70
heterodimers, 63–68
alpha chains, 63–64
beta chains, 64–65
delta chains, 67–68
gamma chains, 65–67
intrathymic maturation of T cells, 22–27
intrathymic selection, 25–26
invariant components, 61
ligand recognition and signaling pathways, 68–70
MHC class I molecules and, 39
MHC class I polymorphism, 45
overview, 60–61
peptide-MHC complex and, 95
T cell epitopes, 179
Theileria infection, 148
tumors, pathogenesis, 71
V region studies, 70–71
T cell recognition, induction of T cell-mediated immunity, 94
T cells, see also αβ T cells; γδ T cells
BHV-1 infections, 161
CD1 expression in cattle, 4
enteric, 215
gut mucosa, 214
helper, see Helper T Cells
memory, see Memory T cells
in nematode infections, 220
recirculation of, 29–30
T cell subsets, see also specific antigens
development and tissue localization of, 27
differentiation antigens, 2–8
fetal development, peripheral pool expansion, 28–30
homing to peripheral lymph node, 115–116
Theileria infection, 148
tissue distribution, 8–11

Tegument protein, BHV-1 infections, 165
TH1 T cells
BHV-1 infections, 165
cytokine production, 75–76
TH2 T cells
BHV-1 infections, 165
cytokine release, 75, 76, 95
Theileria annulata, 48
Theileria colubriformis, 217–221
Theileria parva, 10, 11, 143
CD4+ T cells, 146–147
CD8+ T cells, 147–151
cellular and chemical basis of the antigenic specificity of CTL, 150–151
influence of the immunizing parasite strain on MHC restriction and strain specificities of the CTL response, 149–150
influence of the MHC phenotype on strain specificity of the CTL response, 149
relationship of parasite strain specificity of CTL with cross-protection, 148–149
Theileria-specific CTL, 148
cytokines and *Theileria*-infected cells, 151–152
induction of *Theileria parva*-specific proliferation, 146
mechanisms of protective immunity, 145–146
MHC class II polymorphism, 45
MHC polymorphism and, 48
parasite genome and antigens, 145
pathogenesis, 143–146
vaccination and parasite strains, 144–145
Thiamine, 133
Thoracic duct lymphocytes, in *Theileria* infection, 146
Thromboxane, 221
Thymocyte development pathways, 23–24
Thymus, 8–9
development of, 19–22
lymphocyte migration, 112
T cell emigration from, 27–28
Tissue distribution
migration and homing of lymphocytes, see Recirculation and homing of lymphocytes
T cell subpopulations, 8–11
Tissue-infiltrating T lymphocytes, BHV-1 infections, 164
Tissue specificity, see Recirculation and homing of lymphocytes
T memory cells, see Memory T cells
Transporter associated with Antigen Processing (TAP), 39
Trichinella, 216
Trichostrongylus axei, 213
Trichostrongylus colubriformis, 214, 215
Trichostrongylus vitrinus, 221
Trojan Horse Hypothesis, 204
Trophoblast IFN, 76
Tropism, MVV, 205
Trypanosoma brucei, 10
Trypanosoma congolense, 8
Tumor necrosis factor
and disease diagnosis and prognosis, 235, 236
therapeutic applications, 234
Tumor necrosis factor-α (TNF-α), 43, 79, 81
adjuvant activity, 232
assays, 79
BHV-1 infections, 166
cloning strategies, 77
gene control region, 83
in vitro biology, 84
and neutrophil function, 130–131, 134
receptors, 86
Theileria-specific, 151
Tumor necrosis factor-β (TNF-β), 43, 166
Tumor necrosis factor receptor, types of, 85–86
Tyrosine kinase activation domain (TKAD), 61
Tyrosine kinases, 70

V

Vaccines
BHV-1 infections, 159–160
cytokine activity, 232–234
enteric vaccination against nematodes, 215
FMD, 186–190
in *Theileria* infection, 144–145
Vascular endothelium, 132
adhesion molecules in gut, 119
fetal gut, 119
lymphocyte migration, 112
skin, 117
VCAM-1, 117
VCAM-1, 6, 111, 117
Veiled cells, see Afferent lymph veiled cells (ALVC)
Viral antigens
BHV-1 infections, 159–160
MVV, 205
Virulence factors, and neutrophil function, 133
Virus infections, see also Bovine herpesvirus 1; Foot-and-mouth disease; Maedi-visna virus
cytokine antiviral properties, 230–231
NK activity of infected cells, 135
Vitamin B_{12} binding protein, neutrophils, 129
VLA-4, 6, 111, 115, 116
VP1 141–160 FMDV loop peptide, 187–189
VP8, BHV-1 infections, 165
V regions, 30, 31, 33, 68
EMBL database accession numbers, 71
gene segments, 30, 31, 33, 63, 68
repertoire studies, 70–71

W

WC (workshop cluster) antigens
WC1, 3–4
intrathymic maturation of T cells, 26
lymphocyte recirculation and homing, 111, 120
WC1+ T cells, 5
L-selectin expression, 6
tissue distribution, 9–10
WC3, 8
WC4, 8
WC5, 6
WC10, 8
Western blots, 144, 145

Z

Zeta chain, 61–63